William J. Fay

— DAD —

Dying, We Live

Dying, We Live

A New Enquiry into the
Death of Christ
in the New Testament

Kenneth Grayston

New York   Oxford
OXFORD UNIVERSITY PRESS
1990

Oxford University Press

Oxford   New York   Toronto
Delhi   Bombay   Calcutta   Madras   Karachi
Petaling Jaya   Singapore   Hong Kong   Tokyo
Nairobi   Dar es Salaam   Cape Town
Melbourne   Auckland

and associated companies in
Berlin   Ibadan

Published by Oxford University Press, Inc.,
200 Madison Avenue, New York, NY 10016

Oxford is a registered trademark of Oxford University Press

Library of Congress Cataloging-in-Publication Data
Grayston, Kenneth.
Dying, we live : a new enquiry into the death of Christ
in the New Testament / Kenneth Grayston.
p.   cm.   Bibliography: p.   Includes index.
ISBN 0-19-520789-0
1. Jesus Christ—Crucifixion—Biblical teaching. 2. Jesus Christ—
Resurrection—Biblical teaching. 3. Bible. N.T.—Criticism.
Interpretation, etc.   I. Title.
BT450.G7   1990
232.96'3'09015—dc20   89-35844

2 4 6 8 9 7 5 3 1

Printed in the United States of America
on acid-free paper

Acknowledgments

I take pleasure in acknowledging the support and advice of academic colleagues in the Department of Theology and Religious Studies of the University of Bristol and in its affiliated institution, Wesley College; in thanking Anne Cade, Vicki Jones, and Pam Eglinton, who typed the complex script; and in saying how greatly I have relied on my wife's patience while the book was being written and on her practised skill in proofreading and indexing. The professional care of the publishers—Darton, Longman & Todd in London and Oxford University Press in New York—will attract the reader's gratitude, which will match my own.

Contents

Dying, We Live

1

Introduction

The New Testament writings are concerned with the intentions and activities of God in relation to Jews and Gentiles. He is recognisably the same God as appears in the Old Testament and the writings of early Judaism. Similar styles of language are used throughout, though in different proportions: in the New Testament there is a modest use of cultic terminology and a fuller use of prophetic and wisdom expressions. But it is the same God at work, in his wisdom or by his spirit, through a variety of agents. The decisive *difference* between Jewish and Christian writings appears in the question of agency: the New Testament presents Jesus, at least in some respects, as the unique representative of God. His uniqueness can be considered in two ways: Did he appear as a special person? Did he act at a decisive time? The former question leads to Christology, the latter to soteriology.

If the path of Christology is followed, the Gospels are seen to describe Jesus as a Jewish teacher who gathers followers and debates with other teachers; as a wandering Jewish prophet and performer of miracles who falls foul of the religious leaders. He is referred to in familiar Jewish style as prophet, Lord and Son of God; and is asked whether he claims to be Messiah—but he prefers the modestly low-key expression *Son of man*.[1] Yet after his resurrection *Son of man* is dropped and Christ is the commonest term of reference (though scarcely in any of the Jewish meanings of *Messiah*). *Lord* and *Son of God* are redeveloped to express a new perception of Jesus; the Jewish teacher becomes the Wisdom of God, the prophet and miracle worker becomes the Word of God, and so on.

That path has been thronged with scholars for the last thirty years[2]—since about the time Vincent Taylor completed his trilogy on New Testament Christology (1958). But he had previously produced a trilogy on soteriology, following the other path.[3] His comprehensive work on the death of Christ in the New Testament employed the methods and summed up the conclusions of the first half of the century. In Great Britain he has had rather few followers, though exploration of this path has pressed forward in continental Europe and in the

United States.[4] It therefore seems proper to join in the enquiry, not so much by surveying existing work but by interrogating the New Testament evidence afresh.

If we follow the other path we ask whether Jesus acted for God at a decisive time. We cannot ignore either question, for a special person may be required for a decisive time. Yet we first direct attention to the time and examine the quality of the action. In general the answer is given in terms of eschatology and atonement; that is to say, the period of Jesus' activity is the decisive turning point in God's relations with Jews and Gentiles, the change from the old age to the new age, from one kind of human existence to a new kind. During that transitional period, the things Jesus said and did not only disclose God's intention but (as is the character of prophetic interventions) also set it in motion and made it irreversible. In particular it was the death and resurrection of Jesus that brought all this to the proof, possibly in a manner not clearly foreseen in Jewish expectation. Whatever some Jews may have believed about resurrection, it seems difficult to find any expectation that the resurrection of one person would have saving consequences or would alter the condition of mankind.[5] It is somewhat easier to suggest Jewish models for a saving, life-transforming *death:* the supposed ill treatment and violent death of the prophets, the godly sufferer of the Psalms, the suffering servant of Isaiah, the atoning death of the Maccabean martyrs, and Abraham's offering of Isaac.[6] But none is entirely persuasive, and no Jewish writing gives to a single death the prominence that the death of Christ receives in the New Testament.

It is prominent, of course, with the resurrection, in the four Gospels as well as in Acts, 1 John (but not 2 and 3 John), and Revelation; also in the Pauline Epistles (except for 2 Thessalonians and Philemon), in Hebrews (but not in James), and in 1 Peter (but not in 2 Peter or Jude). Absence from any of the shorter writings is not significant unless the theme appears to call for a reference. The character and style of James exclude so much traditional Christian language that it is scarcely unexpected that Christ's death and resurrection play no part, though the very awkward phrase in Jas 2:1 'our Lord Jesus Christ of the glory' may be a contorted memory of resurrection language. Yet it *is* somewhat surprising that Jas 5:10–11 uses the prophets as examples of suffering and patience and Job as an encouragement to steadfastness, rather than Jesus.[7] As for 2 Thessalonians, views of the Pauline authorship are neither confirmed nor shaken by the absence of explicit reference to the death and resurrection of Christ. The lordship of Christ is prominent (and may presume the resurrection), and the point of the epistle may be to explain (by means of a 'mystery') why that lordship is not yet decisively displayed. Be that as it may, the epistle contributes nothing to this enquiry.

It goes without saying that references to death and resurrection must be taken together. Although I am concerned with the death of Christ and have no intention of making a major study of his resurrection, I cannot study one without the other. Nor can I satisfactorily study the references after removing them from their contexts. Some of the references—for example, those which appear to be credal formulas—may indeed have existed independently before being used in their present position; but to begin with at least, they must be studied in context. Therefore, I invite the reader to join me in an exegetical enquiry, which will often need to look beyond the immediate reference and take in the wider argument. It cannot be assumed (when, for example, Paul refers to Christ's death) that the writer is bringing into play a preformed theology of atonement; he may be working it out as he deals with the ambitions or disorders of a Christian community. I am suspicious of the method that assembles the author's doctrine by fragments taken indiscriminately from his various writings or writings ascribed to him. I am even less attracted by the method that arranges references so as to support one of the views of atonement approved by theological opinion in the Church today. It seems to me necessary to read the evidence as it stands in the existing documents, with our imagination prompted by some knowledge of how people thought and felt in late antiquity.

That is true as much for the Gospels as for the other New Testament writings. The Gospel writers have provided a setting for whatever they chose to reproduce or (as some think) create. Doubtless they expected their contemporaries to make sense of what they wrote, both as being serviceable to the Christian community and as presenting the traditions by which they were made aware of the Lord. The relation between memory, tradition, and the community's self-awareness is complex; no informed reader can now regard the Gospels simply as complementary historical sources. Each Gospel has to be assessed in terms of the theological intentions of its author or editor, and in that respect the enquiry is not impossible. Whether we can go further back is open to question. We know that the Evangelists used sources. If (as I suppose) Matthew and Luke used Mark, we can see what they made of him; if (as I suppose) they used a collection—partly written, partly oral—of dominical sayings, we can more or less see how they used them. Thus, we can venture some conclusions about their theological intentions.[8] If Mark wrote independently of Matthew and Luke, no doubt he used sources, written or oral. We can ask what they were and whether, for example, there was a pre-Markan Passion narrative; we can try to distinguish between tradition used by Mark and his editorial changes and construction. In all three Gospels we can conjecture the earlier history of sayings and episodes and then assign them to probable environments in the primitive church or to conjectural stages in its development. Since we lack first-hand evidence for the earliest stages, the enterprise is circular; and when conjecture is piled on conjecture the results are open to stimulating disagreement.[9] I do not rule out the possibility of discovering something about the process by which sayings reached their canonical form. That would be foolish, for we possess information about the ground where the crop was formed, the seeds that were sown, and the climate that encouraged their growth. But we cannot confidently describe the crossbreeding and genetic engineering. My hope is to gather the crop and grade it for size and quality.

If, then, it is difficult to make confident statements about the earlier stages of the Gospel tradition, it is likewise difficult to be confident about the reported origin of the tradition in the words of Jesus. We need not be intimidated by Bultmann's statement that we cannot know how Jesus understood his own death; but we should not dismiss Marxsen's more cautious formulation: 'Jesus's personal attitude towards his death cannot be discovered *directly,* only by an indirect approach beginning from "what the witnesses testify".[10] Even if the quest for the actual words of Jesus (to which Jeremias devoted his life) is supplemented by the actual facts and his actual intentions (as Schürmann proposes),[11] it has to be admitted that Jesus said very little about his death—nothing in public about its saving significance, something perhaps to the restricted circle of his disciples when death was very close. That does not imply an absence of saving intention, only that it became plain after Easter. To demand that the death of Jesus should have only that meaning he gave it is to ignore the Holy Spirit. The death means not only what Jesus said it meant but also what it was found to mean.

The consequence is that this enquiry cannot begin with the Synoptic Gospels. Instead it begins with the earliest written evidence, namely the Pauline Epistles in the order 1 Thessalonians, 1 and 2 Corinthians, Galatians, Romans and Philippians. Whatever problems there are about the dating, integrity or function of this group, all were written between 50–65 C.E. Colossians and Ephesians are to be placed somewhat later, Colossians possibly while Paul was still alive but no longer fully in command of his missionary work, Ephesians perhaps after his death and so during the period of the Jewish War, 65–70 C.E. During that same period it seems to me probable that the Gospel of Mark and 1 Peter were produced. In the next fifteen years I am inclined to place writings with certain affinities, namely, the Pastoral Epistles, Luke-Acts, Matthew, and Hebrews. Finally, in the last fifteen years of the century there are the Epistles of John, the Gospel of John, and Revelation. In setting out the exegetical material it will be convenient to deal first with the Pauline corpus as a whole; then to consider the Synoptic Gospels, 1 Peter and Hebrews; and to end with the Johannine writings.[12]

At the end of long exegetical treatments I have written summaries. The necessary extent and detail of exegesis may otherwise obscure conclusions that can be drawn; but the summaries are not by-passes round congested areas. In the notes I have provided (1) sufficient justification for statements made, sometimes adding useful detail not strictly required by the argument, and (2) references to selected literature but not a comprehensive bibliography, not even to everything I have read and consulted. The learning and views of other scholars have stimulated me to fresh thought about the New Testament, and I have tried to do the same for others in this enquiry.

In presenting this exegetical enquiry I do not regard myself as either preacher or systematic theologian, even if the language I use may sometimes suggest that I hope to be both. I am inviting the readers of this book to enter, with understanding and imagination, into the thoughts of people whose view of the world was very different from ours. It is of course essential that the Church's preachers should speak today about the death of Christ in a way that engages with modern perceptions. For the integrity of Christian doctrine it is essential that systematic theologians should work out the significance of that death in a comprehensive account of what they propose for belief today. But my task has been to provide them with their basic material. I find that it is richer and more varied than is usually supposed and more closely entwined with other phases of theological, moral, and spiritual discussion than standard technical treatments suggest. If preachers and theologians draw out the consequences of my proposals, I think they should find themselves with room to move, liberated from the tight restraints of stock theories (which now often seem implausible or offensive). No longer under obligation to devise a theory of atonement that is but one part of a doctrinal formulation, they may be able to discover in the death of Christ a critical and interpretative principle for the whole of theology.

2

The Pauline Writings (1)

1 THESSALONIANS*

1 Thessalonians was written to give support and encouragement to the young Christian community of Thessalonica. This purpose can be discerned in the unusual structure of the letter. The introductory sections of letters in the Pauline corpus follow a standard form¹ (most simply displayed in Philemon and Philippians): the writer (1) names himself and any associates, (2) names the recipients, and (3) greets them; then (4) he thanks or blesses God, (5) prays for the well-being of the recipients, and (6) begins dealing with the matters in hand. In 1 Thessalonians, units 1–3 are concisely written but unit 4, the thanksgiving, is enormously expanded. It begins in 1:2, is repeated in 2:13, and is concluded in 3:9 (with a recall in 5:18). Only after the third reference to thanksgiving is the prayer of unit 5 introduced in 3:10–13, after which the writer is ready to move on to matters in hand of unit 6 with the words *loipon oun* (which mean 'and so to other matters', not 'finally' as in RSV).²

Why is the thanksgiving unit so greatly expanded not only in 1 Thessalonians but in 2 Thessalonians as well?³ The answer must be that Paul replied to the Thessalonians by turning their complaints about hardship into matters of thanksgiving. In 1:2–10 he thanks God for their hard work and endurance, their reception of the word with much hardship cheerfully borne that had made them a model to all believing communities in Macedonia and Achaia. In this respect they had become *mimētai hēmon . . . kai tou kyriou,* 'imitators of us and of the Lord' (1:6) in that they received the word (i.e., the gospel) with harsh consequences to themselves, which they bore with joy prompted by the Holy Spirit. This significant statement has two chief difficulties. First, the modern sense of *imitator* is unsuitable for what is being said. If Paul indeed suffered for accepting the gospel,

---

* See the Bibliography for commentaries on each of the New Testament writings discussed in this book.

8

the Thessalonian converts did not set out to imitate him, that is, to contrive a situation in which their conversion would also be painful. No doubt they knew that consequences might be unpleasant and were ready to accept them, but that is not what we mean by conscious imitation. What Paul intended to say was that their experience had been like his;[4] when they became Christians they had followed Paul's pattern. Second, in that case, how can it be said that they also followed the experience of the Lord? Perhaps it can be said in the general sense that like him, they experienced suffering in carrying out God's instructions. As he had endured hardship, so had his apostle and his apostle's converts. If so, the suffering and death of Christ is regarded as providing a model of the experience Christians must expect in this present age.

1 Thessalonians 1:9^b-10

This view, however, is only a subsidiary feature of the section 1:2–10. The main argument acknowledges the hardships but encourages those who bear them by setting out their singular success and the notable example they gave to others. The section ends by describing their condition of waiting for the coming wrath, from which *they* will be preserved and others will not. The description draws on a traditional formula (in 1:9^b–10),[5] probably of hellenistic Jewish-Christian origin, based originally on Jewish preaching to Gentiles. The Thessalonians had turned from their former idolatrous worship (as Jews would regard it) to become servants of a genuine, actively intervening God who was about to bring his wrath to bear. The evidence for this conviction was that he had raised his Son from the dead, for in some forms of Jewish eschatological expectation resurrection of the dead and final judgement belonged together. Thus they were waiting for the risen Son of God to come from heaven and begin the process of judgement. The significant words are these:

> *anamenein ton huion autou ek tōn ouranōn*
> *hon ēgeiren ek [tōn] nekrōn*
> *Iēsoun ton rhyomenon hēmas*
> *ek tēs orgēs tēs erchomenēs.*

> [to wait for his Son from heaven
> whom he raised from the dead,
> Jesus who delivers us
> from the approaching wrath.]

The implications of the formula may perhaps be discovered in the following way.

According to the pictorial language, the Son is in heaven with God. That corresponds closely to the phrase 'God the Father and the Lord Jesus Christ' (1:1, 2 Thes1:2), which reflects the standard hellenistic description of the divine being as *theos kai kyrios*. Hence any expected divine action will be the joint act of the Father and the Lord (see 3:11, where God and the Lord are asked to direct Paul's way to Thessalonica and 3:12, 4:6, and 5:2, where it is not easy, or necessary, to decide whether *Lord* refers to God or Jesus).

In the formula we are discussing, however, *kyrios* is not used, although the parousia of the Lord is often referred to elsewhere in the epistle (2:19, 3:13, 4:15–17, 5:23). It is likely that the primitive Christian confession *Kyrios Iēsous*, 'Jesus is Lord', stands behind the surprising appearance of the personal name Jesus in 1:10 (see *Lord Jesus* in 2:15, 19; 3:11, 13; 4:1, 2). But the formula in fact describes Jesus as 'his Son', for the only time in the epistle.

Here, then, is something similar to the traditional formula incorporated in Rom 1:4, where Jesus is 'designated Son of God in power according to the Spirit of holiness by his resurrection from the dead, Jesus Christ our Lord'. When God raised him from the dead, he overturned the victory of Death and gave Jesus mastery over the powers of Death (to use a Pauline personification). Hence, Jesus shared the life-giving powers of God and therefore became, or was disclosed as, the image or Son of God.[6]

As Son of God, Jesus therefore delivers us from the approaching wrath. The verb *rhyomai*,[7] uncommon in Paul, comes from the piety of the Psalms, which plead with God to deliver his people, and from passages in Isaiah where God is the Redeemer. In Rom 11:26, Paul quotes Is 59:20: 'The Deliverer [*ho rhyomenos*] will come from Zion.' Hence *Iēsous ton rhyomenon hēmas* means 'Jesus our deliverer.' A quotation from Ps 6:1, 4 is apt:

> O Lord, rebuke me not in thy anger,
> nor chasten me in thy wrath
>
> .   .   .   .
>
> Turn, O Lord, save my life;
> deliver me for the sake of thy steadfast love.[8]

When the Lord comes to conduct the final judgement, he does not appear as an impartial judge. He comes to save and vindicate his own. The Thessalonian community can be reassured because they are 'in God the Father and the Lord Jesus Christ' (1:1). They are all sons of light and of the day (5:5). They will be preserved from the disastrous consequences of the wrath because they belong to him who has been raised from the dead if they stand firm in the Lord (3:8).

The death of Christ is naturally presumed by his resurrection; but the death plays no theological part in the formula, which therefore joins a group of sayings ascribing our salvation to the resurrection. Since the formula appears to be a standard pre-Pauline sketch of the missionary proclamation, the absence of any theological function for the death of Christ cannot be remedied by reference to subsequent mention. The resurrection carries the main theological weight and so gives rise to the numerous *kyrios* references in the epistle.[9]

If Jesus was master of the powers of Death, it was indeed perplexing that some Christians should have died. Was it possible that God would not deliver them from death (as the psalmist had pleaded)? Because of such anxieties, Paul had to revise the simple presentation of Christ's lordship (e.g., by inventing the remarkable phrase 'the dead in Christ' [4:16] and by making Death the last enemy to be destroyed [1 Cor 15:26]), and to give more serious thought both to death itself and to the death of Christ.

1 Thessalonians 2:14–16

In the next section of the epistle, 2:1–12, Paul puts forward his own experience of hardship, repudiates adverse judgements on his conduct, and mentions the heavy, demanding work by which he supported himself. In effect he is saying, 'If you have had much to put up with, so have I'. Then he repeats the thanksgiving, in 2:13–16, because in bearing hardships the community is at one with the Christian communities in Judaea:

> You became imitators of the churches of God in Christ Jesus which are
>   in Judaea; for you suffered the same things from your own
>   countrymen as they did from the Jews.
>
> Who killed both the Lord Jesus and the prophets,
> and drove us out
> and displease God and oppose all men
> by hindering us from speaking to the Gentiles that they may be saved.

That surprisingly incidental reference to the death of Christ has a very instructive context.

As in 1:6 *became imitators* means that they had undergone similar awkward experiences, possibly by God's intention.

Mention of the death of Christ belongs to the area of anti-Jewish polemic. To say that the Jews killed the Lord Jesus (presumably meaning the Jesus who was shown by his resurrection to be Lord) must imply that they caused his death. Several New Testament passages distinguish clearly between what the Jews did and what the Romans did, though other passages use the more offensive presentation. The function of the death of Christ in this passage is to reinforce the energies of Christian communities in conflict with their neighbours, partly by an appeal to resentment.

The theological assessment of Christ's death in this passage is determined by a Jewish and early Christian rhetorical formula about the ill treatment and killing of the prophets.[10] This formula, which had its origin in the Old Testament and continued in use both in early Christian and later rabbinic writings, was made to serve several purposes: as part of a confession of sin; a complaint addressed to God; a reproach by, or on behalf of, God with accompanying threats of punishment; polemic against a rival religious group; and an encouragement to heroic loyalty. In no usage is it suggested that the prophetic death does anything more than cause shame, repentance, judgement or endurance. Nor is there any evidence that the prophets were regarded as martyrs whose dying had atoning value. The reference in 1 Thes 2:15 is a matter of reproach, matched by confidence that God's punishment of hostile Jews (and perhaps Gentiles) is now decided (2:16[b]).[11]

Finally, it is notable that Paul's most serious charge against the Jews is not that they killed the Lord Jesus (who is placed in the line of suffering prophets)[12] but that they obstructed the missionary proclamation to Gentiles 'by hindering us from speaking to the Gentiles that they [might] be saved'. That, above all, filled up the measure of their sins. At least in the circumstances of this epistle, the death of Christ has a much smaller theological function than either the resurrection or the evangelization of the Gentiles.[13]

Now that Paul has urged that the hardships of the Thessalonians are a matter for thanksgiving, he feels able to approach a more personally awkward question: Why had he sent Timothy and not visited them himself? In 2:17–3:10 he protests his earnest desire to see them (3:10) and implies that he had suffered much from anxiety about them (3:7) but explains that Satan had hindered him (2:18). When he was with them, he had warned them, 'We were to suffer hardship', for 'to that we are appointed' (*eis touto keimetha*, 3:3). His use of *we* confirms the statement that he is recalling what had previously been said. His teaching must therefore have included the statement that God had so arranged the end-time that his own people would suffer before the Judgement and the transition to the new age took place. It must also have included some reference to Satan (a name unknown to Greek converts) and his activities as tempter (or, better, tormenter [*peirazōn*]). Paul had become anxious in case the tormenter had tormented them and his missionary work had been reduced to nothing (3:5). Once again, the success of his evangelistic commission is primary. Any attack on it, any hindrance to his care for it, is specially wicked; yet Paul knows that *he* must suffer and his converts must suffer, for to that they are appointed. The suffering and frustration is required by God and provided by Satan. Why it should be so is not explained. There is, as it were, a dark side to God's saving activities; and it is possible for salvation to be lost because of suffering and frustration. Fortunately, in this case Timothy brings back an encouraging report, and Paul can safely repeat the thanksgiving in 3:9.

After this enormously extended thanksgiving (unit 4), the prayer (unit 5) is finally announced in 3:10 and written in 3:11–13—a request that the community may be suitably qualified to stand before God at the coming of the Lord with all his holy company, whether saints or angels. So now he comes to the matters in hand (unit 6): they have to do with instructions already given them about their need to avoid 'uncleanness' and their practice of community love (*philadelphia*). In some way, not quite easy to discern, they are related to particular problems within the community. A third matter, however, is plainly related to the community's expectations about the Parousia, and is treated in 4:13–5:11.

1 Thessalonians 4:13–14

If Jesus is their deliverer from the approaching wrath, what is the community to think about members who have died? If they belonged to him who is master of the powers of Death, how could Death claim them? Paul begins his reply by quoting what looks like a credal formula:

> *ei gar pisteuomen*
> *hoti Iēsous apethanen kai*
> *   anestē*
> *houtōs kai ho theos*

tous koimēthentas dia tou Iēsou
axei syn autō. (4:14)

This may be translated,

> Since we believe
> that Jesus died and rose,
> so also [we believe that] God
> will bring with him
> those who have fallen asleep through
> [their fidelity to] Jesus.[14]

The formula bears indications that it is non-Pauline, and that it has been modified from a rather simpler form. For one thing, it is grammatically incomplete so that words must be added to provide a proper link between the two clauses. I have supplied the words *we believe that,* with the intention of making the second clause a consequence of the first. For another thing, the second clause is overloaded, with two prepositional clauses: 'through Jesus' and 'with him'. It may plausibly be argued that Paul was making use of a formula that supported traditional teaching about the general resurrection in the form 'Because Jesus died and rose, so God will bring back the dead with Jesus when he comes from heaven'. That formula, already known to the community, had not suggested to their minds that some community members would have joined the dead before the arrival of Jesus. Therefore Paul qualifies 'those who have fallen asleep' by adding 'through Jesus' (in a sense of *through* already exploited in 4:2). A little later, in verse 16, he changes the rather gauche expression into something simpler ('the dead in Christ') but equally surprising to the Thessalonians. It is true, of course, that 'the dead in Christ' indicates that those who had died were Christians; but it was disconcerting that being Christian, they had died. It became necessary to realise that at least some Christians would die (as Christ had died) before Christ returned to inaugurate the new age for those Christians who were still alive. This position Paul modifies in 1 Cor 15:50, where he admits that 'flesh and blood cannot inherit the kingdom of God'; that is to say, even the living need to be transformed before they take possession of the new age. But already in 1 Thessalonians Paul has taken a step towards the possibility that Christians may share the death of Christ.

Even so, that is an oblique reference to his death. The formula in 4:14 is used to introduce a word of the Lord about the return to life of dead Christians and other events of the Parousia. The statement that 'Jesus died' has no obvious theological function.

1 Thessalonians 5:1–11

In the remainder of the section (5:1–11), Paul repeats what the community already knows about the timing of the day of the Lord. For the rest of mankind, it will come like an unexpected, devastating night attack; for the community—

watchful, armed and ready for the day—it will come as their salvation. They are
to encourage themselves with the conviction that

> God has not destined us for wrath
> but to obtain salvation through our Lord Jesus Christ,
> who died for [*hyper,* v.l. *peri*] us
> so that whether we wake or sleep
> we might live with him. (5:9–10)

This statement, which for the first and only time seems to present the death of
Christ with a strong theological significance, deserves careful examination.

Like the formula in 1:9–10, its context is the divine wrath and the resurrection, with the notable addition that he 'died for us'.

Wrath and salvation are the opposed possibilities when the day of the Lord
arrives. It had been Paul's whole aim that the Gentiles should share salvation
(2:16), which meant living with the Lord Jesus Christ. Those who paid no attention to the gospel (and perhaps those in the community who rejected consecration and practised uncleanness; cf. 4:6–8) would behave drunkenly and be over-
taken by destruction. On the other hand, 'God has not destined us for wrath',
that is, the path marked out for community members leads away from destruction (even though it will include suffering and may include death) and towards
salvation.[15]

'The Lord Jesus Christ who died for us' recalls a fairly common (and pre-
Pauline) New Testament formula. Its sudden introduction implies that the Thessalonians were familiar with it. It goes further than 'Jesus died' in 4:14 by suggesting some purpose in his dying; but the formula is in fact used to suggest a
variety of purposes.

'Christ died *for* us' (*hyper* or *peri,* according to variant readings) probably
lies behind 'the one *for* whom Christ died' in Rom 14:15 (*hyper*) and 'the
brother *for* whom Christ died' in 1 Cor 8:11 (*dia* + accusative). Dying *for* us
means dying for our advantage, namely to make possible our membership of the
Christian community. No explanation is given.

In 1 Pt 2:21, 'Christ also died [v.l. suffered] *for* you, leaving you an example,
that you should follow in his steps' (*hyper* or v.l. *peri*). Our advantage is seen in
the exemplary death, as elsewhere in the New Testament, even if the theological
appraisal is taken further in 2:24.

A more obvious theological expansion of the simple formula is that 'Christ
died *for* the ungodly' and 'that while we were yet sinners Christ died *for* us' in
Rom 5:6, 8 (*hyper*). Thus the death of Christ secured benefits for sinners, notably
the benefit of 'justification'. And Paul continues,

> Since therefore we are now justified by his blood,
> much more shall we be saved by him from the wrath.
> For if while we were enemies
>   we were reconciled to God by the death of his Son,
> much more, now that we are reconciled,
>   shall we be saved by his life.

It is clear that 'justification', or reconciliation, are secured by Christ's death and that salvation from the wrath is secured by his life. Nothing is gained by ignoring that distinction. How his death effects the justification or reconciliation of sinners is not (here) explained; but once they are justified and reconciled to God, they share the life of him who is master of Death. When therefore in the wrath, destruction and ruin fall on all around, those who belong to Jesus are preserved.

The formula 'He died and he rose' is expanded in pre-Pauline tradition recorded in 1 Cor 15:3: 'Christ died for our sins in accordance with the scriptures,' and what follows. The formula points to a scriptural basis (without identifying it) for its statements about death and resurrection, though Paul makes no use of these qualifications and uses the formula only as an introduction to his discussion of resurrection. In the phrase 'for our sins' (*hypertōn hamartiōn hēmōn;* cf. Gal 1:4, *hyper* v.l. *peri*) the preposition *hyper* means not 'for the benefit of' but perhaps 'in order to deal with'. Compare in 1 Pt 3:18 'Christ also died [v.l. suffered] for sins once for all, the righteous for the unrighteous, that he might bring us to God'. In 'for sins', the preposition is *hyper* or *peri*. Neither preposition in itself conveys a precise meaning, but the intention is plain: Christ's death overcomes the alienating condition of sinfulness and restores us to God. The formulation in 1 Peter is immediately followed by a different expansion of the formula 'He died and rose' in 'being put to death in the flesh but made alive in the spirit'.

Lastly, the formula is used to express a common hellenistic theme: the death of one man for the benefit of a whole group. In Jn 11:50–52 Caiaphas is given the ironic inspiration to say 'that it is expedient for you that one man should die *for* the people, and that the whole nation should not perish' *(hyper)*. The evangelist comments that Jesus died not only for the nation but to gather into one the children of God who are scattered abroad, which implies that his death provided the means of identifying those who belong to God. Paul's use of the same principle is different. 'We are convinced', he says in 2 Cor 5:14–15, 'that one has died *for* all' *(hyper)* and he draws the conclusion that 'therefore all have died'. In some way they participate in his death and also in his risen life: 'And he died for all, that those who live might live no longer for themselves but for him who for their sake died and was raised'. He does not stop at saying that they live for him who died for them, which would be a properly grateful response to a martyr; but that they live for him who died and was raised.

This survey does not include similar and equivalent formulas (e.g., Gal 2:20, 3:13; 2 Cor. 5:21), because its intention is to discover what Paul intends in 1 Thes 5:10 by using the precise words *who died for us*. The use in Romans 14 and 1 Corinthians 8 is generally apt but takes us no further; and the use in 1 Peter is apt in 1:6 but not here. The uses in Romans 5, 1 Corinthians 15 and 1 Peter 3 are illuminating except that the Thessalonian formula says nothing of sin and sinners. The use in John 11 and 2 Corinthians 5 is close in meaning except that the Thessalonian formula makes no use of the theme that one dies for all. In fact the closest parallel comes in Rom 14:9 (an expansion of the formula 'He died and arose'), namely

> To this end Christ died and lived again,
> that he might be Lord both of the dead and of the living.

It becomes clear that in 1 Thessalonians, the resurrection of Jesus carries the main theological weight. No doubt any religious group would be distressed by persistent hardship and the death of some members; but this community was both distressed and perplexed because it believed its Lord to be master of Death. Paul seeks to remove their perplexity and to fortify them by introducing the surprising thought of 'the dead in Christ' by constructing an imaginative scenario for the parousia and by explaining that Christ actually died for us so that he could be Lord of both dead and living Christians. Apart from that, however, the death of Christ has no positive theological function. Suffering for the community has to be recognized as willed by God and perhaps engineered by Satan. It can be blamed on those who are hostile to the community, part of a long history of opposition to God, to the prophets, and to the Lord Jesus. Hence in some measure they share Christ's death as well as receiving benefit from it.

1 CORINTHIANS

The situation of the Christian community in Corinth was similar, in some respects, to that of the community in Thessalonica. Both were gifted and energetic communities, both were waiting for the revealing of the Lord Jesus Christ, both were troubled about people who had died, and both were debating questions about the conditions in which Christ would rejoin his people.[16] But the theme of suffering and persecution evident in 1 Thessalonians is absent from 1 Corinthians. In the former, the community is alienated from its neighbours; in the latter, the community seems to be rather closely related to the pagan society from which members had been recruited. That being so, it is not surprising that statements about the death of Christ now become more prominent and take on a critical function in the Pauline instruction. In dealing with questions from the community and with complaints about them, from time to time Paul looks to the death of Christ for a solution; and then devotes the long final dissertation to the resurrection. Thus, whether by design or accident, the epistle is constructed as a development and qualification of the early formula 'Christ died and rose again'.

In a well-developed thanksgiving unit (though short by the standards of the Thessalonian letters), Paul praises the community for its rich equipment *en panti logō kai pasē gnōsei* ('with all speech and all knowledge', 1:4). That in itself was confirmation of the apostolic testimony to the benefits of Christ, so that possession of logos and gnosis is not only proper but to be expected. But what is meant by these two words?

Logos

Paul uses the word *logos* with various meanings. In general it may indicate some kind of statement, sometimes (but not necessarily) a verbal rather than a written statement. Often it means a divine statement—an edict, command, promise or explanation. Clearly this sense does not suit the present occurrence, though it is appropriate for the references in 14:9, 19, 36.

It commonly indicates the apostolic preaching and teaching, as it does in 1:18 ('the logos of the cross') and in 2:4ᵃ where 'my logos' has the same meaning as 'my kerygma'. This does not suit the expression *panti logō* in 1:4.

The meaning 'account' does not appear in the Corinthian correspondence.

Logos can refer to 'speech' in a general sense, sometimes joined with *ergon* ('in word and *deed*') or contrasted with it in a manner familiar in classical Greek ('not merely in words but in *practice*'). The contrast is a prominent feature in this epistle. Paul asserts that the kingdom of God does not consist in logos but in power *(dynamis)*, so that he is not interested in the logos of his arrogant questioners but in their *dynamis* (4:19–20). He had been constrained to preach the gospel not by *sophia logou* (see p. 24) but by the logos of the cross which is God's *dynamis* (1:17–18). His logos and kerygma were not presented in the plausible logoi of wisdom but in demonstration of the Spirit and *dynamis* (2:4ᵇ). He concedes that there is indeed an appropriate wisdom for Christians (12:8, cf. 2 Cor 8:7)—to be communicated, however, not in language instructed by human wisdom but in language instructed by the Spirit (2:13). So he says, 'When I came to you, brethren, I did not come proclaiming to you the testimony of God *kath hyperochēn logou ē sophias*. For I decided to know nothing among you except Jesus Christ and him crucified' (2:1–2). The words left untranslated need care: the genitives 'of speech or wisdom' are joined to an adverbial phrase which means 'in a preeminent manner'. Did Paul mean 'not in a preeminent manner shown in speech or wisdom'? or '[not impressed by] the overriding claim of speech and wisdom'? From 2 Corinthians it is clear that some members of the community were dissatisfied with Paul's speech: 'His letters are weighty and strong, but his bodily presence is weak, and his speech is of no account'. To that Paul retorts that what he is by logos in letters, he will prove to be in action when present (2 Cor 10:10–11, logos contrasted with *ergon*); and, a little later, he claims that though unskilled in speaking, he is not unskilled in gnosis (2 Cor 11:6).

Thus it is clear that Paul resisted criticisms of his own speaking and was reserved about the Corinthian interest in 'speech', preferring the *dynamis* of the cross accompanied by the activity of the Spirit. Yet he admitted logos as a proper gift within the community—indeed, a gift of the Spirit. How, then, do we envisage this logos about which Paul is ambivalent? Fundamentally it is the ability to be articulate, to present an argument or set of convictions by a series of intelligible statements, to examine one's own responses and formulate them in coherent fashion for communication to others. When formally developed, this becomes rhetoric, the art of persuasion. By Paul's day, rhetoric had played a varied and debatable part in Greek life for more than four centuries and was enjoying something of a revival. Its reputation depended largely on what kind of sophist (an itinerant teacher of practical wisdom at some periods, of cynical disbelief in morality at others) was offering, and probably selling, its benefits. It is commonly supposed that Paul was objecting to the more specious kinds of rhetoric, sometimes designated as eloquence. Although eloquence may be defined as the fluent, forcible and apt use of language (as in COD)* and may sometimes be regarded as a pleasing feature of spoken or written style, we often distrust it as designed to mislead, to conceal false and empty promises, or to misrepresent the strength of arguments. It is certain that Paul had been suspected of the worse kind of sophistry (1 Thes 2:5–6); but he is scarcely rejecting that criticism in this epistle, where, if anything, he defends his underuse of eloquence. That he could use it in writing, if not in speaking, is clear from many passages in his letters. Why then does Paul reject it in favour of the cross? Because rhetoric or eloquence can be used not just to please or to dupe the hearers but in a powerful and far-reaching way to form attitudes and fix them permanently. In certain situations rhetorical language escapes the control of those who devise it and seems to take on a life of its own, driving people to actions they had not supposed themselves capable of. Paul had known that experience as a zealous Pharisee persecuting the church of God (1 Cor 15:9, Gal 1:13). When he renounced his former profession, he did not transfer his rhetorical skill from the Jewish to the Christian cause but resolved that the *dynamis* of the gospel should arise from God's action in Christ's death, not from Paul's trained rhetorical experience. When, therefore, he praises the Corinthians for their enrichment *en panti logō* he is expecting them to subject that spiritual gift to critical examination by their fidelity to the cross.

Gnosis

Now we may turn to the second word, *gnosis*,[17] which occurs often in the Corinthian correspondence but infrequently in the other Pauline Letters. It appears as one of the varied spiritual gifts of the Christian community, which are listed in several places: 1 Cor 12:8 as *logos gnōseōs* alongside *logos sophias;* 13:2, 8 with prophecy, tongues, faith and so on, contrasted with love; 14:6 joined with revelation, prophecy and teaching and contrasted with tongues; 2 Cor 6:6, 8:7 with the Spirit, love, truth, faith, and so on. By themselves, however, these references give little information about the nature of gnosis. More instructive is the discussion of food offered to idols in 1 Cor 8:1–11. It begins abruptly with the admission that 'all of us possess knowledge'. That no doubt reflects the manner in which the Corinthians had framed their question; and it suggests that they were looking to gnosis to settle their community problems. Paul at once cautions them against the dangers of gnosis: it makes the possessor conceited, unable to recognize its inherent limitations. But then he states plainly the relevant gnosis—'We know [*oidamen*] that "an idol has no real existence" and that "there is no God but one"' (8:4)—and proceeds to define the second item in relation to God the Father and the Lord Jesus Christ (8:6). Yet (surprisingly) not everyone possesses this gnosis; and a Christian who, by reason of his gnosis, insists on doing what he thinks fit may destroy the brother for whom Christ died (8:11).[18] Thus the only explicit example of gnosis in 1 Corinthians refers to knowledge of the sole being of God and the nonexistence of idols; and it is a piece of knowledge that

---

* For this and other abbreviated sources, see pp. 337–339.

must be combined with love before it can safely be used to decide cultic problems within the community.

Three references in 2 Corinthians confirm Paul's use of gnosis to indicate knowledge of God: 2:14, where the knowledge of God is spread like the fragrance of a sacrificial offering; 4:6, where knowledge of the glory of God on the face (i.e., person) of Christ is a new creative illumination; and 10:5, where knowledge of God is joined with obedience to Christ. If statements in the two epistles are brought together, the following summary can be offered: Gnosis is awareness of God as the sole existent divine being, who is comprehended by means of Christ's lordship and death. This is a spiritual—not a natural—perception, with consequences for the life of the community. It is the apostolic duty to spread this gnosis, even at sacrificial cost. Yet not all members of the community possess it, and those who do can use it to endanger themselves and others. Hence, in practical use it ranks lower than love.

There is one more reference, in 2 Cor 11:6, where Paul, replying to the jibe that his letters are weighty and his speech of no account (10:10), says, 'Even if I am unskilled in speaking [logos], I am not in knowledge [gnosis]'. It is by no means certain that Paul admits the criticism of his speaking (the Greek could properly be translated 'even if I were unskilled') but it is clear that he distinguishes logos from gnosis. Thus, gnosis is the particular Christian awareness of the being of God, and Paul's logos would express that awareness not in a rhetorically attractive form but in displeasing insistence on cross and resurrection. From Paul's manner of argument elsewhere, it seems unlikely that he is siding with the philosophers (gnosis) against the sophists (logos).[19] No doubt trends of discussion within the Corinthian community led him to distinguish form and substance. But as a result, 'utterance of knowledge' in 1 Cor 12:8 means 'gnosis articulated publicly' rather than confined to inward experience.

Sophia

The same passage, however, also contains 'utterance of wisdom' *(logos sophias)*; and the word *sophia* takes the reader back to the early part of the epistle. After the mention of logos and gnosis at 1 Cor 1:5 in the thanksgiving unit, gnosis disappears until chapter 8. In the first major structural unit of the epistle, 1:10–4:21, it is replaced by *sophia*. How are the two to be distinguished?

Ginōskō and Oida. It must first be noted that not everything has been said about knowledge when the use of gnosis has been examined. Two (possibly three) *verbs* are also concerned: *(epi)ginōskō* and *oida*. In their normal, secular meanings they are interchangeable; but Paul uses them with a difference. For one thing, *ginōskō* usually means to know someone or to experience something; only occasionally does it mean to know *that* such and such is the case. (By contrast *oida* can mean both; very frequently it means 'knowing *that*'.) What is more, in 1 Corinthians *ginōskō* is always used in negative constructions: 'The world did not know God through wisdom' (1:21); none of the rulers of this age knew the secret and hidden wisdom of God (2:8). In fact none but the Spirit of God knows what is proper to God (*ta tou theou*, 2:11), and the unspiritual man (the *psychikos anthrōpos*) cannot know what is proper to the Spirit (*ta tou pneumatos*, 2:14). All these disclaimers seem to arise from reflection on a phrase in Is 40:13: 'who has known the mind of the Lord' (the only place in the Septuagint where *nous*, 'mind', represents Hebrew *ruaḥ*, 'Spirit'). Even when Paul admits that at least Christians know God, he argues that such knowing is at present partial: 'If anyone imagines that he knows something, he does not yet know as he ought to know' (1 Cor 8:2). Now we know in part, but even so our knowledge will pass away. Only when the perfect has come shall we know fully as we have been fully known (*epiginōskō*, 1 Cor 13:8–12). Thus our knowing God, possible only within the spirit-moved community, is no more than a dim reflection of his activity in knowing us. If we look for a closer description in 2 Corinthians we find, 'You know the grace of our Lord Jesus Christ, that though he was rich, yet for your sake he became poor, so that by his poverty you might become rich' (2 Cor 8:9). And at the end of the epistle is, 'Do you not know that Jesus Christ is in you?' (2 Cor 13:5). In the same epistle Paul boldly extends his dialectic of knowing God to embrace knowing persons: 'From now on we know [*oidamen*] no one by human perception; even if we once knew [*egnōkamen*] Christ by human perception, we know [*ginōskomen*] him so no longer' (2 Cor 5:16). Since in Paul's statements about Jesus Christ the main event open to human assessment was his crucifixion, his meaning in this statement must surely be that he had been led to revise an earlier understanding of his death. Thus whatever kind of 'knowing' is admitted by Paul, it is qualified by that presentation of Christ that he makes central in his proclamation of the gospel.

There is more to be discovered when we turn to the relevant usage of the verb *oida*. We have already noticed the exegesis of gnosis in 1 Cor 8:4: 'We know *that* "an idol has no real existence",' and so on. About half of the other occurrences of 'knowing *that*' formulate items of specifically Christian information about the condition of the community and its members:

> *1 Corinthians*
>
> that you are God's temple, and that God's Spirit dwells in you (3:16)
>
> that your bodies are members of Christ (6:15)
>
> that he who joins himself to a prostitute becomes one body with her (6:16)
>
> that your body is a temple of the Holy Spirit within you (6:19)
>
> that the head of every man is Christ, the head of a woman is her husband, and the head of Christ is God (11:3)
>
> *2 Corinthians*
>
> that as you share in our sufferings, you will also share in our comfort (1:7)
>
> that while we are at home in the body, we are away from the Lord (5:6)

And half formulate information about eschatological expectations:

1 Corinthians
that the saints will judge the world (6:2)

that we are to judge angels (6:3)

that the unrighteous will not inherit the kingdom of God (6:9)

that in the Lord your labour is not vain (15:58)
2 Corinthians
that he who raised the Lord Jesus will raise us also with Jesus and bring us with you into his presence (4:14)

that if the earthly tent is destroyed, we have a building from God, a house not made with hands, eternal in the heavens (5:1)

All these items of information are used to solve specific problems within the community, and most of them explicitly refer to God, Christ or the Spirit. They may therefore reasonably be described as religious wisdom, if *sophia* is given the standard Greek definition of skill in matters of common life. Hence (to answer the question raised above) wisdom and knowledge are not so much exclusive as differently directed. Both are forms of religious awareness: the one *(sophia)* directed towards the existence of the community as it looks forward to the great transformation, the other (gnosis) directed towards the existence of the divine being who has set the great transformation in train.

1 Corinthians 1–3. It is now possible to consider Paul's preoccupation with *sophia* and *sophos* in the first three chapters of 1 Corinthians (1 Cor 1:17, 19–22, 24–27, 30; 2:1, 4–7, 13; 3:10, 18–20). It is commonly supposed that Paul was attacking *sophia* in two senses: the use of logical and rhetorical devices in preaching and the Gnostic use of *sophia* as the content of salvation.[20] He indeed admitted that there were proper senses in which *sophia* could be, and should be used, but he had probably been pushed to that admission by ways of thought that dominated some part of the Corinthian community. This standard interpretation can best be tested by examining 1:22–25.

When Paul said that 'Jews demand signs and Greeks seek wisdom', was he deriding the religious credulity of the Jews and the intellectual frivolity of the Greeks (or Gentiles, as in verse 23)? Or was he contrasting the predictable and proper demands of these two groups with his own disturbing proclamation? Let us take the Jews first. A 'sign' is an indication from God, not necessarily a prodigy. (Gentiles would read it as an 'omen'.) No doubt there were Jews who hoped that God's approval would be shown by miraculous interventions. Even the rabbis knew such expectations, as the famous story of Rabbi Eliezer shows.[21] Despite the prodigies that supported his halakah, the remaining rabbis took no account of them but relied on Torah given from Sinai. In the hellenistic Judaism of Philo (who took little notice of prodigies), a 'sign' is very frequently a scriptural indication or proof.[22] It is notable that when *sēmeion* in 1 Corinthians is taken to mean a characteristic Jewish demand for prodigies, the only parallel evidence ever cited is from passages in the Gospels where Jesus is asked for a sign. The single Gospel logion (variously developed in Mark, Q, and John)[23] certainly must be taken into consideration in any discussion of the debate between Jews and Christians on the verification of God's intentions, but it can scarcely be regarded (as it stands) as direct evidence for prevailing Jewish attitudes. Moreover, Paul says that Jews regarded the crucifixion of Jesus as a *skandalon,* that is, as something religiously offensive. If so, it could not be neutralised by a prodigy; this would certainly be ruled out by Dt 13:1–3, which prohibits attention to a prophet or a dreamer of dreams who offers a sign or wonder to justify novel religious practices. In any case why should Paul object to prodigies when his own apostolic ministry was accompanied by signs and wonders (Rom 15:19, 2 Cor 12:12, the stock Deuteronomic phrase for saving wonders)? It must therefore be the case that Paul was recording the proper Jewish demand for a scriptural justification of his preaching of the cross. At this stage he was unwilling or unable to provide it (contrast his citation in Gal 3:13), and offers only God's saving *dynamis* (verse 24).

Paul therefore leaves the Jews and turns to the Greeks who are easier to deal with because they are familiar with individuals who sacrifice their lives to preserve the community.[24] It is useful to remember that Corinth in Paul's day was little more than a hundred years old, more a Roman than a Greek city. Many of its families had risen rapidly in social position and wealth. It provided opportunities for advancement in commerce, banking, artisan skills, and civic administration.[25] At least it was so for those who exercised good practical judgement, who were experienced, shrewd and calculating—in other words, who possessed the virtue of *sophia.* When the Gentiles heard the crucifixion of Jesus being offered as a religious proclamation, they regarded it as 'folly' *(mōria)* or 'weakness'. The conjunction in verse 25 of folly and weakness indicates that (in this context) folly means not 'silliness' but socially destructive stupidity.[26] *Sophia* is whatever makes for social cohesion and advancement; by contrast, *mōria* implies ill-considered and rash conduct, which may flout received views and will tend (rather through stupidity than wickedness) to destabilise society. Naturally, I am not suggesting that the Gentiles who heard Paul's announcement of the gospel expected him to confirm or advance their social position; but I think it likely that they would reject a proclamation that ran contrary to their standard social attitudes. Their criticism of Paul's gospel was not intellectual but social.

This social interpretation of *sophia* is confirmed by two considerations. First, it corresponds well enough to the items of religious wisdom in the 'knowing *that*' formula, namely, information about the social status and relationships of the Christian community and about the function of the community in the expected period of transformation. The first of these may be regarded as the distinction between the community and the *kosmos,* the second as the community's relation to the present *aiōn* and the coming one. That leads at once to the second confirmation, namely the description of the desired wisdom as 'wisdom of the world' (1:20–21, 3:19) or 'wisdom of this age' (2:6, cf. 3:18). The word *kosmos* is used with exceptional frequency in 1 Corinthians, and this fact arouses the suspicion that some of the community were saying, 'What will the world say?' in the sense of society at large.[27] In this context, *world* means that social grouping to which Christians belong but from which in important respects they differentiate themselves. But the shrewdness necessary for successful existence in society at large has (according to Paul) been shown by God to be in fact destructive folly; for God has chosen what society regards as stupid, weak and contemptible to embarrass society's shrewd and powerful place holders (1 Cor 1:27–28). The charge against social wisdom is not that it is man-centred, even though it is called human wisdom *(anthrōpinē sophia)* in 1 Cor 2:13. After all, what else could it reasonably be, since God is surely concerned with the well-being of humanity? No, the charge against social wisdom is that it becomes overconfident and fails to observe its own inherent limitations: 'The *sophia* of the *kosmos* is *mōria* with God. For it is written, "He catches the wise in their *panourgia*"' (1 Cor 3:19), that is, in their readiness to attempt anything and to use any means to their chosen ends. Consequently the *kosmos* by the exercise of its own *sophia* cannot know God in his saving activity, which not only goes beyond the limits of social wisdom but also operates by means that must appear to the *kosmos* as demeaning and imprudent (1 Cor 1:21).

But it is not yet clear why God must overturn social wisdom and condemn it as folly. The answer lies in Paul's indentification of social wisdom as 'wisdom of this age'.[28] At once that gives the social structure a temporal qualification. Its aims and its skills—indeed its wisdom—have only relative authority when compared with some other age, especially with the new age (an expression commonly used by New Testament exegetes but not by Paul). In fact, 'the rulers of this age ... are doomed to pass away' (1 Cor 2:6) when Christ delivers the sovereign power to God the Father after destroying every rule and every authority and power, even the power of death itself (1 Cor 15:24–26). Equally doomed, it would seem, are human physical existence (1 Cor 6:13) and spiritual activities such as prophesying, speaking with tongues, and gnosis (1 Cor 13:8). The vivid image contained in the statement, 'When I became a man I did away with the affairs of childhood' (which uses the same forceful verb *katargeō* used previously for 'destroy', 'doomed to pass away'), suggests that the destruction is not wanton or condemnatory but a necessary action in moving from immature to mature social existence. Granted, however, that 'this age' was destined to pass away, it was inevitable that the wisdom of this age would resist its dissolution. The very purpose of social wisdom is to maintain the present situation and to preserve the power of those who are socially dominant. Hence the oracle becomes relevant in which God says, 'I will destroy the wisdom of the wise, and the cleverness of the clever I will thwart' (1 Cor 1:19). The quotation from Is 29:14 referred originally to the practical understanding claimed by the king's advisers who were responsible for Israelite foreign policy. Paul gives the words a wider reference in a larger crisis. In this crisis wise men, whether Jewish scribes or Greek debaters, no longer have any clear function (and in this perception lies not only disenchantment with social wisdom but also far-reaching reservations about the scribal use of Torah;

see 1 Cor 1:20). In the process devised by the divine Wisdom for the dissolution of the *aiōn*, social wisdom has perversely become self-destructive, and the *kosmos* therefore cannot know God in his saving intention (1 Cor 1:21).

Hence, Paul insists that he was sent to preach the gospel but not with *sophia logou* which, in view of my earlier discussion, may be rendered 'not with verbal wisdom'[29]—with an implicit contrast between *word* and *power*. Nor was his speech and message *en peithoi[s] sophias [logois]*, but in demonstration of the Spirit and of power (1 Cor 2:4).

The untranslated words present well-known problems of text[30] (as the square brackets suggest) and vocabulary (since the adjectival form *peithois*, presumably from *peithos*, is otherwise unknown). The noun *peithō* has the sense of 'persuasion', either the means of persuading others or a persuasion in one's own mind, hence 'conviction' or 'confidence'. Nowhere else does Paul use the noun or the normal adjective (*pithanos*, 'persuasive'), though he uses the verb, generally its perfect form, in the sense 'I am confident'. In this section of the epistle he has just argued that no human being should boast in the presence of God (1:29), and he now refers to his own weakness and insecurity as God's messenger. In so doing he makes the point that he could not encourage himself by his inner confidence in some proposed human or social wisdom.[31] He had to make the best he could of a demonstration of Spirit and power, which came from preaching Christ crucified and was not under his control. If the word *logois* was indeed an original part of the text, I suspect it of being an afterthought, as if to say, 'no confidence in ('no persuasiveness about', if *peithoi* has the active meaning) social wisdom—expressed in mere words'.

At this point, however, often to the perplexity of exegetes, Paul abruptly inserts a passage in commendation of wisdom.[32] Instead of moving promptly from 'demonstration of the Spirit and power' to his regret at being unable to address his readers as spiritual men (1 Cor 3:1), he concedes that there *is* a wisdom to be discussed within the community among members who are described as *teleioi*. He introduces (or reintroduces, see 2:1), the word *mystery*,[33] associates it with a scheme of thought (God's age-long intention, previously hidden and now revealed) that appears again in the deutero-Pauline writings, and supports it with an unidentifiable citation, presumed to be scriptural. And before resuming the account of his manner of speaking to the community, he draws a distinction (and argues for it) between *pneumatikoi* and *psychikoi* (spiritual and unspiritual persons, as RSV translates). The distinction appears again with a different application in 1 Cor 15:44, 46, but is taken up in 1 Cor 3:1 by a contrast between *pneumatikos* and *sarkikos* (RSV translates 'men of flesh'). Paul ingeniously regains the direct line of his argument, but clearly something has driven him off course in 1 Cor 2:6–16. The difficulties may be solved, however.

In 1 Cor 2:6–16 Paul is writing somewhat defensively. He is withdrawing from an overexposed position. Corinthian readers would not hear (as we commonly do) the consistent preacher of Christ crucified battling against purveyors of wisdom; they would hear Paul, who himself had offered wisdom, withdrawing to his primary preaching of Christ crucified. That Paul should have offered wisdom is not surprising. If the preaching of death and resurrection had the powerful

effect of creating a believing community, it would soon become necessary to provide—or encourage the community to provide—an understanding of its existence and an expectation for its future. Thus *sophia* would develop and perhaps
get out of hand if the inspiration of 'resurrection' outran the sobering thought
of 'death'.

The dissensions within the community[34] convinced Paul that teaching associated with himself, Apollos or Cephas was proving destructive (1 Cor 1:12; 3:4,
22). They were in dispute about divergent Christian life-styles, claiming the
authority of named persons and using slogans of social success: 'Already we are
filled. Already we have become rich. We have become kings' (1 Cor 4:8). Their
kinds of *sophia* had led them to a position of self-confidence and satisfaction
within the community but also to rivalry and dissension. Therefore it was necessary to return to the foundation, namely Jesus Christ (1 Cor 3:10–15)—hence
(in Paul's understanding) to cross and resurrection.

It is clear from the various words used to describe those who heard Paul's
instruction—*teleioi, pneumatikoi, psychikoi, sarkikoi* (1 Cor 2:6, 13–15; 3:1,
3)—that no fixed terminology was in mind. The controlling image is of maturity
and immaturity. The *sarkikoi* ('men of the flesh') are immature Christians who
by jealousy and strife show that they are still held by ordinary human desires.
The *psychikoi* are unreflective non-Christians to whom the divine spiritual energies seem unsocially stupid. The *teleioi*[35] are mature Christians (not the 'perfect')
who can be entrusted with explanatory myths of the divine wisdom; and the
pneumatikoi are experienced Christians in whom the divine Spirit is at work to
create the structure and life-style of the community. Hence, part of the anxiety
about *sophia* is that the community has accepted it from named leaders rather
than discovering it by their own endowment of Spirit; and Paul is partly to blame,
though he says apologetically, 'I could not address you as spiritual men' (1 Cor
3:1).

In principle, of course, they were spiritual: 'Do you not know that you are
God's temple, and that God's spirit dwells in you?' (1 Cor 3:16). The ultimate
life-creating energy was at their disposal, as becomes abundantly clear in the discussion of spiritual gifts in 1 Cor 12. They have no need to dispute about human
teachers of wisdom, 'for all things are yours, whether Paul or Apollos or Cephas
or the world or life or death or the present or the future, all are yours; and you
are Christ's; and Christ is God's' (1 Cor 3:21–22). For a second time the four
names of 1 Cor 1:12 are recited, but this time to make clear the true offence of
their former mention where Paul, Apollos and Cephas are raised to the same level
as Christ, or Christ is brought down to the level of Paul, Apollos and Cephas. But
it is not these three or any others but only Christ whom God has made their
sophia, which Paul expands by three words: righteousness, sanctification and
redemption. Each word occurs only here in 1 Corinthians, so that this is likely
to be a previously formulated sentence, not simply devised for this discussion.
Their relevance to this discussion can, however, be seen if they are released from
their conventional translations. *Dikaiosyne* means 'acceptability' to God, *hagiasmos* means 'consecration' to God's purposes, and *apolytrōsis* means 'liberation'.
Hence those who are 'consecrated in Christ Jesus' (1 Cor 1:2) are set free by the

gifts of the Spirit of God to make a proper judgement in all matters and are not subject to the appraisal of others (presumably Paul, Apollos and Cephas, who are put forward as rival authorities). After all, according to scripture no one can give instructions to the Lord, but we have the mind of Christ (1 Cor 2:14–16). The conclusion of this discussion is that Paul is talking to the whole community about its powers and responsibilities. He is not talking about different grades of community membership (as if in Gnostic fashion) but about a community that is tempted to avoid its responsibilities by seeking *sophia* from named persons of authority.

There is another sense, however, in which Paul speaks of *sophia,* the sense that prompted the diversion at 1 Cor 2:6. It is not a wisdom of this age *(aiōn);* hence it is directed neither to maintaining the present structures of social life nor to creating structures, however provisional, for the Christian community. It is a divine wisdom, decreed by God before the ages but hitherto hidden, intended for the glorification of those who love God. The extent and benefits of that glory are now revealed by the Spirit. Indeed the main thrust of 2:6–16 is directed towards those benefits. Paul's reference in verse 4 to 'demonstration of the Spirit and power' is expanded and explored in verses 9–16.

His negative mention of 'the wisdom of men' in verse 5 is offset by verses 6–8, which contain an explanatory myth. By finding a myth in verses 6–8, I mean that they draw on a familiar stock of images, imbued with emotional force and traditionally used to portray the activities of the divine being, especially in relation to human beings. By describing this as an explanatory myth, I mean that the verses are not intended to provide objective information but to locate certain convictions within a particular framework of understanding. Nor indeed are they intended to convey saving information by which mature Christians, the *teleioi*—those, perhaps, who have passed beyond the stage of initiation and are now ready to advance to a higher stage—may begin the Gnostic journey. Paul's concern in this section of the epistle is with the spiritual benefits that are the common property of the community. Verses 6–8 are no more than an aside for the benefit of those who wish not simply to accept a mystery but in some measure to understand it. Paul in effect likens God to a king intending to benefit his people but surrounded by corrupt or hostile members of his court. He cannot proceed directly to do what he would do but must appear to yield to their opposition. But he sets and conceals a trap that will use their lack of loyalty and perception to destroy them, as David did when forced by Absalom's rebellion to flee (2 Sm 15:32–37, 17:1–14). This imagery lies behind the opening chapters of Wisdom, which address the rulers of the earth and commend wisdom to them, although ungodly men lie in wait for the righteous man and test him with insult and torture:

> Thus they reasoned, but they were led astray,
> for their wickedness blinded them,
> and they did not know the *mysteria* of God,
> nor hope for the wages of holiness,
> nor discern the prize for blameless souls. (Wis 2:21–22)

The mysteries of God seem to include the conviction that death is not necessarily a punishment but may be a test of goodness, to be rewarded with immortality. Later, in Wis 6:22, the ideal wise ruler promises to disclose the mysteries of wisdom from the beginning of creation. Paul seems to be using wisdom language and the deception myth to suggest that the crucifixion of the Lord of glory can be understood as a proper activity of the originating divine wisdom, devised to trap the rulers of this age[36] by their own incomprehension, to break their power, and to deliver promised benefits to God's people. This piece of wisdom mythology is interesting precisely because it is an interlude in Paul's argument. It is introduced to qualify Paul's otherwise depreciatory references to wisdom, and outside the framework of a polemical discussion, it indicates how Paul could bring the death of Christ into relation with God. The crucifixion, though carried out by God's enemies, was the means used by God for bringing benefits to his people. There is no suggestion, at least in this imagery, that the death of Christ is a means of influencing God's attitude towards his people.

Nor indeed is it implied that God directly commanded the crucifixion or that Christ directly sought it. The doubtless primitive statement that 'Christ died and rose again' is here represented by passive verbs—in 1:13 (by implication) the aorist, in 1:23 and 2:2 the perfect (embracing both the death and the proclamation of it)—or by the neutral phrase, 'the cross of Christ' in 1:17–18. In 2:8 the active verb has the rulers as subject, and 'the Lord of glory' as object. Elsewhere only Gal 6:14 associates *Kyrios* with the cross, and the designation 'Lord of glory' is unique in the New Testament, though common enough as a reference to God in 1 Enoch.[37] The designation refers back to the promise of glory for God's people (verse 7) and implies that Christ is the master and provider of that glory. But is he regarded as a divine being who possesses the glory that constitutes God himself?[38] If so, his crucifixion was no more than a formal make-believe. Is it not necessary to regard 'glory'[39] as a reference to the resurrection state, to remember that the title *Lord* is primarily a resurrection title, and to read verse 8 as meaning that the rulers crucified him who was to become the Lord of glory? If so, the statement expands both parts of 'He died and he rose', and God's primary action is once again displayed in the resurrection.

The Power of the Cross

In the section 1 Cor 1:10–4:21, which deals with internal conflicts of the community, there are no more explicit references to the death of Christ; though it seems likely that Paul's description (in 4:9–13) of the apostles' destiny to become 'fools [*mōroi*] for Christ's sake' picks up an echo of the folly *(mōria)* of the word of the cross in 1 Cor 1:18. In contrast to the ebullience of some competing Corinthian Christians, Paul expects the apostolic commission to be marked by deprivation, social rejection, and self-sacrifice. In that measure, the apostolic norm was the crucifixion of Christ.

It is necessary, however, to say more about references that have already been noticed in the discussion of Corinthian wisdom. From 1 Cor 2:8 it can be con-

cluded that the crucifixion is God's secret wisdom, sufficiently powerful to
thwart and defeat the 'rulers of this age'. This conclusion agrees with the statements in 1:18, 23–24 that the preaching of Christ crucified is 'the power of God
and the wisdom of God'; and from 2:2 it seems that the main thrust of Paul's
preaching at Corinth must have been 'Jesus Christ and him crucified'. In offering
this proclamation, religiously offensive to Jews and socially destructive to Gentiles, Paul was operating with the power of God which was not only to over-
throw 'the rulers of this age', but also to set believers on the road to salvation
(1:18), to give them mastery of the world, life, death, the present and the future
(3:22). Those who were prematurely claiming that they had become kings (4:8)
should indeed look forward to the power of the kingdom of God (4:20). So the
preaching of the cross is power to transform, equip and no doubt preserve believers. But the cross can be 'emptied of its power' if the dominant purpose of believers is the achievement of (social) wisdom. Hence 'the word of the cross' is not a
power formula that can be used and controlled by anyone and in any fashion: it
remains always the power not of the preacher but of God. Later in the epistle
instances will be given of the disastrous misuse of the formula, whereby its power
is not merely lost but becomes destructive.

Something similar may be said of baptism, which is mentioned more frequently in 1 Corinthians than in any other epistle. From 1 Cor 1:13 it is clear
that the proclamation 'Christ died for you' (see 1 Thessalonians) is coordinate
with baptism into Christ's name. No doubt this reference to baptism implies that
'by one Spirit we were all baptized into one body' (1 Cor 12:13), for in both places
the mention of baptism is part of a remedy for dissension among Christians. In
1 Cor 1:13–17 Paul certainly expresses reservations not about baptism itself but
about *his* baptizing and the consequent danger of associating the Christian condition with his name (or the name of Apollos or Cephas) instead of the name of
Christ. If baptism is valued chiefly for the merits and wisdom of the baptizer
rather than for the name of Christ into which candidates are baptized, the proclamation of the cross is deprived of its power to transform, equip and preserve
believers. Whatever power is attached to baptism derives strictly from its relation
to the death of Christ.[40] How and why that is so receives no explanation at this
point except that the proclamation of crucifixion—and so presumably that of
baptism—overthrows the wisdom of this age. We therefore need to look further.
The reference in 1 Cor 15:29 to 'being baptized on behalf of the dead'[41] strongly
suggests that baptism was regarded in Corinth as a saving and protecting ritual.
Moreover, Paul's midrash on the Exodus in 1 Cor 10:1–2, which says that 'our
fathers . . . were all baptized into Moses' exploits the imagery of deliverance and
protection.[42] It must be taken as very probable that baptism was regarded as a
kind of ritual talisman, corresponding to the preaching of the cross as a verbal
talisman. To some readers the word *talisman* may be offensive as implying a
magical power or as pushing baptism and preaching in the direction of the widespread ancient and modern use of amulets and nostrums. True indeed, and a
better word might be *seal* if it could claim to be a New Testament designation
of baptism.[43] All the same, even if *talisman* must be written with reservations, it
serves to remind us (1) that nothing precise is alleged when something is rejected

as magical unless an attempt has been made to assess what magic implies in the relevant culture; (2) that some forms of magical powers and effects were acceptable within primitive Christian circles; (3) that the Christian practice of preaching and baptism has not seldom managed to make the theme and the action superstitiously trivial; and (4) that Paul's reservations about baptism and his disputation about the Christian kerygma show that he was aware of the dangers.

Sexuality

With the end of chapter 4 the direct discussion of the nature of the gospel is, for the time being, at an end. In 5:1–6:20 Paul turns to questions of sexual freedom and aggressive litigation within the community; and he applies the Christian confession of death and resurrection to these moral problems. The community itself must take responsibility for preserving its own purity: it must therefore withdraw itself from the 'unfaithful' (in the moral sense) and must certainly not take its disputes for judgement to the 'unfaithful' (in the unbelieving sense). Paul is thus encouraging the community to develop its own kind of social wisdom for managing its internal and external relations.

First he deals with a case of sexual deviance *(porneia)* which he regards as particularly scandalous: 'A man is living with his father's wife' (1 Cor 5:1). He argues for the offender's exclusion from the community on the grounds that his action is a defilement that, unless removed, will affect everyone. They must remove him for the same reason that leaven is removed at Passover (thus taking for granted that the Corinthian community has knowledge of Passover customs). The sudden change, at first sight surprising, from *porneia* to leaven is perhaps to be explained from M. Kerithoth 1.1 which lists the thirty-six offences punishable by cutting off from the community: they include 'if a man has connexion with his father's wife' (from Lv 18:8 and 29) and if he 'eats leavened bread during Passover' (from Ex 12:15). Then Paul moves from the second personal imperatives of verse 7[a] to the first personal statement of verses 7[b]–8:[44]

> For Christ, our paschal lamb, has been sacrificed.
> Let us, therefore, celebrate the festival,
> not with the old leaven, the leaven of malice and evil,
> but with the unleavened bread of sincerity and truth.

The words *sincerity* and *truth* (which translate *eilikrineia* and *alētheia*) suggest moral elevation; but since Paul is speaking of festival rites, it would be preferable to translate 'purity and authenticity'. He is urging the community to celebrate the festival (either literally or as a figure for the Christian life) in a blameless manner. He does so because, like Israel of old, the community lives within the protection provided by the Passover sacrifice of Christ. When the Passover lamb was first sacrificed in Egypt, the lintels and doorposts of their houses were smeared with the blood, so that the Lord would pass over the doors and would not allow the destroyer *(ho olethreuōn)* to enter and slay them. In a similar way, the Corinthian community was protected by Christ's blood from the destructive forces outside its boundary, but only if it maintained a well-recognized boundary (which is the aim of Paul's explanation and rebuke in 1 Cor 5:9–6:8) so that a serious offender could be handed over to Satan for the destruction *(eis olethron)* of the flesh.[45] Thus, even more explicitly than his previous references, Paul's present moral use of the death of Christ regarded as a Passover sacrifice indicates that he is thinking of it as a seal or talisman. It perhaps needs to be said that he was not speaking literally. Jesus was not a lamb; he was not offered sacrificially to God by a Jewish family; he was not sacrificially killed as lambs are killed; nor was his blood drained off, collected and disposed of in the prescribed manner. If Christ's death can indeed be described as a sacrifice (despite Jewish aversion to human sacrifice), it is only because Paul's argument can be understood in the form, 'When we are protected from the retribution that falls upon pagan wickedness, it is as if Christ's death were equivalent to the Passover sacrifice; and alas for us if we allow impurity to render the sacrifice ineffective'. Since Paul makes figurative use of the Passover to set the mind moving in the direction of his thought, he is not offering an explanation of the apotropaic effect of the cross; but he is quite clearly trying to say that a Christian community is protected from the moral collapse of pagan society not by religious self-confidence but by attitudes and actions moulded by the death of Christ.

The comparison of Christ to the paschal lamb is commonly thought to underlie other New Testament passages, though nowhere else is the identification explicit: 1 Pt 1:19; Rv 5:6, 9, 12 and 12:11; and Jn 1:29, 36 and 19:36. Only in one of these passages, Jn 1:29 ('Behold, the Lamb of God, who takes away the sin of the world', where the reference is a notoriously perplexing problem) is there any connexion between the lamb and the removal of sin. Yet it is not uncommon to find 1 Cor 5:7 interpreted as a sacrifice dealing with sin. For example, 'Christ as the Lamb of God summed up God's action for the deliverance of his people; and the context suggests (though much less explicitly than John 1:29) that he delivered them by bearing for them the burden of their guilt and thus removing their sin' (Barrett on 1 Cor 5:7). The first part of that statement can be amply supported by the celebration and expectation of deliverance that runs throughout the Passover Haggadah; but what in the context of 1 Corinthians 5 suggests that deliverance had been effected (or perhaps would be effected) by Christ's bearing of their guilt and hence removal of their sin? Some connexion, it is true, must be made between the description of Christ as 'our Passover sacrifice' and the community's tolerance of scandal (1 Cor 5:1–2) and its injudicious consorting with disreputable people (1 Cor 5:9–10). A little later it is said that some Corinthian Christians had themselves been disreputable but had been washed, sanctified and justified in the name of the Lord Jesus Christ and in the Spirit of God (1 Cor 6:9–11). But what has that to do with Passover? The Mishnaic tractate Pesahim contains not a word about sin and guilt; nor does the traditional Passover Haggadah, except in one prayer that mentions the gift of the Temple 'for the atonement of all our sins'.[46] As Dalman argued long ago, the Passover was not expiatory. The Mishnaic tractate Zebahim 4.6 (animal offerings) says that an offering must be slaughtered while mindful of six things, beginning with the offering (i.e., someone intending to make an offering under one sacrificial category may not slaughter under another category). Then it adds, 'If it is a Sin-offering or a Guilt-offering, also of the sin'. Thus, Passover is excluded from the sacrifices for sin. As the Haggadah itself says, 'This Passover is . . . because the Holy One . . . passed over the houses of our fathers in Egypt'; and this was still understood by Christians in the second century. Justin[47] accused the Jews of falsifying parts of the scriptures: they excised certain passages, including one from Ezra. Since this passage is not found in any surviving text or translation of the Old Testament, it is probably early Christian targum: 'This passover is our saviour and our refuge'. That, too, is Paul's meaning: 'We claim his saving protection; we must not mar it by tolerating impurity'.

As Paul continues (in 6:12–20) to probe the Corinthians' sexual morality, he turns in verse 14 from the death of Christ to his resurrection. The community, it would seem, uses the catchphrase 'All things are lawful for me'—perhaps originally commended by Paul—to claim their freedom in matters of sex and food. Paul makes only a brief reference to liberty in choice of food (1 Cor 6:13ᵃ) before discussing sexual liberty, which, in the end, occupies him until 7:40, after which he deals at some length with food and eating in 1 Cor 8, 10–11.

In trying to follow Paul's argument in 1 Cor 6:13–20 it is necessary to remember that his use of *sarx* and *sōma* does not always suit what would naturally be suggested by *flesh* and *body* respectively. This may be shown from the occurrences of the two words in the epistle. In standard Hebrew fashion—reflecting the use of *basar*—'all flesh' in 1:29 (somewhat hidden in modern translations) means 'all mankind'.[48] In 15:50 'flesh and blood' means humankind in its vulnerability; and the flesh, as the area of hardship in 7:28, is our material existence. 'Wise according to the flesh' in 1:26 means 'in human terms', and 'Israel according to the flesh' in 10:18 means 'in its human aspect'. Thus *flesh* means the general existence to which all human beings belong, to be distinguished (in the rather special argument of 15:39) from the kinds of existence to which other living beings belong. But in 6:16 there is a less general use of the word when Paul quotes from Gn 2:24, 'The two shall become one flesh', which means that they become one in kinship.[49] Since, as Paul surely knew, the Hebrew word not infrequently meant 'body' in the ordinary sense, he substituted *sōma* for *sarx* when he gave his own targum, though using *sōma* somewhat more freely than the ordinary sense would suggest.[50]

A soma is, of course, a physical body, but it is differentiated from the *koilia* (the body cavity, either the digestive or reproductive organs). Food and belly are self-regulative, merely components of the material existence called *sarx*. God will bring both to an end, either when we die or at the parousia. But the soma and its sexual activity are not self-regulative: the soma is not for *porneia* but for the (risen) Lord, and the Lord for the soma. Why should *porneia* be given this prominence? If fornication is injurious, why not gluttony? Because (Paul may suppose) gluttony belongs to the world of flesh, whereas fornication has its consequences on the body. The world of flesh is to be done away, but the body is to be raised by that power of God which raised the Lord.[51] For that reason Paul can make the curious ruling that other sins are outside the body, but that someone who fornicates sins against his own body. Without being explicit, he is relying on the ancient Hebrew view that acts of fornication establish an alien kinship which is destructive of true kinship (and of course gluttony and other sins have no such consequences).[52] So, then, soma is indeed a physical body but more than a physical body. The bodies of Christians are *melē* of Christ: limbs, organs, even perhaps component parts (verse 15ᵃ, cf. 1 Cor 12:27: 'You are *sōma Christou,* each part [of the community] being a *melos* [limb or organ]').[53] Christians are therefore not autonomous, they do not belong to themselves but to the Lord, a conviction driven home by what looks like a familiar dictum: 'You were bought with a price' (verse 20ᵃ, repeated in a different context in 1 Cor 7:23). Neither use of the dictum implies that they were bought to be set free.[54] Hence comparison with the inscriptions of sacral manumission is not appropriate.[55] There is a general affiliation with Old Testament language of redemption (e.g., Ru 4:4-5, where the next of kin is invited to buy Ruth with the parcel of land; and Is 43:1, 3-4, where the Lord gives other people in exchange for the life of his own people). But in neither passage does the verb *agorazō* ('buy') appear; indeed, both *agorazō* and *exagorazō* are without theological significance in the Septuagint. Nor does the word *timē* ('price') in the redemption rituals of Ex 34:20 and Lv 27 help explain Paul's dictum. Further, in contrast to Gal 3:13 and 4:5 (where *exagorazō* is used) there is no reference, explicit or implicit to the death of Christ; and it can scarcely be proper to interpret the intention of Paul by referring to Rv 5:9 ('by thy blood didst purchase for God men of every tribe'). It is not to be denied that Christ's death can be compared to a ransom, to a price that was paid; but in this context Paul's intention is to insist that members of a Christian community belong to the Lord not because they have chosen so to belong but because the Lord has acquired them by his own free decision. This is not a dictum about redemption but about Christian obedience, as it is in 1 Cor 7:23. This strong metaphor asserts that Christians belong exclusively to God, who obtained them at great cost—perhaps (though we cannot be sure) at the cost of sacrificing his Son to the rulers of this age.

Christians are what they are chiefly because God has exercised his power in raising Christ from the dead (verse 14). Now, when someone is united to the Lord, he is one spirit with him (verse 17). Hence, 'your body is a temple of the Holy Spirit within you, which you have from God' (verse 19ᵃ). According to the context, this applies to individual Christians what is said in 3:16 of the whole community. Even so, Paul wrote *to sōma hymōn* where *body* is singular (except in a well-supported variant reading *bodies*) and *your* is plural, exactly as in the closing words of verse 20: 'glorify God in your body'. He varied his expression from 'your bodies' in verse 15 to 'your body' in verse 19 probably because he already had in mind that 'by one spirit we were all baptized into one body' (1 Cor 12:13). Hence 'your bodies' belong to the body of Christ and in so belonging are the temple of the Holy Spirit.

This discussion of *sōma* may now be restated systematically and then taken a little further. *Sarx* is, as it were, the undifferentiated human condition, whereas *sōma* is a particular personal identity. Since it belongs to the world of *sarx* it is of course weak, vulnerable and mortal. Sharing psyche, the life of the *sarx,* it is a *sōma psychikon,* naturally perishable, dishonourable and weak; but sharing, as it does, pneuma, the spiritual life of God, it will be raised imperishable, glorious and powerful (1 Cor 15:42–44). Hence Paul's response to the Corinthian assertion of moral freedom (verse 12) contains these elements: (1) you belong not to yourselves but to the risen Lord, and you are free to do what he bids you; (2) in your bodily existence you belong to the body of Christ in such a way that sexual deviance can ruin your Christian status; and (3) as belonging to the body of Christ you are the temple of the Holy Spirit, but you are not yet experiencing existence as spiritual bodies; hence the statement that 'God raised the Lord and will also raise us by his power' (verse 14) makes the resurrection both a promise for the future and a warning for the present.

Paul is now ready to deal with the first matter referred to him (1 Cor 7:1–40), namely, relations between marriage partners. If 'your bodies are members of Christ' (1 Cor 6:15), consequences should follow for the state of marriage. So Paul works out the proper consequences without further reference to resurrection teaching, though he reinforces obedience by repeating his dictum, 'You were bought with a price', in 7:23 and insists strongly that 'the form of this world is passing away' in 7:31.

Eating

Then he moves to a second referred question, namely, 'food offered to idols'; and the discussion turns upon the extent and limits of Christian freedom in a pagan world (1 Cor 8:1–11:1). For the first time gnosis plays a leading part in the argument (after its initial mention in 1 Cor 1:5); and the indifference of food problems, briefly indicated already in 1 Cor 6:13, is established: 'Food will not commend us to God. We are no worse off if we do not eat, and no better off if we do' (1 Cor 8:8). It is clear, however, that opinion in Corinth was divided; and the debate introduced conscience *(syneidēsis)* into Paul's vocabulary. The majority of Corinthian Christians had no doubt 'turned to God from idols' (1 Thes 1:9; cf. 'When you were heathen, you were led astray to dumb idols', 1 Cor 12:2) and had been forced to consider whether they could still consume sacrificial meat or must avoid it. Those who were aware that 'an idol has no real existence' were not pained by conscience when they ate sacrificial meat. By thus having a clean conscience they escaped any consequences of believing in the gods. Others, however, had not been released from familiarity with the gods, their conscience was but weakly aware of the sole Lordship of Jesus Christ, and if they ate they feared that their conscience would not be clean (1 Cor 8:7). What conclusion would such a 'weak' Christian draw if he saw another Christian, untroubled by conscience, at table in an idol's temple? Might he not be encouraged to eat sacrificial meat, despite the judgement of his conscience, supposing that the gods as well as the Lord Jesus Christ have real existence? In that case he both lacks the gnosis of Christ's sole Lordship and has abandoned the protection of his conscience. Therefore the untroubled Christian, by thoughtless use of his gnosis, has caused the destruction of the 'weak' Christian, 'the brother for whom Christ died' (1 Cor 8:11). In so sinning against his brother by striking at his conscience (however weakly instructed by gnosis), he sins against Christ.

Such seems to be Paul's meaning, though it is elliptically expressed. It is partially concealed from modern readers since Paul's use of *conscience* is different from ours. To us, a 'weak conscience' implies a conscience that gives only feeble guidance about moral problems and weak condemnation of wicked actions. For Paul, a weak conscience was but weakly instructed by gnosis. For us a 'defiled conscience' (if we used the expression) would be a conscience that gave perverted guidance and judgement. For Paul it is a conscience that has been impaired by deliberate disregard. (Dionysius of Halicarnassus[56], whose main writings fell in the period 30–8 B.C.E., says that deliberate falsehood defiles [*miainein*] the conscience of the historian). Again, we would suppose that an active conscience might protect us from moral wrongdoing, whereas Paul supposes (in this instance) that a clear conscience prevents Christians from being drawn back into pagan religion. He treats conscience with respect even when it is wrongly informed and regards it as one of the safeguards of human existence. His view corresponds to the description contained in a fragment attributed (wrongly?) to Epictetus: 'When we were children our parents handed us over to a nursery-slave [*paidagōgos*] who should watch over us everywhere lest harm befall us. But when we are grown up, God hands us over to the *syneidēsis* implanted in us, to protect [*phylattein*] us. Let us not in any way despise its protection, for should we do so we shall be both ill-pleasing to God and have our own *syneidos* as an enemy'.[57]

It is now possible to draw out the significance of Paul's reference to the death of Christ in 1 Cor 8:11. 'The brother for whom Christ died' has already been discussed in reference to Thessalonians as a variant of a fairly common pre-Pauline formula, implying that his death secured benefits for believers. Since the Christian brother was a former idolater, the benefit of Christ's death (better, his death and resurrection) must have been release from allegiance to the gods and transference to the Lordship of Christ. On this point the uncertainty at Corinth was whether Christ had triumphed over the gods or whether he had demonstrated their nonexistence. Whichever view was held, this was a far-reaching conclusion about the death of Christ.

It is clear that Christians who felt free, as Paul did, to eat sacrificial meat with a good conscience were indeed protected from further worry about the gods, but they were not protected from harming fellow Christians. That second protection was provided once more by the death of Christ. If they destroyed those who had been obtained by the Lord's death, they would be sinning against their Lord. Hence awareness of the death of Christ was a protection when conscience itself gave no help.

Paul's exploration of Christian freedom in regard to sacrificial meat is qualified by the conviction that a freedom need not be exploited, and indeed must not be exploited to the injury of others (hence the excursus in 1 Corinthians 9). In 1 Corinthians 10 a further qualification is developed to guard against overconfidence about Christian freedom and the source of it. Paul approaches this extension of his exposition by a midrash on the Exodus. Israel, having been protected by the Passover from the destroyer, was equally protected by the divine cloud from the pursuing Egyptians and by divinely provided food and drink from death in the wilderness. Their passage through the sea was, as it were, a baptism, and their spiritual food (manna) and spiritual drink (water from the rock) was, as it were, a participation in Christ akin to the Lord's Supper. But the protection of God and the evidence of his favour were not irrevocable: 'With most of them God was not pleased; for they were overthrown in the wilderness' (1 Cor 10:5). The journeying people gave way to idolatry and immorality. God's protection was withdrawn, and they were destroyed by the destroyer (*olothreutēs*, 1 Cor 10:10, cf. 5:5). These things took place to warn the Israelites, and they were written down to instruct Christians living at the end of the ages.

This ingenious exercise kills three birds with one stone. It indicates Paul's continuing anxiety about the effects of idolatry, his way of understanding the Lord's Supper, and his need to rebuke some members of the Corinthian community for unwisely presuming on the protection they enjoy. The argument takes a step forward when Paul says 'shun the worship of idols' (1 Cor 10:14). In all, he uses idolatry words fifteen times in 1 Corinthians and only four times in his other epistles.[58] Therefore his anxiety about idolatry in the Corinthian community at this time was most pressing. Since he was not greatly concerned about the eating of sacrificial meat even in idol temples, except for the disturbing consequences for less-instructed Christians, he was probably not moved by the need to prevent Christians from actually taking part in pagan cults.[59] He was protesting against the continuation of a pagan manner of life. Some of the community had been fornicators, idolators, and so on; but they had been cleansed, consecrated and made acceptable to God in the name of the Lord Jesus Christ and in the Spirit of God.[60] They were not therefore in danger from the gods (which had no real existence) but from the demons (1 Cor 10:19-20). In so speaking Paul was echoing ancient Jewish polemic against the gods, as in Dt 32:17: 'They sacrificed to demons which were no gods'. What did Paul imply when he used *daimonion* for the first and only time in his epistles, and opposed it to *theos*? The change from *theos* to *daimonion* (as used in the Hebraic tradition) is a demotion.[61] A god has independent existence and powers peculiar to his being: a demon is entirely dependent upon other beings (sometimes divine beings but especially human beings and animals), has no creative but only a destructive existence, and possesses no power except what is drawn from its victims. For the sake of a strong visual contrast between two cups and two tables, Paul allows in name the real existence of demons, but what he means by demons are the self-destructive and socially destructive forces of human existence. He is in the tradition of the Testament of Reuben, which speaks of the seven spirits of deceit which turn out to be fornication, insatiableness, fighting, obsequiousness and chicanery, pride, lying and injustice.[62] That is very much the kind of thing that Paul associates with idolatry; but when he now associates it with demons, it ceases to be a religious cult that people may or may not choose to follow and becomes instead a dangerous threat that must be avoided: 'I do not want you to be partners with demons', that is, helplessly join in their destructive work. If people drink the Lord's cup and share the Lord's table, they are members of the Lord's household; they cannot at the same time be members of the demonic household.[63]

If they try to belong to both households they will provoke the Lord to withdraw his protection—unless they think they are already stronger than the Lord and can protect themselves! (1 Cor 10:22).

The contest between the claims of the Lord and the idols is, of course, unequal, because the Lord represents God and the idols are no more than demons. But it is certainly a religious, not simply a moral conflict. Paul makes his point by reference to the public sacrifices of Judaism, whereby an animal is killed, the blood is drained off and applied to the altar, and the flesh is either wholly burnt (as an offering to God) or partly burnt and partly eaten by the priests and other worshippers. In a communion sacrifice *(zebaḥ šelamin)* the worshippers may naturally be described as 'partners in the altar' as they eat the sacrificial portions; but it is not easy to suppose that the Corinthian community would be familiar with the varieties of Jewish sacrifice.[64] They would simply know that Jews did not consume sacrificial blood. They killed an animal and in effect asked God to accept both the blood and the intention of the killing. In that way they were doubly 'partners in the altar', both in the sacrificial action and in the subsequent eating. Hence to claim the protection of the Lord by eating and drinking at his table is to be involved in the most profound and ancient human relations with the deity.

The Lord's Supper

That conviction must underlie Paul's rhetorical questions in verse 16:

> The cup of blessing which we bless,
> is it not a participation in the blood of Christ?
>
> The bread which we break,
> is it not a participation in the body of Christ?

The cup and the bread obviously refer to the tradition of the Lord's Supper written down in 1 Cor 11:23–26, but it is necessary to interpret verse 16 on its own terms. No doubt the tradition of chapter 11 was in Paul's mind when he wrote verse 16 and in his readers' minds when they read it, but the fact remains that he needed the tradition in chapter 11 but not here. Indeed, in verse 16 he reverses the order, putting cup first and bread second, as also in verse 21. This order can scarcely be regarded as evidence for a variant Corinthian liturgy[65] but may well represent a teaching sequence corresponding to the standard sacrificial order of first offering the blood and then eating the meat—not at an altar but at a table. Wishing to allow the words of the Christian cult to function at this point in a particular way, Paul has moulded them to a simple and familiar Jewish type. The opening words of each question are strongly Jewish. In particular, 'the cup of blessing' is the standard Jewish phrase for the cup of wine that ended any meal (or the third Passover cup), and it meant the cup for which God was blessed (i.e., thanked).[66] Bread and wine were natural components of a cultic meal, as may be seen in sources roughly contemporary with Paul. *Joseph and Asenath* (an Egyptian-Jewish legend, possibly of the first century C.E.) speaks of having 'eaten the bread of life and drunk the cup of immortality'.[67] The Community Rule of Qumran makes provision for the communal meal thus: 'When the table has been prepared for eating, and the new wine for drinking, the Priest shall be the first to stretch out his hand to bless the first-fruits of the bread and new wine'.[68] In these examples, as well as in the Lord's Supper tradition, bread precedes cup. Hence, for this argument, Paul not only reduces the cult to its simplest (and therefore presumably its essential) form but also inverts the common order so that first emphasis is given to the cup and main development to the bread.

The key word is *koinōnia* ('participation'), which was used significantly in 1 Cor 1:9: 'You were called into a *koinōnia* of [or with] his Son, Jesus Christ our Lord'. Here it appears twice more (and not again in the epistle) as a *koinōnia* of the blood and of the body of Christ. Translation is not straightforward.[69] The verb means to have a share of something or to have a share with someone in something or to contribute a share to some enterprise. Correspondingly, the noun means (1) the sharing of something, participation in; (2) association with others in something, participation with and in; and (3) contribution. Here the choice is between (1) and (2). Since verse 17 develops a strongly corporate image, the meaning most probably is participation with fellow Christians in the blood and the body of Christ. In a somewhat similar way, the emphasis on 'calling' in 1 Cor 1:9 suggests that the phrase means 'into a participation with fellow Christians in the community of his Son'. Further elucidation depends on the meanings suggested by *blood* and *body*.

'The blood of Christ' is not a common expression in Paul. Apart from this reference and the Lord's Supper tradition in 1 Cor 11:25, 27, there are only two more references—Rm 3:25 and 5:9, both of which read like formal statements. Furthermore, Rm 5:9–10 equates 'his blood' with 'the death of his Son' and contrasts it with 'his life'.[70] Hence it may be agreed that all four references come from pre-Pauline traditions and that *blood* signifies violent death, as it does throughout the Old Testament,[71] where the shedding of blood always brings dangerous consequences. The dangers may be averted if animal blood is shed by appropriate atoning rituals that attract God's protection by pouring the blood around or smearing it upon his altar. If human blood is at risk, there lies an appeal ('Save me from bloodshed', Ps 51:14 NEB) to God who is 'the avenger of blood' (Ps 9:12 NEB). Sacrificial blood and blood guilt are dominant themes in the Pentateuch ('Blood pollutes the land, and no expiation can be made for the land, for the blood that is shed in it, except by the blood of him who shed it', Nm 35:33). Violent death (for which *blood* becomes the stock term) is seldom absent from the rest of the Old Testament and naturally gives rise to violent images of vengeance. According to Is 49:26, the Lord says, 'I will make your oppressors eat their own flesh, and they shall be drunk with their own blood as wine'. Ez 39:17–20 describes a grotesque banquet when the Lord God summons birds of every sort and all beasts of the field to 'eat the flesh of the mighty, and drink the blood of princes' at the great sacrificial feast he prepares. It is, and was meant to be, repulsive. To judge from within Paul's own culture (where he himself stands, as shown by his reference to Israel in verse 18 and his repudiation of pagan cults in verses 20–21) he was moved into a religiously offensive area by insisting on a participation in the blood of Christ. It is true that he avoids words for eating and drinking (though they must be presumed and are explicit in 1 Cor 11:26), and he says *body,* not *flesh;* but what kind of interpretation is possible for a Pharisaic Jew to put forward? The answer may in part be found in an ancient story of David. 2 Sm 23:13–17 recounts an incident in his campaign against the Philistines, then encamped in Bethlehem. Three men of this company risked their lives to get water from the well at Bethlehem for David to drink; but he refused to drink it, with the words, 'Shall I drink the blood of these men who went at the risk of their lives?'[72] The incident is reproduced in very similar words in 1 Chr 11:19 but more importantly in 4 Mc 3:6–19: 'But David, though still burning with thirst, considered that such a draught, reckoned as equivalent to blood, was a grievous danger to his soul. Therefore opposing his reason to his desire, he poured out the water as an offering to God'. 4 Maccabees, perhaps written in 40 C.E., is a hellenistic Jewish treatise arguing that martyrdom is in line with reason, that martyr deaths expiate the sins of the people and purify the land, and encouraging readers to follow the example of the heroic Maccabean martyrs. Thus, in a manner of speaking, the people of God drink the blood of the martyrs, that is, they themselves survive because the death of others has effected purification, ransom, and propitiation (4 Mc 6:29, 17:22). If such a thought is at home in a hellenistic writing much concerned with the life of wisdom ('the culture acquired under the law, through which we learn with due reverence the things of God and for our worldly profit the things of men' 4 Mc 1:15–17), is it not equally at home in Paul's letter to the wisdom-seeking Corinthians? Paul is at least saying that Christians participate together in receiving the benefits of Christ's death; but he may be implying even more: that Christians are involved with God in the responsibility for Christ's death and the consequences of it. The thought of that death as a protective talisman is not absent (the protection of Christ is incompatible with the protection of the gods, verse 21), but Paul is now moving to a participatory understanding—a participation in the blood of Christ and also in the body of Christ.[73]

For Paul the name Christ signifies two existing realities: the consequence of his death and the consequence of his resurrection. With participation in the body of Christ we come to the consequences of his resurrection. So far in this epistle the word *body* has appeared twelve times, almost always as a human body, a particular personal identity, experiencing the life of *sarx* and to some extent the life of pneuma (see pp. 32–33). But already in 1 Cor 6:19–20, the change from 'your bodies' to 'your body' suggests that the community is being regarded as a body, in anticipation of the metaphorical exploration of *body* in chapter 12 and its conclusion: 'Now you are the body of Christ and severally members of it' (1 Cor 12:27, cf. verse 12). In itself *the body of Christ* is not common in Paul or elsewhere:[74] apart from a specific indication to the contrary, it is unlikely that it would bear different or unrelated meanings. Hence, participation in the body of Christ, with particular emphasis in verse 17 on the oneness of belonging, means sharing in such measure as unchanged human beings can (cf. 1 Cor 15:51) in the risen life of Christ. Reference to the one loaf clearly prepares the way for dealing with improprieties and dissensions at the Lord's Supper. In effect, Paul is saying,

'If your community is torn apart into self-regarding groups, you are not living as the body of Christ'. What it would mean to go further and say that Paul regarded Christ as being sacramentally present is not clear to me. That Christ's protection was available to those who shared his sufferings and Spirit-filled life should not be in doubt; nor that protection was replaced by judgement if protection was claimed from the sacrament without participation in the blood and body of Christ.

A similar warning is sounded after Paul's direct citation of the Lord's Supper tradition. At 1 Cor 11:17 he turns his attention to unsatisfactory behaviour when the community is assembled for the common meal. All his complaints arise from the discovery that it has ceased to be a *common* meal: 'It is impossible for you to eat the Lord's supper, because each of you is in such a hurry to eat his own, and while one goes hungry another has too much to drink' (1 Cor 11:21 NEB).[75] Thus, they despise the *ekklēsia* of God and shame its poorer members. That is what Paul means by eating the bread and drinking the cup of the Lord unworthily. There is no complaint about liturgical impropriety or defective theological understanding of the cult but that they will not allow the formative words of the cult to shape their communal behaviour. Hence, Paul recites the tradition he had already conveyed to them, having himself received it 'from the Lord' (verse 23). The tradition begins historically with 'the night when he was betrayed' and therefore is a tradition about *Jesus*. Paul cannot mean that he received it from Jesus himself or in an appearance or vision; he must mean that he received it as he took part in Christian worship when the risen Lord was in some sense made present by the recital of his words and actions.[76] Hence, it is a tradition about the *Lord* Jesus. The early Christians called on the Lord Jesus (Christ) (1 Cor 1:2, 12:3); they were a *koinōnia* of the Lord Jesus Christ (1:9); they experienced his grace (16:23) and knew his power to forgive (6:11) and to condemn (5:4), to support and to give victory (15:31, 57). He had already been seen by the apostles in his resurrection (9:1, 15:5–8), was expected to return (1:7, 8; 5:5), and was already the reigning Lord (15:25), to be spoken of alongside God the Father (1:3, 8:6). The common meal of the church was the *kyriakon deipnon*. The adjective *kyriakos* (used in inscriptions and papyri to distinguish the imperial service) was also applied to the first day of the week, *kyriakē hēmera*:[77] John was in the Spirit on the Lord's day and saw him who had died but was alive for evermore (Rv 1:10–18). So also the Lord's Supper was the common meal for acknowledging the power of the risen Lord. Here then is the tradition:[78]

> The Lord Jesus
> on the night when he was betrayed [*paredideto*]
> took bread,
>
> and when he had given thanks,
> he broke it,
> and said, 'This is my body
> which is for you.
> Do this in remembrance of me'.

> In the same way also the cup,
> after supper,
> saying, 'This cup is the new covenant
> in my blood
> Do this,
> as often as you drink it,
> in remembrance of me.' (1 Cor 11:23c–25)

This is obviously one form of the tradition, to be placed with the Lukan tradition and alongside the tradition in Mark and Matthew and to be compared with the Johannine tradition. Paul and the synoptic Gospels testify to a meal on the evening when Jesus was betrayed (or handed over). He performed the two common Jewish actions of taking bread and a cup (of wine), saying the standard thanksgivings, and giving the signal to partake. But then he diverged from common practice by designating the broken bread as his body and by associating the (contents of the) cup with a covenant and his blood. The Fourth Gospel also recounts an evening meal at which the betrayal of Jesus played a major part; and elsewhere a feeding miracle followed by a discourse in which Jesus identified himself as the living bread from heaven, and invited his hearers to consume his flesh and drink his blood.

Within this general agreement there are many divergences. Mark and Matthew place the statements about the bread and the cup side by side; Paul separates them by the meal, and Luke (if the longer text is original) follows Paul after Luke's own preliminary references to eating and drinking. John, surprisingly, introduces sayings about eating flesh and drinking blood into a discourse mainly concerned with the heavenly bread. The synoptic Gospels describe a supper for Jesus and the Twelve, John directs the discourse to the Jews, and Paul says nothing about the supper company. The synoptic Gospels present the supper as a result of Passover preparations; though Luke says that Jesus did not in fact eat the Passover with the disciples, and no account is a recognizable description of Passover. John's feeding miracle takes place when Passover was near, but the symbolism of the discourse relies on manna in the wilderness; and John's final supper is explicitly placed before the Passover. Paul's tradition makes no reference to the Passover despite the exodus references in 1 Corinthians 10 and the description of Christ as our Passover in 1 Corinthians 5. And finally Paul's tradition has a double command to repeat the ritual action with a special intention, present also in the Lukan statement about the bread (in the longer text) but absent from Mark and Matthew. Instead, Mark attaches to the bread the ritual instruction 'Take' (which becomes 'Take, eat' in Matthew); and for the wine Matthew provides the instruction, 'Drink of it, all of you'. The Johannine form is not a liturgical text and provides no such instructions.

From these observations it follows that each form of the tradition had its own particular function and that interpretation of the form in 1 Corinthians 11 and Paul's use of it can claim help from other forms only with caution. No doubt all the forms derive from one source, but it is notoriously difficult to discover the original and to be confident that the intention of the original form is determinative for subsequent forms.

The double ritual instruction tells the community to 'do this' *(touto poieite)*, namely, to take bread, say the thanksgiving, and break the bread; and correspondingly *(hōsautōs)* to take a cup, put wine in it, and say the thanksgiving.[79] Thus, they have 'the cup of blessing which we bless' and 'the bread which we break' in 1 Cor 10:16. In the second instruction, 'as often as you drink it' do not make the drinking optional but emphasise it, as their amplified repetition in verse 26 makes plain. Not 'Break bread and, if wine is available, use the cup-thanksgiving also' but 'Whenever, having received the bread, you drink the wine, do so in remembrance of me; indeed whenever you eat the bread and drink the cup, you proclaim the Lord's death'.

The actions are done *eis tēn emēn anamnēsin*, 'to promote remembrance of me' (the possessive pronoun *emēn* is objective); and it is necessary to decide what was intended by *remembrance*. The many words associated with memory produce a wide range of possible interpretations. It is important to notice that biblical writings give more prominence than does our culture to one feature of remembering or reminding, namely, bringing something to a person's attention so that action will be taken. For example, promises are brought to attention in the hope that they will be fulfilled, sins that they may be punished or forgiven, good deeds that they may be rewarded, needs that they may be relieved, commands that they may be carried out. Very often God is asked to remember and act appropriately. Hence, it has been suggested[80] that 'in remembrance of me' should be taken to mean 'that God may remember me'. If so, the intention would be something of this kind: 'By these actions we remind God of what he did for us in Jesus Christ and beg him to complete his promised help, and [continuing with verse 26] we will repeat these action-prayers until Jesus actually comes'. In itself the interpretation is not out of the question, especially for a mind accustomed to rabbinic exegesis; but it would scarcely have been explicable to the Corinthian community unless Paul had written 'remembrance of me before God'. It must therefore have meant 'remembrance of me before the community', that is, bringing to their attention what needed to be done about the body and death of Christ. Other meanings are possible but unlikely. Why does one remember a person who has died? To avenge his death, to perpetuate a name that otherwise would perish, or to claim a share in his achievements: the first is inappropriate, the second is not wholly excluded by belief in the resurrection, and the third is not far from the point. The remembering of the Lord by performing these actions was indeed to bear his name and to share his benefits by using the means he himself had provided for claiming his power.

The statements about the bread and the cup should be considered separately since the tradition places one at the beginning and the other at the end of a meal (even if, as some suppose, the Corinthian *ekklēsia* no longer followed that liturgical pattern). The statements are both similar (in using *this is* to identify the two actions) and divergent (in using *cup* in the second but not *bread* in the first; and in making *covenant,* not *blood,* in the second parallel to *body* in the first). The meaning of neither statement is self-evident. Both require interpretation, and it seems likely that the grammatically awkward phrase *which is for you* was a second-stage explanatory addition. The double formula moves from simplicity to complexity in a liturgically acceptable manner. Thus, the two statements, held together in the context of a meal, make different but complementary assertions. If *sōma* is allowed its meaning as a body of people, and *covenant* implies the people with whom it is made, both statements are applicable to the Christian community, pledged by the presence and power of Christ's body and sealed by his blood. It is now proper to take each statement in turn.

The statement about the bread requires a decision on the meaning of *body*. Whatever Jesus may have intended by his words at the Last Supper, we must ask what *body* means in the setting of Corinthian worship and Paul's discussion of it. From the previous examination of 1 Cor 10:16^b–17 it is natural to presume that *my body* means the particular personal identity of the risen Christ within which the Christian community can be and needs to be included. Is that understanding, however, advanced or destroyed by the additional words *to hyper hymōn,* 'which is for you'? This text is given by excellent Alexandrian witnesses (papyrus 46 Codex Sinaiticus (first scribe) A B C 33 1739 Origen) with a little additional support; but it is enigmatically abrupt and is variously supplemented: by adding *thryptomenon* ('broken in pieces') in D, *klōmenon* ('broken', from *eklasen,* at the beginning of the verse) in correctors of Sinaiticus, C, D and in the majority of other manuscripts, and *didomenon* ('given', from the longer text in Lk 22:19) in the Vulgate and the Coptic versions. If the scribes were right in wishing to supply some other verb than *is,* they give a choice between regarding the body as broken or given up in death and regarding it as broken (i.e., distributed) or given (i.e., made available). It depends on the meaning of the preposition *hyper.* If it has the same meaning here as in the 'Christ died for us' formula (see Thessalonians), it has been detached from the blood where it would aptly belong and appropriated for the body; but one must not too quickly assume that it has by itself an atoning or sacrificial meaning. It can carry such meanings if they are clearly indicated by accompanying words; but *hyper* is very commonly used over a wide range of meanings: for the benefit of, for the sake of, on behalf of, and concerning (equivalent to *peri*). If we consider only the Pauline uses that refer to Christ's actions, the following phrases appear:

1. Christ died *for* you (Rm 14:15, 1 Cor 1:13 [in effect, with v.l. *peri*] 1 Thes 5:10 [v.l. *peri*]); he died *for* the ungodly, *for* us (Rm 5:6, 8); one (he) died *for* all (2 Cor 5:14, 15); Christ gave himself *for* me (Gal 2:20).
2. Christ died *for* our sins (1 Cor 15:3); he gave himself *for* our sins (Gal 1:4 [v.l. *peri*]) cf. Rm 8:32.
3. God made him to be sin *for* us (2 Cor 5:21); he became a curse *for* us (Gal 3:13, cf. 1 Cor 5:7 [v.l. hyper]).
4. Christ intercedes *for* us (Rm 8:34).

In every passage the meaning of *for (hyper)* is determined by the connected words, and sometimes *peri* will do as well as *hyper.* Under (1), (3) and (4) *for* can be expanded into *for the benefit of;* under (2), perhaps into *make atonement for;* but it is the presence of the word *sins* that suggests the meaning. The same conclusion follows an examination of the Septuagint, where *hyper* is used (though not very frequently) of people, sins, and religious virtues. The theme of dying for others is evident in the Maccabean writings, for instance, 'Let our punishment be sufficient on their behalf [*hyper autōn*, v.l. *peri*]; make my blood their purification, and accept my life in return for theirs' (4 Mc 6:28–29) and 'Gladly for God's sake [*hyper tou theou*] we give the limbs of our body to be mutilated' (10:20). But such passages scarcely suit Paul's letter. Rather more appropriate is the description of Judas as ever the protagonist in body and soul for his fellow citizens *(hyper tōn politōn)* in 2 Mc 15:30 and the remark that the God of heaven surely protects the Jews as a father acts on behalf of his sons *(hyper huiōn)* 3 Mc 7:6. Since the protective value of the sacrament is not far from Paul's thoughts in this letter, 'my body which is for you' means 'for the benefit and safety of your community'.[81]

Such an understanding of *body,* even if Paul himself had not intended it, would doubtless have suggested itself to the Corinthian community; for it would have strengthened confidence in their protected, unassailable condition. Before Paul corrects such overconfidence, he sets down the statement about the cup.

The cup statement is close to Luke 22:20 but not identical: 'In the same manner also the cup [Lk, Also the cup in the same manner], after supper, saying,'

"This cup is the new covenant	Lk omits *is*
in my blood [*en tō emō haimati*]	Lk, *en tō haimati mou*
Do this,	Absent from Luke, which has, 'which
as often as you drink it,	is poured out for you'
in remembrance of me".	

Thus, Luke agrees with Paul in placing the cup after supper, in making the (contents of the) cup signify a covenant sealed by blood (whereas Mark seems to identify the wine with Christ's blood of the covenant), and by referring to the *new* covenant; but disagrees with Paul in omitting the command to repeat the action and by adding a parallel to 'which is given for you' in Luke 22:19[b], a form that substitutes *you* for Mark's *many*. It is possible that Luke offers a hybrid form, in some way suited to his presentation of the Last Supper as a meal of which the disciples alone partake, since Jesus abstains from both eating and drinking. Luke has only one command to repeat the action, and Paul may have added the second command in order to emphasise the significance of Christ's blood, that is, his violent death, whenever the Lord's Supper was held. In verse 26 he explicitly associates eating this bread and drinking (this) cup with proclaiming the Lord's death. Yet the death of Christ is not presented here as a sacrificial or martyr's death for others or for the community (as in the synoptic traditions) but equally with the body as a means of recalling the Lord Jesus. The thought may be reformulated in this way: when the community repeats the actions singled out by Jesus, he is present not only in the power of his risen body but also in the power of his death.

In the Markan tradition 'my blood of the covenant' may refer to Ex 24:8, where the blood of sacrificial animals binds the parties to a covenant. Half the blood is thrown against the altar (which represents the Lord), and half is thrown upon the people; and so the covenant is sealed. In Luke and Paul 'the new covenant' refers to Jer 31:31–34, where it replaces the Exodus covenant, which Israel and Judah had broken. Hence 'days are coming' when God will forgive their sin and lead the exiled people back as he once led them out of Egypt. He will make a new covenant with them, different from the Exodus covenant in this respect, that they will now have both the will and the ability to obey.[82] Although this prophetic conviction should be of great interest to Paul in his teaching on knowledge of God and on the Spirit, he makes no use of it at this point. Indeed neither *new* nor *covenant* occurs again in this epistle—though in Rm 11:27 he mentions a promised covenant in words taken from Is 59:20–21[a] and 27:9 (LXX), like a scrappy version of Jer 31:33–34; in Gal 4:24 he distinguishes two covenants corresponding to flesh and spirit; and in 2 Cor 3:6, 14 he discusses the new and old covenants. Even if Paul had that kind of thing in mind, he was not disposed to share it with the Corinthian community at this stage. The word *new* is even less frequent: in addition there is only *new creation* in Gal 6:15.

In citing the Lord's Supper tradition, therefore, Paul fixes his mind on the blood of Christ. Although he avoids an explicit commandment to drink what is designated as blood, he cannot and does not wish to avoid the imagery of participation. Just as the community signifies by sharing the one loaf that it participates in the risen presence of Christ, so by drinking from the one cup it signifies its participation in the sufferings of Christ: 'If one member suffers, all suffer together; if one member is honoured, all rejoice together' (1 Cor 12:26). Jeremiah's new covenant made no mention of blood: the earlier covenant provided this necessary word. For it was blood that united the deity and the community, specifically the Lord's death. Every Lord's Supper is a proclamation (*katangellō*, a verb most commonly used for proclaiming the gospel and its benefits) of the Lord's death, and therefore of the community's participation in it. True as it is that the community shares his risen presence, until the Lord comes they must also proclaim and share his death.

The community's experience of suffering comes from two sources: from God and from their own sin. The word *paredideto* ('was betrayed' or, better, 'handed over') in 1 Cor 11:23 is part of the passion tradition. In Pauline use the verb refers not to Judas or the Jewish authorities but to God (Rm 8:32) or to Jesus himself (Gal 2:20). Hence, Paul's *paradosis* of the supper implies that the death of the Lord Jesus was contrived or permitted by God; and the conclusion can easily be drawn that the community's share in Christ's death is equally required by God. But their suffering may in part be self-inflicted. If they eat the bread and drink the cup of the Lord unworthily, they are guilty of (an offence against) the body and the blood of the Lord (verse 27). Eating and drinking in obedience to the Lord's command brings benefits, unless anyone 'eats and drinks without discerning the body', when his actions inevitably bring judgement upon himself. The meaning of 'without discerning the body' (*mē diakrinōn to sōma*) is much disputed and turns on the translation of the verb *diakrinō* and the sense to be given to *sōma* in this sentence. The verb here and in verse 31 means 'to assess properly'.[83] It cannot mean 'fails to distinguish [a proper meaning of the verb] the bread of the Lord's Supper from ordinary bread', because it is clear from 1 Cor 10:1–22 that the community already makes an injudicious distinction; and in any case the phrase speaks of 'body', not 'bread'. It is difficult to see what sense of *diakrinō* would be appropriate to *sōma* interpreted as 'the Christian community', though disregard for the proper values of the community is not far away. It is best to interpret *sōma* as meaning the risen body of the Lord. Those who eat unworthily (as described above) fail to make a proper assessment of the power of the risen presence, which can bring not only benefits but also judgement—the latter seen in weakness, sickness and even death (verse 30).

The Gospel of Death and Resurrection

Paul's exposition of Christ's death and resurrection in the Lord's Supper therefore brings to attention the death of Christians, as also in 1 Cor 5:5. In both places he attempts some reassurance: the erring member's spirit may be saved in the day of the Lord Jesus; and present sufferings, though merited, are a chastening so that Christians 'may not be condemned along with the world' (1 Cor 11:32). Then Paul turns his full attention to things of the Spirit but finally returns to the death of Christians in chapter 15. He begins with another explicit quotation of the tradition he had received (presumably in the same manner as the tradition recited in chapter 11):

> that Christ died for our sins
> in accordance with the scriptures.
>
> that he was buried,
> that he was raised on the third day
> in accordance with the scriptures,
>
> and that he appeared to Cephas,
> then to the twelve. (1 Cor 15:3–5)

The simple pre-Pauline formula 'Jesus died and rose again' (1 Thes 4:14) has already been examined in the discussion of Thessalonians and compared with similar statements. In the present *paradosis* it has been much developed, and it is possible to suggest a process by which the enlargement took place.[84] It is first to be noted that 'he was buried' *(etaphē)* stands as a simple statement of fact, evidence that Christ had really died. Unlike the other leading phrases, it is not glossed, presumably because at this stage it lacked any religious application (contrast Rm 6:4). This suggests that the original formula was simply,

> Christ died, he was buried;
> he was raised, he appeared,

where 'he appeared' is equally evidence that Christ had really risen. Then the (un-Pauline) phrase, 'in accordance with the scriptures', was added to each limb of the formula to assert with some emphasis (it may be supposed) that both the death and the resurrection were rooted in the ancient oracles and were not a way of repudiating them. The addition has the same wide generality as the statement in Acts 17:2–3 that Paul 'argued with them from the scriptures, explaining and proving that it was necessary for the Christ to suffer and rise from the dead'. That, of course, is a familiar Lukan theme (Luke 24:27, 'Beginning with Moses and all the prophets . . .') and is also present in a similarly oblique fashion in Mk 12:24 (the Sadducaic question about resurrection) and in Mt 26:54, 56 (the acceptance of arrest in Gethsemane). These passages notably lack precise indications, and the explanation is probably simple. As a result of the shocking death and astonishing resurrection of Jesus, the first Christians had experienced a shift in their perception of the scriptures. They had to say not 'Here are clear predictions of what should happen, and it has now happened exactly so' but 'Now this has happened, we read the scriptures again and see that it had to happen so'. Thus, in Acts 4:10–11, Peter discovers in 'Jesus Christ of Nazareth, whom you crucified, whom God raised from the dead' a new understanding of 'the stone which was rejected by you builders, but which has become the head of the corner' (as in Ps 118:22). In somewhat the same way, the words 'in accordance with the scriptures' must have indicated that the general tenor of scripture provided a firm base for Christian belief in the death and resurrection of Christ.

The next additions, however, caused a displacement of the references to scripture. The words *for our sins* were added to the verb *died,* and *on the third day* to the verb *raised.* Since the former addition gives a theological significance to the death, it is likely that the latter addition was similarly intended to indicate a theological aspect of the resurrection. Both additions therefore hint at the kind of scriptures to be examined for an understanding of the death and resurrection of Christ. The theological importance of 'the third day'—originally perhaps an eschatological symbol—has long been obscure to us; but the relation between death and sins is easier to grasp.[85]

The word *hamartia* ('sin'), in the plural, occurs here for the first time in the epistle. It is picked up in verse 17 and reappears in the more characteristic Pauline singular in verse 56. Indeed, the rather extensive Pauline vocabulary for wrongdoing is largely absent from 1 Corinthians. The verb *hamartanō,* 'to act in a religiously offensive way', occurs seven times. *Hamartia* is such an act. Paul uses *hamartia* language rather sparingly (except in Romans) and normally in the singular; here he is following traditional usage as in Gal 1:4 (he 'gave himself for our sins').[86] 'Sins' may be looked at in two ways: (1) they may be accidental, careless or deliberate contraventions of the rules by which God governs mankind and the universe, in which case, they introduce instability and danger into normal life and a powerful remedy is needed to restore the divine system; (2) they may be actions that destroy belief in the power, goodness and presence of God, in which case, they strike at the root of human existence and a powerful remedy is needed to restore confidence. It is not a matter of appeasing an offended deity or of satisfying an abstract principle of justice; it is a matter of repairing what the sinner has broken down and making credible what the sinner has defiled. Paul seldom deals with sin in such terms, but here he accepts a traditional formulation (which is not necessarily alien to his perception of the death of Christ as a 'talisman') and builds it into his own exploration of the resurrection.

The statement 'that Christ died for our sins' is not easy to interpret, even when put in the form 'to deal with our sins' (see the discussion of *hyper*, pp. 42–43 and n.). It depends on how sins are viewed (Are they occasions of defilement or social injustice or defiance of God?), on the kinds of persons involved (e.g., devout and repentant individuals or unheeding and selfish communities), and on the kind of imagery used for relating sinners to the deity (is he described by some such analogy as king or judge or father?). Since this is traditional language, we can only look at the source from which it comes; but that itself is not easy to discover. Broadly speaking, there is a choice between judicial, self-sacrificial and cultic contexts. In a *judicial* context, the death of an innocent person would be required as a kind of transferred retribution for sins committed by others, in itself an odd conception. It is unsupported by any enquiry in New Testament writings about the conditions of atonement (such as is found in M. Yoma 8.8) or by any passage describing God as judge and judging and so may be dismissed from consideration. In certain desperate situations it seemed possible that the *self-sacrifice* of a few persons, or even of a single person, might save a whole people. Thus, in the Maccabean conflict it might be hoped that the divine Providence would deliver hard-pressed Israel, the martyrs having 'as it were become a ransom [*antipsychon*] for our nation's sin' through their blood and the propitiation (or atonement) of their death (4 Mc 17:22). That such thoughts were active in Christian circles is shown by the recording in Jn 11:40 of Caiaphas' remark, 'It is expedient for you that one man should die for the people'.[87] This context, certainly appropriate to some parts of the New Testament, seems to belong to a period of major conflict not evident in the formal expansion of 'He died—he rose' and so may be put aside. Therefore we are left with the *cultic* context where 'for our sins' is obviously at home. *Ta peri hamartias* is a stock phrase for the sin offering, and the great cleansing ritual of Leviticus 16 makes atonement for the people of Israel 'once in the year because of all their sins' (Lv 16:34). Hence, it is necessary to give some consideration to the Temple cult.

In the long history of the Temple, from its dedication by Solomon to its destruction by the Romans, there is frequent alternation between elaborate practice of the cult in an existing Temple and substitute practices for the cult in the absence of a Temple and between adherence to the cultic life and criticism of it. In Solomon's prayer at the consecration of the Temple (recorded in 1 Kings 8, where the king has his preexilic priestly status, and reproduced with priestly alterations in the postexilic 2 Chronicles 6) this is already clear. With detailed emphasis the Deuteronomic compiler in 1 Kgs 8:27–43 makes the Temple the sole centre of forgiveness for every Israelite and for all Israel, and from verses 62–66 it can be seen that the Temple is the place of offering sacrifices and celebrating festivals. Yet the last part of Solomon's prayer, verses 44–53, is the work of an exilic reviser modifying the prayer for the captives who can no longer take part in the cult.[88] Some such device was necessary whenever the Temple was beyond the worshippers' reach or had fallen into disuse or had been defiled by the Gentiles—or indeed when the Temple had fallen into disrepute by reason of the character of those who served it. Two contrasted responses were made: (1) Ezekiel and the exilic scribes coordinated and developed ancient traditions to create a revised cultic system, which eventually was put into operation in the postexilic

Temple; (2) other cultic, prophetic and wisdom writers sought forgiveness by direct appeal to God, taught that atonement could be achieved by religious devotion and social goodness, and even spoke derogatively of the sacrificial system.[89] There were some, however, who regarded the various atoning activities of type (2) as acceptable equivalents of the directly sacrificial activities of type (1) if valid sacrifices were not possible in the Temple. That line of thinking may suggest an origin for the fourth servant song in Is 52:13–53:12 or at least a way it might be understood in the Septuagint. The beginning and constant theme of Deutero-Isaiah's message is that God had sufficiently punished his people and now had wholly forgiven them. In that case, how could they understand their not only suffering but suffering in an alien land? If Israel had been wicked ('the sins of my people', verse 8[d]), the nations had been far worse. The answer appears to be that God's people had been punished for their wickedness by the nations in order that the nations, observing the manner of their punishment and subsequent exaltation, should realise their own condition and perceive the truth of God. Both God's servant and the nations will come to understanding (52:13[a], 15[d] LXX), and the nation's understanding is expressed in a ritual confession that 'he was wounded *because of* our wicked actions and made weak *because of* our sins [verse 5, *dia* + acc.]. In fact 'he carries [the consequence of] our sins' and 'the Lord handed him over to [the effect of] our sins' (verses 4[a] and 6[c]). At that point in the Septuagint *we* statements cease, and presumably the prophetic or divine voice resumes with the defence of the servant (verses 7–9), whose death is now explicitly mentioned. Finally the divine verdict is pronounced in verses 9–12, and cultic language now appears: 'Their sins he will take away' (11[e]), 'He has taken away the sins of many' (12[e]), and 'Because of their sins he was handed over' (12[f]). In other words, sufferings caused by the sins of others are now pronounced by God as a death that atones for their sins (cf. *peri hamartias* in verse 10[b]). The servant song moves in the direction of sacrificial equivalent for evaluating the death of the servant, whether he is Israel or some other person.

That movement of thought remained isolated in Jewish religion. Even the New Testament writings make scanty reference to the fourth servant song by way of quotation and allusion and (apart from 1 Pt 2:22–25) seldom fix attention on the sacrificial phrases.[90] Hengel's statement that Isaiah 53 may have had some influence on the formula 'Christ died for our sins' is as much as can be said;[91] but it does not greatly matter. The situation of the early Church led some part of it to find acceptable equivalents for Temple worship. Like the community at Qumran, the early Christians could regard themselves as the true temple of God; and indeed Paul makes mention of such a view in 1 Cor 3:16: 'Do you not know that you are God's temple and that God's Spirit dwells in you'. If the community itself was the true temple, some feature of the community's life would correspond to the sacrificial cult of the old temple.[92] In Qumran the community's strict practice of the law replaced the bloody sacrifices of the Jerusalem Temple, and the sacrificial living of the community members made 'atonement for guilty rebellion and for sins of unfaithfulness' (1 QS 9.3–4). Christian tradition also knew about spiritual sacrifices. Paul urges his readers to present their bodies to God as a living sacrifice, which is their spiritual worship (Rm 12:1). 1 Peter 2:5

describes the community as a holy priesthood offering spiritual sacrifices, acceptable to God through Jesus Christ. But the spiritual sacrifices of Christians are not atoning sacrifices, unless perhaps they are part of Christ's sacrifice. The Qumran Teacher, who composed the Hodayoth, endured much suffering and called himself God's servant but nowhere suggested that he was suffering in place of sinners.[93] The distinctive conviction that 'Christ died for our sins', which (as Hengel has shown[94]) would be readily comprehensible in hellenistic culture, can therefore be expounded in the following way: as the death of an animal in the sin offering joined deity and people together in an atoning ritual, so the death of Christ made possible an atonement that did not simply restore the status quo but opened up the possibility of a new age.

That all-important extension of atonement is required by the double formula, 'He died for our sins, . . . he was raised on the third day', and indeed by Paul's closely associating death and resurrection in other epistles.[95] In this chapter his development of the primitive formula is entirely devoted to the resurrection, so much so that the reference to 'sins' is detached from Christ's death and added to his resurrection: 'If Christ has not been raised, your faith is futile and you are still *in your sins*' (1 Cor 15:17). That final phrase *(en tais hamartiais hymōn)* is without parallel except for Jn 8:24, 'Die in your sins', which in turn reflects the judicial pronouncements of Ez 18:24, 26: a formerly righteous person shall die by reason of the sins he has now committed. Paul's use of the phrase presumably means that if Christ has not been raised, Christians are still condemned for their sins (which is not yet the same as 'in the power of sin' as explored in Romans). Yet he quickly replaces 'in your sins' with a more vivid image, namely 'in Adam', contrasted with being 'in Christ'. Adam is associated with death, Christ with life. If Christ has not been raised, they cannot be in Christ; they must still be in Adam. If they cannot be in Christ, they cannot share Christ's present reign or his fight against the enemies of God and finally against death itself (verses 21–26). They are condemned to bearing 'the image of the man of dust', for they cannot bear 'the image of the man of heaven' (verse 49). Hence, it is plain that Paul has burst through the traditional imagery of the atoning death of Christ. The predicament of the Corinthian community is not that they can be charged with specific sins that may or may not be remedied but that they may remain untransformed human beings who cannot enter the Kingdom of God (verse 50). For their transformation both the death and the resurrection of Christ are needed.

Summary

Thus it becomes clear that Paul had come a great distance between the writing of 1 Thessalonians and 1 Corinthians. The resurrection no longer bears the main theological weight, though its influence within the life of the community is profoundly important. There is no doubt that Christ is the Lord of Glory, that his resurrection *dynamis* is present within the community that participates in (or actually is) his risen body, that the Lord is their protector against the demons, and that his *dynamis* conveys to them a remarkably effective social wisdom—or at least that some of them thought so and saw no other way in which the Lord's resurrection could sensibly be regarded. With much of this Paul could scarcely disagree, but his experience of the Corinthian community left him deeply dissatisfied with the development of their wisdom. He therefore set himself a complex double task. On the one hand he had to undermine the community's confidence in its resurrection *dynamis* without destroying its reliance on the risen Lord; on the other hand he had to develop in them an awareness of the crucifixion as the critical principle for assessing their manner of life.

He set about the former task by restoring to resurrection teaching its eschatological function, the primary purpose of 1 Cor 15. The last enemy, Death, has not yet been destroyed; and for that reason alone the community can neither ignore the resurrection appearances of Jesus nor dismiss from their mind his expected return. But during the period of waiting (which the Thessalonians had already found troublesome), how were they to meet the pressures arising within the community or coming from outside, by what standard were they to formulate and control their responses?—By developing an understanding of the death of Christ.

In only one passage of the epistle death and resurrection are spoken of together,[96] and there Paul explicitly quotes a *traditional formula*, 1 Cor 15:3–5. He was not the inventor of the relation between Christ's death and our sins, which has a cultic origin as one variant of the numerous substitutes for the Temple ritual. Paul's justification for balancing excessive claims for the resurrection by drawing out the meaning of the death is to be found in the pattern of the traditional formula in 1 Cor 15:3–5 and indeed in the shorter formula which appears in 'the brother for whom Christ died' (1 Cor 8:11), which is used to restrict the insistent and inconsiderate use of Christian gnosis. The same formula is implied by 'Was Paul crucified for you?' in 1 Cor 1:13, and is at once given a cultic association by 'Were you baptized in my name?' Since baptism was regarded as at least a saving and protecting ritual (at this point intended by Paul as saving the community from dissension), it follows that proclamation of the death of Christ was of similar effect. An equivalent result is obtained when, to save the community from sexual deviance, Paul compares Christ's death to the sacrifice of the Passover lamb. This contains no elaborate theory of sacrifice but simply a conviction that God wills to protect a community whose life is responsive to Christ's death. The same thought of God's protection lies behind a midrash on the Exodus, warning the community that their cultic protection can be withdrawn if they are presumptuous and unfaithful, particularly as regards the Lord's Supper. Thus, the cup of blessing signified that the community jointly shared the protective benefits that Christ's death had obtained, and the broken bread signified their share in Christ's risen life. The risen Lord himself was *recalled* by means of the anamnesis (especially of his death), which claimed a share in what he had achieved. That claim, however, would be valid only if the community accepted the new covenant in Christ's blood which (in Paul's use of the tradition) implies modelling their communal behaviour on the self-giving of Christ.

In thus working out the implications of Christ's death, Paul is properly relying on the imagery, traditional sayings, and cultic formulas already at home within the community. In the opening chapters of the epistle, however, he breaks new ground. He opposes the word of the cross to Corinthian wisdom and knowledge for the simple reason that wisdom and knowledge, however justifiable in themselves, are directed towards the preservation of this present age and therefore resist the radical transformation God has inaugurated in Christ. The word of the cross is religiously offensive to Jews and therefore undermines Jewish self-confidence. It is socially destructive to Gentiles and therefore subverts Gentile social awareness. To those who are caught up in the dissolution of present society it is shockingly stupid, but to those who are sharing in the process of transformation it is God's powerful agent of change. Indeed, it is the hidden mystery of the divine will, which entraps and overthrows the ruling powers of this age and opens up the possibility of God's rule through the crucified and risen Lord.

2 CORINTHIANS

The notorious literary and historical problems of this epistle offer little hindrance to studying the references to Christ's death and resurrection. The structure of the epistle as it stands is commonly agreed on:

1:1–2:13	Greeting, Paul's recent troubles, frustrated plans, anxiety about an offender, and relief at a solution.
2:14–7:4	A polemical discussion of the apostolic ministry of reconciliation.
7:5–16	Reversion to Paul's plans and relations with Corinth.
8–9	A (double) treatment of the collection.
10–12	An attack on rival missionaries.
13	Final words to the community.

All direct references occur in chapters 1–5 and 13, though indirect references may perhaps be detected elsewhere.

1:1–2:13

The extended thanksgiving unit 1:3–11 has consolation *(paraklēsis)*[97] as its theme, the divine response to affliction. The affliction may be Paul's hardships endured for the community's sake or his despair in a situation of deadly peril or the sufferings of the community. The consolation may be renewal of confidence, recovery from illness or improvement of spirits on hearing better news from Corinth (2 Cor 7:4, 6–7, 13). In Paul's mind, this experience of great distress and recovery of hope is interpreted as 'sharing abundantly in Christ's sufferings' (verse 5) and relying 'on God who raises the dead' (verse 9). If the primitive formula 'He died—he arose' stands behind the thanksgiving unit, it has now become an interpretative pattern. The death and resurrection of Christ are indeed saving events insofar as Christians participate in them. The precise meaning of *ta pathēmata tou Christou* (2 Cor 1:5)[98] has been variously understood. If *Christ* is taken to mean 'Messiah', Paul may be calling to mind the widespread belief that severe hardships will precede the messianic kingdom, the so-called travail pains of the Messiah.[99] In 1 Thes 5:3 he compares the sudden destruction preceding the day of the Lord to birth pangs and in Rm 8:18 speaks of the *pathēmata* of this present time in contrast with the glory that is to be revealed. But the present passage is scarcely part of the conventional cosmic catastrophe; and when Paul details his sufferings (as he does rather frequently in this epistle), he is not giving voice to dreadful apprehensions but mentioning what has actually happened to him. It seems likely that the phrase *ta pathēmata tou Christou* had a certain currency in the early church, for it occurs again in Phil 3:10 and (with the verb *sympaschō*) in Rm 8:17 immediately before 'the sufferings of this present time'; in 1 Pt 1:11, 4:13, 5:1; and in Heb 2:9–10. No doubt Paul had the suffering and shocking disgrace of crucifixion in mind, the exclusion of Jesus from the company of the godly ('He made him to be sin', 2 Cor 5:21), and the rejection of his work by Israel. But the aptness of the phrase in this context arises from the sufferings of Paul and the community: it means the sufferings of those who are members of Christ's body and belong to Christ.

The situation of the community has changed sharply since Paul wrote 1 Corinthians. At that earlier time they were exploiting the possibilities of social wisdom and so were held in honour (1 Cor 4:10); they were carrying on an active social life with their pagan neighbours (1 Cor 5:9–10, 6:1, 8:10) and encouraging unbelievers to attend their enthusiastic assemblies (1 Cor 14:23–24). But now they need consolation. No direct indication is given about their sufferings, but Paul judged that they could be partly mitigated if he told them about his own sufferings. They must have said, 'Why should we suffer for being Christians?' In effect, he replied, 'We all share in sufferings for Christ's sake'; deliverance depends not on ourselves 'but on God who raises the dead'. When Paul moves beyond the thanksgiving unit, he may give a hint about the cause of their sufferings: he, at least, had behaved in the *kosmos* (the social world of Corinth), especially towards them, not with fleshly wisdom *(sophia sarkikē)* but with the the grace of God (see pp. 22–24 for *sophia*, pp. 31–33 for *sarx*). The grace of God gives strength, even in desperate situations; social wisdom—at one time so much prized by the Corinthians—is weak, vulnerable and mortal. It is to be suspected that the community's earlier confidence in their social skills had turned out disastrously.[100] In any case, 'wisdom' is not mentioned again in the epistle; and the balance between sharing Christ's sufferings and experiencing his life is prominent in the polemical discussion of the apostolic ministry in 2:14–7:4.

2:14–7:4

In this unit of the epistle Paul's mind leaps from one image to another without attempting to present a consistent symbolic scheme; but it is usually possible to follow his meaning and to discern why the particular symbols seemed appropriate. He writes throughout with opponents in mind whom he probably puts with 'those who are perishing' (2 Cor 2:15, 4:3), and appears to be answering a variety of charges, namely, that he persistently commends himself instead of getting letters of commendation from others (2 Cor 3:1, 5:12); that in disgraceful, under-

handed ways he tampers with God's word (2 Cor 4:2) and so offers a veiled gospel
instead of the clearly written scripture with its ancient splendour (2 Cor 4:3, 3:6–
7), that he does this to persuade his hearers (2 Cor 5:11); that he is out of his
mind (2 Cor 5:13) and is always in some kind of trouble (2 Cor 4:8–9). To such
an attack, Paul launches his reply with exceptional vigour: his humiliation and
suffering is the means by which God is victorious, and God (as it were) leads the
apostle a captive in his triumphal procession (NEB; Collange, p. 24). Not only
that but it is 'God who in Christ always leads us in triumph' (2 Cor 2:14). The
words *in Christ*, an essential qualification of this image and the one that succeeds
it, are a very familiar feature of Paul's writing. In this instance (as in three other
passages of 2 Corinthians) it refers to the means by which God performs his saving activity, namely, by the death and resurrection of Christ.[101] On this, Paul's
sufferings and the effectiveness of his apostolic ministry (in diffusing the fragrance
of the knowledge of God) are modelled.

The knowledge of God is knowledge of Christ crucified and risen. In 2 Cor
2:15–16 it is as pervasive as a fragrance *(euōdia)*, but by no means all can tolerate the smell of it. It is an 'aroma' *(osmē, which may be a fragrant or a foul
smell)*. In Tob 8:3 it is the smell of burning that drives the demon away; very
frequently in Lv and Nm *osmē euōdias* is the soothing smell of sacrifices acceptable to God. The high priest of later times, Simon b. Onias, 'reached out his hand
to the cup and poured a libation of the blood of the grape; he poured it out at
the foot of the altar, a pleasing odour to the Most High, the King of all' (Sir
50:15). The two words *osmē* and *euōdia* are not joined together here (as they
are in Phil 4:18), but it is scarcely to be denied that Paul was playing on the
inevitable sacrificial imagery when he compared himself to 'the aroma of Christ'
in verse 15. Compare Ez 20:41, where God says, 'As a pleasing odour I will
accept you, when I bring you out from the peoples . . . and I will manifest my
holiness among you in sight of the nations'.[102] Yet this sacrificial ministry, however pleasing to God, meets with opposed responses. To some it arises from death
(namely, the death of Christ) and leads only to death; to others, it arises from life
(namely, the resurrection of Christ) and leads triumphantly to life. 'Those who
are perishing' and 'those who are being saved' have still not shifted from their
estimates of the cross as either folly or the power of God (1 Cor 1:18).

Paul admits (though grudgingly) that his gospel may be veiled, that is, the
glory and the lordship of Christ is not plainly visible to all who hear the message
of Christ crucified. Why not? Because 'the god of this world has blinded the
minds of the unbelievers', alongside whom (Paul perhaps insinuates) his opponents take their stand. But Christ is the likeness *(eikōn)* of God. Just as the first
Adam was made in the image and likeness of God when God separated the light
from the dark (Gen 1:3, 26), so the glory of God is seen in the face of Christ
when he has been brought from the darkness of the crucifixion to the light of
resurrection (2 Cor 4:3–6).

Before writing this bold midrash on Genesis 1, however, Paul has already
attacked his opponents with an equally venturesome midrash on the shining face
of Moses and the veil that covered it in Ex 34:27–35. He distinguishes between
the written code *(gramma),* on which it seems his opponents base their case, and

the Spirit, meaning the spiritual interpretation of scripture that he develops. He presumes (1) that the written code kills, for it killed Jesus, and the Spirit gives life, for it is the Spirit of the living God and raised Jesus from the dead (cf. 1 Cor 15:45, 'the last Adam became a life-giving spirit') and (2) that he is the minister of a new covenant (2 Cor 3:6), which supersedes the old covenant of the written code. He therefore argues that since the Mosaic covenant possessed its own glory (which he does not dispute), the new covenant must have even greater glory. By means of the Spirit, we behold the glory of the Lord and we are transformed into his likeness from glory to glory *(apo doxēs eis doxan)*—from the veiled glory of the crucifixion (cf. 1 Cor 2:8) to the open glory of the resurrection (2 Cor 3:7–18).

That Paul does indeed regard the knowledge of Christ crucified and risen as being pervasive as a fragrance is shown by his self-disclosure in 2 Cor 4:7–9. He introduces it with the overfamiliar and rather perfunctory image of treasure in earthenware containers in order to maintain that his concern is with the power of God, not with his own experiences, however instructive and insistent they may be (see Collange). *They* are represented by four sets of contrasted passive participles, all indicating that he was under exceptional but never overwhelming pressure. Such experiences are then interpreted as 'carrying in the body the death of Jesus, so that the life of Jesus may also be manifested in our body'. Such statements are sometimes ascribed to Paul's mysticism, presumably in the sense that he experienced the sufferings of Christ in thought, imagination and sympathy.[103] But that is unlikely. He not only insists on bodily experiences, of himself and of Jesus, but says that he is constantly being handed over to death (by God; the verb is *paradidōmi,* familiar in the Passion tradition) for the sake of Jesus. In fulfilment of his apostolic mission (neatly indicated in verse 12 by the change of pronoun, from the expected 'in *us*' to 'in *you*'), Paul is required to undergo mortifying experiences, where *mortifying* is understood in one of its antique senses. He does not simply risk death; he actually experiences the power of death (the last enemy to be destroyed, 1 Cor 15:26), so that some areas of his immense energy are devastated and are never usable again.[104] The death and resurrection of Jesus provide Paul with a grid, as it were, for locating his own experiences within the purpose of God and for interpreting them when they seem most frustrating. At the same time, his own experience of being handed over to death (of dying daily as he put it in 1 Cor 15:31) in order that others should experience new life gives him his clue to the theological meaning of the crucifixion and fills out the proverbial words of 1 Cor 15:36: 'What you sow does not come to life unless it dies'.

Yet that is not all. Not only dying for others but also dying in order that the life of Jesus should be displayed in our body (verse 10) and in our mortal flesh (verse 11). It is not satisfactory to assume that here body and flesh *(sōma* and *sarx)* mean the same. Paul is not repeating the same statement in alternative words but making one complex statement by two parallel but not identical phrases. He intends to say that the life of Jesus (that is, the risen life of him who was crucified) is to be displayed not only in our body but also in our flesh. It is understandable that our body (which, besides belonging to *sarx,* belongs to the Lord) should display his life. It is more remarkable that *sarx* (which is dishonourable, weak and mortal) should display the risen life. But, according to Paul, it does. And he brings this section to an end in 4:14 by quoting a formula and a deduction from it:

> He who raised the Lord Jesus
> will raise us also [so that we are] with Jesus
> and will bring us with you into his presence [at the judgement, see
>   2 Cor 5:10].

Thus, the death of Jesus and what God made of it stretches from the apostolic responsibility to the final purpose and justification of the Christian community.

4:16–5:10. Paul has not yet finished either with the death of Christ or the apostolic sufferings; but before he continues with those themes, he develops an argument in 2 Cor 4:16–5:10 that presents notorious problems of interpretation. It is not necessary for me to offer a detailed exegesis, but I am required to indicate why Paul placed it between his treatments of the death of Christ in 4:10–11 and 5:14–15 and to show its connexion with both.

The section begins 'So we do not lose heart', repeating what was said in 4:1 and therefore taking a step forward. Readers of the letter should now be able to understand what Paul means when he says that 'our outer nature is wasting away, our inner nature is being renewed every day', especially as he uses familiar hellenistic phrases, common in contemporary philosophical and religious discourse, namely, *ho exō* and *ho esō anthrōpos* (which RSV translates as 'nature').[105] Some of the scholarly disagreement about the passage has arisen between those who are determined to welcome and others who are determined to resist the intrusion of hellenistic ideas. But Paul is doing little more than rephrasing in conventional words his own teaching about soma, already discussed on pp. 31–32. In distinction from *sarx,* which is the undifferentiated human condition, soma is a particular personal identity. It belongs both to the world of *sarx* and to the world of pneuma (spirit), whether it indicates the personal identity of an individual or a group. In a much more familiar convention, *anthrōpos* (meaning an individual or mankind) can also suggest a personal identity. Hence by *ho exō anthrōpos* is meant a personal identity relating to external things, which is wasting away (i.e., soma as belonging to *sarx*); and by *ho esō anthrōpos* is meant a personal identity related to the internal concerns of the Christian community, which is being renewed every day (i.e., soma as belonging to pneuma). So for Paul the hardship of his life, and indeed of the Christian life, is offset by the renewing force of the Spirit. His present bearable hardship[106] will produce an eternal fulness of glory beyond all comparison as he fixes his attention not on the (external) temporary things he can see but on the (internal) things which are unseen but eternal.

He now sets about persuading the Corinthian community to think in the same way, to evaluate present hardship in terms of future glory. He offers a persuasive demonstration, as he says plainly in 2 Cor 5:11: the demonstration takes place in 2 Cor 5:1–5, and the desired result is presented in 2 Cor 5:6–10. A good deal of exegetical confusion arises if commentators fail to notice the purpose of this passage; but of course even more arises from Paul's compressed use of imagery and his compulsive play on words.

He is talking about some *somatic* existence: the bodily experience of the life and death of Jesus in 2 Cor 4:10; presence in, or absence from, the body in 5:6, 8; and actions performed in the body in 5:10. Moreover, the word *skēnos* (the neuter form, not the more familiar feminine *skēnē*) in 5:1, 4 is here translated 'tent', although its almost invariable application in Greek was a metaphor for the human body.[107] Paul picks earthly house(hold) and heavenly building as a provocative image for the dual relations of soma (5:1), and then (5:2-4) forces verbs familiarly used of putting on and taking off clothing to do duty for an exchange of metaphorical habitations. An attempt to represent his riddling speech in verses 2-4 might run as follows:

> For indeed in this [earthly household] we are groaning, longing for our heavenly building to be superimposed

> with the hope that,[108] when it [earthly house or heavenly building] has been [disposed of or imposed, respectively], we shall not be found naked [found defenceless by God at the judgement].

> For indeed we who exist in the *skēnos* are groaning,
> weighed down [by hardships],
> because we have no desire to be exposed
> but to have [a dwelling] superimposed
> that what is mortal may be overwhelmed by life. (5:2-4)

Paul is trying to persuade a Christian community to abandon the security it obtains from the relations of soma with *sarx* for the promised but unseen securities of pneuma (for God has already given them the Spirit as a guarantee, 5:5); but, very understandably, they are still holding on to their visible securities as a prisoner clutches a few rags when led away to interrogation.

Paul has encouraged at least himself by this verbal dexterity and so moves towards a conclusion, with another metaphorical shift.

> So we are always of good courage;
> we know that while we are in residence in the body
> we are non-resident with the Lord,
> for we walk by faith, not by what we see.

> We are of good courage
> and we resolve rather to be non-resident in the body
> and in residence with the Lord.

> For which reason we make it our aim,
> whether in residence or non-resident,
> to be pleasing to him.

> For we must all appear before the judgement seat of Christ,
> each one to receive the reward of what was done in the body,
> in proportion to his deeds, whether good or bad.[109] (5:6-8)

Thus it begins to appear how difficult a task Paul has set himself. Christian existence in the body pulls him in two directions. In one direction lies unhindered communion with the Lord, towards which he is driven by the activity of the Spirit. In the other direction lie the comforting securities and persistent hardships of common human existence, where life has to be carried on by faith, not by what can be seen. And whichever way he turns, he has to satisfy the Lord. It is not perhaps surprising that he let slip the curt remarks of verse 13: 'If [you think] we are out of our senses, [it is our response] to God; if [you think] we are talking sense, [it is our service] to you'.

5:14–15. At this point, however, Paul launches his most striking argument as he gives the reason for what has just been said:

> For Christ's love controls us[110]
> because we are convinced
> that one has died for all;
> therefore all have died.
>
> And he died for all
> that those who live might live no longer for themselves
> but for him who for their sake died and was raised. (5:14–15)

By *Christ's love* he means Christ's action in dying for all, stated and repeated in verses 14ᶜ and 15ᵃ. By *all* he means all who can say 'Christ died for us' (Kümmel); he is not, in this argument, uttering propositions about all mankind; for he is making use of one variation of a familiar formula (see p. 15). Almost certainly, however, he goes beyond the formula in the surprising conclusion that 'therefore all died'. To assert that 'Christ died for all' must at least mean that he died for their benefit. But if the consequence of his death was that all died, where was the benefit? If we ask in what capacity Christ died, it is commonly answered that he died as their substitute so that what should have happened to them in fact happened to him alone. As in martyr or hostage theology, he died that others should not die—the opposite of Paul's startling deduction. Nor is it quite satisfactory to say that Christ died as their representative, so that he spoke and acted on their behalf, saying and doing what they wanted said and done, though it was beyond their ability. Christ died for all, Paul appears to say, to show what all must do. He is what Heb 6:20 calls 'a forerunner on our behalf', an idea Paul elsewhere expresses by calling Christ the image of God and the last Adam.

It is clear that Paul arrived at this special sense of *hyper pantōn* ('for all') by extending the thought of 2 Cor 4:14 from the second half of the formula 'He died—he arose' to the first half. If we share the life of him who was raised, we also share the death of him who was crucified. But the actual words used ('all died') are more absolute than the now-familiar thought of participation (in daily life all share some aspects of his dying). Can they imply that all are deemed to have died and so begin life anew as if they had been raised from the dead? But such a premature resurrection is ruled out by the future tense in 2 Cor 4:14; and the legal fiction of deeming something to be so when it is not is useful for exceptional cases but not for the generality. Nor is it satisfactory to read back an interpretation from verse 15[b] and say that 'all died' in that they ceased living for themselves and began living for him who for their sakes died and rose again, for two reasons: (1) since it is not indicated why the death of Christ should prompt or make possible that result, the problem remains unsolved and (2) it is not clear what living for him who died and rose for them could mean. Barrett refers it to 'the realm of actual obedience and unselfish living', but that seems to be no more than the acceptable face of Pharisaism. What we need is something that is typified by the death of Jesus and brings us into relation with the risen Jesus as Lord. Now when Jesus was crucified he became religiously offensive to Jews and socially destructive to Greeks (see pp. 21–22). What is typified therefore in his death is a negation of Jewish and Greek identities. If the cross is a permanent symbol of God's intention, if Christ crucified is irremovably part of the gospel, then the death of ancient social structures is the aim of Paul's preaching. When you became members of the Christian community, he said in effect, you ceased to be Jews and Greeks and became the Church of God (1 Cor 10:32). Paul himself was intensely proud of his Jewish identity and, speaking as a fool to offset Jewish claims, boasted of it (2 Cor 11:22). But elsewhere he renounced his Jewish identity for the surpassing worth of knowing Christ Jesus his Lord (Phil 3:4–11); and thenceforward to the Jews he became as a Jew, and to those outside the law he became as one outside the law—though in fact in-lawed to Christ (1 Cor 9:20–22). As he said in Gal 3:28, 'There is neither Jew nor Greek, there is neither slave nor free, there is neither male nor female; for you are all one in Christ Jesus'—three revolutionary remarks of great potential scope, however restricted in actual effect. As far as the Corinthian community was concerned, the powerful self-consciousness of Greeks is evident in the first epistle's treatment of social wisdom and knowledge; the equally strong Jewish self-consciousness stands behind the second epistle's early concern for Moses and its later polemic against Paul's opponents. Thus, the death of Christ was a blow against the existing structures of the world, and his resurrection a claim for loyalty to a new structure. After all, the form of this world was passing away (1 Cor 7:31).

5:16–17.   Yet for the time being life had still to be lived in the flesh. Strategies had to be devised for this new in-between kind of existence, and they are suggested in verses 16–17: 'From now on we regard no one [who has confessed Jesus as Lord] as bound by the existing structures. Even if we [Christians] once regarded Christ [because of his seemingly offensive or destructive crucifixion] as rejected by the existing structures, we regard him so no longer. Therefore if anyone is [the object of God's action] in Christ, he is a new creation; the old has passed away, behold, the new has come'. Those two notorious verses are not too difficult to grasp if appropriate supplements are added (as so often must be done) to Paul's compressed rabbinic style of writing and if they are interpreted as constituent parts of Paul's consistent argument in this epistle. Exegesis of chapter 5 in particular has been much hindered by an inclination to interpret remarkable assertions apart from their context.

It is therefore necessary to ask why Paul immediately says of the new creation in Christ that 'all this is from God, who through Christ reconciled us to himself and gave us the ministry of reconciliation' (verse 18). Why does he repeat (in verse 19) that he is equipped with 'the message of reconciliation' and then (in verse 20) on Christ's behalf beseech the community to be reconciled to God? Why did they need reconciling to God, for these were the very people to whom Paul had said, 'From God you have your existence in Christ Jesus' (1 Cor 1:30)? It is true that some within the community needed reconciliation to others (perhaps the offender discreetly mentioned in 2 Cor 2:5–11, as well as Jewish and Gentile factions); but this is about reconciliation *to God*, not to fellow Christians. Reconciliation means the exchange of enmity for friendship. It could easily be supposed that God was at enmity with some members of the community for their negligent or defiant activities (the 'trespasses' of verse 19) and needed persuasion to be reconciled to them. Paul might then be pleading with the community (in a loose phrase) to seek reconciliation with God, to do whatever was necessary to turn God's enmity into friendship. The situation would then be similar to what is found in 2 Mc 7:33: 'If our living Lord is angry for a little while, to rebuke and discipline us, he will again be reconciled with his own servants', who, in fact, suffer martyrdom.'[11] But of course Paul exactly reverses that situation: the reconciliation was effected not by the death of the sinners but by the death of the sinless Jesus, and it is not God who is reconciled to sinners but sinners who are reconciled to God. At least two things are clear. First, this process of reconciliation is not mutual. It is not as if two parties in conflict decided to be friends and not enemies. What needs to be done is done by one party (namely God) and accepted by the other party (namely sinners). So when Paul again uses this manner of speaking in Rm 5:11, he says, 'We have now accepted our reconciliation'. God, who has good reason to be hostile to sinners, in fact reconciles them to himself. Second, sinners need to be reconciled to God because they see him as their enemy.'[12] Their own enmity towards God is aroused by the threat which God offers or seems to offer to their security, intentions and hopes. It would be no reconciliation to persuade them that God poses no threat; because a loving God must indeed attack whatever subverts his goodness toward us. Now the death of Christ, religiously offensive to Jews and socially destructive to Greeks, is a damaging attack by God (if it is indeed God's doing) on the existing structures of the world and therefore on our securities, intentions and hopes. Therefore, Paul with all his heart beseeches the community to be reconciled to God; for only if they will accept Christ's death as the act of a friend, not an enemy, will they experience the consequences of his resurrection.

5:18–21. It is noticeable that verses 18–21 are written in repetitive style, which in fact suggests a deliberate structure:'[13]

> All this [the new creation] is from God
> who *through Christ reconciled us to himself*
> and gave us the ministry of reconciliation;

granted that [*hōs hoti* AGB s.v. *hoti* 1.d.] in Christ God was
   reconciling the world to himself,
not counting their trespasses against them,
and entrusting to us the message of reconciliation.

So we are ambassadors for Christ,
with God as it were [*hōs*] making his appeal through us.
We beseech you on behalf of Christ,
be reconciled to God.

For our sake he made him to be sin who knew no sin,
so that in him we might become the righteousness of God.

The italicized statements are Paul's chosen way, in this part of the argument,
of talking about the divine activity in Christ. Reconciliation was an uncommon
theme in the ancient world. Paul himself seldom uses it (to 2 Cor 5:18, add 1
Cor 7:11 of man and wife; Rm 5:10–11, 11:15; and, less important, Eph 2:16
and Col 1:20, 22); and it is otherwise absent from the New Testament. Even here
it is used with some diffidence: witness the phrases 'granted that God' and 'with
God as it were'. Paul knows that he is stretching what can be thought proper of
God, but he is concerned with a problem that is not greatly illuminated by traditional imagery. In verse 19[b], 'not counting their trespasses [*paraptōmata*]
against them' is modelled on the language of Ps 32:1–2: 'Blessed is he whose
transgression is forgiven, whose sin is covered. Blessed is the man to whom the
Lord imputes no iniquity'. The verb for 'impute' in the Septuagint and for
'reckon' in Paul is *logizomai,* though the Septuagint does not use *paraptōma.*
The psalm rejoices in a clear conscience that comes from frank confession of sin
and the Lord's willingness to accept repentance and cancel the adverse report.
But with some practice and a trained sensitivity, that is the easiest problem of
sin. It is in no way a prescription for dealing with other consequences of sin
beyond the distressed individual conscience or with the sin of those who fail to
see that they are sinning. As Paul himself well knew, the gravest rejection of the
will of God may arise from devotion to what the will of God is thought to be. 'I
am the least of the apostles', he said, 'unfit to be called an apostle, because I
persecuted the church of God. But by the grace of God I am what I am, and his
grace towards me was not vain' (1 Cor 15:9–10). The problem is even worse
when the assumptions of a whole culture take for granted the support of religion
and cannot make room for any questioning of the existing institutions in God's
name. That, no doubt, is why 19[a] speaks of God's reconciling the world to himself, where 'world' *(kosmos)* means the social world surrounding the community
(see pp. 22–23), as is shown by *their* and *them* (a *constructio ad sensum* in 19[b]),
which plainly refer to people. It must be held that by submerging the image of
sin as a criminal record in talk about reconciliation, Paul had made an important
advance in the understanding of salvation.

    Having thus established the context in which a discussion of salvation may
take place, Paul now redefines 'in Christ God was reconciling' by using a pregnant interchange formula:[114] 'He made him to be *hamartia* that we might become

dikaiosynē in him'. Its meaning does not lie (and was not intended to lie) on the surface. Interpretation must begin from the meanings of the contrasted words *hamartia* and *dikaiosynē*. It has been suggested that they are not contrasted but complementary—that *hamartia* perhaps means 'a sacrifice for sin' (as twice in Leviticus) and would then be the necessary action before our righteousness. But the first *hamartia* in the sentence cannot be 'a sacrifice for sin' and so rules out that interpretation. Moreover, elsewhere in the epistle *dikaiosynē*, meaning God's saving goodness, is contrasted with God's judgement, with lawlessness, and with Satan's work (2 Cor 3:9, 6:14, 11:15). Hence, if we supplement the final phrase (for simply to repeat Paul's words indicates a failure in understanding), it must have some such rendering as, 'that we might become the object of [or the beneficiaries of] God's saving goodness in Christ'. By logical contrast *hamartia* should mean 'the destructive malice of Satan'. In the Corinthian epistles, Satan makes his most frequent appearances in Paul's writings (elsewhere only Rm 16:20, 1 Thes 2:18, 2 Thes 2:9). Satan tries to provoke Christians and to take advantage of them by false representations (1 Cor 7:5; 2 Cor 2:11, 11:14) and, in a rather unpleasant way, is quite useful for disposing of obstinate sinners and preventing excessive pride (1 Cor 5:5; 2 Cor 12:7). In the hierarchy of spiritual beings he has no higher status than a revenue officer tracking down tax evaders. In other words, already in Paul's day *Satan* was an archaic term that tended to conceal the seriousness and power of sin, as is shown by Paul's remarkable development, especially in Romans, of sin personified. That development is already present in the Corinthian epistles. In 2 Cor 11:7 *hamartia* perhaps means little more than a wrong action, and *sins* of 1 Cor 15:3, 17 is the stock term of pre-Pauline formulas; but there is something different in 1 Cor 15:56: 'The sting of death is sin, and the power of sin is the law'. Here sin is compared to a fatal sting (from LXX of Hos 13:14) that causes death, and sin gets its deadly power from the law—a theme extensively treated in Romans. So Paul, who is too sensible of the majesty of God to have him unequally yoked with Satan, opposes the destructive power of Sin to the saving goodness of God. So we may interpret, 'God made him [the victim of the destructive power of] sin that we might become [beneficiaries of] God's saving goodness in him' (the 'in him' formula being used, as so often, of the divine activity).

So far I have examined only the central contrast of this interchange formula. But it will be noticed that the concluding 'in him' corresponds to the initial 'him who knew not sin'. The meaning of that description seems not to have been clearly perceived. Paul does not say 'he did no sin' (as in 1 Pt 2:22) or even that he had no sin (cf. 1 Jn 1:8) but that he did not know sin. Therefore Paul's meaning is ill represented by 'Christ was innocent of sin' (NEB). That could be taken for granted and was not particularly important. If Paul himself could be 'as to righteousness under the law blameless' (Phil 3:6), so, presumably, could Jesus. What happens if we give full weight to '*knew* not'? Surely it is irrelevant to quote (as Bultmann does) passages from Hermas and the rabbis about babes who do not know what wickedness is, or a child one-year old who has not tasted sin. Paul cannot have imagined that Christ died in a state of infantile morality. He might mean that Christ knew no charge against him or any feeling of guilt, but Paul knew very well (1 Cor 4:4) that such confidence was without significance. If these interpretations are cleared out of the way, the nearest parallels are in Romans: 'If it had not been for the law, I should not have known [the power of] Sin' (7:7) and 'Through the law comes knowledge [*epignōsis*] of [the power of] Sin' (3:20). The verb *ginōskō* is being used in the sense of 'recognise' (AGB s.v. 7), either in the sense of recognizing the claims of something or recognizing its authority. The Hebrew *yada* (BDB s.v. 2, p. 394) can be used in statements about knowing the Lord and not knowing alien gods (often in Deuteronomy and Jeremiah); Hosea denounces the spirit of idolatry within them, for 'they know not the Lord', even though they cry, 'My God, we Israel know thee' (Hos 5:4, 8:2). Thus, 'knew not sin' means 'did not recognize Sin as a power'. No doubt Christ was well aware of the extent and strength of Sin's power, but he yielded nothing to it. Hence, if we let go the epigrammatic form of verse 21, an interpretation can be offered in these terms:

> Christ gave no recognition to the power of sin;
> but for our sakes God made him a victim of sin's power,
> that we might receive the benefits of God's saving goodness,
> which was active in Christ.

Some comments are necessary on that formulation.

The understanding of this passage, and similar passages, is often confused by an individualistic conception of sin. Attention is fixed on the will of the individual sinner and on particular wrong actions, especially those which contradict known rules of religious behaviour, even though everybody knows that sinning is a complex affair, that the individual is almost always responding to pressures from his community, and that his will may be the least effective contributor to his sin. Even when the godly person is intentionally carrying out the divine will, resisting immoral pressures, and seeking the benefit of others, the consequences may, in greater or lesser degree, be damaging. The power of sin is not simply to prompt wrong actions, but to take good endeavours and make them destructive. Thus, it may be said of Christ that he was truly the godly man, the image of God, so that all his actions reflected God and he made no concessions to Sin; yet he, too, was made a victim of the power of Sin in that all his goodness was so turned as to destroy himself and to create unbelief among those who were most zealous for God.

Among more instructed students of theology, sin is often defined as 'that which separates us from God'. Since that view is obviously self-centred, it can be argued that the definition is itself sinful. It would be better to say that sin is any human activity that makes it difficult for people to believe in the goodness, the power and the presence of God. In that sense the crucifixion of Jesus, to both Jews and Greeks, looks like a denial of God and we can better understand Paul's harsh form of words, 'made him to be sin'. Correspondingly, the *dikaiosynē theou* ('righteousness, or saving goodness, of God') is what persuades us of the goodness, power and presence of God. In fact, what we become in Christ is the most persuasive testimony to God.

The whole thing is God's doing, and it is for our benefit (*hyper hymōn;* see pp. 42–43). Christ is not caused to act as our substitute. He does not become the victim of Sin's power in order to release us from that power; for clearly, even with the best Christian intentions, we are caught up in it. He does not bear the punishment of our wickedness, the dreadful consequences of our involvement with Sin, in order that we should go scot-free. We are painfully aware of the consequences that cling to us. Christ is not the victim of a divine legal system which makes it easy for the authorities to punish somebody even if the guilty go free. In any case, human affairs are far too serious to be left to any legal system concerned not with truth and justice but with legality. Christ is not our substitute but our representative or fore-runner. He dies (and indeed rises again) as the image of God, the true Adam, what we are created to be. God, as it were, looks at him and sees us. Christ therefore marks out the path that we must follow: if we share his victimization, we may then share the saving goodness that is operative in him.

With such confidence, Paul again urges the community to accept God's grace as it has just been described; and also, returning to the theme of 2 Cor 4:7–9, justifies his apostolic ministry in terms of its sufferings. In the earlier reference he was hard pressed but never overwhelmed; now, however, his apostolic existence is altogether more ambiguous. First comes the tale of hardships (2 Cor 6:4^b–5), then the Christian resources that can be deployed (verses 6–7), leading to an unsystematic but very effective array of nine pairs of words in contrast (verses 8–10). For this study the most striking is AS DYING AND BEHOLD WE LIVE, and the rhetorical pressure suggests a deep involvement in the apostle's emotions, security, reputation and influence. Death and resurrection is at the heart of his theology and of his self-awareness—and also of his relations with the community: 'You are in our hearts, to die together and to live together' (2 Cor 7:3).

7:5–16

When in 2 Cor 7:5–16 Paul resumes what he began to say in 2:12–13 he returns to his theme word 'comfort' *(paraklēsis)*. Its meaning has perhaps been deepened by the thoughts set down in 2:14–7:4, and even emphasized by the ambiguity revealed in 6:8–10. It is by no means unlikely that Paul deliberately broke his train of thought at 2:13 in order to resume it at 7:5 after a major theological argument. His teaching about cross and resurrection has its application to practical decisions and difficult relations; nor can it be properly understood unless studied (as far as possible) against the situation that called it forth.

Chapters 8–9

The possibly double treatment of the collection in 2 Cor 8–9 follows very suitably the concerns dealt with so far, and the two chapters may contain echoes of the previous argument. Thus the quotation in 2 Cor 9:9—'His righteousness endures for ever'—whether intentionally or not, is an apt comment on our aim of 'becoming the righteousness of God in him' (2 Cor 5:21). Paul's intention that no one should blame him for the administration of the gift (in 8:20) clearly reflects his attempt to avoid blame in 6:3; so that his aim for 'what is honourable not only in the Lord's sight but also in the sight of men' (8:21) echoes 'honour and dishonour' in 6:8. It is more notable that 8:9 presents another interchange formula:

> You know the grace of our Lord Jesus Christ,
> that though he was rich,
> yet for your sake he became poor,
> so that by his poverty you might become rich.

The formulation in terms of rich and poor echoes 'as poor, yet making many rich' in 6:10.

It is not fanciful to relate the metaphorical enrichment of the Corinthian community with their earlier pursuit of social wisdom (see p. 22).

The word *grace* is prominent not only in the formula but also throughout chapters 8 and 9 and elsewhere in the Corinthian epistles. *Grace (charis)*[15] has a set of linked meanings, which Paul likes to use in stimulating association. At its simplest, it is a generous action or gift, such as the Corinthian community's money for the relief of Christians in distress (2 Cor 8:4, 6, 7, 19). That kindness is prompted by God's gifts freely bestowed on the communities in Corinth and Macedonia (1 Cor 1:4; 2 Cor 6:1; 8:1; 9:8, 14). They are, in fact, properties of the new life that the communities receive from the risen Lord: 'He who raised the Lord Jesus will raise us also with Jesus and bring us with you into his presence. For it is all for your sake, so that as grace [*charis*] extends to more and more people it may increase thanksgiving [*eucharistia*], to the glory of God' (2 Cor 4:14–15; the word play is deliberate). The gift of new life is God's unforced generosity to those who have no claim but their need, often marked by perversity and ruin. Paul's words at 1 Cor 15:9–10 (quoted above, p. 60) speak of the resurrection's power and indicate the effectiveness of grace. In another self-revealing passage in 2 Cor 12:5–10, he tells how he sought escape from the pressures and sufferings of apostleship but learnt from God that 'My grace is sufficient for you, for my power is made perfect in weakness'. Thus the experience of grace is participation in suffering and new life in Christ—but willingly or under compulsion? The Lord's answer to Paul's insistent request for relief from harassment may be read as either reassuring or peremptory—inevitably, since grace always means generosity from the greater to the lesser, from the powerful to the weak, from the holy to the unholy. For that very reason the interchange formulas are of the greatest importance. In 2 Cor 6:1 the formula of 5:21 is described as the grace of God. Now, in the formula of 8:9, the grace of God becomes the grace of our Lord Jesus Christ (the very name and title indicates the crucified one who is risen) to indicate that the conferment of grace is not by superior condescension but by a poor man in his poverty to us in ours, the last Adam to the first Adam.

The imagery of poverty and riches is determined by Paul's charitable appeal (2 Cor 8:2), and is somewhat similar to Jewish stories about reversal of fortune at the hands of God. The poor man, who is seldom far away in the Psalms, praises the Lord: 'Who is like the Lord our God, who is seated on high, who looks far

down upon the heavens and the earth? He raises the poor from the dust, and lifts
the needy from the ash heap, to make them sit with princes, with the princes of
his people' (Ps 113:5–8). That might well be called the grace of God; but nowhere
is it suggested that God's agent of grace abandoned, or was deprived of, his own
riches that the poor might profit by his poverty and become rich. It should be
noted that the adjective *poor (ptōchos)* is not confined to poverty; in the Septuagint it often represents Hebrew *'ny* meaning poor and wretched, afflicted, subject to others. When did Christ change from a well-provided, independent position to a condition of wretchedness and dependence? Most of the commentators
say *at the incarnation* and take for granted that Paul thinks of a preincarnate
heavenly existence for Christ, who became immeasureably impoverished when he
became man. But does Paul really take such a depreciatory view of human existence, almost as if it sundered communion with God (like the ancient Sheol)? And
could he expect the Corinthian community to pick up the reference? Where else
in his letters to them would they have found (regardless what they themselves
may have expected) instruction about a preincarnate existence of Christ or even
the becoming man of a divine being. Surely not in the midrash of the travelling
rock (1 Cor 10:4). If we look outside the Corinthian letters, Paul mentions
Christ's birth in Rm 1:3: 'descended from David according to the flesh' (a distinguished rather than a humiliating beginning), and in Gal 4:4: 'God sent forth his
Son, born of woman, born under the law' (assuming without question that Jesus
could be respected for his Jewish birth as Paul could for his). There is nothing
humiliating or leading to wretchedness in the birth, but there is in the death of
Jesus. In this passage it is premature to introduce thoughts of incarnation and
'preexistence': we are still concerned, as we were in 2 Cor 5:21, with the benefits
of his Passion, for God's 'gift beyond words' (2 Cor 9:15 NEB).[116]

Chapters 10–12

In chapters 10–12, Paul turns with surprising sharpness (though he pleads 'the
meekness and gentleness of Christ', 2 Cor 10:1) to attack rival missionaries who
appear to be Jews, claim to be apostles of Christ, and (in Paul's view) preach
another Jesus and offer a different spirit and a different gospel (2 Cor 11:4, 13,
15, 22). They have sought to undermine reliance on Paul's competence, and he
gives all his energies to repudiating their criticisms. It is impossible either to discover (except in general terms) what different views or convictions were put forward by Paul's rivals[117] or to feel confident that the contending parties can be
properly identified. The whole section is written in emotional, polemical language; and the syntax is sometimes impenetrable. Pauline irony plays its part but
too often turns (as it seems to modern readers) into heavy sarcasm. For these
reasons it is not easy to be patient with the apostle; and when there are strong
reasons for doubting the connexion of chapters 10–12 with the rest of the epistle
and for hesitating about placing them earlier or later in the sequence of letters,
it is tempting to abandon the section and move on.

But that would be a mistake. The repellent features of this section are a measure of Paul's frustration (he knows he is making a fool of himself) in trying to

convince the Corinthian community that his way of behaving comes not from incompetence or craftiness but from policy. To simplify, *they* say that Paul behaves in an embarrassingly inferior manner because he is an inferior agent, whereas the rival apostles behave in an agreeably superior manner because they are superior agents. Paul replies that he is just as much a superior agent as they are and *therefore* behaves in an embarrassingly inferior way. In effect, though he is rich, yet for their sakes he becomes poor. Without entering into unnecessary detail, that schematic statement may readily be justified.

When face to face with the community, Paul is humble *(tapeinos)*, that is, feeble and slavelike (10:1) as well as crafty (12:16), though he is typically aggressive at a distance. His bodily presence is weak, his speech contemptible (10:10). Compared with the rival apostles he is so inferior that he dare not claim support from the community (11:5, 7; 12:13–14). He lacks the marks of an apostle (12:12), notably the power to impose his authority (11:20). His oft-repeated sufferings are a tale of weakness, typical of someone living the life of feeble humanity *(kata sarka,* 10:2, 11:23–27). Rival apostles can command visions and revelations 'out of the body' (12:1–5). All these charges Paul admits, though he qualifies some and explains others. On the question of visions, he lays claim to a visionary experience (either once fourteen years ago or as long ago as fourteen years and by implication [see 12:7] since then continued) but whether 'out of the body' or not, who can say? He could, no doubt, make boastful claims for such a person and his experiences. But embodied existence, such as we all must live, is always subject to weakness (since it is partly *sarx,* see p. 31), so that its strength comes from the Spirit. Paul has no confidence in self-commended experiences raising a believer beyond the weakness of human existence. God had given him a *skolops tē sarki,* a 'thorn in the flesh' (as it is usually translated, though *skolops* chiefly means a pointed stake and can be equivalent to *stauros* [cross] AGB s.v.). Its purpose was to subdue overconfidence in revelations and to make him understand that God's grace was sufficient—that when he was weak, he was strong (12:9–10).

In response to these depreciatory comparisons, Paul claims power, given him by God, to punish and destroy if need be (10:4–6) but insists that his authority is properly used for building up, not for destroying or frightening (10:8–9, 12:19). The Lord has assigned Corinth as his apostolic province (10:13, 14), that he may promote increasing faith in that city and extend the scope of the gospel beyond it (10:15–16). He has, as it were, betrothed the community and must complete the betrothal by marrying them to Christ (11:2). And they are not his only responsibility: there is his anxiety for all the churches (11:28). He will most gladly spend and be spent on their behalf (12:15). It is clear that he limits the exercise of his own authority in order that the community should take up its own responsibility rather than confer authoritative positions on rival apostles. The really effective stroke of irony is that God may humiliate the apostle when he finds the community so morally discreditable, the apostle whom the community thought *tapeinos* for suffering so much on their behalf. In all this Paul speaks before God in Christ (12:19). What comes out of these tormented pages is an attempt to work out, in the hellenistic expansion of the Church, the response of an apostle when he is despised and rejected of men.

Chapter 13

From these impassioned words Paul turns to his forthcoming visit, when he promises them an exercise of authority, namely the authority that Christ exercises in him:

> He is not weak towards you, but powerful among you.
> For indeed he was crucified by reason of weakness
> but he lives by reason of God's power. (2 Cor 13:3-4)

'By reason of' represents *ek* (AGB s.v. 3-4). The purpose of this neatly constructed couplet is not to make weakness and strength antithetical (as if strength followed weakness or vice versa), but to hold them together as constituents of the same divine activity *in Christ*. Compare the body, which is sown in weakness and raised in power (1 Cor 15:43), with the divine assertion, 'My power is made perfect in weakness' (2 Cor 12:9). Weakness and power correspond to flesh and spirit (and hence to Paul's conception of soma, the body), as well as to the old age and the new. But it must not be forgotten (according to 2 Cor 8:9 and passages that identify Christ as the image of God) that the weakness is not inherent weakness but chosen weakness. Therefore it is also grace. Hence when Paul immediately reapplies the couplet to himself and the community, he is defining how the grace of God operates for us.

> For we also are weak in him
> but we shall live with him by reason of God's power towards you.

In this way he combines a version of the standard formula ('He died—he arose'), which speaks of a completed double action, with a participatory appropriation of the formula whereby 'He died' is represented by our willing acceptance of hardship and rejection, and 'He arose' is left open as an object of faith and experience.

Summary

The two Corinthian epistles are complementary in their exploration of the death of Christ. Although the situation in Corinth has changed by the time that 2 Corinthians was written, certain convictions established in the earlier epistle (particularly the function of the death of Christ in criticising social wisdom) are presumed in the later writings.

2 Corinthians contains ample evidence of a sustained attack on Paul's own kind of wisdom and knowledge. The apostle (it is said) is weak and continually in trouble. He behaves abjectly and shows no attractive benefits of being in Christ. Paul does not deny the weakness—after all, he had chosen to be weak, not domineering—but he accounts for it and interprets it. In part that weakness is God's

doing: Paul was given a stake in his flesh to subdue overconfidence and to show the sufficiency of God's grace. It is the means by which God triumphs, though not everyone so understands it: to some, Paul's sacrifice is the stink of death; to others, the fragrance of life. Indeed, God constantly hands him over to death for Jesus' sake, a daily exposure so that others may experience new life. 'Dying and behold we live' is the pattern of his precarious apostolic work, as indeed it is the pattern of all Christian living in the body, which belongs both to the flesh (which is weak, dishonourable and mortal) and to the spirit (which communicates the life of the risen Lord). If we are resident in the body as belonging to the flesh, we are not resident with the risen Lord; if we are not resident in the body as belonging to the flesh, we are resident with the risen Lord.

The gift of new life is God's unforced grace, the spontaneous generosity of all-sufficient Deity to weak and fallible mortals. But grace, as shown in Christ, is not condescension; it is the act of a poor man in his poverty for the benefit of us in ours. Thus the death and resurrection of Christ are saving events insofar as Christians participate in them. The primitive formula 'He died—he rose' provide an interpretative pattern: experiencing distress, we share the sufferings of Christ; receiving consolation, we rely on God, who raises the dead. The characteristic formula 'in Christ' indicates the means by which God performs his saving activity, namely, by the death and resurrection of Christ. Christ's love is his action in dying not chiefly as a martyr—not solely as our representative—but as our forerunner, to show the way that all must go. When Paul says memorably 'One has died for all; therefore all have died' he means that former identities as Jew and Greek have been lost, to be replaced by a new identity in Christ.

That proposed change, of course, would be regarded as a threat—as indeed it was a threat—to the securities, hopes and intentions of the *kosmos*. So Paul begs his readers to be reconciled to God, to accept the death of Christ as the act of a friend (not an enemy), and so to experience the consequences of his resurrection; for although Jesus gave no recognition to the destructive power of sin, God made him a victim of that power so that we (interchanging places, as it were) might become beneficiaries of God's saving goodness displayed in Christ. And something of that kind lies behind Paul's talk about dying daily, about carrying around the dying of Jesus in his body. The resultant polemic, though too insistent and too angry, is not morbid: it is redeemed by the use of creation and covenant imagery. What God does in Christ is a movement from darkness to light, from the veiled glory of the crucifixion to the open glory of the resurrection, from the written code that kills to the spirit that gives life.

GALATIANS

This epistle, despite its numerous problems of interpretation, has a remarkably clear argumentative structure; and references to the death of Christ are made at significant points of the argument. The first comes in the epistolary greeting, another in Paul's description of the gospel, a third as the precondition for receipt of the Spirit. Other references appear in the scriptural proof of the gospel, at the heart of the consideration of sonship, in the repudiation of circumcision, and in the description of life in the Spirit. The final three were written by Paul's own hand in the conclusion of the letter. These observations suggest that the death of Christ is identified by Paul as one of the main matters in dispute between himself and his opponents, even if the opponents had no intention (or no overt intention) of repudiating the cross.

The Greeting, 1:1–5

In Gal 1:1–5 the stock phrases of a Pauline greeting are carefully joined with a variation of the standard confessional formula. The greeting is immediately polemical, intended to repudiate objections to Paul's apostleship:

> Paul an apostle—
> not from [the decision of a group of] men
> nor through [the action of] a man,
> but through [the action of] Jesus Christ
> and of God the Father who raised him from the dead.[118]

There is no other explicit reference to the resurrection in the epistle (though frequent mention of the living Christ), and the standard formula 'He died—he rose' is reversed, the resurrection in verse 1 and the death in verse 4. The resurrection probably stands here to serve a double purpose, to indicate that Paul's apostleship has proper divine authority (verse 12) and to stress the lordship that is perhaps flouted by the objectors to Paul's gospel, though the *kyrios* title plays little part in the epistle except in the greeting and in the conclusion (Gal 6:14, 18).

> The Lord Jesus Christ gave himself for our sins
>
> > to deliver us from the present age, bad [as it is],
> > according to the will of our God and Father.

'*For* our sins' (*hyper*, or *peri* v.l.) belongs to the type of formulaic variant shown in 1 Cor 15:3 (see p. 42). But if in that passage the cultic imagery seemed most appropriate, in the present passage the reflexive verb *dontos heauton* ('he gave himself') lends weight to the self-sacrificial imagery. It is supported by 'He gave himself [*paradontos*] for me' in Gal 2:20 and elsewhere in the wider Pauline corpus by 'He gave himself [*dous heauton*] as a ransom for all' in 1 Tim 2:6 and 'He gave himself [*edōken heauton*] for us to redeem us from all iniquity' in Tit 2:14. (Cf. also Mk 10:45, 'He gave his life [*dounai tēn psychēn autou*] as a ransom for many'.) It is possible that the self-giving may recall the Jewish image (derived from the Maccabean martyrs) of the godly man who seeks to persuade God, by his own death, to take compassion on the rest of God's people who are in extreme need because of their sins or their persecutors. But the Galatian situation scarcely requires such imagery (which Paul does not adopt elsewhere) and it is better to interpret the phrase *gave himself* as indicative of the grace of Christ (see 2 Cor 8:9), which is made explicit in verse 6. The aorist participle (*dontos* from *didōmi*) points to the historic action of the crucifixion, and it is taken for granted that readers will make that identification (see Gal 3:1).

The formula continues, however, with what seems like a complementary but implausible purpose of Christ's death, namely, our delivery from this present evil age. The verb *exaireomai* ('deliver') is not elsewhere used by Paul, though it occurs five times in Acts and frequently in the Septuagint. The Greek phrase *ek tou aiōnos tou enestōtos ponērou* is awkward (as the above translation suggests)[119] and has several variants. When Paul elsewhere speaks of the present age he uses *houtos* ('this'), not *enestōs* ('present'); he uses *ponēros* ('bad') elsewhere only five times and only once as an adjective. Hence, it is possible that here he uses another section of a non-Pauline formula. But even if he does, he is obviously willing to countenance these words and to bring them within the scope of his own thought about the death of Christ. But what does he mean by what he writes? He cannot mean that the present age is no longer present (cf. 1 Cor 1:20, 3:18) or that Christians have been transferred into a situation where the present age cannot touch them (for in Rm 12:2 they are urged not to be conformed to this present age). For Paul this age had its rulers and its god (1 Cor 2:6, 8; 2 Cor 4:4); therefore, he must mean that Christ's death delivers us from the powers at work in the present age. If the formulaic elements in verse 4 are not just cultic words that sound right and operate effectively but also have coherent meaning, the interpretation is that the death of Jesus sets us free from the disabling consequences of our sinful actions and thus delivers us from our helpless addiction to evil forces that pervert our present social situation. It is as if we had put ourselves in the wrong and become guilty persons by a serious misjudgement or a stupid folly or a malicious action—with the result, however, that we are not only guilty and subject to penalty if caught but are entrapped in shady dealing and criminal activity quite beyond our intention, but now inescapable. We are not only sinners committing sins but are also the unintended but corrupted agents of sin. In Paul's tradition and teaching we are delivered from that hopeless situation by the death of Christ, though no direct indication has yet been given of how that may be possible. It is simply 'according to the will of our God'; and Paul ends the greeting section liturgically by ascribing glory to God, which is a Hebrew way of saying, 'Let us submit ourselves to the majesty of the divine will'.[120]

The Gospel Restated, 2:15–21

Immediately after the greeting Paul makes an indignant response (but still in hellenistic epistolary fashion) to the Galatians' shift from the grace of Christ (outlined in verse 4), and he puts the evangelists of a variant gospel (though it is no true gospel at all) under a curse (*anathema*, cf. 1 Cor 12:3, 16:22), which replaces the normal thanksgiving of a Pauline letter. Then, after a defensively transitional verse in Gal 1:10,[121] he asserts the divine origin of his gospel and its independence of human authority, source or teacher in 1:11–12.[122] Gal 1:13–2:14 is Paul's *cursus vitae* arranged as a proof of his apostolic independence and hence of the divine authority of his gospel.[123] Now, in Gal 2:15–21, comes his restatement of the gospel as perceived and experienced by a Jew.

It is important to realise why Paul chooses to restate the gospel at *this* stage of the argument and why he first presents it as a question of 'justification' before turning, in 2:20, to a participatory application of 'He died—he rose'. In the previous section he has portrayed himself as an active and devoted Jew who was called to preach the Son of God among the Gentiles. In those early days his contacts with Jerusalem and Judea were scanty (Gal 1:18, 22), for his work lay in Arabia, Damascus, Syria and Cilicia (Gal 1:17, 21). Not for many years did he pay a second and very significant visit to Jerusalem, when it was agreed that he had been entrusted with the gospel for the uncircumcised, just as Peter had been entrusted with the gospel for the circumcised (Gal 2:7–9). Then at Antioch (Paul's territory rather than Peter's) the decisive encounter took place to decide the question how Jews and Gentiles should live and whether Gentile Christians should be compelled to live as if they were Jews (Gal 2:14). It is in that context— and almost entirely in that context—that the question of 'justification' arises in Paul's epistles. The 'justification' group of words has minimal significance unless the relations between Jews and Gentiles is of central importance, as it is in Galatians and Romans and to a smaller extent in Philippians.[124] Paul argues thus: 'If I, a Jew and not a Gentile sinner [meant ironically], cannot be acceptable to God by acting as required by Torah, how much less can Gentiles be acceptable to God by performance of Torah'.

We Jews know, says Paul, that a person is not 'justified' (i.e., declared acceptable to God[125]) by performing requirements of the law but through faith in Christ. That is the common conviction of Jewish Christians, including Cephas and Barnabas. Paul activates the appropriate religious response (here as in Rm 3:20) by reciting key words taken from Ps. 142:2 (LXX): 'Enter not into judgement with thy servant for no living person shall be *justified* before thee'. The psalmist is often taken to mean that no one will be acquitted of sin in the divine court, thus invoking the imagery of a criminal trial. But in fact the psalmist says not one word about sin; the whole psalm is a plea for help—and a tactfully reproachful plea at that. It could be read as an attempt to saddle God with responsibility for the success of the psalmist's enemies: therefore, when he begs for God's saving goodness (in verses 1 and 11), he disavows any intention of blaming God: 'Do not take proceedings against thy servant, for in thy presence no one can claim to be in the right as against God'. That is precisely the conviction of Job when he says 'No man can win his case against God'. The overwhelming divine power and authority make Job's complaint against God a hopeless cause. 'Though I am in the right, he condemns me out of my own mouth; though I am blameless, he twists my words' (Jb 9:2, 20). Despite what his friends say, Job is asking for God's favour, not his forgiveness; and finally God answers Job: 'Will you put me in the wrong in order to claim that you are in the right?' (Jb 40:8). Both Job and the psalmist are making a plea for God's help, knowing that they cannot claim a right to that help. When, therefore, Paul sets down his doctrine that 'by works of the law shall no one be justified' (Gal 2:16), he is reviving ancient Jewish piety for the special situation of Jewish and Gentile admissions to the Christian community.

In a different context, what he says can be recognized in the Community Rule and the Thanksgiving Hymns of the Qumran sect, for instance, 'Thine, thine is righteousness, and an everlasting blessing be upon thy name. [According to] thy righteousness let [thy servant] be redeemed [and] the wicked be brought to an end' (1 QH 17, DSSE, p. 205).[126] The notable differences between Qumran and Paul are that sectarian 'knowledge' has a dominating role in the Qumran ḥymns, and that some of the hymns are given to excessive protestations of sin and self-loathing:

> Yet I, a shape of clay kneaded in water, a ground of shame and a source of pollution, a melting pot of wickedness and an edifice of sin, a stray-ing and perverted spirit of no understanding, fearful of righteous judge-ments, what can I say that is not foreknown, and what can I utter that is not foretold? ... What shall a man say concerning his sin? And how shall he plead concerning his iniquities? And how shall he reply to righ-teous judgement? For thine, O God of knowledge, are all righteous deeds and the counsel of truth; but to the sons of men is the work of iniquity and deeds of deceit. (1 QH 1:21–27, DSSE, p. 167).

It is, of course, not necessary to regard all such protestations as praiseworthy consciousness of sin. They may be the proper ritual abasement of oriental devo-tion (even or especially in Ps 130); but it is significant that in Gal 2:15, 17 Paul takes pains to push aside the thought of sin: 'We ourselves are Jews by birth and not Gentile sinners; nor is Christ an agent of sin, nor do we become sinners if we seek to be acceptable to God in him—though I (or anyone, in my position) would indeed be at fault if I restored what I had demoted (namely, Torah) as a ground of acceptability with God (Gal 2:17–18).

The demotion of Torah is sharply indicated in verse 19: 'I through the law died to the law, that I might live to God'. Here we encounter the first example of Paul's strange use of a verb connoting dying followed by *law* in the dative (cf. Rm 7:4, 6 and an equally surprising construction in Gal 6:14).[127] The difficulty of the phrase is shown by the reluctance of translators to abandon 'died to the law', which no one can ever have supposed to be intelligible English. Even 'live to God' is not fully comprehensible when changed to 'live for God' (NEB), though the phrase is now familiar. To live for God means to live so as to serve God's purposes; hence to die to the law means to cease living so as to serve the purposes of the law. More plausibly, since Paul thinks of the law (like sin) as a power, the contrast is between living (or ceasing to live) by the power of the law and living by the power of God. The distinction would scarcely seem meaningful to Paul's Pharisaic contemporaries; but Paul, unlike them, had been forced by the law itself to demote the law. Presumably he means that his zeal to assert and defend the law had made him a persecutor of the Church of God and therefore an enemy of God. When it pleased God to reveal his Son in him, Paul recognized his own enmity to God and could no longer trust the law as a true indication of God's intentions and a genuine source of his power. As Jesus had died, sentenced by the law, and had been raised by God, so Paul could recognize his own experience as the death of his most powerful, law-defending impulse, accompanied by a new life-giving initiative from God:

> I have been crucified with Christ;
> it is no longer I who live,
> but Christ who lives in me;
> and the life I now live in the flesh
> I live by faith in the Son of God,
> who loved me
> and gave himself for me. (Gal 2:19ᶜ–20)

When Paul says 'I', what does he mean? Does he mean 'I myself' or, more generally, 'anyone in my position' (as may well be possible in Gal 2:18); or is he offering a hypothetical example using the rhetorical device of the first person?[128] Clearly, he is not writing hypothetically: his phrasing is not equivalent to 'Supposing I have been crucified with Christ, then it is no longer I who live'. He may perhaps imply that others in his position would say what he says, but Gal 1:13–14 suggest that few could have been in his position. Surely the whole direction of the argument requires Paul to speak for himself: 'I, a Jew exceedingly zealous for the law and experienced in it, have demoted the law in order to be identified with Christ. Why then do you Galatians, who are not Jews, wish to adopt the law in order to be identified with the crucified Christ?' Even a proper suspicion of 'mysticism' and anxiety about the exegesis of Rm 7:7–25 ought not to prevent the exegete from recognizing Paul's self-disclosure in this passage, granted the implication that his readers should imitate him as far as their different circumstances permit.

What then is Paul saying about himself,[129] about the ego that makes him what he is and not someone else? He is saying that his former ego has been crucified with Christ and is no longer alive. Yet since he is obviously a living person—living indeed in the world of 'flesh'—how does he now regard himself? Not as a new ego, raised from the death of his former ego, but as a person indwelt by (the risen) Christ. When Paul speaks reflectively about himself, as he often does, it must be taken for granted that a person's ego is not an inward self-awareness that persists unchanged in all circumstances. The formation and activity of the ego is determined by the forces that press on an individual life, by the power that claims authority over it. Hence Paul's ego, formed and stimulated by Torah in 'the present evil age', has been destroyed; and a new ego, fully formed by 'the powers of the age to come' (to use a non-Pauline phrase, Heb 6:5), has not yet replaced it. Paul hoped that he might attain the resurrection of the dead but did not consider that he had already attained it (Phil 3:11–13). By the grace of God he was what he was; his apostolic energy came not from himself (*ouk egō*) but from the grace of God which was with him (1 Cor 15:10). In this epistle he is explicit: 'He who had set me apart before I was born, and had called me through his grace, was pleased to reveal his Son in me [*en emoi*] (Gal 1:15, cf. 1:12). Hence it was wonderful but not surprising that the Galatians had originally received him as an angel of God, indeed as Christ Jesus (Gal 4:14). Moreover it is clear that Paul had expected to encounter Christ Jesus in them and, in disappointment, was again in travail until Christ was formed in them (Gal 4:19 _mechris hou morphōthē Christos en hymin_). He himself is indwelt by the risen Christ, but it is unthinkable that he should boast in anything but the cross of our Lord Jesus Christ or claim consideration for anything but the stigmata of Jesus that he bore in his body (Gal 6:14, 17). The statement 'I have been crucified with Christ' contains one kind of participation language (_with_-language), and 'Christ lives in me' contains another (_in_-language). Both invite study.

With Christ and In Christ

The _with_ statements[130] appear in two rather diverse groups. The first group contains eschatological expectations: 'Since we believe that Jesus died and rose again, even so, through Jesus, God will bring _with him_ those who have fallen asleep. . . . The dead in Christ will rise first; then we who are alive . . . and so we shall always be _with the Lord_' (1 Thes 4:14–17, cf. 5:10); 'He who raised the Lord Jesus will raise us also _with Jesus_ and bring us with you into his presence' (2 Cor 4:14); since God handed over _(paredōken)_ his Son for us all, 'will he not also give us all things _with him?_' (Rm 8:32); when Paul thinks he has a kind of choice between death and life, his desire is 'to depart and be _with Christ_' (Phil 1:23). Thus, infrequently but over the whole period of Paul's letter writing, being with Christ is a hope for the transformed future. The second group of _with_-statements is more restricted. Paul uses the verb _sysstauroō_ ('concrucify') in the passive in Gal 2:19 and again in Rm 6:6 (elsewhere the verb is confined to references in Matthew, Mark and John to the crucifixion of Jesus with two others). In Galatians Paul has been _crucified_ with Christ; in Romans, when he draws out the significance of baptism, the thought is carefully extended to _living_ with Christ: 'We were buried therefore _with him_ in baptism into death, so that as Christ was raised from the dead by the glory of the Father, we too might walk in newness of life. For if we have been united _with him (symphytos)_ in a death like his, we shall certainly be united _with him_ in a resurrection like his. We know that our _palaios anthrōpos_ [old human identity] was crucified _with him_ so that the sinful body might be destroyed. . . . But if we have died with Christ, we believe that we shall also live _with him_' (Rm 6:4–8). It looks as if Paul's startling use of _concrucify_ was prompted by his dispute with the Galatian communities and that in Romans he amalgamated that new thought with the earlier, independent use of _with Christ_ in eschatological statements. In Rm 8:17 he brings both together concisely: 'fellow heirs _with Christ,_ provided we suffer _with him_ in order that we may also be glorified _with him_'. These three passages (one in Galatians and two in Romans) are not quite isolated, but there are no more true parallels. Thus 2 Cor 13:4 says, 'He was crucified in weakness, but lives by the power of God. For we are weak _in him_ [v.l. _with_], but we shall live _with him_ [v.l. _in_] by the power of God for your sake'. And Phil 3:10 speaks of knowing him and the power of his resurrection, and the sharing of his sufferings, being conformed to his death. The direct assertion that we both die _with_ Christ and are raised _with_ Christ seems almost elaborately avoided. To be raised with Christ and to be with him—both as an eschatological hope—causes no serious hesitation; but to be crucified *with* him is something rather different. What Paul intends by it in Galatians must wait until we have examined what is said about the crucifixion itself. Meanwhile it may be noted that 2 Tm 2:11–12 reproduces the teaching of Romans in a faithful saying and that Colossians remoulds it by reinterpreting the meaning of dying with Christ and by presuming that the expected resurrection has already taken place (Col 2:12, 13 [cf. Eph 2:5], 20; 3:1, 3–4).

In-language is used by Paul of God, of Christ and of the Spirit, as well as of Sin and the law of Sin (Rm 7:17, 20, 23). This last manner of speaking supplies the clue for understanding the rest. When Paul says, 'It is no longer I that do it, but Sin which dwells within me', he implies that Sin is a hostile power that has taken possession of him, with or without his consent, as a base for its destructive work. Correspondingly the divine being takes possession of the believer or the community as a basis for the divine work (so 1 Cor 3:16, 6:19 the temple of the Holy Spirit). Hence, God's intention of disclosing his Son *in* Paul is something more than the disclosure of the Son *to* Paul (as Gal 1:16 is usually translated);[131] it is the intention of displaying the power of the gospel of Christ among the Gentiles, the gospel as already outlined in Gal 1:1, 4. The power is God's power of raising the dead, as is made plain in the comprehensive formulation of Rm 8:9–11. In that passage Paul speaks not only of 'Christ in you' but also of 'the Spirit of him who raised Christ from the dead being in you'. The two phrases are not merely stylistic variants. *Christ* means the counterpart to Adam, the last Adam, who became a life-giving spirit (1 Cor 15:22, 45) because he was brought to life from the dead and sustained in being by the Spirit of God, just as God formed the first Adam and breathed into his nostrils the breath of life (Gn 2:7). Therefore Paul's faith now derives from the conviction that he has been crucified with Christ and that his own ego has not yet been recreated as a wholly new identity but that the power of the last Adam is now displayed in the life he still lives in the flesh (cf. Phil 1:21, 'For me to live is Christ'). His train of thought is, God in Christ and Christ in me.

Son of God

He does not, of course, use the words *last Adam* in this passage, but *Son of God*.[132] Why he uses *Son of God* is a fruitful question once the title is rescued from premature considerations of nature, deity and incarnation. It should first be observed that the title, which occurs fifteen times, is not uniformly distributed (There are seven in Romans, four in Galatians, two in 1 Corinthians, and one each in 2 Corinthians and Thessalonians: seven are pre-Pauline formulaic references [Rm 1:3–4; 8:3, 32; Gal 2:20, 4:4; and 1 Thes 1:10]; and eight are Pauline [Rm 1:9, 5:10, 8:29; Gal 1:16, 4:6; 1 Cor. 1:9, 15:28; and 2 Cor 1:19].) This distribution suggests that the arguments of Romans and Galatians had a special interest in *Son of God* language, no doubt because in these two epistles Christians, in their nature and destiny, are described as 'sons of God'. The references may be set out in three groups:

'God sent his [own] Son' (Gal 4:4, Rm 8:3) in the standard conditions of human, indeed of Jewish, existence: born of woman, born under law, and in the (precise) likeness of sinful flesh. He sent him when he judged the time to be fully ripe for his eschatological act *(to plērōma tou chronou),* which is the presumption of the *future* sending expected in 1 Thes 1:10. When the sending is still future, its purpose is to save from the coming wrath; when the sending is transferred to the past, its purpose is to deal with sins *(peri hamartias),* 'to redeem those who were under the law, so that we might receive adoption as sons'. Hence, God sent his Son as his agent and representative, equipped with plenary powers, as did the vineyard owner in Mk 12:6. There is no need to make judgements about the authenticity of the parable or its appropriateness to the teaching of Jesus; its very existence as a popular story shows that hearers would understand what the master meant when he sent his 'beloved son'. Paul makes at least half the meaning plain when he says that in the Son of God is the *yes* to all the promises of God (2 Cor 1:19–20).

The strongest associations of Son of God are with the resurrection and its consequences, from the expectation of the Son from heaven, whom God raised from the dead (1 Thes 1:10) to the Son who delivers the kingdom to God the Father after destroying every rule and every authority and power and is himself then subject to God (1 Cor 15:28). According to the formulaic statement in Rm 1:3–4 (referred to in Rm 1:9) the gospel of God concerns his Son who was descended from David according to the flesh (which points to the weakness of human existence) but was designated Son of God in power according to the Spirit of holiness (which points to the strength of divine existence) because of the resurrection of the dead. When, therefore, God decided to reveal his Son in Paul, it was the risen person he revealed, by faith in whom Paul lived (Gal 1:16, 2:20)— though not only Paul but all community members were called to participate in God's Son, our Lord Jesus Christ (1 Cor 1:9). They receive adoption as sons: God sends the Spirit of his Son into their hearts, and they cry out *Abba!,* 'Father!' (Gal 4:4, 6). In other words, those who are foreknown to God are predestined to be conformed to the image of God, namely his Son (Rm 8:29). As the first Adam was created in the image of God, so the Son of God is the divine image and it is our proper destiny to be conformed to that image.

The connexion between the title Son of God and the death of Christ is much less common. In only one passage is it made explicit: 'We were reconciled to God by the death of his Son; much more . . . shall we be saved by his life' (Rm 5:10). It must be presumed that our salvation by the life of the Son (which belongs to the resurrection group) has induced a reference to the death of the Son. Two other Pauline references are so by implication: 'He who did not spare his own Son [a reference to the Binding of Isaac][133] but gave him up [*paredōken*] for us all, will he not also give us all things with him [presumably, with him as the risen Lord and Son of God, Rm 8:32]?' and the present passage, where crucifixion is explicitly associated with Christ, where Christ is said to live in Paul, and (it continues) 'the life I now live in the flesh I live by faith in the Son of God, who loved me and gave himself [*paradontos*] for me' (Gal 2:20). This delicacy in attaching the death of Christ to the title Son of God,[134] evident elsewhere in the New Testament, can be explained if Son of God in the first place was a designation of the *risen* Christ. Even when the divine sonship was predicated of Jesus at his baptism or at his birth, it must still have seemed awkward to say that the Son of God had died by crucifixion. Here then is a further piece of evidence that the resurrection carried most theological weight in the earliest preaching and in traditions that stemmed from it. In writing to the Galatians, Paul lays great stress on the crucifixion: to abandon it in favour of the law, would be to nullify the grace of God (Gal 2:21). Hence, Paul states reliance on the death of the Son of God in a highly personal form in order to bring out the grace of that self-giving (cf. 2 Cor 8:9) and to prepare for his attack on the Galatian position by implying, 'If I, who am so deeply versed in Torah, set aside Torah for the grace that the Son of God showed in dying for me, why do you wish to move away from Christ crucified to dependence on Torah?'

Proof of the Gospel, 3:1–4:7

The attack is launched in 3:1, which in fact begins an elaborate 'proof of the gospel' extending from 3:1 to 4:7 and containing important references to the death of Christ in 3:1, 13 (and 4:5?). In the initial preaching to the Galatians, Christ crucified was publicly proclaimed in the manner of an official announcement—if, that is, the verb *proegraphē* has that meaning. But Betz has drawn attention to the endeavor of an ancient orator 'to deliver his speech so vividly and impressively that his listeners imagined the matter to have happened right before their eyes.'[135] In that case the translation would be 'vividly portrayed', a statement not out of keeping with Paul's very personal testimony in Gal 2:20. The result of this preaching was the gift of the Spirit to those who received with faith the message of Christ crucified. Paul argues that where faith was so effective, law was not necessary to the gospel; and he takes for granted a relation of cause and effect between the death of Christ and the gift of the Spirit. The implication is twofold. First, those who heard with faith were not simply listening and thus preparing themselves to adopt cultic practices of the Jewish religion; they were sharing the death of Christ and losing their old identity. Second, the Spirit they received was the Spirit of him who raised Christ from the dead (as Paul later explained in Rm 8:11). They were taken over, as Paul himself had been, by the last Adam, who is a life-giving Spirit (1 Cor 15:45). The general resurrection had not yet taken place. But they had received the Spirit as a pledge (2 Cor 1:22, 5:5), and so they lived by the Spirit (Gal 5:25); for the written code kills, but the Spirit gives life (2 Cor 3:6).

Nevertheless, the written code has its attraction for some members of the Galatian communities; and in any case it is the solid ground on which the gospel of Christ crucified is erected. Therefore, Paul moves in Gal 3:6–14 to a discussion of Torah as both blessing and curse. He begins from the axiom that God's original, primary relation with his people is marked by his promise and their trust. The leading proof is provided by the story of Abraham. God promised him innumerable descendents (Gn 15:6): he trusted God and that was regarded as good evidence of his acceptability (i.e., of his qualification to receive the divine blessing). Since Abraham was the founding father of Israel, it follows that the people of Israel are sons of Abraham in the sense that they are related to God by faith (Gal 3:6–7). Moreover, the Abrahamic blessing was extended to all the Gentiles who equally receive God's favour by faith. So if inheritance (i.e., entering into possession of God's blessing) is by the law (as some people are saying), it is no longer the blessing promised to Abraham (Gal 3:18).

In that case, why were commandments given to Israel by God through Moses? Why does Torah contain not only the foundation stories but also the legislation? In Galatians Paul does not fully set out the stages of his argument. He says that the law 'was added because of transgressions' (Gal 3:19). Perhaps he implies some such question as, How could people distinguish between the divine promise, which would produce blessing, and the spurious promises, which would produce destruction? After all, as he later said in Romans, the power of Sin was at work in human society and the power of decay in the created world (Rm 5:13 8:20–21). Therefore, God introduced the legislative Torah, not to cause sinning but to identify sins;[136] and he placed it alongside the benedictional Torah. Yet God's *promise* was still primary. The legislation given four hundred and thirty years after the promise to Abraham is secondary and does not cancel the promise (Gal 3:15, 17).[137] Indeed the prophet, speaking long after the Mosaic covenant, reaffirms the primacy of faith, or trust in God: 'He who through faith is acceptable shall live' (Gal 3:11 quoting Hb 2:4). Nor is the legislation contrary to the promise: if life-giving legislation were possible, acceptability with God would certainly be found in the legislation of Torah (Gal 3:21). But legislation is not life-giving. By setting forth commandments you can require people to do or not to do something. In that way you can in some measure guard and protect their life; but you cannot easily repair their life when it has been injured nor produce life from the dead. Living by trust in God's promise is one thing; living by adherence to Torah legislation is quite another thing because it is always under threat: 'Cursed be everyone who does not abide by all the things written in the book of the law, and do them' (Gal 3:10) quoting Dt 27:26, the final clause of the ancient Dodecalogue of Shechem. Communal life is indeed possible under a legislative covenant with God: 'You shall therefore keep my statutes and my ordinances, by doing which a man shall live: I am the Lord' (Lv 18:5, quoted in Gal 3:12). But that places all responsibility on the person who performs or fails to perform the commandment. Those who fail become bearers of the curse and are thrust out of the cultic community; as proscribed persons, they can no longer abide anywhere.[138] Life and relations with God become disturbingly precarious and therefore wholly unlike life that trusts in the divine promise. If, however, such is the effect of Torah legislation, why was it introduced? Paul replies, As a corrective device: 'The scripture shut up all things under sin' (Gal 3:22).[139] The meaning of those words has been disputed. The verb *synkleiō* ('shut up') appears in this and the next verse, where it is said that 'before faith came we were guarded by the law, shut up until faith should be disclosed'. It is true that the verb can mean 'to hand over' (parallel with *paradidōmi*), but that meaning requires the preposition *eis*, not *hypo*. The scripture has not handed everything over to sin but confined everything by the threat of sin and guarded us by the fence of Torah, a usage not remote either from the man whom God has hemmed in in Jb 3:23

or from the advice of the men of the Great Synagogue to 'make a fence to the Torah' (Aboth 1.1).[140] Paul presupposes the image of a community existing within its protective fence but always threatened by the power of sin if they stray carelessly beyond the boundary or if they rashly break down the fence. Paul had no doubt that it was possible to live acceptably to God within the confines of Torah ('blameless as regards such acceptability as is contained within the law', Phil 3:6); but it was not possible to claim the Abrahamic promise by staying within the boundary—precisely because that promise was made to all the Gentiles. It was necessary to venture across the boundary by faith in Jesus Christ (Gal 3:22[b]). Now that 'the time had fully come' (Gal 4:4) the legislative Torah had served its important but provisional purpose. To change the imagery, it was like the custodian *(paidagōgos)* who protected and corrected school boys in the ancient world (Gal 3:23–25). Or it could be compared to the guardians and trustees who supervised the heir to the family property until he came of age (Gal 4:1–3). Thus legislative Torah, for all its great merits and its holiness, is limited and constrictive.[141] If, however, you do not agree with Paul that the decisive time has come to break out of its constrictions, you will invoke or dread the curse of the law.

3:13–14. It should be noted that whereas the thought of blessing is congenial to Paul, the thought of cursing is not—and it is less common. The noun *katara* (a 'curse') and the adjective *epikataratos* ('under a curse') appear only in Gal 3:10, 13, where they depend on Dt 27:26 and 21:23. Their meaning may be filled out by *anathema*, which has already been used in Gal 1:8–9 as an exclusion formula for those who proclaim a different gospel. To be anathema or under a curse is to be excluded from God's people (cf. Rm 9:3, where Paul is himself willing to be excluded if that would save Israel), to lose the benefit of God's blessing, and to be handed over to the consequences of one's own choice (as in the threefold handing over in Rm 1:24–32). In that context Paul says,

> Christ redeemed us *from the* curse of the law,
> having become a curse for us—
> for it is written,
> 'Cursed be everyone who hangs on a tree'—
> that in Christ Jesus
> the blessing of Abraham might come upon the Gentiles,
> that we might receive the promise of the Spirit
> through faith. (Gal 3:13–14)

That carefully structured statement resembles 2 Cor 5:20 in two respects: in the startling assertion that Christ became a curse (as in 2 Corinthians he was made to be sin), and in the pattern whereby Christ takes upon himself our disadvantages that we may receive his advantages. The two formulas, however, differ notably: in one Christ is subjected to the power of Sin, in the other to the curse of God. The matter needs closer examination.

Redeemed represents the very uncommon verb *exagorazō*, which simply means 'to buy', or occasionally 'to buy back'. It is not a technical word for purchasing the freedom of slaves, nor does it carry the sense of ransoming. At the most it means bringing back (what has been lost, given away, or sold) at some cost to one's self.[142] Presumably Paul uses it here and in Gal 4:5 to indicate that Christ, at cost to himself, brought back the reliance on God's promise that had, in part at least, been surrendered in the course of devoted service to legislative Torah.

The cost to Christ was that he became a curse.[143] Paul does not say, as we might expect, that he became 'accursed'; but follows the Hebrew of Dt 21:23 which says 'a hanged man is God's curse'. The point at issue is this: if a member of the covenant community has been put to death for a capital offence (probably covenant breaking) and his dead body has been publicly exposed by hanging on a tree (no doubt as a sign of repudiation and as a warning), the dead body itself becomes the instrument of a curse upon the people unless it is removed and disposed of before sunset. By citing that Deuteronomic provision, Paul indicates that (according to his reading of Jewish religious law) Jesus had been put to death for a capital offence and that thereby he had become a destructive threat to the covenant people of Israel. Because Jesus had died under the curse of the law and therefore under the curse of God, he was excluded from the community of Israel.

Since he died under a curse, his death could not atone for his own offence; even less could it atone for the offences of others. Mishnah Sanhedrin 6.2 requires that a convicted person on the way to execution must be given an opportunity of confession, supplying (if necessary) the words, 'May my death be an atonement for all my sins'. In Tosephta Sanhedrin 9.5, 'Those who are put to death by the court have a share in the world to come [the resurrection life], because they confess all their sins'. In the fourth century C.E., Rabbi Abbaye said that he who died in his sins, if put to death by a Jewish tribunal, had no atonement; but martyrs executed by the government had atonement (Sanh. 47a *fin*); and of course it was held that the martyrdom of Israel at the hands of the Gentiles is an atonement in the world to come (Rab Anth 226). But Jesus did not die as a convicted man who had finally sought forgiveness; nor, like the martyrs in 2 Mc 7:32 did he admit to suffering for his own sins in the hope that God would bring to an end the wrath that had justly fallen on the whole nation. This description of the death of Jesus deliberately makes it a *skandalon* to Jews.

There are two ways of dealing with a curse: by weakening or removing its power and by counteracting it with a stronger blessing. Both methods are in use here. First, the power of the legislative Torah is removed by God's action 'in Christ Jesus', that is, in his death and resurrection. If Christ died under a curse, he had no place in the world to come; but since God raised him from the dead, God had overruled the curse and cancelled the law that uttered it.[144] In that case, Christ was not a threat to the covenant people insofar as they were brought back (at personal cost to Christ) to the covenant with Abraham. Moreover, the covenant no longer needed the protection of 'the curse of the law' because the blessing of Abraham had come to the Gentiles. In other words, legislative Torah had been replaced by foundational Torah, which can even be called 'the law of Christ' (Gal 6:2).[145] Christ is the *yes* to all the promises of God (2 Cor 1:20).

The Function of Torah

The divine energy that raised Christ from the dead is the Spirit, and that Spirit is the first fulfilment of the promise to Abraham (Gal 3:14). As Paul develops his 'proof of the gospel' (see p. 77), he takes trouble to disarm objectors and to give a positive function to Torah. The section Gal 3:15–18 argues that the later covenant of the law does not supplant the earlier covenant of the promise. In the course of that argument, Paul introduces the notoriously perplexing comment on the seed of Abraham; 'It does not say, "And to seeds", referring to many; but "And to thy seed", referring to one, who is Christ' (Gal 3:16; 'seed(s)', i.e. *sperma(ta),* which RSV translates by 'offspring(s)'). The point, hinted at rather than plainly made, is simple enough: God's promise to Abraham was not made for the benefit of numerous and separate descendants but for his numerous descendants in mankind regarded as a whole—and represented by Christ. Hence, legislative Torah was the valid way of dealing with transgressions until the seed should come for whom the promise had been intended, as is stated at the beginning of a section (Gal 3:19–25) that defines legislative Torah by its constrictive and protective function. This Torah was ordained (by God for Israel and was delivered) through angels by the hand (i.e., agency) of a mediator. At this point also, Paul throws another hasty glance at the deficiency of Torah when compared with promise, so hasty as to put his meaning almost beyond comprehension. None of the numerous attempts at explanation can make much of the words as they stand: 'Now the mediator of one is not, but God is one'. If however we remember Paul's interest in the unity of mankind at this point and abandon the suggestion that he throws in an inept definition of mediation, it is possible that the mediator of the legislative Torah, namely Moses, does not appear as a mediator for the one constituency of mankind (but for the particular constituency of Israel), whereas (in Torah itself it is said that) God is one (and is therefore concerned to speak to mankind as one). The implication would be that legislative Torah had a necessary function for Jews but not for Gentiles until the promise could be fulfilled by faith in Jesus Christ. Then there is reestablished the direct line from God to Abraham, from Abraham to Christ, and from Christ to mankind. To sum up the whole operation,

> When the time had fully come,
> God sent forth his Son,
>   born of woman,
>   born under the law
> to redeem [*exagorasē*] those under the law
> that we might receive adoption as sons.
> And because you are sons,
> God has sent the Spirit of his Son into our hearts
> crying, 'Abba! Father!'
> so thou art no longer a slave, but a son, (Gal 4:4–7)

Thus it plainly appears, from this second use of *redeem,* that the effect of Christ's death was to make the decisive transition between the necessary function of the law and its obsolescence, from the suspension of the promise to its operation, from slavery under the law to freedom of the Spirit, and from the old identity threatened by the curse to a new identity marked by the indwelling of Christ Jesus.

Freedom, 4:8–5:24

When Paul has ended his 'proof of the gospel', he naturally makes an appeal to the communities. He urges them not to fall back into a kind of enslavement (Gal 4:9–11) but, for the sake of their association with the apostle (Gal 4:12–20), to understand their heritage of freedom (displayed in a midrash on Sarah, Gal 4:21–31) and to grasp the kind of freedom for which Christ set them free (Gal 5:1). That statement, echoed in 5:13, supports what has just been said about the effect of Christ's death: it was to set believers free. That freedom would be lost if they accepted circumcision as a necessary condition of receiving the promises of God (Gal 5:1–12). What is more, by the words 'loved me and gave himself for me' in Gal 2:20, Paul has prepared the way for 'faith working through love' in 5:6, and for the two remaining references to *agapē* in the next section, Gal 5:13, 22.[146] The section 5:1–12 is concerned with protecting Christian freedom from incursions of the law; and it contains an explicit reference to the cross, although its interpretation is difficult:

> But if I, brethren, still preach circumcision,
> why am I still persecuted?
> In that case the stumbling block of the cross has been removed.
> (Gal 5:11 RSV)

For my present purpose it is not necessary to solve the problems of this verse; it is sufficient to note that Paul opposes the preaching (*kēryssō*) of circumcision to the preaching of the cross. There is a clear choice between presenting circumcision (and the accompanying intention of obeying Torah) as the necessary entrance into the divine blessing and presenting Christ crucified (and the accompanying activity of faith) as the only entrance. The cross is religiously offensive (*skandalon* cf. 1 Cor 1:23) precisely because the crucified person comes under the curse of the law (Gal 3:13). The Galatian Christians might indeed be under pressure to be circumcised in order to avoid persecution because of the cross (Gal 6:12–13); but acceptance of circumcision led back again into the world of flesh, that is the competing cultural structures of Jews and Gentiles, whereas faith in Christ crucified admitted one to a new existence structured by the Spirit and marked by freedom.

Therefore Paul finds it necessary to protect Christian freedom not only against incursions of the law but also against the insidious influence of the flesh, (to which the Galatian Christians belonged), namely, the habits and impulses of their native hellenistic culture. Hence the section Gal 5:13–26, which leads up to an instruction to 'walk by the Spirit' because 'those who belong to Christ Jesus have crucified the flesh with its passions and desires' (Gal 5:24–25). In recent study the Pauline use of *flesh* has excited a good deal of interest.[147] Every instructed reader has long known that the association of sin with the flesh is no exclusive or even special indication of sensuality. In standard Hebraic fashion, the transferred sense of *flesh* indicates human beings in their frailty, fallibility and mortality. But there are important Pauline uses where *flesh* seems to appear as a power that, in defiance of God, holds human beings in its grip. That use is present in Galatians 5. Up to that point, most references to the flesh have caused no difficulty; 'flesh and blood' (i.e., human beings, 1:16) and 'not ... all flesh' (i.e., nobody, 2:16) are stock Hebraisms; 'a weakness of the flesh' is a bodily disorder (4:13, 14). 'Child born according to the flesh' (4:23, 29) means one born according to the normal, though unpredictable, possibilities of human birth,— contrasted with one born according to the abnormal possibilities of the divine power—so that the contrast to flesh (which was close at hand in the references already cited) becomes explicit. *Flesh* often implies a contrast with God in his power, perfection and eternity. So Paul's present life in the flesh (2:20) is contrasted with the indwelling of Christ and the new life that will finally be his. In 3:3 the Galatians have begun their Christian existence in the Spirit and now are tempted to complement that experience by adopting Jewish requirements of the flesh, namely, circumcision in accordance with the law. Paul, I imagine, was not displeased at making that obvious point about flesh and circumcision. But his intention may have been wider: not only circumcision, but the whole area to which the law applies can be called flesh. With that in mind, what now happens in Galatians 5? Christian freedom can provide an opportunity for the flesh to accomplish its desires, which are opposed to the divine Spirit (5:13, 16, 17). The flesh is active in socially damaging ways, amply suggested in the so-called works of the flesh (5:19–21). The condemnatory words mostly express social judgements against various kinds of group exploitation, rivalry, and conflict; just as the contrasted 'fruit of the Spirit' is expressed in social virtues (5:22–23). At this point it is customary to ask how the same word can be used to describe both (fallible) human existence and a (malign) universal power. But, as Käsemann (*Perspectives on Paul* p. 26) says, it is a fictitious problem.

There are probably two reasons why students of Pauline vocabulary have manoeuvred themselves into a difficult situation from which they hope to escape by high-flown language and appeals to Gnosticism: they take it for granted that Paul is dealing only with individuals and that he applies to them an abstract concept called man. If Paul (who certainly makes use of conventional hellenistic morality), were really a conventional hellenistic moralist, he might indeed reflect in somewhat abstract fashion on human nature. In that case *flesh* might translate as 'our lower nature' (NEB) and be described as the 'self-regarding element in human nature which has been corrupted at the source, with its appetites and propensities, and which if unchecked produces the "works of the flesh" listed in vv. 19f.' (Bruce on Gal 5:13.) But 'the extraordinary contours of Paul's thought' (Jewett p. 94) suggest that he was unconventional in this matter; and his references to the Spirit separate him from the moral teacher who tries to improve the conduct of so-called libertines by analysing the abstract qualities of human nature. Moreover, as Käsemann also says (*Perspectives on Paul* p. 27), 'There is no such thing as man without his particular and respective world'. The individual

who can be isolated from the culture in which he grows ceases to exist as a
human being. Even the individual characteristics that distinguish one person
from another are regarded as important or unimportant according to the culture
in which they occur. Therefore it is necessary to recognize that Paul extends the
use of *flesh* to cover social organizations (as the Qumran community had already
done[148]). When he claims that he has good reason for confidence *in the flesh*, he
explains (in Phil 3:4–6) his meaning; flesh includes circumcision, membership of
God's people Israel, belonging to the particular tribe of Benjamin, brought up in
a Hebrew-speaking household, by conscious choice a Pharisee, a defender of the
faith and an irreproachable observer of Torah. Every phrase makes clearer the
picture of a social culture to which the writer once belonged, which he has now
abandoned. This is the being crucified with Christ, the loss of an old identity and
the search for a new one, the departure from one cultic community and the entry
into another. Paul's use of *flesh* moves understandably from bodily existence to
one's immediate family, then to one's remoter family ('my kinsmen according to
the flesh [by race]' Rm 9:3 RSV) and all the members of one's major social group
('I may provoke to jealousy them that are my flesh', Rm 11:14 RV; 'to make my
fellow Jews jealous' RSV). It ends with *all flesh* meaning mankind as a whole.
Flesh means not corporeality but sociality.

Social groups can develop immense power in fighting for survival and in moti-
vating their members. When Paul talks about the desires and the activities of the
flesh, he has in mind the impulses present in social groups. Although in Romans
he can give some credit to worthy Gentiles and can find some virtue in the
authorities, in general he has a low opinion of hellenistic society; and in Galatians
his opinion is at its lowest. Those who act out the desires of the flesh cannot
possess the Kingdom of God (Gal 5:21, the social image of *kingdom* opposed to
the social image of *flesh*). Therefore Christians are those who have crucified the
flesh with its passions and desires: they have renounced their involvement with,
and dependence on, hellenistic culture just as drastically as Paul had behaved
towards his own Jewish culture. Participation in the crucifixion of Jesus was not
simply an experience that transformed the inner life of a believer; it affected all
the believer's social relations as well, and through the resurrection power of the
Spirit gave a new social existence.

Conclusion

This understanding of Gal 5:24 is confirmed by the even more striking statements in Paul's own hand-written conclusion to the epistle, which serves to sum
up the argument and fix its central contention (Betz). He concentrates on circumcision and the cross. Each word both indicated and symbolised a wounding of
the human person, and each symbol acted as the necessary ritual for admission
either to the Jewish or to the Christian community. Crucifixion was offensive to
both Jews and Gentiles; circumcision was ridiculed by Gentiles. If Gentiles could
be persuaded to accept circumcision, members and supporters of the Jewish community would register a social victory and to that extent counteract the negative
effect of the public portrayal of Christ crucified (Gal 3:1) with its social offen-

siveness and its repudiation of Torah. The social victory is represented (in a typically Pauline wordplay) by 'make a good showing in the flesh' and 'glory in your flesh' (Gal 6:12–13). In response Paul says

> Far be it from me to glory
> except in the cross of our Lord Jesus Christ,
> by which [or, through whom] the world has been crucified to me,
> and I to the world.
>
> For neither circumcision counts for anything,
> nor uncircumcision,
> but a new creation. (Gal 6:14–15)

The Mishnah tractate Nedarim 3:11 contains a saying of Rabbi Eleazar b. Azariah (c. 80–120 C.E.): 'Hateful is uncircumcision, whereby the wicked are held up to shame, as it is written, For all the nations are uncircumcised (Jer 9:26)'. Then follows a catena of sayings from the rabbis praising circumcision and ending with 'Great is circumcision: but for it the Holy One, blessed is he, had not created his world, as it is written, Thus saith the Lord, but for my covenant day and night, I had not set forth the ordinances of heaven and earth (Jer 33:25)'. To that kind of glorying, which regards circumcision as the necessary foundation of all existence (perhaps included among the *stoicheia tou kosmou,* Gal 4:3, 9) Paul opposes his glorying in the cross. Yet nothing could disguise the fact that crucifixion was religiously offensive to Jews and socially destructive to Greeks (see section on 1 Cor 1:23); and Paul makes the point sharply in a reciprocal epigram, using datives of relation with the verb *crucify* in the passive. As earlier in the epistle, to be crucified means to be under a curse and hence rejected, so Paul's epigram says

> By means of the cross of our Lord Jesus Christ,
> the *kosmos* has been rejected as far as I am concerned,
> and I am rejected as far as the *kosmos* is concerned.

The *kosmos* means the present social structure, and is equivalent to the flesh. It is to be replaced by a new creation; and it might be said, 'Great is the cross: but for it the Holy One would not have recreated his world'.

In the penultimate sentence of the letter Paul curtly bids his readers to cease troubling him, for he bears in his body the *stigmata* of Jesus. He was of course circumcised (Phil 3:5), but that mark has been superseded by the wounds of his apostolic ministry: 'always carrying in the body the death of Jesus' (2 Cor. 4:10). Being crucifed with Christ had meant a radical questioning of his identity and a cutting adrift from the dual cultures in which he carried out his commission from God. It had also left its marks on his body.[149]

Summary

In this epistle, Paul makes little use of formulas apart from the initial quotation, which founds his apostolic commission on the resurrection and stresses the self-giving of Christ as the means of transition from the present evil age to a new era inaugurated by the gift of the Spirit. From this beginning he develops a remarkably consistent account of the crucifixion as the act that undermines and destroys a person's existing social identity and opens the way for a new identity. This may be expressed in a strikingly personal manner, as when he says, 'I have been crucified with Christ' and finds himself indwelt by the risen Son of God. This participation in Christ's death implied a correspondence between Paul's experience and the death of Christ, which, by its disturbing nature, threatened the Jewish community, cancelled the legislative Torah, restored the promise to Abraham, and bestowed the resurrection power of the Spirit. The death of Christ is not in any ordinary sense an atoning act but a response of God's grace when he hears the appeal for help from people who know they are acceptable to him only by faith. The result is a transition from the necessary function of the law to its obsolescence, from the suspension of the promise to its operation, from enslavement by the powers at work in this age to the freedom of the Spirit, and from an old identity always under threat to a new identity promised by the indwelling of Christ Jesus.

Even if all this can be read in intimately personal terms, it is also to be understood in social terms. The crucifixion destroys the *kosmos;* those who are crucified with Christ suffer the loss of their social identity (which is the significance of flesh) and move towards a new social identity prompted by the Spirit. Participation in the death of Christ not only changes the inner life but social relations as well, and the Spirit (which raised Christ from the dead) brings into being a new social existence.

ROMANS

The interpretation of Romans depends to a large extent on a judgement about its structure. Is it to be viewed as a 'treatise' on the gospel and its various consequences (1:16–15:13), which happens to be enclosed with a letter in which Paul commends himself to the Roman community (1:1–15 and 15:14–16:27, except that chapter 16 may have been intended for Ephesus)? Does the 'treatise' (if such it is) extend only from 1:16 to 8:39, leaving chapters 9–11 and 12:1–15:13 as useful but perhaps not essential additions? If 1:16–8:39 is the core of the epistle, is it an expansion of the argument of Galatians or a fusion of Jewish legal thought from Galatians with hellenistic participatory thinking from the Corinthian correspondence? Or, to move away completely from the 'treatise' description (which finds much of its support in Western theological debates), should we presume that Romans is much like other Pauline letters, where the theological argument and the ethical advice is worked out in relation to the situation of the readers? If so, it would appear that Paul, very conscious of his apostleship to Gentiles (Rm 1:5, 13; 15:16, 27), is concerned that Gentile communities should be interested in the survival of the Jewish Christians in Jerusalem (Rm 15:25–27). What is more, the conflict between the weak and the strong in Rm 14:1–15:13,[150] which Paul treats with close knowledge and care, leads up to an appeal that Jews and Gentiles should welcome one another in the Christian community. Since the theological argument begins with the assertion that the gospel is for the Jew first and also for the Greek (Rm 1:16), it is likely that Paul had a consistent aim throughout the letter.[151] Two consequences follow: (1) statements about the death and resurrection of Christ are likely to be part of the whole argument, and (2) the argument is about *communities* of Christians, Jewish and Gentile, not simply about individuals. Once an excessively individualistic reading of Paul's argument is abandoned, it is easier to see why chapter 8 is not the proper ending and why chapters 9–11 are essential to the main argument.

It is clear that Romans was written when Paul perceived that he had reached a turning point in his apostolic mission. He had completed his responsibility 'from Jerusalem and as far round as Illyricum' and had no intention of taking over the Roman community's responsibility for preaching the gospel (Rm 15:19–20). Therefore he would move towards Spain. Treating his readers as a competent, autonomous community of Christians, he writes to involve them in two important endeavours: the extension of his Gentile apostleship to the West (Rm 15:24, 28) and the preservation of the Jewish Christian mother community in Jerusalem (Rm 15:25–27, 30–32). Before resuming the onward course of his Gentile mission (Rm 1:5, 13; 15:16, 18, 27; 16:4, 26),[152] he must not only return to Jerusalem and present the collection to the Jewish Christian community there—both as a charitable gift and as a symbol of the Christian unity of the Jewish and Greek communities—but he must also settle once for all (if he could) any misunderstandings about the central importance of the Jewish heritage within the gospel of Jesus Christ as it spread into the hellenistic world.

The Gospel

The introductory section follows the standard form. The writer (1) names himself (1:1), (2) names the recipients (1:7ª), (3) greets them (1:7ᵇ), (4) thanks God for them (1:8), (5) in his apostolic character prays for them and hopes to visit them (1:9–10), and (6) justifies his intention of visiting them, a theme that continues in 15:14–33. The notable feature, however, of this introduction appears in unit 1, which Paul expands by announcing his apostolic qualification and by defining the gospel he proclaims:

> the gospel of God
>
> which he promised beforehand
> through his prophets
> in the holy scriptures,
>
> [the gospel] concerning his Son
> who came to be from the seed of David
> according to the flesh,

> who was designated Son of God in power
> according to the Spirit of holiness
> by his resurrection from the dead,
> Jesus Christ our Lord.

Most commentators now agree that this is a pre-Pauline formula, perhaps somewhat adapted (Wilckens). For our purpose it is necessary only to make a few comments. First, the gospel is defined in terms of the resurrection. As in 1 Thes 1:9–10, the death of Christ plays no independent part. The formula, which is probably early (since it attaches the *divine* Sonship to the resurrection), leads on to an identification of Jesus Christ our Lord. It is therefore an elaboration of the primitive confession 'Jesus is Lord', not so much a Christological formula as a declaration of the lordship under which Christians have placed themselves. Second, the formula is strongly Jewish, partly in its wording, partly in its assertion that the gospel is not novel but ancient, partly in its 'Pharisaic' focus on the resurrection (cf. Acts 23:6, where Paul is portrayed as splitting the council on the issue of the resurrection). It is no doubt hellenistic Jewish, as is its *Son of God* language, but the formulation is perhaps intended to be eirenic in a community where Jews and Greeks are in uneasy contact. At least, for the time being, it avoids the scandal of the cross. The only thing that could reasonably be said, on the basis of this formula, is that Christ is Lord of death.

Objections to the Gospel

At the end of the introductory section Paul cites his responsibility to Greeks and barbarians, to wise and foolish, as the reason for his eagerness to preach the gospel to the Roman community. Then, instead of continuing with plans for doing so (which are postponed till 15:14), he turns to implied objections against the gospel. Somebody was saying that he ought be be ashamed of it. Why they said so can be discovered from his thematic reply in Rm 1:16–17. The gospel as he presents it is offered 'to everyone who has faith, to the Jew first and also to the Greek'. Seen from a less sympathetic, Jewish position, that would be rephrased, 'Paul offers the gospel to Gentiles without first requiring them to fulfill even the minimum conditions for becoming God's holy people; and he offers it simply in response to faith without requiring obedience to Torah. Therefore Paul's gospel is at once destructive of Judaism and of morality.' With such objections in mind, Paul has framed his thematic statement, which provides four indications of his forthcoming argument. First, the gospel about God's Son is not only the offer of a new lordship; it is also salvation (i.e., rescue) from the oppressive lordship that gives no release. It is 'power . . . for salvation'. Second, it is God's power, just as it is 'the gospel of God'. The thought is none other than the promise of deutero-Isaiah: 'I am the Lord your God, the Holy One of Israel, your Saviour. . . . I am the Lord, and beside me there is no saviour' (Isa 43:3, 11). Thus, in the forefront is the salvation of God's people (i.e., not simply needy individuals), and it is precisely in chapters 10 and 11 that *salvation* appears again (see 9:27; 10:1, 9, 10, 13; 11:11, 14, 26). Hence, in traditional Jewish—and particularly Isaianic—

terms, 'the power of God for salvation' can be described as God's *dikaiosynē,* his saving goodness. This is what the gospel contains. Third, it was first offered—and is still first offered—to the Jew; but it is also offered to the Greek—in fact, to everyone who exercises faith. Fourth, the central requirement of faith—*ek pisteōs eis pistin,* from the faithfulness [of God] for the fidelity [of mankind]—is derived from the prophetic word of Hb 2:4 (cf. Rm 9:33, where Is 28:16 is cited in support of faith). When the argument is well advanced, faith will be proved by Gn 15:6 from the heart of Torah, with a supporting text from the writings in Ps 32:1–2 (Rm 4:3, 7–8).

Salvation for Jews and Gentiles

In those four indications Paul declares his intention of proceeding wholly within Jewish presuppositions to subvert the Jewish conviction that the promise of salvation belonged only to Jews and was maintained exclusively by obedience to Torah. The demonstration that salvation is for both Jews and Gentiles occupies his attention from 1:18 to 5:21. Thereafter he turns to the discouragement of sinning and to the place of the legislative Torah. He begins with a standard Jewish polemic against human ungodliness and wickedness, against which the wrath of God is directed (Rm 1:18–23). In line with Jewish teaching of his day, he explained the archaic phrase *wrath of God* as a conviction that God did not destroy sinners (though they were worthy of death, Rm 1:32) but handed them over to the unpleasant consequences of the wrong choices they had made (Rm 1:24–32). It is not therefore 'an inevitable process of cause and effect in a moral universe',[153] because it is not always obvious that unpleasant results follow from wicked choices. Nor, again, is it the retributive activity of God venting his displeasure on sinners and making the punishment fit the crime. In Paul's view of the wrath God hands sinners over to the unpleasantness they have obstinately chosen, no longer protects them from the consequences, and leaves them bound to do what they have chosen to do. At this point making use of the diatribe style, Paul attacks the person who condemns wrongdoing while doing the same himself (Rm 2:1–11). This is a standard argument against moral hypocrisy. He uses the Jewish dogma of reward against his critics: wicked people are handed over to their wickedness, good people receive the goodness they are seeking. This is a dogma of faith, of course, but once believed, it defines God's impartiality. What is true for the Jew is equally true for the Gentile. Hence Jews with Torah and Gentiles with the unwritten law are in parallel situations when God judges the secrets of mankind, according to Paul's gospel, by Christ Jesus (Rm 2:12–16).

Only now does Christ appear in Paul's polemic against ungodliness and wickedness. Nor will he appear again until the indictment has been completed at Rm 3:20. At this point, perhaps with some irony, he draws attention to the gospel of which he should be ashamed, not simply a gospel that might be found in Jewish missionary propaganda but a gospel that contained Christ Jesus. And it becomes clear, almost incidentally, that Christ is neither an agent of the wrath nor someone who persuades God no longer to exercise the wrath. As Paul sees it, the 'handing over', God's treatment of obstinately wicked choices, is his necessary way of perserving *dikaiosynē* (cf. Rm 3:5–6). Hence, to anticipate, the saving function of Christ will be not to pacify the wrath but to rescue those who are bound by the consequences of their sins.

Resuming the diatribe style in Rm 2:17–24, Paul now addresses Jews directly and challenges their social confidence (*kauchēsis,* of which 'boasting' is no longer a suitable translation) in God, his law, and its many benefits. In Rm 2:25–29 he concludes that the circumcised but sinful Jew comes off worse than the uncircumcised but law-abiding Gentile. If this argument seems to deprive the Jew of every advantage, it cannot deny his chief advantage, that Jews have been entrusted with the divine utterances (Rm 3:1–9ᵃ). In that case they know well a string of quotations from the Psalms and elsewhere (i.e., Torah in the wider sense) that complain that 'none is righteous, no, not one'. Despite the mark of circumcision and the guardianship of Torah, the divine benefits have not produced an obedient community. The legislative Torah has not restrained them from wrongdoing. If that is true of Jews, how much more true of Gentiles. By the activities of Torah *all flesh* (i.e., all mankind) is not made acceptable to God. What Torah produces is awareness of (the power of) Sin (Rm 3:20).[154]

God's Saving Goodness

That however is not the final dismissive judgement on Torah; for the law and the prophets bear testimony to a disclosure of God's saving goodness (*dikaiosynē*) that is not dependent on legislative Torah (*chōris nomou,* Rm 3:21). It is in fact God's saving goodness (1) through faith (2) in Christ Jesus (3) for all who believe (Rm 3:22). That threefold statement is immediately repeated in reverse order: 'for all who believe' becomes 'For there is no distinction; since *all* have sinned and fall short of the glory of God' (Rm 3:22ᵇ–23); 'in Christ Jesus' is expanded to

> They are justified by his grace as a gift,
> through the redemption which is *in Christ Jesus,*
> whom God put forward as an expiation
> by his blood, to be received by faith (Rm 3:24–25ᵃ);

'through faith' immediately takes up that clause and vindicates God who 'justifies him who has *faith* in Jesus' (Rm 3:25ᵇ–26). Having thus stated his theme and reiterated it, Paul now develops it at some length, but this time in the original order. Item (1) is treated in Romans 4 by means of Abraham's faith; item (2) in Rm 5:1–11, where the central theme is that 'Christ died for the ungodly'; and item (3) in Rm 5:12–21, which speaks of the condemnation and the acquittal of all men.

It is clear that God's *dikaiosynē* dominates Rm 3:21–26 (the noun in verses 21, 22, 25, 26; the verb in verses 24, 26; and the adjective in verse 26). Paul has returned to the theme he announced in Rm 1:17: the saving goodness of God *ek pisteōs eis pistin,* 'from the faithfulness [of God] for the fidelity [of mankind]'. Indeed, the wording of verse 22 recalls that earlier formulation: 'God's saving goodness *dia pisteōs Iesou Christou* for all *tous pisteuontas*'. Doubtless *Iesou*

Christou is an objective genitive, so that the phrase should be understood to mean, '[to be received] through [our] faith in Jesus Christ'; but it could perhaps be expanded differently as, 'God's saving goodness through [his] faithfulness [displayed] in Jesus Christ and [available] for all who believe [on him]'. In any case Paul now moves beyond what I have called the 'Pharisaic' focus on the resurrection and introduces the death of Christ as a second essential component (alongside the resurrection) of the divine *dikaiosynē*. He does so in formulations of his own choice, but scarcely of his own composition.[55] When Paul is still addressing a Jewish Christian audience, it is reasonable to suppose that his argument advances by the tactful adaptation of a formula from the Jewish Christian section of early Christianity.

Through Faith. I will consider first his approach to the theme of *faith*. Verses 25[b]–26 offer a kind of theodicy: there is a double demonstation of the divine *dikaiosynē* in relation to God's forbearance in passing over former sins and to his present activity. Whatever the original formula meant, Paul presumably intends it to refer *first* to God's forbearance in handing sinners over to what they desired (instead of destroying them) and therefore providing for them a way back to his favour and *second* to God's saving activity at the present time, namely, the time of proclaiming the gospel before this age gives way to the new age. So God is right in doing what is right (for his people) and—here Paul interposes his constant theme—he treats as right (i.e., accepts) anyone who derives his confidence from faith in Jesus.

Once that has been said, Paul briskly opposes faith to the operation of Torah, uses the Jewish confession of the one being of God to assert that both Jews and Gentiles are acceptable to him by faith, and claims that thereby Torah is upheld (Rm 3:27–31). Then follows the demonstration from (foundational) Torah that Abraham's acceptability with God and the possibility that his descendants should receive the 'many nations' promise depended solely on faith. The nature of faith is described entirely in Jewish terms, even when he says that Abraham believed in God, 'who gives life to the dead and calls into existence the things that do not exist' (Rm 4:17). But that makes the opening for him to conclude his midrash with a Christian confessional formula. The statement of Gn 15:6, 'It was reckoned to him' applies to us who believe

> on him that raised Jesus our Lord from the dead
> who was handed over for our trespasses
> and was raised for our justification. (Rm 4:24–25)

The thrust of the discussion requires the formula to be introduced entirely for its reference to the resurrection, that is, for its second clause. That is made plain by the preliminary words, which echo the confession 'Jesus is Lord', itself derived from Christian testimony to the resurrection. In principle, Paul is moving no further than his definition of the gospel in Rm 1:2–4.

But when the formula is examined in detail, more is implied. The word *dikaiōsis* ('justification'), used again at Rm 5:18 (*dikaiōsis zoēs*, 'acquittal and life') but nowhere else in the New Testament means 'an act of setting right'. The second clause might be more comprehensibly rendered 'God raised him in order to set us right'. It is a Jewish formulation, though not obviously from Paul, even if it conveniently suits Paul's general discussion of God's *dikaiosynē*.

Both clauses use *dia* with the accusative but in different senses. In the first clause it is causal, 'because of our trespasses'; in the second it is final, 'in order to'. The formula was constructed to be verbally memorable rather than theologically precise; yet it testifies to an intention of making parallel and complementary statements about the death and resurrection of Christ.

If we now turn to the first clause, we have *paredothē dia ta paraptōmata hēmōn*. The verb *paradidōmi* ('to hand over') is used in a variety of senses in the New Testament, but it has a special significance in the passion tradition. In Gal 2:20 (cf. Eph 5:2, 25) it is said that Christ handed himself over for the speaker's benefit. In Rm 8:32 it is said that 'God did not spare his own Son but gave him up for us all [*hyper hēmōn pantōn*]; the same thought must be present in the aorist passive of the verb here. The noun *paraptōma* now occurs for the first time in the epistle. It is in effect not easily distinguishable from *parabasis* (which has already been used in Rm 2:23, 4:15) except that *parabasis* always carries the thought of infringing a command and *paraptōma* can have the broader meaning of 'blunder'.[156] Thus the first clause can be written out more explicitly in this fashion: 'God handed Jesus over [to death] because of our disastrous misjudgements'. That translation does not reduce the meaning of *paraptōmata* but strengthens it and gives it force (for the offence of 'trespass' is scarcely regarded seriously in these days). Granted this translation, however, the question must be asked, Who could sensibly have made such a confession? The formula must surely be attributed to the Jewish contemporaries of Jesus[157] or to such as wished to associate themselves with his repentant contemporaries. When (according to Acts 2:36–38) Peter said to the men of Israel, 'God has made him both Lord and Christ, this Jesus whom you crucifed', and then advised them to repent, presumably he expected them to respond with something like Rm 4:25.

When it is asked where the wording of the formulation came from, it is possible to reply with echoes of Isaiah 53. In Is 53:6ᶜ (presumably spoken by the 'nations' of 52:15), it is said, 'The Lord has laid on him the iniquity of us all'. In the Septuagint this appears as 'The Lord handed him over for our sins' (*paredōken tais hamartiais hēmōn*). And in 53:12ᵈ it is said, 'and made intercession for the transgressors', which in the Septuagint appears as 'because of their sins he was handed over' (*dia tas hamartias autōn paredothē*). It is clear that the Septuagint translator was either using a somewhat different text or making the best sense of what he did not fully understand. In 53:11ᵈ he was even more perplexed and not without cause. The Massoretic Text seems to say, 'By his knowledge shall the righteous one, my servant, make many to be accounted righteous' (RSV; though see NEB); but the Septuagint translator seems to mean that God intends to justify the just one who well serves many people. It may be possible that *paredothē* and *dikaiōsin* of Rm 4:25 could have been suggested by the Greek version of Is 53;[158] but at most this perplexing song about the suffering servant provides an emotional background for the primitive Christian confession. Without further information, it is useless to speculate about any process of thought by which

Jewish or Gentile Christians could have selected two or three words from Is 5 3 and then constructed a confession in antithetical parallelism. It is even more precarious to suppose that recitation of Rm 4:2 5 would call to mind the dominant features of Is 5 3.

'In Christ Jesus'. Now we can turn to the significance of *Christ Jesus* by studying the formulaic words in Rm 3:24–25, which can be set down thus:

dikaioumenoi	being made acceptable [to God]
dōrean tē autou	
chariti	freely by his grace
dia tēs apolytrōseōs	through the liberation [accomplished]
tēs en Christō Iesou	in [the death and resurrection of] Christ Jesus
hon proetheto ho theos	whom God put forward [or intended]
hilastērion	as an atonement
dia tēs pisteōs	[effective] through faith
en tō autou haimati	by means of his blood

Even if, as seems likely, Paul is here making use of an existing formula, it is not a simple task to remove his additions and recover the original. The words in verses 24[b] and 2 5[c] can easily be recognized as Pauline emphases, whose removal simplifies the structure. But *in Christ Jesus* in 24[d] is even more characteristically Pauline but it cannot be simply removed since some reference to Christ is necessary as the antecedent of *whom* in verse 2 5[a]. With that reservation in mind, it is possible to simplify the exegesis by first noting the special Pauline phrases.

The word *grace*[159] has already appeared in the epistolary greeting (Rm 1:7, cf. 16:20) and as a reference to Paul's apostolic commission in Rm 1:5 (cf Rm 12:3). Now it is used to describe the quality of God's saving action in Christ, namely, God's spontaneous generosity to undeserving and unthankful people. We know from 2 Corinthians that this meant God's eschatological activity in the death and resurrection of Christ. So also it does in Romans: 'Grace immeasurably exceeded the power of Sin in order that, where Sin had established its reign by way of death, grace might establish its reign through God's saving goodness, and issue in eternal life through Jesus Christ our Lord' (Rm 5:20–21). But in the present passage it is specifically associated with Christ's death and opposed to the power of Sin. It is marked out as God's free gift (*dōrean;* cf. the noun *dōrea* associated with grace in Rm 5:15, 17), not his due payment for actions required by his authoritative law (Rm 4:4; 6:14–15; 11:6).

It is commonly agreed that *through faith* must be separated from what follows.[160] It qualifies the manner in which the *hilastērion* is presumed to operate. Just as *grace* emphasises the spontaneity of God's generosity, so *faith* emphasises the unforced, uncalculated human response to that generosity. The words incorporate within the formula (though in a manner that is stylistically awkward) the insistence on faith that appears in verse 22 and is explored in chapter 4.

Once the force of these Pauline qualifications has been noted, it is necessary to regard the simplified formula and discover the direction in which its symbolic language is pointing: 'We are made acceptable to God

through that liberation
accomplished in [the death and resurrection of] Christ Jesus
whom God put forward [or intended]
as an atonement
by means of his own blood.

The discussion of *apolytrōsis*[161] has already been anticipated by rendering it as 'liberation'. The noun is rare, and it is therefore necessary to determine its meaning in each context where it occurs. It belongs, of course, to a group of words derived from the simple verb *lyō*, meaning 'to release', sometimes to release on receipt of a ransom. Hence, *lytron* is a means of releasing, or ransom money, and *apolytrōsis* can describe the release of prisoners when a ransom or compensation is paid (though it need not refer to any payment). It occurs again at Rm 8:23 and in eight other New Testament passages. In none of them is the thought of ransom necessary or appropriate (not even in Eph 1:7, which is closest to Rm 3:24: 'In him we have redemption [*apolytrōsis*] through his blood, the forgiveness of our trespasses'). It cannot be denied that elsewhere Christ is said to have bought benefits for those who believe in him (using the verb [*ex*]*agorazō*), and 1 Pt 1:18–19 says that 'you were ransomed from the futile ways inherited from your fathers, not with perishable things such as silver or gold, but with the precious blood of Christ'. But the payment is made explicit, as it is not in Rm 3:24; and it is unsound exegesis to transfer what is explicit in such passages (and perhaps also in Mark 10:45, *lytron* and 1 Tm 2:6, *antilytron*) and attach them to one occurrence of *apolytrōsis*. In any case, it is no great loss. The ransom metaphor wilts before the question, To whom was the ransom paid? and becomes nothing more than a dramatic way of indicating the demand made on Christ. In the present context that is indicated more effectively in 'by means of his own blood'. An attenuated metaphor is less significant than a real liberation.

The arguments for translating *proetheto* as 'put forward' or 'intended' are nicely balanced, and the matter is not worth anxious debate. With either rendering, the important point is clearly made: that God himself has provided Christ Jesus as an atonement. The presumption must be that God made such provision for the benefit of sinners rather than to satisfy his own dignity, status and reputation.

'As an atonement' represents the single word *hilastērion*,[162] and avoids the debate between 'expiating sins' and 'propitiating the deity'—chiefly because that debate is wrongly conceived. The word *hilastērion* may be used here in one of three possible meanings. First, it can mean the golden cover for the ark (containing the tables of the law), which, together with the cherubim was the place of the divine presence. In Hebrew it is called *Kapporeth* (from the root *kpr* meaning 'to cover over, pacify, make propitiation') rendered by the Septuagint as *hilastērion* (and *hilastērion epithema* at its first mention, in Ex 25:17, which looks like a double translation: 'a propitiatory, a cover' [see Josephus *Ant.* 3.135]). On the great day of atonement it was partly sprinkled with blood to secure renewal of sanctity for the priesthood and the sacred enclosure. This obscure sacrificial ritual has been seized upon to provide a proper religious symbol for the death of Christ, but it is unsuitable once the cultic intention is understood. Moreover, it is symbolically confusing (if Paul's readers were familiar with the Levitical ritual) to compare Christ himself to the cultic object and his blood to that which is sprinkled on and towards the cultic object (though admittedly there is no limit to the confusion that religious minds will sometimes tolerate). The best that can be made of this meaning of *hilastērion* is to say that the death of Christ is the locus where atonement was mysteriously performed—this time publicly instead of hidden away within the most holy place.

Second, the popular hellenistic meaning is a votive offering. The people of Kos set up a votive offering to influence the gods to deal favourably with the emperor's welfare. It is an inducement for the gods to do what, if they choose, they may properly do. This depends on the conviction that when someone approaches an eminent person in hope of a favour, he does not go empty-handed. If one of the meanings of *propitiation* is 'inducement', it would perhaps fit Rm 3:24.

Third, there is one well-known hellenistic Jewish example of *hilastērion* in 4 Mc 17:22. The dying martyr, a holy devotee in the sacred war against tyranny, offers his violent death on behalf of sinful Israel, so that 'through the blood of these godly men and the *hilastērion* of their death the divine Providence delivered the previously ill-treated Israel'.

If meanings (2) and (3) are put together, they suffice to explain Rm 3:24: God himself has made plain the one inducement he will accept from suppliants who appear before him, namely their trust in Christ Jesus who shed his blood at God's behest. The correctness of this interpretation is shown by *prosagōgē* in Rm 5:2, a word belonging to ancient court ceremonial, meaning the privilege of access to the great king: 'Since we are made acceptable by faith, we have peace [instead of enmity] with God through our Lord Jesus Christ, through whom we have received this access to the divine grace in which we [firmly] stand, and we exultantly hope [to recover] the glory of God'. A similar perception occurs in the formulaic phrases of 1 Pt 3:18: 'Christ died for sins once for all, the righteous for the unrighteous, that he might bring us to God', using the verb *prosagō*.

1 Pt 3 speaks of sins, which Rm 3:24 does not. All indeed have sinned (Rm 3:23) with the consequence not so much that they are guilty of particular faults as that they are deprived of the glory of God. The business of atonement is less the finding of charitable and (if possible) legal means of forgiveness as of changing the condition of sinners. The Greek root from which *hilastērion* derives regards sin as a dangerous contamination that can be removed only by calling on an even more powerful protective device, in itself not without danger. The process of atonement (as expressed in this group of words) is concerned with securing God's favour, averting danger, restoring functions that have been lost, and reequipping people to take part in worship. If Paul used *hilastērion* for any other reason than its presence in the formula he adopted, he would probably have thought of it as an inducement for God (not to suspend the wrath, for how then could he judge the world? [Rom. 3:5–6] but) to intervene on behalf of sinners and rescue them from the consequences of what they had obstinately chosen to do.

The atonement is effected 'by means of his own blood' (*en tō autou haimati*), which is repeated in Rm 5:9 and recalls *en tō emō haimati* in 1 Cor 11:25 (cf. 1 Cor 10:16 and 11:27), the only Pauline references to the blood of Christ (see pp. 43–45). It may be conjectured that the formula of Rm 3:24–25ᵃ had its original setting at the Lord's Supper and that the apparently objective statement about God's action in Christ Jesus was complemented by a participatory understanding when the cup was shared. Paul's further exposition of our being made acceptable to God by means of Christ's death draws on the resources of the divine love. For the first time in the epistle we meet agape in Rm 5:5, 8, and the first major section of the argument closes with the powerfully rhetorical references to the love of Christ and God in Rm 8:35, 37, 39. It must be taken for granted that all matters discussed between the beginning of chapter 5 and the end of chapter 8 fall within the scope of God's agape.

According to the structure proposed above (p. 90.), Paul first presents a formulaic statement of God's saving goodness in Christ Jesus in 3:24–25ᵃ and then expands it in Rm 5:1–11. He begins with a two-sided description of Christian self-confidence *(kauchōmetha)*,[163] which is the opposite of disappointment (verse 5): We confidently hope to recover the glory of God, and we maintain our self-confidence through suffering, endurance, testing experiences and expectation. In characteristic fashion Paul joins a genuine awareness of acceptance by God to an experience of suffering that makes the awareness only provisional. Having peace with God, which is the opposite of being at enmity with him, does not imply an instantaneous transformation of our condition. God's love is poured out (suggesting great abundance, in contrast to the pouring out of divine wrath) into our hearts (our wills and intentions) through the stimulus of the Holy Spirit bestowed on us. Not that we are made loving but that God's love for sinners takes possession of all we do and, by the operation of the Holy Spirit, continually removes evidences of the wrath and restores us to godly hope. But how can Paul say that we are overwhelmed by God's love if we must endure suffering? Because God demonstrates his love for us in Christ's death for the weak, the ungodly and the sinful rather than for the upright and the good. Paul feels his way from a preliminary statement in verse 6, via a tentative counterstatement in verse 7, to a confident affirmation in verse 8 of the standard primitive confession:

6   For moreover [*eti gar*] Christ—when we were weak—
    yet at that time [*kata kairon*] died for the ungodly.

7   For one would scarcely die for an upright man—
    though indeed for a good man possibly someone
      would venture to die—

8   But God demonstrates his particular love for us,
    namely [in that] while we were still sinners
    Christ died for us.

It is necessary to remember that Paul seldom speaks of God's love; that when he does, the thought is normally taken from the Old Testament, and that God's love for Israel receives special emphasis in Torah.[164] God's rescue of Israel from bondage in Egypt is attributed solely to the fact that God set his love upon them:

'It was not because you were more in number than any other people that the Lord set his love upon you ... but it is (simply) because the Lord loves you ... that the Lord has brought you out with a mighty hand, and redeemed you from the house of bondage. ... Know therefore that the Lord your God is God, the faithful God who keeps covenant and steadfast love with those who love him and keep his commandments ... and requites to their face those who hate him, by destroying them' (Dt 7:7–11). The image of the Egyptian bondage is before Paul's mind (prompted perhaps by *apolytrōsis* in Rm 3:24: note 'redeemed' [*elytrōsato*] in the above quotation), and he returns to the thought of 'bondage' *(douleia)* in chapters 6, 7 and 8. What he does here, however, is emphasise God's love for his people and at the same time remove God's enmity towards those who hate him. Not only when we were weak (which might result from misfortune or malice) but when we were ungodly and sinful, Christ died for us. Nor did Christ die like a heroic Maccabean martyr in order to induce God to rescue sinful Israel. The death of Christ was the action of God's own love to make a way back to him for those who had been handed over to the consequences of the wrath. This symbol of love that rescues from bondage shapes Paul's understanding of the formula 'Christ died for us' (see pp. 14–15).

From God's love, he now turns to our expectation with a standard 'if so ... all the more' argument:

> All the more therefore,
> now that we are acceptable by means of his blood,
> we shall be saved by him from the wrath.
>
> For if, when we were [God's] enemies
> we were reconciled to God through the death of his Son,
> all the more, now that we are reconciled,
> we shall be saved by his life.
>
> Not only so, we have confidence *(kauchōmenoi)* in God
> through our Lord Jesus Christ,
> through whom we have accepted this reconciliation.

That we are acceptable to God by means of Christ's blood picks up a phrase from the formula in Rm 3:25 and extends its consequences to salvation (cf. Rm 1:16, God's power for salvation) form the wrath. From the parallel phrase in verse 10, our salvation from the consequences of the wrath is by sharing in his (risen) life. It is not implied that God is persuaded no longer to operate the wrath (in the sense explained in Rm 1:24–28) but that he provides rescue from our bondage to self-chosen punishments so that at the final day we are not condemned to destruction. To guard against any impression that God has been induced to do what he properly should not do, Paul rewrites the statement in terms of reconciliation: we are reconciled to God and accept this reconciliation from Christ (see p. 59).

'For All Who Believe'. Finally, Paul develops his theme that God's saving goodness is for all who believe (Rm 3:22): without distinction *all* have sinned and fall short of the glory of God (Rm 3:22ᵇ–23). The development takes place in Rm 5:12–21: the notorious words of verse 12ᵈ plainly refer back to what was already said in 3:23ᵃ 'in that all sinned'.¹⁶⁵ The argument is written in a notably abstract and symbolic style, in order (no doubt) to give it the widest possible reference: to mankind, not simply to Jews. Two single human figures are contrasted: Adam and Jesus Christ. Both can be called symbolically 'the one man', and both represent 'all mankind' or 'the many'. The same device had already been used in 1 Cor 15:22, 45–49. There, however, it was applied to questions about the nature of the resurrection life, and therefore the nature of Adam as 'the man of dust' and the nature of Christ as 'the man of heaven' are in the forefront. But here the symbols are used to elucidate the death of Christ; therefore the actions of Adam and Christ are foremost, though in a seemingly general form. The misjudgement *(paraptōma)* of Adam resulted in death for mankind (Rm 5:15); the right judgement *(dikaiōma)* of Jesus Christ resulted in acceptability and life for mankind (Rm 5:18). Hence Adam *ex contrario* was a 'type' of him who was to come, that is, he was a symbolic figure who indicated the universal significance of Christ. Behind these two figures can be discerned a fundamental opposition between Sin and death on one hand and God and life on the other.

Sin and Death

In Rm 5–8 there is an unusual concentration of words for sin and death. So far Paul has not clearly indicated *why* the death of Christ can make us acceptable to God and reconcile us to him. There have been references to the blood of Christ (Rm 3:25, 5:9), and he has been handed over for our trespasses (Rm 4:25); but these are no more than quotations of pre-Pauline formulas. What he now does is explore the connexion and consequences of Sin and death. For the moment he confines himself to the period between Adam and Moses and the period since Christ. The period of Torah is parenthetically dismissed (Rm 5:13, 20ᵃ) until he treats it more seriously in Rm 7. The reason for this temporary neglect is plain: he is totally preoccupied with what is authoritative for all mankind.

In these chapters, Sin is represented as if it were in conscious rivalry to God. Sin entered the world (5:12–13), it increased (5:20), it reigned in death (5:21). It can be an indwelling power (7:17, 20), presumably in the sinful flesh (8:3) or the sinful body (6:6) and therefore can have a law of its own (7:23, 25; 8:2). Sin can use your members as instruments of wickedness (6:13). Of itself Sin is powerless, but it can seize opportunities presented by God's commandment and create unlawful cravings (7:8–9). It deceives and destroys (7:8, 11, 13; 8:10). It entices people into its service and in reward pays them with death (6:23). Correspondingly, people may be under Sin (3:9), sold under Sin (7:14), slaves of Sin (6:16, 17, 20), they may remain in Sin (6:1), and with their flesh they may serve the law of Sin (7:25). This symbolic manner of speaking should not mislead us into supposing that Paul was describing a rival deity. If that had been the case, there would have been little difficulty, for Jewish propagandists were expert in discrediting rival deities. But with Sin there was nothing to get hold of: no cult to deride, no claims to disprove, no convictions to attack. Sin has no creation of its own;

it simply destroys what is presented to it. It lives by feeding on what is good. It resembles a dangerous infection like rabies, which evades a country's protective devices through some act of carelessness or defiance. Once it is introduced, there is no certain cure. The result is usually death for animals and humans, and further contamination. The only way of dealing with it—even then not guaranteed—is to kill animals who may become infected. If the comparison with organic disease and animals is unsatisfactory, we might compare Paul's conception of Sin to the introduction and spread of fascism, anti-Semitism, and racism— with the added difficulty that moral perversions cannot be removed by killing immoral people. However that may be, it is clear that 'the forgiveness of sins', in any ordinary use of the phrase, is no answer to the problem that Paul perceives.[166]

Moreover, Paul sees that death is closely linked with Sin (Rm 5:12, 14, 15, 17, 21). When he says that 'through one man's trespass sin came into the world, and death came in through sin, so that death spread to all men' (Rm 5:12), his intention ought not to be misunderstood. He is not providing interesting though speculative information about the origin of mortality or drawing simple-minded conclusions from an ancient myth. Why, in such a discussion as this, should he attempt that kind of thing? He is indeed using mythical language, probably for two related purposes: (1) to say something about individual sinning:—if Adam symbolically represents all mankind, I also am Adam and the consequences of my sinning spread out to mankind; and (2) to suggest that the most damaging consequence of sinning is the extension of death (not merely in the physical sense): '*We* determine the state of being dead by the moment of extinction of physical life, but for Israel death's domain reaches far further into the realm of the living. Weakness, illness, imprisonment, and oppression by enemies are a kind of death'.[167] That manner of thinking cannot be dismissed as a pious fancy because for Jewish people death was the weakest form of existence, premature death was regarded with dismay, and in the end death severed communion with God. 'For Sheol cannot thank thee, death cannot praise thee; those who go down to the pit cannot hope for thy faithfulness. The living, the living, he thanks thee, as I do this day' (Is 38:18–19).[168] So for Paul, sinning diminishes our awareness of God and our possibility of praising him and begins to move us towards the area where God no longer exercises his saving goodness. Death reigned (i.e., dominated human life) from Adam to Moses, even over those who had not sinned in Adam's manner by breaking a specific commandment of God (verse 14): 'Because of one man's trespass, death reigned through that one man' (verse 17), perhaps in the sense that 'each of us has become his own Adam' (2 Bar. 54.19); 'Sin established its reign by way of death' (verse 21 NEB). In all this, death is clearly not simply— or even chiefly—termination of physical life but it is destructiveness, especially self-destructiveness.

But Christ's godly, saving act was not just equal and opposite to Adam's ungodly, destructive act. There was a great imbalance. Even though Adam's fault was universally disastrous, Christ's act of grace was overwhelmingly greater. The theme is developed with increasing emphasis in verses 15–17, with the conclusion that as death established its reign through one man, how much more will those who have received the abundance of grace and the gift of acceptance live and reign in [the death and resurrection of] the one man Jesus Christ. There is a change in sentence structure (deliberate, since Paul could easily balance death reigning with grace reigning in verse 21) to suggest that the impersonal dominance of death gives way to the personal effectiveness of living people. Here two sovereignties are in mind, one in the past, one in the future. The old sovereignty was initiated by an act of wrongdoing when Adam acted for himself in disregard of God. The new sovereignty was initiated by an act of overwhelming generosity when Christ acted for others in obedience to God. The code words for the old humanity, namely, *ungodly activity, condemnation, disobedience* and *sinners,* are replaced by other words for the new humanity, namely, *godly activity, acquittal, obedience* and *acceptable to God* (Rm 5:18–19). In verse 19 the verb twice used of the many is *kathistēmi* (to make, constitute). Adam's act made mankind sinners (i.e., slaves of Sin), Christ's act will make mankind acceptable to God. Hence, the death of Christ, in this argument, is not presented as an atoning act but as a constitutive act. As Adam constitutes mankind in the present age, Christ constitutes mankind in the coming age.

The law is still not allowed to enter the discussion (verse 20), partly because mankind is under consideration and not simply the Jews, partly because law is not constitutive for mankind in the coming age (as indeed it was incapable of constituting an obedient mankind in the present age). The constitutive feature of mankind is not obedience to the law but (as Paul will say in the opening verses of Romans 6) faith in him who died and rose again. But before taking that next step, let us consider the implications of the argument so far. The problem to be solved was not the inevitability of death for mankind, for Paul as a Pharisee knew that God was able to reach into Sheol. It was the problem of *Sin and death.* How had God dealt with a dangerously pervasive power hostile to his own sovereignty and destructive of his people? His answer may be stated in five stages:

1. To the sovereignty of Sin and death God opposed overwhelming power, not simply power sufficient to rescue sinners but power large enough to thwart Sin.
2. This power had its source in the grace of God. Thus it was spontaneous and unmotivated,[169] called forth not by some demand on God but by his own generosity and love.
3. This power was put to work in Jesus Christ who entered the sovereignty of death, so that Christ's death ended his work and the manner of his death (religiously offensive to Jews and socially destructive to Greeks, 1 Cor 1:23) discredited it. Indeed, God made him a victim of Sin's power (2 Cor 5:21).
4. For that very reason the overwhelming sovereignty of God's grace raised him from captivity to death and in him constituted a new mankind for the coming age.
5. Hence, eternal life is now a possibility for all who have faith in Jesus Christ (i.e., in his death and resurrection), who for them is both forerunner and Lord.

It is clear (especially from the change of sentence structure in verse 17) that this conflict of sovereignties, set out in a modest myth, is intended to apply to the needs and hopes of Christians.

Dying and Rising in Baptism

In what immediately follows in Romans 6 it is so applied, making use of the cultic metaphor of dying and rising in baptism. It may reasonably be supposed that Paul regarded baptism as the common ritual for admission to a Christian community even though, at Corinth, he distanced himself from the activity of baptizing. Moreover, when baptism was coordinated with Christ's death, it could be presented in argument and teaching as a saving and protecting device (see pp. 28–30). In what is perhaps a pre-Pauline formulation, 'being baptized into Christ' can be rephrased as 'having put on Christ' and so having become 'sons of God by faith in Christ Jesus' (Gal 3:26–27).[170] That will be complementary in effect (though divergent in metaphor) to being baptized in one Spirit and being caused to drink one Spirit (1 Cor 12:13). Yet Paul is never entirely at ease with such baptismal language, perhaps because it was too easy, to the satisfaction of Greeks and the scandal of Jews, to assume that baptism in fact conveyed the virtues that it only promised. In any case, Romans 6 and 8:1–17 are a vigorous encouragement for Christians who have been baptized into Christ's death to work out the possibilities of walking with him in newness of life.

In diatribe style, Paul returns at Rm. 6:1 to a malicious objection to his position and turns it away by showing it to be unreasonable: 'How could we, who have ceased living in the power of Sin, continue so to live?' (Rm 6:2).[171] Throughout the chapter, Sin is represented as a hostile power from which Christians have been liberated. Formerly they were enslaved to Sin and were bound to serve it; now they are no longer so bound and need not—indeed must not—serve Sin. The argument in chapter 6 is set out in a rapid alternation of indicatives and imperatives (verses 3–10 indic., 11–15 imp. or equivalent, 16–18 indic., 19 imp. and 20–23 indic.).

The first set of indicative verses, Rm 6:3–10,[172] appealing to baptism as a starting point, establish a relation between the death and resurrection of Christ and something corresponding in the experience of Christians: 'Surely you know that all of us who have been baptized into Christ Jesus were baptized into his death'. The phrase 'baptized into Christ Jesus',[173] which is unique in the New Testament, must be regarded as Paul's adaptation of a less-unfamiliar baptismal formula, 'to baptize into allegiance to someone' *(baptizein eis to onoma)*, akin to the similar faith formula *(pisteuein eis to onoma)*. Instead of taking the (traditional) formula 'baptized into Christ' and explaining it by another metaphor (as in Gal 3:27), he changes to the faith formula of Gal 2:16 ('We have believed in Christ Jesus') and allows it to fill with the meaning of his *en Christo Iēsou* formula. Hence, he argues, 'When you were baptized, you were indeed brought into allegiance to Christ—but as participators in his dying and rising again.' They were not only serving a new Lord, but their lives were now patterned by his experience of death and resurrection. Baptism was more than the ritual entry to a new religious community and less than a mystical or magical transformation into a new nature. It was the act that initiated the transfer of the pattern of death and resurrection from Jesus to the person being baptized. Not for a moment, however, does this transfer minimise the singularity of Christ: it would not do to take some other heroic or tragic person and use his example as a universal pattern. What is said about the death and resurrection of Christ has significance precisely because he is the last Adam and gave himself for us. What we appropriate in baptism are the benefits of his self-giving and a share in the new constitution of mankind.

Since immersion in water is not an appropriate analogy for burial (Tannehill, p. 34), it is clear that Paul is reflecting on an existing credal formula of the type that lies behind 1 Cor 15:3–5, namely, He died, he was buried, he arose, he appeared (see pp. 45–46). Without hesitation he applies the first two clauses to believers: they have died and been buried with him, which he rephrases in verse 5: 'We have been united with him in a death like his'.[174] In what respect is our dying like his dying? The answer follows in verse 6: 'Our old identity has been con-crucified with his in order to make an end of a bodily life that belongs to Sin, so that we are no longer in servitude to Sin'. None of this is unfamiliar, for it has been worked out in principle in Galatians (see pp. 72–73) and developed in the previous chapter of Romans: 'The kind of death that Christ died put a stop, once for all, to the condition in which, whether he wished it or not, he could be subject to the power of Sin' (verse 10[a]). Exactly that may be said of baptized Christians. So much for 'He died, he was buried'. But what of 'He arose, he appeared'? It is scarcely to be expected that 'He appeared' would apply to them, and indeed in Rom 8:18–19 Paul says that the creation eagerly awaits the revealing of the sons of God. But Paul also displays caution in referring to whatever, in Christian life, corresponds to 'He arose'. The statement 'As Christ was raised from the dead by the glory of the Father, we too may walk in newness of life' (in verse 4) is made unambiguous in verse 8: 'Since we have died with Christ [*ei* with indic.], we believe that we shall also live with him'. Thus a Christian is not yet living the resurrection life (despite what some Christians believed and wrote), and the most that could be asserted of a Christian is 'The kind of life he now lives is fully open to the power of God' (verse 10).

From that assertion the imperatives follow: 'So you also must take account of yourselves as being what you are, no longer living in the power of Sin but living in the power of God in [the death and resurrection of] Christ Jesus.'[175] They are no longer to go on putting their capabilities at Sin's disposal, but they must offer them for God's use *as if* they were risen from the dead. For they are no longer dominated by Sin; they are not under the demands of law but the influence of grace. But Paul does not say, and cannot say, that they are no longer under the power of death. Christians are still mortal. They must not allow Sin to dominate their mortal bodies (Rm 6:12), and they can be confident that God's indwelling Spirit will bring to life their mortal bodies (Rm 8:11). They await the liberation of their bodies (Rom 8:23) but must contend with 'the sufferings of this present time' (Rm 8:18). It might be said that the gospel had changed nothing because Christians were still subject to suffering and death; but it had changed everything, because these things could now be interpreted as a share in *Christ's* suffering and death. In the death and resurrection of Christ Jesus, God was changing the consequences of death, destroying the power of Sin, and establishing the new constitution of mankind. When Christians died, they died not as slaves of Sin but as slaves of God—to speak in human terms (Rm 6:16–23).

Sin, Law, and Death

The same interest in death continues in Rm 7:1–6, which, picking up a preliminary reference in Rm 6:14–15, turns attention directly to the law.[176] Death releases not only from the power of Sin (Rm 6:7) but also from the authority of the law and thus opens up new possibilities of relationship (Rm 7:1–3). The application in verse 4 is highly compressed and allusive but not otherwise difficult: 'You, my [Jewish] brethren, have been put to death with regard to the law through [your baptism into] the bodily life that belongs to Christ [for you were baptized into his death]. In consequence you can become the servant [not of Sin but] of another who in fact was raised from the dead that we [all] might bear fruit for God'. In other words, they have been discharged from the law, they are dead to that which held them captive. The effect of being baptized into the death of Christ is that they are no longer at the mercy of the flesh; they now serve in the new possibility of the Spirit, not the old necessity of the written law (Rm 7:5–6).

But what could 'put to death with regard to the law' possibly mean? When Paul says that the law dominates a person during his life (Rm 7:1), he is saying, whether consciously or not, something more than the innocuous legal principal that the validity of law ceases at death. For those of us who are not professionally concerned with the operation of law, the legal system is taken for granted, and we normally adhere to it within reason and for ordinary purposes avoid thinking about it. But to Pharisees such as Paul (who was extremely zealous for the tradition of the fathers, Gal 1:14) the law was a passion from which only death could separate them. The evidence is plentifully available in Aboth. Rabbi Hananiah b. Teradion (d. 135 C.E.) said 'If two sit together and words of Torah are spoken between them (in religious duties or worldly business), the Shekinah rests between them' (3.2). A saying attributed to Akiba (also d. 135 C.E.) goes, 'Beloved are Israel, for to them was given the precious instrument; still greater was the love, in that it was made known to them that to them was given the precious instrument by which the world was created, as it is written, "For I give you good doctrine; forsake not my Law"' (3.15). Rabbi Nehunyah b. Ha-kanah (c. 70–130 C.E.) said 'Whoever takes upon him the yoke of the Torah, from him shall be taken away the yoke of the government and the yoke of worldly care; but he that throws off the yoke of the Torah, upon him shall be laid the yoke of the government and the yoke of worldly care' (3.5). If human existence is a struggle between the good and evil inclinations, 'the evil *yetzer* has no power over against the Law, and he who has the Law in his heart, over him the *yetzer* has no power' (Midr. Ps. on 119:10).[177] But Paul, precisely because of his devotion to Torah,

had been led to a very different judgement and indeed to a very different understanding of mankind's problem.

In Rm. 7:7 he asks, 'Shall we say that the law is sin?'[178] The question, thus
translated, is not logically constructed; but the Greek is itself very abrupt: *ho
nomos hamartia?*—not merely 'Is Torah sinful?' but 'Is the power which Torah
has over us the power of Sin?' Is this precious gift which seems to be bestowed by
God no more than a deceptive agent of Sin? No indeed! But Torah is involved
with Sin in two damaging ways that are indicated, first in Rm 7:7[b]-13 with
respect to Torah's prohibitions and second in Rm 7:14-25 with respect to positive commands. In the former passage (as is normal) verbs in the past tense are
used; in the latter passage, verbs in the present. In both the first person singular
is striking and obviously deliberate (as also the second person singular in Rm 8:2
if *se* is orginal there), but its significance is much discussed. As a result of modern
study, most exegetes no longer hold that Paul is offering illustrations mainly from
his own religious experience. To suppose that the Tenth Commandment, against
coveting, caused him peculiar temperamental difficulties is to concede too much
to the popular homily. To search for a period in his young life when he once
lived apart from the law is fruitless, and to imagine that he confesses his
anguished failure to practise the Torah is to misunderstand Pharisaism. The present use of the first person singular is not wholly similar to its use in Gal 2:19[c]-
20 (see p. 73), where I hold that Paul is indeed speaking for himself and urging
others to become as he is. In Romans 7 the first person is used, at least in considerable measure, as a rhetorical device to suggest 'some particular person' or perhaps 'someone in my position'. The situations described apply not merely to Paul
but to anyone who, zealous for the traditions of the fathers, encounters the death
and resurrection of Christ and has to rethink his understanding of Torah. Hence,
even if we admit that these accounts in the first person are not a mere transcript
of Paul's experience, we must also acknowledge that they are not divorced from
it. As we examine the accounts, it will be necessary to remember that Paul is not
really engaging in psychological analysis (as proposed by Lietzmann and by
Dodd) but is discussing the theological problem of Torah.

The Prohibitions

'If it had not been for the law', he says, 'I should not have known Sin' (Rm 7:7[b]).
He means 'the power of Sin' (see Rm 3:20, and above, n. 154). Paul is not offering
an abstract discussion of sin, morality and religious law but demonstrating his
awareness of Sin as a dormant adversary that can spring into life when it has the
opportunity of perverting a divine commandment. Sin has no independent existence; apart from God's holy law, Sin is lifeless (verse 8). When, however, God
utters a prohibition, Sin seizes upon it as a base of operations (*aphormē*, verse 8).
For an example of such a prohibition Paul turns to the Ten Commandments
(which are mostly formulated in negative fashion) and chooses the tenth according to a convention, already present in the Judaism of his day, by which the
prohibition of 'coveting'[179] sums up the Decalogue. The word *coveting* has fallen
out of use in modern English, and it poorly conveys the meaning of *epithumia*

(in a bad sense). Philo speaks of *epithumia* 'pulling and dragging someone in spite of himself towards what he has not got' (*Mos.* 2. 139). It is an obsession with acquiring or appropriating. It is 'that fount of injustice, from which flow the most iniquitous actions, public and private, small and great, dealing with things sacred or things profane, affecting bodies and souls and what are called external things. For nothing escapes. . . . Like a flame in the forest, it spreads abroad and consumes and destroys everything' (Philo *Decal.* 173). The author of 4 Maccabees[180] (perhaps written c. 40 C.E.) says 'Reason is proved to subdue the impulse not only of sexual desire, but of all sorts of covetings. For the Law says, "Thou shall not covet thy neighbour's wife, nor anything that is thy neighbour's." Verily, when the Law orders us not to covet, it should, I think confirm strongly the argument that the Reason is capable of controlling covetous desires, even as it does the passions that militate against justice' (4 Mc 2:4–6). The author of this moralising work rewrites the story of the martyrs to demonstrate how religious reason, informed by the law, governed and subdued the emotions that would have led the martyrs to violate the law. His intention, in fact, is exactly opposite to Paul's. For him the law is a protection against *epithumia;* for Paul the law's prohibition of *epithumia* is the cause of it.

What was Paul's obsession? It can scarcely have been an exorbitant, acquisitive temperament in any vulgar sense. The reported words of Acts 20:33–34 ring true, despite edifying exaggeration (Haenchen on Acts 20:33–34): 'I coveted no one's silver or gold or apparel. You yourselves know that these hands ministered to my necessities, and to those who were with me'. Paul's excessive demands were not directed against fellow Christians but, in a devoutly perverse way, against God. He desired to excel in performance of Torah, and it was the very existence of Torah that gave him his obsession. Well could he say that his Jewish compatriots had a zeal for God without proper knowledge of its consequences (Rm 10:2); for with that kind of zeal he persecuted the church and counted himself blameless according to the law (Phil 3:6). It was God's precious gift of the law that had opened the way for him to become obsessive about the acquisition of his own righteousness.[181]

But it may be objected that God's gift could never have such a damaging effect. Paul's reply to the objection in verses 9–11, is to cite the case of Adam. He gives Adam a voice, as it were, and we hear him speaking with the pronoun *I*.[182] In the beginnings of creation, Adam lived in a spellbound world. If in the midst of the garden there was a magic tree that needed to be protected from Adam (because it was the tree of life and therefore God's exclusive property) or from which Adam needed to be protected (because it was the tree of the knowledge of good and evil and therefore dangerous to mankind), it was kept inviolate by means of the *divine spell*. To put it in more familiar manner, creation and primal mankind lived within the sevenfold blessing 'and behold it was very good' (Gn 1:4, 10, 12, 18, 21, 25, 31).[183] But God uttered a prohibition: 'Of the tree of the knowledge of good and evil you shall not eat' and the spell was broken. The magical protection was withdrawn, and now the tree was protected (or Adam was protected) only by *Adam's will* to obey God. Adam himself became responsible for good or evil, blessing or curse. The commandment, in this case a prohibition, was holy and just and good (verse 12). It was intended to protect or enhance Adam's life (verse 10), in particular to allow Adam a will of his own to exercise in relation to the will of God. In the primal spellbound existence, before any prohibition was uttered, Sin was dead (inoperative). Since there was no transgression (Rm 4:15), there was no occasion to take account of sin (Rm 5:13). But once the prohibition had been uttered—the prototype of law coming in to increase the trespass (Rm 5:20)—Sin sprang into action, found a base of operations in the prohibition, deceived Adam (cf. 2 Cor 11:3, 'as the serpent deceived Eve by his cunning') and killed him. God had warned Adam that 'in the day that you eat of it you shall die' (Gn 2:17). But the serpent had said, 'You will not die. For God knows that when you eat of it your eyes will be opened, and you will be like God, knowing good and evil' (Gn 3:4–5). In fact, the serpent was right.[184] The result of Adam's transfer from a spellbound world to a rule-bound world was that at one and the same time, he was exalted and abased—exalted to exercise his will in relation to the will of God and to share the divine responsibility for good and evil, abased by becoming subject to the correct but deceptive arguments of Sin. According to Paul, sharing the responsibility of God when deceived by Sin is nothing more or less than death ('I died', 'the commandment . . . proved to be death to me', 'sin . . . killed me'). 'Death inflicted on me by sin is not just a dying expected at the end of life, but is even now the mark of my life destroyed by the deceit of sin' (Bornkamm, "Sin, Law, and Death," p. 92). The only hopeful feature in this desperate predicament, when Sin contrives death by using what is good, is that the awakened spirit can be brought to see, by means of the commandment, how exceedingly sinful Sin can become. This is not a question of commonplace disobedience or even of vicious wickedness but of the deceitful corruption of what is good. Hence what Paul has demonstrated by reference to the Adam myth can be applied to Torah as a whole, specifically to Torah given through Moses. It is all related in the past tense because this part of the argument is logically prior to what follows.

The Commands

Paul now turns from the prohibition to the positive command and changes to the present tense.[185] He sets aside disobedience and explores the devout person's attempt at obedience. Disobedience stretches back to Adam's disobedience in the remote past. What, then, are the chances of success *at this present time* for someone like the devout Pharisee who earnestly desires to obey God as disclosed in Torah? Very slender, according to Paul, for it is a question of relations between a being who is Spirit, and human beings who are flesh and sold as slaves to Sin. For the first time (except perhaps for verse 5) *sarx* ('flesh') appears in Romans in its specially Pauline sense, meaning not corporeality but sociality (see pp. 83–84). Flesh can of course indicate that weak, vulnerable and mortal existence to which mankind belongs; but when Paul speaks of 'my flesh' in verse 18 he means that social culture to which he belongs, he and the Jewish Christians with whom he is arguing. The statement 'Nothing good dwells in me, that is, in my flesh' is as startling as being crucified with Christ (Gal 2:20) or as suffering the loss of all

Jewish prerogatives (Phil 3:8). Yet these references in Rm 7:14, 18 are no more than a preparation for what will be said in Romans 8.

The course of the present argument, however, is decided by the rueful admission, 'I do not understand what I am doing' (verse 15). He fails to do what he intends and succeeds in doing what he rejects (verses 15, 19). Here is no questioning of God's commands, but a deep-seated conviction that the law is good (verse 16). Not only does he intend what is good (verses 18, 21), but his inmost self delights in God's rule. This is the characteristic posture of the *ḥasid* (the man of exemplary piety) and of the devoted Pharisee. Even more, as far as language is concerned, it is the posture of the upright man of hellenistic morality, with his 'inmost self' *(esō anthrōpos)* and 'law of the mind' *(nomos tou noos)* (verses 22–23). But the only consequence is that he fails to do the good and finds himself doing the bad. The intention of doing good is present, but performance is not (verses 18–19, 21). This is not a well-intentioned person of weak will who means to do good but, under pressure or through carelessness, does bad instead. It is the person who strenuously does what is good, only to find that in his hands it turns out unsatisfactorily, damagingly, disastrously. This is certainly not the confession of a Jewish *ḥasid* or a hellenistic upright man. Both could, and did, admit sins and failures but took it for granted that what they did well was good, and what they did badly could be amended. Nor is this a reflexion of Qumran's teaching about the good and bad spirits within mankind (and, most likely, within each human being), the spirits of truth and perversity. According to the Community Rule 4.23–25, 'Until now the spirits of truth and falsehood struggle in the hearts of men and they walk in both wisdom and folly. According to his portion of truth so does a man hate falsehood, and according to his inheritance in the realm of falsehood so he is wicked and so hates the truth. For God has established the two spirits in equal measure until the determined end, and until the Renewal, and He knows the rewards of their deeds from all eternity'. That naive moralism is certainly not Pauline, because (1) Paul's constrast is not between one God-given spirit and another but between flesh and spirit; (2) indwelling Sin is hostile to God and by no means a tolerated permanent tenant until the end; and (3) the ways of the spirit of falsehood ('greed, . . . haughtiness, falseness and deceit, cruelty and abundant evil, ill-temper, . . . abominable deeds [committed] in a spirit of lust, . . . a blaspheming tongue' Community Rule 4.9–11) are not suggested, by the slightest hint, as belonging to the failure of someone delighting inwardly in the law of God.[186] Nor, again, is there much relation between Rm 7:14–25 and the rabbinic teaching about the good and bad impulses.[187] A well-known rabbinic passage attributes the following assertion to God: 'I created within you the evil *yetzer* [impulse], but I created the Law as a medicine. As long as you occupy yourself with the Law, the *yetzer* will not rule over you. But if you do not occupy yourself with the Law, then you will be delivered into the power of the *yetzer*, and all its activity will be against you'. But Paul is describing a situation in which devotion to the law is no safeguard and where repentance cannot remove unintended bad consequences. He portrays someone confronted by two powers: one of them he respects and consciously serves; the other he despises and rejects, yet he finds himself doing what it demands. He is quite explicit and emphatic: 'It is no longer I that do it, but Sin which dwells within me' (verses 17, 20). When he considers what he is able to do [his abilities, *melé,* usually translated 'members'] he detects that another rule [*nomos,* usually translated 'law'] is in operation that fights against the rule accepted by his mind and makes him captive to the rule of Sin that pervades his abilities (verse 23). The statement sounds a little extreme, sharing the same exaggeration as 'Nothing good dwells within me' (verse 18). Surely some good things are both intended and performed; but Paul is reflecting the disenchantment with religious goodness and even success that overtakes even the most sanguine temperaments when confidence is shaken by unexpeced failure and good initiatives produce disastrous consequences. According to Phil 3:6, like any strict Pharisee, Paul had no difficulty in fulfilling Torah; but his achievement was (as he was led to see) disastrous. His zeal led him to persecute God's *ekklēsia,* and by his devotion to God's cause he became an enemy of God (cf. Rm 5:10). The problem was not the wickedness of his sins but the destructiveness of his devotion. Paul writes out this testimony (as in part it must be regarded) not for theoretical reasons but to warn his fellow Jews about the devotion they are proposing to, or demanding from, Gentile Christians. He therefore speaks as a Christian looking back on his Jewish past; but he cannot have been unaware that even if Christians are no longer captive to the power of Sin, that power is not broken. Therefore Christians are free to fight Sin (Nygren p. 242). They must remember the slavery from which they were set free, for Sin will still ruin (if it can) their best religious initiatives—as, for example, Paul's relations with the Corinthian community. It is not surprising that this passage ends with a dramatic cry of anguish from the inmost experience of the devout, upright person: 'Who will rescue me from the body of this death',[188] that is, from the bodily existence threatened by such death from the power of Sin. It is answered equally dramatically by 'Thanks be to God through Jesus Christ our Lord!'[189] This (perhaps pre-Pauline, liturgical) formula (cf. Rm 1:8) for the first time brings Christ into the discussion and by itself is probably sufficient to evoke the benefits of salvation, namely, peace with God, exultant confidence in him, and the dominance of grace (Rm 5:1, 11, 21). If so, it is noteworthy that Paul's remedy for the law's unsatisfactory performance is the lordship of Christ, which in turn points rather to his resurrection than to his death.

What God Has Done

The argument, however, sweeps on; and its momentum is helped by the rhetorical force of Rm 7:25[b] and 8:1:

> So, then, I of myself
> serve the law of God with my mind,
> but with my flesh I serve the law of Sin.

> Thus as things now are
> for those in Christ Jesus
> there is no penalty [of that kind].

The perplexed and despairing figure of 7:15–25ᵃ ('I of myself') is opposed to 'those in Christ Jesus' for whom there is no such penalty (see AGB, MM on *katakrima*) as desiring to serve God while being enslaved to Sin. That assurance can be given because

> the rule of the life-giving Spirit has set thee free
> from the rule of Sin and of death (Rm 8:2),[190]

though not, it would seem, from the disadvantages of belonging to the flesh. In my treatment of Galatians (pp. 83–84) I have already argued that, when flesh appears as a power that seems to defy God and imprison mankind, it refers to the social structure of human existence. The same meaning is present in this passage of Romans. The contrast between flesh and spirit is not between the physical and the mental aspects of human life (that is clearly excluded by the popular, untechnical use of pneuma for the human spirit in verse 16); nor is it between conflicting forms of an abstract thing called human nature. The contrast is between human social existence—weak, fallible and corruptible—and a corporate experience that in some measure displays the power, assurance and permanence of God.[191] Even the holy law of God, weakened by having to operate in the flesh, could not achieve its life-giving intention, nor destroy the power of Sin over human beings (cf. the contrast in Rm 2:29, 7:6 between Spirit and written code). But, Paul continues, what the law could not do

> God [has now done, by his gift of the Spirit]
> having sent his own Son
> in the likeness of sinful flesh
> and [to act] concerning Sin.
> He condemned Sin in the flesh
>
> that what the law claims to do
> may be fulfilled in us
> who live our lives, not according to the flesh,
> but according to the Spirit. (Rm 8:3–4)

This elaborate but imperfect formulation requires careful study.

'What was impossible for the law' in verse 3ᵃ requires something like 'God has now made possible', which can be inferred from the general movement of the sentence. *How* God has remedied the failure of the law can be deduced from verse 2 and the subsequent references to the Spirit. In Gal 3:1–2 it was taken for granted that the preaching of Christ crucified prompted the gift of the Spirit (see p. 77), which has brought Christ from the dead (Rm 8:11). Hence the supplementary words in square brackets in verse 3ᶜ.

The sending of the Son, which does not explicitly refer to his death, is rare in Paul (elsewhere only in Gal 4:4, where the verb is *exapostellō*, not *pempō* as here). Not even the hymn of Christ's descent and ascent in Phil 2:5–11 uses sending language. The two passages in Galatians and Romans indicate that the Son of God was sent, at the appropriate final time, as God's agent and representative. He was equipped with plenary powers to act in the standard conditions of human existence (see p. 81).[192]

The purpose of the sending was to deal with Sin. The Revised Standard Version translates 'for sin', with the variant 'and as a sin offering' in the margin. The New English Bible favours 'as a sacrifice for sin', and puts 'and to deal with sin' in the margin. The Greek is *peri hamartias*. It is sometimes misleadingly stated that these words are common in the Septuagint for a sin offering. It is indeed true that Leviticus frequently uses *peri hamartias* for animals required for the sin offering, and for the actions of atoning or sacrificing; but the sacrificial idea is always provided by the cultic context. The same is true of the solitary references in Ps 39:7 and Is 53:10; but the cultic reference is entirely absent from this passage of Romans and is very uncommon in Paul. It is scarcely to be expected that Sin could be prised from its tenacious hold on the flesh by a stock cultic sacrifice intended to please God, not to dislodge Sin. If something more precise is required than 'to deal with Sin',[193] I suggest 'with authority concerning Sin' by analogy with 1 Cor 7:37, 'He has authority concerning his own intention'.

God dealt with Sin by 'condemning Sin in the flesh', interpreting *flesh* as human social structures. *Condemn* does not mean (as it commonly does in today's English) to express disapproval or to censure but to restrain or to punish. Sin is condemned to be dispossessed, to be driven out from its apparently impregnable position in human society. That is the significance of *sinful flesh:* not that human social structures are inherently wicked but that they are so weak, fallible and corruptible that Sin can easily take them over and find a secure base of operations there. The effect of God's action in Christ is to make Sin's position untenable if people will now use their newly given freedom to fight Sin. The phrase *sinful flesh* was already known at Qumran: 'I belong to the Adam of wickedness and to the assembly of wicked flesh' (1 QS 11.9; Leaney, p. 255) where *flesh* is already a social concept. Thus the meaning of verse 3[d] is that God sent as his agent someone required to live in the assembly of wicked flesh, hence to be subject to death and to the perversion of his best intentions. He was entrapped by Sin, exactly like all human beings. That does not imply that he sinned by choice or by carelessness or by a defect of temperament (the sinlessness of Christ can still be maintained, though in this context it scarcely comes into consideration), but it recognizes the belief that Sin exists by perverting what is good and strengthens its hold by destroying what is best. Verse 3[d] is constantly discussed as if Paul were laying down Christological axioms, and it is debated whether Christ took upon himself fallen or unfallen human nature.[194] This readily wanders into a world of technical fantasy, for it scarcely realises that human beings live in social structures that decide what their nature shall be. 'In the *likeness* of sinful flesh' has caused much perplexity. *Homoiōma* ('likeness') may indicate virtual identity (as in Rm 5:14, 'those who did not sin *in the same way* as Adam sinned') or what corresponds (as in Rm 6:5, 'united in a death *corresponding to* his death) or similarity with a difference (as in Rm 1:23, where a pagan cultic object is 'the likeness of an image'). *Homoiōma* is one of a group of words with a similar range of meaning: likeness, form, appearance, structure (see n. 174). In the one remaining Pauline use in Phil 2:7 it sits between *morphē* and *schēma*, 'form' and 'structure'. Thus God sent his own Son to live, like all of us, in the social structures that are the operating center of Sin, with authority to act against Sin and to clear the way for Sin to be brought into the open and fought.

The Gift of the Spirit

How Sin can be fought is indicated in what follows. Christians cannot, of course, cease living in the flesh even though they are not bound to Sin; but there are consequent disadvantages. Indeed, the human scene is most disparagingly treated. The whole direction of the flesh is death and enmity towards God; it is not ruled by God, indeed cannot be. Those who walk or exist according to the flesh, those who are in the power of the flesh cannot please God. If this sounds unduly pessimistic, it is saying no more than what is said in Rm 1:18–2:24, where Jewish and hellenistic social structures are exposed in their offensiveness to God. But because Paul has reduced his former denunciation of particular social evils to a radical criticism of the flesh, he can now present the Spirit as the positive counterpart. Indeed, Paul seems carried away when he says 'You are not in the flesh' (cf. Rm 7:5, 'while we were living in the flesh'), you are in the Spirit if in fact the Spirit of God dwells in you' (verse 9). But it is important not to misread him. Because theologians seldom resist the temptation to discuss questions of 'nature', it is taken for granted that his use of *in*-language *(in the flesh, in the Spirit)* refers to a change of nature from flesh to Spirit, either from material to immaterial or from communal to sectarian. But Paul talks much more about human situations than about human nature. Flesh and Spirit are powers, and to be 'in' one or the other is to be in the power (or under the control) of flesh or Spirit. Paul is therefore saying, Although you cannot withdraw from the social world of your contemporaries (cf. 1 Cor 5:10), you need not be dominated by the destructive pressures it brings to bear; for you can be moved by the power of the life-giving Spirit if the Spirit of God dwells in you.

Paul here makes use of not uncommon symbiotic metaphors—those in which one organism achieves its aims by living inside another to the benefit or the disadvantage of the other organism. He adopts the simple Hebraic view that human beings have a spirit within them that is their motive force (1 Cor 2:11, 'the spirit of the man which is in him'; 'my spirit', 'our spirit' in Rm 1:9, 8:16). It may be instructed by gnosis, illuminated by the knowledge of Christ, or given confidence by the truth of Christ (1 Cor 8:7; 2 Cor 4:6, 11:10). But it may be envisaged as cooperating with God: 'Work out your own salvation . . . for God is at work in you, both to will and to work for his good pleasure' (Phil 2:13; so also Rm 8:28 if it could be translated, 'In all things he cooperates with them for good'). The inward activity of God is not compatible with some purposes of the human spirit, as in 1 Cor 3:16 and 6:19, which use the variant metaphor of the sacred enclosure: 'You are God's temple . . . and God's Spirit dwells in you' and 'Your body is a temple of the Holy Spirit which you have from God'. In Rm 8:9 the indwelling Spirit of God stands in obvious contrast to indwelling Sin and its destructive activity (Rm 7:5, 8, 17, 18 (no good impulse), 20, 23). This indwelling Spirit, in cooperation with our own spirit, makes Christians what they are, namely chil-

dren of God (verse 16); but precisely because it is a life-giving energy that stimulates new activity in the Christian set free from Sin's captivity, two qualifications are required. First, there is what Käsemann (p. 224) calls 'the eschatological
reservation': we possess the Spirit only as the first fruits (verse 23; cf. *guarantee*,
2 Cor 1:22); for we have entered the new life genuinely but by no means completely. Second, we possess the Spirit of him who raised Christ Jesus from the
dead. It is not the Spirit of triumphant mankind or even of transformed mankind
but 'the Spirit of Christ', that is, the Spirit that led him to his death in confidence
that God would do what he would do. Consequently Paul could go a step further
and say, 'If Christ is in you' (verse 10), in effect drawing upon his own conviction
in Gal 2:20, that 'it is no longer I who live, but Christ who lives in me'.[195] The
prior condition in Galatians is that Paul had been crucified with Christ. Something similar is said here: 'If Christ is in you, the body indeed is dead because of
Sin but the Spirit is life because of righteousness'.[196] If the reader's mind is following the outlines of Paul's thought, there is no danger of supposing that Paul cannot distinguish between the indwelling Spirit and the indwelling Christ. In examining Gal 2:20 (p. 73), I argued that Paul regards himself not as a new ego raised
from the death of his former ego but as a person indwelt by (the risen) Christ,
awaiting a new ego that will be fully formed by the powers of the new age. So
here the ego (represented by *body* rather as the word operates in the English
compound *anybody*, SOED s.v. *body* III) is dead because of Sin's destructive
power but has been given life by the Spirit of him who raised Christ from the
dead, and has been replaced by the indwelling Christ (the first stage towards being
'conformed to the image of his Son', Rm 8:29). This, of course, is not new: it
takes up references to 'the body of sin' (Rm 6:6, 12; 7:4, 24), release from which
is cautiously associated with baptism (see p. 101). It looks forward to 'the
redemption of our bodies' when 'the sufferings of this present time' are replaced
by 'the glory that is to be revealed to us' (Rm 8:11, 13, 18, 23).[197] Present sonship, made known to our spirit[198] by the Spirit of Christ, holds before us the
prospect of becoming 'fellow heirs with Christ, provided we suffer with him in
order that we may be glorified with him' (verse 17). With that sobering thought
of participating in Christ's sufferings, the enthusiastic acceptance of Spirit possession is put in proper perspective.

In the following section, Rm 8:18–27, suffering with Christ is merged with
the even larger theme of the suffering of the creation that affects even those who
have the first fruits of the Spirit. But everything is according to the will of God
(Rm 8:28–30), and this phase of the argument ends with a confident peroration
in Rm 8:31–39, which sounds like the statement of pleas on behalf of Christians
in their conflict with hostile powers:

> What then shall we say about these matters [about Rm 5–8, which
>   begins and ends with God's love]?
>
> If God is for us [recalling the whole theme of 'justification'] who is
>   against us?
>
> He who spared not his own Son [perh. glancing at Abraham's
>   willingness to sacrifice Isaac, Gn 22:16]

but gave him up for us all (*paredōken,* as in Rm 4:25; 1 Cor 11:23]
will he not also give us all things with him [cf Rm 5:9–10]?

Who will bring a charge against God's chosen?
It is God who justifies [accepts us].

Who condemns? [cf. 8:1, 3]
It is Christ Jesus who died,
even more who was raised,
who indeed is at the right hand of God
who also intercedes for us [cf. the Spirit's intercession, Rm 8:26–27].

It is clear that Christians are not on trial but are pleading the case for redress against their tormentors. The three distress cries of verses 18–27 become more substantial in verses 35–36, where, in the midst of triumphal rhetoric, Paul instances the real dangers for a Christian (taken from his apostolic experiences). It is possible to remind God of the cry of his chosen—'For thy sake we are being killed all the day long; we are regarded as sheep to be slaughtered' (verse 36, Ps 44:22)—with, of course, the same implication: 'Rise up and come to our help! Deliver us for the sake of thy steadfast love'.[199] Because God loves his people, they can confidently expect the overwhelming success of their appeal (verse 37). And so Paul finally evokes, in resounding words, the disorder of creation with its competing rival powers, which make God's lordship seem impossibly remote; and he concludes that nothing can separate us from God's love in Christ Jesus our Lord.

The conclusion of the matter is therefore the lordship of Christ, which stems from his resurrection. The old formula 'He died—he rose' is prolonged in verse 34[d,e] ("who indeed . . .") to Christ's session at the right hand of God (not elsewhere mentioned by Paul, though no doubt presumed in Phil 2:10–11) and to his intercession for the suppliants (cf. Heb 7:25). Christ behaves like the *paraclete*[200] of 1 Jn 2:1 except that he uses his influence with God in favour not of forgiveness but of deliverance. In verses 26–27 the Spirit also intercedes with God. If Paul's imagery belonged to a consistent pattern, presumably the indwelling Spirit (which was, of course, the Spirit of God) prompted requests the community itself was incompetent to formulate; and the indwelling Christ (who was of course the Son of God) represented the community and its needs before the Father. The only claim for help Christians can present is God's love for them, and the evidence of that love is that God did not spare his own Son. This imagery of pleading for God's help is an oriental convention that in no sense implies that God has to be persuaded, propitiated or satisfied. With ceremonial courtesy it pleads the people's need and God's known generosity but claims no right to his favour. It allows God to be God and to be God in his own way.

God's Dealings with Israel

Thus ends Paul's response to the first main criticism of his gospel: that it is indifferent to sin because it relies on grace, not on obedience to Torah. He has been dealing with the human predicament. But now (in Rm 9–11) he deals with what I may call the divine predicament and turns to the second main criticism: that Paul is hostile to God's Jewish people and destroys God's credibility by allowing him to abandon the Jews. In Rm 9:1–5 he retracts his confident statement that nothing can separate him from the love of God in Christ Jesus. One thing could: he could almost pray to be accursed and cut off from Christ for the sake of his brothers, his kinsmen 'according to the flesh'. Paul has turned from the problem of Sin to the problem of God.

Paul conceives the problem thus: 'Can the word of God be thought to have failed?' By *word of God* he may mean God's promises to Israel or more probably (as six times elsewhere in Paul) the proclamation of the gospel. Has God failed to carry out his promises to Israel, or does the proclamation of Christ's death and resurrection fail the necessary condition of relating God to his people? There are two main discussions, both inconclusive.

The first, in which Christ does not appear, considers how God carries out his purpose (Rm 9:6–29). In brief, Paul argues that God keeps his saving power within his own control and operates by a process of choice (or election). Hence he is not under obligation to *all* Israel. He even goes outside Israel for agents of his purpose (which certainly has dark aspects as well as light) and ventures, in quoting passages from Hosea, to describe what God has done for Gentiles in language that belongs to his promises for Israel (Rm 9:25–26). If God does so much for hostile Gentiles, how much more will he do for disobedient Israel!— though the response of Israel is unpromising.

The second discussion is a restatement of earlier arguments about what God requires. Is it faith or observance of Torah? (Rm 9:30–10:21). How can Jews be saved if they fail to realise that 'Christ is the end of the law, that everyone who has faith may be justified'? (Rom 10:4). The continuing debate about that ambiguous statement is not directly relevant to an enquiry into the death of Christ, but it has some bearing on Paul's midrashic development of a passage from Deuteronomy that is used to justify a Christian credal formula. So it is suitable to give a little attention to the meaning of *telos* ('end'). The word has a wide range of meanings, but only two are relevant. Is Christ the end of the law as being its fulfilment, expressing the true meaning of Torah, or as the goal to which Torah directs God's people? Or does Christ put an end to Torah, which in any case he would do if he were its goal and fulfilment?[201] Since in Rm 7:6 we are 'discharged from the law, dead to that which held us captive', since the law belongs to the old epoch and was in any case a supplementary addition to the promise, it looks as if Christ makes an end of the law as far as acceptance with God is concerned. Not, of course, that Paul was dispensing with the scriptures *(hai graphai):* he is clear that 'whatever was written in former days was written for our instruction [*didaskalia*], that by steadfastness and encouragement of the scriptures we might have hope' (Rm 15:4). He has abandoned the scriptures as law *(nomos)* and has restored to them their original purpose (as he sees it) as instruction (*nouthesia*, 1 Cor 10:11). Scripture is no longer the source from which a limited group of people (namely, Israel) takes the symbolic rituals that define and preserve its identity (cf. the conflict about circumcision in Galatians and the levelling of circumcision and uncircumcision in Rm 2:25–29, 4:9–12). Neither the obligation of these rituals nor the performance of them gains or guarantees or preserves acceptability with God—only faith in the God whose saving activity is displayed in the death and resurrection of Christ.

We now come, in Rm 10:5–8, to a use of passages from Deuteronomy for which Paul has often been reproved. He seems to be making fanciful or disingenuous play with scriptural demands for the practice of Torah so that they are made to commend faith. But that reading misunderstands his intention. He begins with an explicit statement by Moses that 'the man who practises the righteousness which is based on law shall live by it', alluding to Lv 18:5 (repeated at Ez 20:11, 13, 21; Rm 10:5; Gal 3:12). With that quotation we may compare the conclusion of Josephus' words in praise of the laws of the Jewish people, when he says: 'For those . . . who live in accordance with our laws the prize is not silver or gold. . . . Each individual, relying on the witness of his own conscience and the lawgiver's prophecy, confirmed by the sure testimony of God, is firmly persuaded that to those who observe the laws and, if they must needs die for them, willingly meet death, God has granted a renewed existence and in the revolution of the ages the gift of a better life' (*c.Ap.* 2. 218). That combines a strongly Jewish self-consciousness with willingness to suffer in defence of the laws and a Pharisaic belief in the resurrection life. Paul has no quarrel with suffering and resurrection. Nor is he indifferent to Jewish self-consciousness, but he cannot agree that the preservation of God's people depends positively on performing Torah and negatively on excluding Gentiles. Indeed, he detects another strain in Torah, which is first indicated by the cue words for the well-known warning conveyed in the sermon in Deuteronomy 8:1–9:6. He writes only the introductory phrase 'Do not say in your heart' (Rm 10:6), but the continuation was presumably present to him and his Jewish readers: 'Beware lest you say in your heart, "My power and the might of my hand have gained me this wealth". You shall remember the Lord your God, for it is he who gave you power to get wealth; that he may confirm his covenant which he swore to your fathers.' (Dt 8:17–18). 'Do not say in your heart . . . , "it is because of my righteousness that the Lord has brought me to possess this land" (Dt 9:4)'. Thus, Moses himself agrees that whatever the advantages and merits of practising Torah, the outcome still depends on the unrestricted will of God.

So much, as it were, for a sighting shot. Paul now makes use of a common Jewish image—the heights of heaven and the depths of the abyss (e.g., Prv 30:4; Baruch 3:28–29; 2[4] Esd 4:8; cf. Philo, *Quod omn. prob. liber sit* 68, *Virt.* 183), which became a symbol for the pursuit of wisdom or virtue. He finds it in a very odd section towards the end of the Deuteronomic Torah, Dt 30:11–14, which implies that the community requires no persons of magical (or other) skills to obtain from the above-world or from the below-world the controlling secrets; but that the commandments are within the oral tradition of the community and the community's will to maintain them. There is no secret, exclusive tradition; and this section comes immediately before the dramatic choice between life and death, good and evil, blessing and curse. For Paul, the true significance of this assurance that the word is near—is revealed by Christ. The word is 'the word of faith which we preach' (verse 8), that is, the death and resurrection of Christ. Just as Moses has declared that God has done all that is necessary to make his word available, so the Christian preacher can assure his listeners that their efforts are not required to bring the Redeemer down from heaven or to rescue him from his descent into the underworld. Thus, the principle of faith in the saving activity of God is established from the heart of Torah. If knowledge and possession of the constitutive Torah by which Israel became God's people was independent of their efforts, how much more was knowledge and possession of the saving goodness of God by which Christ brought this age to its end and extended God's dominion to the Gentiles independent of their efforts (verse 12). Thus, it is notable that the descent and ascent scheme (which appears again in Phil 2:6–11, Eph 4:9–10, Jn 3:13) makes no theological use of the death of Christ.[202] His death is mentioned simply as the precondition of his resurrection. And when (in verses 9–10) Paul brings in a confessional formula, it is wholly related to the lordship and resurrection of Christ:

> If you confess with your lips
> that Jesus is Lord
> and believe in your heart
> that God raised him from the dead,
> you will be saved.

The conclusion must follow that insofar as we are concerned with the problem of God, that is, with asking questions about what God has done or has not done and why God has acted in this way or that, Paul brings Christ into his answer by referring to the significance of his resurrection, not to the significance of his death (which dominated his handling of the problem of Sin). This difference perhaps appears because the death of Christ, however extensive its consequences, is a completed event of the past; whereas his resurrection is only the beginning of what is yet to be done. However that may be, Paul concludes this second discussion by asserting that faith comes from hearing the word of Christ and by admitting that most Jews are disinclined to listen (verses 16–21). But Christ does not appear in the third, concluding section, where Paul assures his Jewish readers that God has not rejected his people, warns Gentile Christians against despising the Jews, and discloses his *mystērion*—that when the full number of the Gentiles has come in, the Jews will relent and all Israel will be saved (Rm 11:1–36).

Community Needs

Christ, however, reappears when Paul works out community procedures in Rm 12–15. In particular, references to his death and resurrection play a significant part when the problems of the 'strong' and the 'weak' are under discussion. The essential problem is a lack of confidence among some members of the community that the right cultic action is being taken (about matters of eating and drinking and festival days), if not as regards the whole group, at least as regards themselves. In verses 14:7–12, Paul introduces the fundamental Christian awareness of God—which is a matter of life and death. It cannot be self-regarding but must be directed to, and supported by, the Lord, who is defined as 'Christ who died and lived again, that he might be Lord both of the dead and of the living'. Hence it is no business of theirs, if they are lacking in confidence, to judge others, or, if they are brimming with confidence, to despise others. For all must stand before God's judgement seat with Christ as Lord (supported by Is 45:23), and each must give account of himself to God. Once more the resurrection lordship of Christ is centrally important. Nevertheless, if you injure your brother by the cultic decision you have taken in response to Christ's lordship, you are no longer walking in love: 'Do not let what you eat cause the ruin of one for whom Christ died (Rm 14:15). This standard formula (see pp. 14–15), meaning that Christ died to make possible their membership in the Christian community, permits Paul to provide a boundary for unrestricted cultic freedom from the very same source (namely, the Christ who died and rose) that gives the freedom in the first place. Finally, Paul's ruling that 'each of us is to please his neighbour for his good' is supported by the constitutive example of Christ who did not act to please himself (Rm 15:3). That example is described with the help of Ps 69:9 where reproaches against God fall on the godly sufferer. In effect, Paul is saying, 'We who are full of confidence may be right, and those who lack confidence may (unknowingly) be reproaching God. But Christ could suffer in that way: so can we'.

Summary

When we survey the result of working through Romans, one conclusion is strikingly obvious: so many of the references to the death and resurrection of Christ are derived from formulas. Either they bear the marks of pre-Pauline tradition (sometimes adapted by Paul), or Paul has expressed what he needs to say in formulaic fashion.

There are formulas that testify exclusively to Christ's resurrection, and therefore to his lordship (sometimes expressed as his Sonship):

> Christ died and lived again,
> that he might be Lord
> both of the dead and of the living. (Rm 14:9)

The epistle begins with a pre-Pauline formula confessing him who is 'designated Son of God in power according to the Spirit of holiness, Jesus Christ our Lord'. No doubt, thereby he is Lord of death, but the formula is in fact used to justify Paul's apostolic commission (Rm 1:3–5). The primitive confession 'Jesus is Lord' appears to stand behind the triumphant eucharistic cry in Rm 7:25ᵃ, 'Thanks be to God through Jesus Christ our Lord', with the implication that resurrection has replaced the unsatisfactory lordship of Torah by setting up the new lordship of Christ. Similarly, when Torah, made ineffective by the flesh, became incapable of rescuing mankind, God destroyed the opposing lordship by the Spirit that raised Christ from the dead: 'If the Spirit of him who raised Christ from the dead dwells in you, he who raised Christ from the dead will give life to your mortal bodies through his Spirit which dwells in you' (Rm 8:3–4, 11). Finally, when Paul uses the imagery of ascent and descent in Rm 10:6–13 it is directed entirely to the resurrection:

> If you confess with your lips
> that Jesus is Lord
> and believe in your heart
> that God raised him from the dead,
> you will be saved. (Rm 10:9)

He is the one Lord of Jew and Gentiles, and everyone who calls on the name of the Lord will be saved. Thus, the risen lordship is directly responsible for the overcoming of death, the replacement of Torah by Spirit, the unity of Jew and Gentile in the new humanity, and the apostolic spread of the gospel.

Rather more characteristic are formulas that speak of both death and resurrection, with stress sometimes here, sometimes there. Rm 4:24–25 contains the double formulation and stresses chiefly the resurrection:

> We 'believe on him that raised from the dead Jesus our Lord,
> who was put to death because of our trespasses
> and raised for our justification'.

The function of that formula, as Paul used it, was to conclude an exposition of faith, which is confidence in God who raises the dead. In Rm 6:3–4, baptismal teaching is used to describe the Christian's participation in Christ before his new responsibilities are pressed home. Hence, being crucified with Christ is strongly asserted; but even so the thrust of the argument is expressed by 'If we have died with Christ, we believe that we shall also live with him' (verse 8). We are to consider ourselves dead to sin and alive to God in Christ Jesus (verse 11). Hence, the dying is a preparation for the subsequent living—but a very important preparation, for it is dying to Sin, that is, ceasing to live in Sin's power. That liberation is effected by being crucified with Christ and buried with him; and Paul's formulation seems to be Paul's transference to the believer of the primitive formula 'He died, he was buried' and (in anticipation) 'he arose, he appeared'. It is to be noted that participation in Christ's death changes the condition of the believer rather than the attitude of God. In my discussion of Rm 5:12–21, where Adam and Christ represent mankind under two rival lordships, I have already remarked that the death of Christ—his *dikaiōma* corresponding to Adam's *paraptōma*— is displayed as a constitutive, not an atoning, act: 'As by one man's disobedience many were made sinners [agents of Sin], so by one man's obedience many will be made righteous [agents of God]' (Rm 5:19). Finally, in the elaborate plea on behalf of Christians against the powers that cause them such distress (Rm 8:31– 34), two features are noteworthy: (1) the death of Christ appears indeed in the pleading—but even more his resurrection, his presence at the right hand of God, and therefore his excellent position for interceding on their behalf; and (2) God's handing over of his Son to death on their behalf implies that God is on their side and with Christ will give them all things—the Christian plea for help confidently relies on God's proven willingness and ability to rescue in the greatest extremity.

Last of all, there are formulas that speak only of Christ's death, beginning with the basic formula that lies behind 'that person for whom Christ died' in Rm 14:15 (see pp. 14–15). In 5:6, 8 Christ died for the weak, the ungodly, the sinful (i.e., people in the power of Sin). We may compare Rm 15:3 where Christ did not please himself but endured reproaches intended for others. It must be noted (1) that this action is attributed entirely to God's love and (2) that the consequence of Christ's death is to make us acceptable to God *(dikaiōthentes)* and to reconcile us to him and that being saved from the wrath (i.e., the vileness we have chosen to be and to do) depends on living the life of Christ. So even here the resurrection is a necessary complement of the death.

'Through Faith in Christ Jesus for All Who Believe'. We come finally to the formula in Rm 3:24–25[a] (probably pre-Pauline but modified by Pauline additions), which plays an important part in the carefully structured section 3:21–5:21. Paul's presentation of God's saving goodness *(dikaiosynē)* contains three elements: *through faith, in Christ Jesus,* and *for all who believe.* The formula provides the first definition of *in Christ Jesus* (which is then expanded in 5:1–11) and is explicit in some respects though ambiguous in others.

It is entirely clear in asserting God's initiative and activity: God put forward or intended Christ to do what he did. This is strengthened by the characteristic Pauline emphasis in 'by his grace as a gift'.

The meaning is less certain when Christ's action is described as a *'hilastērion* by his blood'. Since Paul nowhere else uses *hilastērion* or any related word, we can adopt any of its common meanings suited to the context. I have suggested 'inducement' as the most probable meaning, when qualified by Paul's addition of *dia pisteōs* ('to be received by faith', NEB). When God regards someone who relies not on his personal merits or offered inducements but entirely on the blood of Christ (a phrase from the Lord's Supper), that is, on his self-giving for others, God is induced to do for that person what he alone can properly do for him. When a person shows that kind of faith, he is in a savable position.

What is offered in Christ Jesus is a liberation *(apolytrōsis).* The same word is used at Rm 8:23 of the liberation of our bodies, a variation of the motif of the future revelation of glorious freedom that takes up the thought of creation's waiting for liberation from decay (Käsemann). In chapter 3 the kind of freedom intended is not explained, but in later chapters (using the *eleutheroō* words) Paul begins with liberation from the power of Sin and ends with 'the glorious liberty of the children of God' (Rm 8:21).

Those who exercise faith receive this liberation and are acceptable to God. They are *dikaioumenoi* (verse 24), that is, they are regarded as *dikaios* by God, who indeed treats as *dikaios* anyone who bases his appeal to God on faith in Jesus (verse 26). The meaning of *dikaios* and its associates is complex and ambiguous to all those who are convinced that Paul is struggling with a legal system, but they misread his intention. If we ourselves are set free from the authority of the law, it is unlikely that God would still be entrapped by it. It is not to be supposed that God needed a good lawyer to get him out of trouble and so engaged Paul. The truth is that Paul does not talk about legal court proceedings but about appeals to God from despairing human beings, and he tells them how access to him can be obtained. That is put beyond doubt by Rm 5:1–2.

No doubt Paul attached his teaching to existing formulas in order to establish, in line with accepted Hellenistic Jewish views, the lordship of Christ and the signficance of his death. When Christ becomes Lord, other powers—law, Sin, and Death—are forced to give up their victims. By Christ's death therefore, which is an act of God, those victims are freed from the consequences of the wrath (which handed them over to the powers they had chosen to serve), and they become acceptable to God and learn to participate in Christ's death, which thus becomes the constitutive act of a new mankind that will share in Christ's resurrection.

PHILIPPIANS

It is not easy to give confident answers to the questions, Was the Epistle written in one piece or was it composed out of several letters? and From which period of Paul's ministry did the Epistle, or its component parts, emerge? I take the view that it was probably written as a single letter when Paul was imprisoned in Rome.[203] Even if I am wrong the consequences will not be damaging. Because of the apparently fragmented character of the epistle, uncertainty about the answers does not greatly hinder exegesis, and confidence about the answers does not greatly advance it. Even if it were permissible to shuffle the various sections of the epistle, it would make it no easier to identify the people denounced in Phil 3:2–4:1.[204]

It is, however, necessary to decide what the epistle is about. Every reader knows that *rejoicing* is a constant refrain. The noun and the verb appear at Phil 1:4, 18, 25; 2:2, 17, 18, 28, 29; 3:1; 4:1, 4, 10—eight times more frequently than in all the rest of Paul's letters. 'Rejoice . . . again I say rejoice' (Phil 4:4) is all too true. Why should the apostle be so insistent unless he himself and the Philippian community were thoroughly depressed? It is notable that no word of joy is spoken between Phil 3:1 and 4:1. In 3:2–21, when attacking dangerous opponents of the gospel, Paul is both angry and miserable. Thus, there are two sets of distressing circumstances: one must be met by resolute opposition, the other must be accepted and turned into joy. In both, appeal is made to the death of Christ.

The circumstances in which Paul can rejoice and urge rejoicing concern himself and the Philippian community. Paul's imprisonment (Phil 1:7, 13–14, 17) is recognized by the *praetorium* and others to be a consequence of his loyalty to Christ; but it has divided the (Roman) Christian community. They have not lost confidence and are boldly speaking the word of God. But some preachers show good will and love to Paul, and others are moved by envy and rivalry, 'thinking to afflict me in my imprisonment'; that is, they think him a failed apostle, whose predicament discredits the gospel. Such a judgement might indeed prove depressing; but Paul rejoices that for whatever motives, Christ is being proclaimed (Phil

1:18). Even if—to consider the most depressing possibility—he is put to death, that will simply be a libation added to the sacrificial offering of the Philippians' faith; so he is glad and rejoices with them (Phil 2:17). This rejoicing is fed by the relation between them and the apostle (Phil 1:4; 4:1, 10), and he hopes that his cause will succeed and that he will rejoin them and add joy to their loyalty (Phil 1:25). For the Philippian community was also upset and divided, not only by their concern for Paul but also by their embarrassment about Epaphroditus. He had been sent, with money, to take charge of Paul's affairs; but he had broken down with what sounds very like a depressive illness. So Paul has to send him back, with generous excuses and instructions to welcome him gladly (Phil 2:25–29). To which he adds a very involved paragraph about his gratitude for the Philippians' initiative (Phil 4:10–19), for he must say both that it was a wonderful gift and most timely and also that its failure was of no consequence. Thus, all the rejoicings are hemmed in by depression and failure.

There is yet more: 'Complete my joy by being of the same mind' (Phil 2:2). The community is threatened by opposition and is likely to suffer as Paul himself is suffering—presumably from the civil authorities (Phil 1:27–30). Paul either foresees the possibility that they will be driven into factions or has information that they are already divided (Phil 2:3–4, 14–16), especially Euodia and Syntyche (Phil 4:2). It is another notable feature of the epistle that the apostle often goes out of his way to address 'you all' or the equivalent (Phil 1:1, 4, 7, 8, 25; 2:17, 26; 4:21). This is a fairly common phrase in the beginnings and endings of Pauline letters; but Philippians uses it five times as frequently as the rest. When it is used in the body of Romans, 1 and 2 Corinthians, Galatians, and Thessalonians, its function is to overcome potentially dangerous divisions. So too in Philippians. The epistle was written out of depression and embarrassment to a deeply loved but now disturbed community.

It was surely written with Timothy listening to Paul's dictation. Paul hopes that he can shortly send Timothy to get news of the community, to be Paul's direct representative ('as a son with a father') and therefore to take action as Paul himself would have acted had he been free to travel (Phil 2:19–24). Timothy, I suppose, wishes the letter to contain some indication of what he is authorized to say about the religious dangers to the community. Not all distressing circumstances can be dealt with by resignation and rejoicing; some must be sharply attacked. Hence Phil 3:2–21 is added in a different tone and with a variety of arguments that can be developed or set aside when Timothy sees for himself what the situation really is. These are sighting shots that may perhaps bring the enemy to action. At the moment the Philippians are only asked to be very careful, to be true to what they have already reached and to stand firm (Phil 3:2, 16; 4:1).

We shall, however, give attention first to the situations that call for resignation and rejoicing. As far as concerns references to the death of Christ, they are contained in Phil 1:1–3:1ᵃ. In standard manner, the writer (1) introduces himself and Timothy, (2) names the Philippian community with its bishops and deacons, and (3) greets them (1:1–2). Then he (4) thanks God for their particularly close association with his preaching, defence and confirmation of the gospel (1:3–8),

and (5) prays for their well-being until the day of Christ (1:9–11).[205] Then he
turns to the business in hand by informing his readers about his own situation
(1:12–17) and reflecting on it in a remarkable manner (1:18–26).

'To Live Is Christ, To Die Is Gain'

His confusing situation—whereby the *praetorium* knows that his imprisonment
is for Christ and the Christian community finds his imprisonment so divisive as
to promote rival proclamations of Christ—will turn out for his deliverance (to
use a phrase from Jb 13:16). Awareness of Christ is made available by two
depressing facts: the imprisonment of the apostle and rivalry within the community. Paul wards off depression by remembering the praying community at
Philippi and by the support of the Spirit of Jesus Christ, meaning the Spirit that
raised Christ from the dead (the nearest parallel is Rm 8:9–11) and proclaimed
him Lord. Therefore he can eagerly expect his deliverance, not necessarily the
favourable outcome of a Roman trial but the hope that God will not shame him
by repudiating his work and testimony. He has reason to hope that God will
magnify Christ in Paul's person *whether Paul lives or dies*. The use of the verb
magnify, which echoes 'The Lord be magnified' in Ps 34:27, 39:17 (LXX), makes
it clear that he is thinking not about his own fate or reputation but about the
lordship of Christ. That observation must control the interpretation of the gno-
mic sentence:

> *Emoi gar to zēn Christos kai to apothanein kerdos.*
>
> [For to me to live is Christ, and to die is gain.] (Phil 1:21)

The first clause has an affinity with Gal 2:20, 'It is no longer I who live, but
Christ who lives in me'. Hence, Paul's meaning is, 'For me to live is Christ magnified as Lord in my life; for me to die is Christ even more magnified as Lord in
my death'.

Verse 21 is usually regarded as a reference to communion with Christ, supported by 'to depart and be with Christ' in verse 23. That is most unlikely, for
two reasons. First, can we really suppose that Paul is weighing the advantage of
an increase in his individual communion against the advantage of conferring
more benefits on the Philippian community (verses 23–24)? What on earth is such
a display of personal piety doing in this context? Second, why should someone
in whom Christ dwells, who is himself in Christ, who has been crucified *with*
Christ, describe a closer communion as 'departing to be with Christ'? On the
contrary, that phrase belongs to the eschatological theme (see p. 74), and cannot
be made to support the communion interpretation. The nearest approach to that
would be to read verse 21 as 'For me living means serving Christ, and dying
means serving him more fully'.

Paul is not reflecting on the death of Christians in general but on his own
death, that is, on the death of God's apostle to the Gentiles (Rm 11:13). Whereas
in Thes 4:17 and 1 Cor 15:52 he presumes that he himself will be alive when
the Lord returns, in Philippians he is prepared to withdraw from the scene to be

with Christ, that is, although the Lord is at hand, Paul may well be among those who have fallen asleep, whom God will bring *with* him (1 Thes 4:14). He is resigned—or almost resigned—to his own dispensability. He is contemplating the possibility of sharing the crucifixion of Christ in a manner he had not previously thought possible, of being set aside, of being no longer required, as the least of the apostles, who by the grace of God worked—and suffered—more than all the others (1 Cor 15:9–10).

Of course we find him expressing the conviction that he will continue for the benefit of the community and will visit them again (Phil 1:24–25, 2:24). But what is the real function of these words? In 2:24 they do no more than authenticate the authority of Timothy. In the present passage they are the result of a debate and a choice. But it is a formal debate and certainly not Paul's choice. What Paul has done is provide the Philippian community, depressed by the possibility of his death, with a formula for accepting what may well happen. If he dies, they can say, 'He wanted to come and intended to come; but it was not the divine will that he should, because by his death Christ's lordship is even more effectively proclaimed than by his life'.

In an elementary form belief in the efficacy of the testimony of martyrdom is present in this passage.

In view of these considerations it should be clear that Paul is not producing novel information about the condition in which Christians who have died exist between their death and the return of Christ. In whatever condition dead Christians are, Paul (expecting to be with Christ) will share that condition.

So much for Phil 1:18–26. But Paul has not yet finished with the prospect of his death: he returns to it in 2:17, at the end of the next section (1:27–2:18), where he is giving guidance and encouragement to the community, which is likely to be frightened by its opponents and suffer in much the same way as Paul is suffering (see p. 121). When he encourages them 'to stand firm in one spirit, with one mind striving, side by side for the faith of the gospel', he implies that they are—or are likely to become—divided about the necessity of suffering for Christ's sake. He sets before them the supportive resources they have in Christ (Phil 2:1) and pleads with them to bear the common interest in mind (Phil 2:2–4). He forbids them to grumble and question and presses them hard by a string of religious sanctions: they are to be unblemished children of God, to be light-bearers in a dark and unpleasant world, to be the cause of Paul's pride in the day of Christ, and to be in sequence with the outpouring of his own life (Phil 2:14–17).[206] This all hangs together, and it is not at first obvious why verses 5–11 are present. We must therefore enquire about the function of the Christological hymn which apparently interrupts these powerful instructions about community behaviour[207] (Phil 1:27 *politeuesthe*, 'discharge your obligations as citizens' [AGB s.v. *politeuomai* 3; but see Martin, p. 82]).

The Hymn to Christ

The hymn is written into the letter in order to persuade its readers that under God, Jesus Christ is exclusively Lord at every level of existence, despite the

appearances. Both Paul and his readers are suffering at the hands of the authorities 'on earth'. He—and perhaps they—may be thrown to the powers 'under the
earth', that is, they may be put to death. Their depressing condition may be
caused by hostile powers 'in heaven' or may even be evidence (as some in Rome
are asserting, Phil 1:17) that God himself rejects them. But whatever the anxieties
and insinuations, Jesus Christ is Lord of all the powers 'in heaven and on earth
and under the earth', and they must eventually yield their authority to him (Phil
2:10–11). When, therefore, Christians make the confession 'Jesus is Lord' they
are invoking the most powerful of all names that God has bestowed upon him
(verse 9), which is made available to those in the community who are disposed
to regard each other as 'in Christ Jesus' (verse 5). Once the Philippians have
accepted this massive reassurance, they need not be daunted by Paul's religious
sanctions (see p. 122). Whether Paul is present or not, they can work out their
salvation for themselves (with fear and trembling, it is true), for God is at work
in them to will and to work for his good pleasure (Phil 2:12–13). This is another
way of saying what is said in Rm 8:37–39.

By thus identifying the function of the hymn as Paul uses it (not necessarily,
of course, its original function), it has been possible to give full weight to the
second and more comprehensible part of the hymn. This rescues the second part
from the comparative neglect it receives when attention is concentrated on the
provocative and perplexing first part. And it also opens up possibilities of understanding the function of the first part. It says that Jesus Christ received the
supreme name because he was distinctively related to God and to mankind. In
his proper existence *(hyparchōn)* he was 'in the form of God *(en morphē theou,*
verse 6ᵃ), but he took 'the form of a slave' *(morphēn doulou,* verse 7ᵃ). The latter
means 'the [powerless] condition of a slave'; hence the former means 'the
[powerful] condition of a god'. That does not say that he was or is a divine being.
Whatever Greeks might have thought, a Jew would have realised that the phrase
in the condition of had been chosen to exclude identity and probably to bring
Adam to mind who had been made in the image and likeness of God. Thus verse
6ᵃ is the indicator of an Adam Christology: Jesus Christ is the true, heavenly man,
which is the underlying assumption of *in Christ Jesus* in verse 5ᵇ (see Appendix
C). Therefore the Philippian community can take courage because the appointed
Lord is not a secondary deity like the many gods and many lords (1 Cor 8:5) of
Greek religion but the origin and model of their own community life, through
whom God had chosen to exercise his ruling power.

It is to be hoped that the Philippians grasped that high-minded thought, but
it may have made them uneasy. It may be well enough to be rescued by overwhelming power from outside; but the very exercise of external power begins to
destroy the internal situation, just as the 'liberation' of small countries by the
great powers undermines the self-confidence of the liberated. If God from on high
should liberate the Philippians from their opponents, how could they be urged to
work out their salvation for themselves? The same consideration is enough to
rule out the familiar exemplary understanding of the first part of the hymn. In
Phil 2:3 Paul has said, 'In humility count others better than yourselves' and said
in verse 8, 'He humbled himself'. To the plea 'He humbled himself; so must you'

it is proper to answer, 'But he was a very special person. Anyone with great power can easily be humble without damaging his interests; we cannot'. The transfer from the condition of a god to the condition of a slave is not to provide an example but to make it possible for God to exercise his power from within the human situation and at its lowest point. The transfer may be seen in two stages:

> Being in the [powerful] conditions of [a] God
>
> he did not think equality with God had to be seized [or retained]
> but made himself nothing,
> taking the [powerless] condition of a slave,
> and appeared in the likeness of mankind.
>
> Being found in form as a man,
> he humbled himself,
> and became obedient to the point of death,
> namely death on a cross.

"He did not think equality with God had to be seized [or retained]" is the torment of the exegete because (1) it is not clear whether 'equality with God' is equivalent to or more than 'the condition of a god' in 6ª; and (2) the meaning of *harpagmon* (related to a verb meaning 'to steal or snatch') is almost impossible to decide: does it mean something not possessed and therefore to be seized or something already seized and therefore to be held fast? If we keep clearly in mind the movement from the condition of a god to the condition of a slave, the required meaning is, 'He did not think that the condition of a god, now described as equality with God, was to be retained at all costs'. Interpreters who find this too simple imagine Jesus Christ debating with himself whether he should exploit his special position for his own advantage (Whatever could that be?) or considering whether he should seize by force the greater privilege of equality with God—and of course in the end deciding he should not. But such interpreters have surely been thinking too much about the temperamental aggression of the Greek gods or reading too much about the teenage chaos of the Gnostic pleroma. Many interpreters, indeed, are impressed by the implicit contrast with Adam who by an absent-minded act of defiance (scarcely of violence) ate from the sacred tree that he might be like the gods, knowing good and evil (Gn 3:5). Christ, they suggest, resisted the temptation of becoming equal with God by an act of defiance and instead took the path of humiliation, for which he was rewarded with lordship. But this improbable temptation is not necessary for the contrast with Adam, which may be rewritten in the following way: Adam, created in the image and likeness of God, was wholly dependent on the divine protection in matters of good and evil but sought to become independent of God by knowing good and evil for himself. Therefore he became subject to sin and death. By contrast, Jesus Christ, being in the form of God, renounced his independent existence as Lord and voluntarily took the powerless position of slave; and so appeared in the likeness of mankind. The first Adam and the last Adam arrived at the same point, one by self-assertion, the other by self-renunciation (which again is discouraging for the 'exemplary' view). This understanding brushes aside elaborate discussions of 'emptied himself' (for 'made himself nothing': the verb *kenoō* means what it does elsewhere in Paul: he deprived himself of lordship to become a slave);[208] and it regards *likeness* as equivalent to *form* in the next line (he appeared not *resembling* a human being but *as* a human being). The heavenly Man has become a man, like other men.

The second stage, marked by the resumptive words "Being found in form as a man," goes further than the exchange of a lord's power for a slave's powerlessness. He exposed himself to humiliation in that he became obedient to the point of death—in fact to the slave's death of crucifixion. To a Greek it would be a sufficient humiliation to be forced into the condition of a slave, let alone to accept that condition by free choice. But Christ goes even further by accepting death in the repulsive manner of crucifixion[209] which is described as becoming 'obedient'. To whom was he obedient? It is commonly said, *to God;* but is it proper to regard obedience to God as a humiliation, even for the heavenly man? Leaving aside the view that no answer is required because obedience of any kind is a virtue,[210] the answer must be, 'He became obedient, that is, subject, to the destructive forces that dominate mankind, of which the last and most obstinate is death' (1 Cor 15:26). This verse does not contain the characteristic Pauline personification, Death, but it comes close. Nor does it give any hint that the supremely humiliating crucifixion was necessary to persuade God to save mankind or to effect atonement. If it is thought that the background of the hymn must be found in the servant song of Is 53 because he took the form of a servant (though the Septuagint used *pais,* a title of honour, not *doulos,* a word of humiliation) and because 'he poured out his soul to death' in Is 53:12 (though the Septuagint used *paredothē,* not *ekenōsen* as in Phil 2:7[a]), it is surprising that coincidence in wording is not found and that vicarious suffering, the striking central theme of the Servant Song, is absent from Philippians 2. It ought to be clear that in origin, this is not a salvation hymn. Of course, it has a saving significance from the context in which it is here placed: the community can be confident of God's power to save them from their troubles because God has asserted his universal lordship from within the most shameful and distressing human situation. If the hymn was modelled on a Gnostic myth describing the descent of the heavenly man, it is perhaps the greatest service that gnosis has done for Christian theology to suggest that the divine being saves from within, not from without, the human situation. But in origin this hymn, whether composed or adopted or adapted by Paul, was a hymn in praise of the lordship of Christ, an articulation of what it meant to be 'in Christ'. If the apparently naive question is asked 'Since Christ is Lord at the beginning of the poem and returns to lordship at its end, what difference has his renunciation and humiliation achieved?' the answer is equally simple: 'He becomes possessor of the supreme name to which all must yield and that the church may use even when completely humiliated'. A somewhat more complex answer will be available at the end of the next section, when Paul has completed his case against the religious dangers the community must resist.

Paul's Gnosis

The section Phil 3:2–21 (see p. 121) contains, as Houlden (p. 96) says, 'probably the most succinct statement we have of Paul's central doctrine of the work of Christ, formulated in some of the same terms as in *Romans* and *Galatians*, where it is treated more extensively'. The section is in places sharply polemical and is directed against 'enemies of the cross of Christ' (Phil 3:18) and perhaps against others.[211] Since Paul speaks emotionally it is impossible to go confidently behind his contemptuous words and recover even a prejudiced view of the opponents. But it can be seen that his contempt was called forth because a group of Gentiles were (1) setting up a kind of pseudo-Judaism by promoting circumcision as the initiatory rite to a mystery religion, thereby (2) promising a way that led to 'perfection' in this present earthly, social existence; (3) perhaps claiming that they were already raised from the dead and so were hostile to sharing the cross of Christ or even that the blood of circumcision was the only necessary symbol of Christ's blood; and (4) probably putting this forward as their kind of gnosis.[212] Paul retorts by presenting his own gnosis: 'the surpassing worth of knowing Christ Jesus my Lord'. For knowledge he had renounced the privileges of a properly circumcised, passionately devoted Jew, irreproachable to his fellow Jews and to God. In this Epistle *gnosis* appears only in verses 8 and 10 and therefore does not revive the dominance it had in 1 Corinthians (see pp. 18–21); but here as there, it was probably introduced to counter the Gnostic claim of the rival propaganda. It has the happy consequence that Paul is able to say plainly what this gnosis is and why it is important. As often in Christological passages, his language is formal and carefully structured:

> I count everything as loss
> because of the surpassing worth of knowing Christ Jesus my Lord.
> For his sake I have
>
> suffered the loss of all things,
> and count them as refuse,
> in order that I may gain Christ
>
> and be found in him,
> not having a *righteousness* of my own, which is from the *law*,
> but that which is through *faith* in Christ,
> the *righteousness* from God in response to *faith;*
>
> that I may know him
> and the power of his *resurrection*,
> and the sharing of his *sufferings*,
> being conformed to his *death*,
>
> that if possible I may attain the *resurrection* from the *dead*. (3:8–11)

Examination of the structure helps the task of exegesis: the first line is repeated and strengthened in the third and fourth. Consequently, the second is repeated and extended in the fifth and sixth. Since 'knowing Christ Jesus' in the second line means 'knowing him as my Lord', so gaining Christ in the fifth means 'gaining him as my Lord.' Moreover, this key theme is repeated in the tenth line, where his lordship naturally evokes the thought of resurrection. Thus, Paul's gnosis is developed in verse 9 by the words *righteouness, law* and *faith;* and in verses 10–11 by *resurrection, sufferings* and *death.* This difference points to two situations in which gnosis is effective.

Paul gains Christ as his Lord and is found—that is, treated or regarded by God as being—in Christ. When Paul stands before God, his plea is heard and decided not on the grounds of the genuine claims listed in verses 5–6 but simply on the grounds that he is a servant of his Lord who vouches for him. The word *righteousness (dikaiosynē)* means what makes him acceptable to God, what qualifies him to receive God's forgiveness and salvation. As in Mt 6:8, *dikaiosynē* can mean 'religious obligation', such as circumcision; and it can also mean blameless and passionate devotion to Torah. When Paul counts all these things as loss (even, in his irritation, as *skybala*), he is not despising the genuine purposes they serve (for he could still behave as a Jew to win Jews, 1 Cor 9:20); but he is renouncing the attempt to use them as a claim to be accepted by God. What he said was open to misunderstanding and still is. If, it may be asked, blameless performance of at least the moral demands of Torah is not the supreme qualification for God's favour, what is the purpose of Torah? The purpose of Torah is to make life possible and enjoyable. Let me compare Torah to a musical score: then performance of Torah is like the instrumentalist's technical training, hard work and skill. But even the best-equipped player may produce lifeless, "perfect" music unless his imagination is at work as he plays, unless he can say, like Tortelier, 'I told a story as I played'.[213] The story Paul told was the death and resurrection of Christ: that story set his imagination to work and allowed God to communicate not simply to Paul but to any who might hear the gospel. Of course, when Paul was a Pharisee he told a story (Passover) as the basis for this practice of Torah (whereas the Christian propagandists had, in Paul's eyes, no story but only a religious fantasy). But whereas the Passover story limited God's benefits to Jews and proselytes, Paul was an apostle to the Gentiles, and the death and resurrection of Jesus was a universal story. Moreover, even blameless performance of Torah and a suitable story could not guarantee Paul's acceptability to God, for in his zeal for God's Torah and God's people he had become a persecutor of God's *ekklēsia.* Therefore, Paul's acceptability to God can never derive from Paul's achievement: the initiative must come from God. Faith, conceding lordship to Christ, allows to him the sole responsibility for obtaining that initiative.

My reference to Paul's 'story' has already anticipated the second situation in which his gnosis is applied. Paul stands not only before God but also before his fellow apostles and missionary colleagues, his converts, critics and rival propagandists, the Jewish people and the Roman state, as well as the unevangelized Gentiles as far off as Spain. Time may be running out, and the Lord is at hand. Who is sufficient for these things? If Christ is his lord, Christ must exercise his power in and through Paul—either the power that Christ has because of his resurrection or the power by which he was raised from the dead. This would correspond to 'It is no longer I who live, but Christ who lives in me' (Gal 2:20), just as Paul's renunciation of his Jewish identity corresponds to 'I have been crucified with Christ'. It cannot, however, as a consequence be assumed that the dying part is dead or that the living part is secure. The structure of the last four lines cited above (often remarked on with some perplexity) warns us that those who are in Christ are likely to share his sufferings (see p. 52) and be conformed (cf. Phil 3:21) to his death. For the gnosis Paul speaks of is not simply the experience of divine power but the experience of resurrection power, that is, the power of God to produce new life when a previous life (with its intentions, proposals, hopes and so on) has collapsed in frustration and perhaps shame. When Paul hopes 'that if possible I may attain the resurrection from the dead', he is maintaining the possibility of resurrection life but leaving the decision entirely to God. For that reason alone Paul says nothing about baptism in this passage, even though identification with Christ in his death and prospectively in his resurrection can elsewhere be explained from baptism (Rm 6:3–4). Baptism, like circumcision, cannot guarantee what he hopes for. He has only faith and a modest *if perhaps.*[214]

The Resurrection Hope

Paul's reserve in making a claim on the resurrection hope is rephrased in verses 12–16 ('Not that I have already obtained'), leads into an instruction to the community and a renewed attack on the opposition (see p. 122) and finally to a restatement of the resurrection hope in verses 20–21:

> Our commonwealth is in heaven,
> and from it we await a saviour,
> the Lord Jesus Christ,
>
> who will transfigure the body of our humiliating condition
> and conform it to the body of his glory,
> by the power which enables him indeed
> to subject everything to himself.

What remains to be said can be set out as comments on this passage. This obviously structured passage is similar in many respects to the Christological hymn of Phil 2:6–11. The subjection of everything recalls 'Every knee shall bow . . .', and the formal confession of Jesus Christ as Lord is present in both. *Form (morphē)* and *likeness (schēma)* in 2:6–8 are taken up in *transfigure (metaschematizō)* and *conform (symmorphos)* here. Our humiliating condition here *(tapeinōsis)* reflects his self-humiliation *(tapeinoō)* in 2:8, and both passages end with *glory* (Martin on Phil 3:21). Thus the lordship hymn of chapter 2, intended to encourage the community in its need, now reaches forward to its proper end in the final transformation of human existence.

The Lord Jesus Christ is expected from heaven as a *saviour,* a description nowhere else used by Paul.[215] He comes to save his servants not from sin but from their humiliation. The stock words for sin do not appear in this epistle: such unpleasant characteristics as partisan spirit and conceit are mentioned (Phil 2:3) but they are not cause for theological alarm. Christians, it is true, are living among a crooked and perverse generation (Phil 2:15) and may be tempted to put confidence in the flesh (Phil 3:3); but even so they can count on whatever is honourable, just, pure, and so on (Phil 4:8). Their problem is not sin but humiliation.

It is easy to assume that Paul's resurrection hope refers to the bodily existence of individual persons; and so, in a way, it does. But Paul actually says, 'the body of *our* humiliation', which, without setting aside personal existence, could well refer to the community. Human existence is humiliating not simply because we possess physical bodies but also because we are hard-pressed by our social failures. That the Philippians knew; but Paul reminded them that their commonwealth was in heaven and therefore depended on faith as much as did their acceptance with God.

Summary

The epistle assumes and celebrates the universal lordship of Christ, which derives from his resurrection or exaltation by God. But it must also take account of personal and social humiliation for the apostle and the community. Therefore, the lordship of Christ is to be celebrated by both life and death; and the humiliating death of Christ provides a way of 'reading' their own humiliation. The 'universal story' of Christ's death and resurrection is the image by which God communicates his truth and remakes life, even if (or especially when) present existence has collapsed in frustration and, perhaps, shame. When Christ is awaited as saviour, he is known as one who exercises his supreme lordship not from outside the human condition but from within the most shameful and distressing human situations.

3

The Pauline Writings (2)

COLOSSIANS

This epistle is oddly situated in the larger Pauline corpus. It is pulled away from the central style of Pauline letter writing by the phraseology it shares with Ephesians yet is tied closely to Paul by much of its thought and vocabulary and by its personal links with Philemon. If it is directly Pauline, Paul has, in some points, revised his theology and changed his manner of argument. If it is only indirectly Pauline, it cannot have been composed much later than the letter to Philemon. Neither can it be placed in the middle of Paul's career (in the famous alleged Ephesian imprisonment) if the Epistle to the Ephesians is regarded as a post-Pauline writing. It is best to treat Colossians like a painting that cannot confidently be attributed in its entirety to one of the masters and to describe it as 'from the school of Paul'. It will then be unnecessary to align it at all points with other Pauline writings.[1]

It begins in standard fashion with (units 1–3) the address in 1:1–2 (though its wording is not quite normal)[2] and (4) the thanksgiving 1:3–8, which is formally constructed in terms of faith, love and hope. The stress falls on *hope*, laid up for them in heaven as the truth of the gospel; this is not elsewhere in Paul so explicitly set in the forefront. It may be compared with 'you have been raised with Christ' and 'you have died, and your life is hid with Christ in God' in Col 3:1, 3; so that the Christian hope is the forthcoming disclosure of the resurrection life they already have reserved for them (Col 3:4). That is not remote from what Paul says elsewhere (e.g., the revealing of the sons of God, Rm 8:19), but it is more explicit and clearly linked to the resurrection. The thanksgiving thus identifies the gospel, removes the isolation of the community by assuring them that this gospel is 'bearing fruit and growing' in the whole *kosmos* and authenticates Epaphras as an approved teacher.[3]

Then, (unit 5) in the prayer that follows, the main substance of the epistle is indicated: they are to be filled with *knowledge* of the divine will in complete wisdom and spiritual perception, so that their *behaviour* is appropriate to the Lord and wholly satisfactory. They are to bear *fruit* in all kinds of good activity, and to grow in the *knowledge* of God (an expansion of verse 6). They are to be strengthened with full *power* (= knowledge) corresponding to the might associated with the divine glory and therefore to become able to exercise maximum *endurance and patience* (Col 1:9–11). The pattern ab-ba-ab establishes the two themes: knowledge of God and conduct appropriate to such knowledge; and the repetitions are used to introduce the catchwords of the epistle. One such is *fulness, being filled.* It uses—no doubt more conventionally than literally—the image of an external power that takes possession of the believer and his community and fills them. So they are to be filled with knowledge *(epignōsis)* of the divine will (Col 1:9), they are to possess the total richness of the fulness of understanding in their knowledge *(epignōsis)* of the divine mystery (Col 2:2).[4] Since in certain circumstances the mystery can be defined as 'Christ' (because in him the whole fulness of deity dwells bodily, Col 2:9) or as 'Christ in you' (Col 1:27), it can be said that the community has come to fulfilment in him (Col 2:10). The apostle's aim is to present everyone mature *(teleios)* in Christ (Col 1:28; this 'maturity' is outlined in 3:12–14) in such a way that maturity is equivalent to being fully possessed (Col 4:12). When to this is added the excessive use of *all* (all wisdom, power, endurance and patience, riches, treasures of wisdom and knowledge and so on), it becomes clear that the epistle is a rather heavy-handed attempt to present the gospel as if it contained (in our terms) every prophylactic and vitamin for any circumstance of life. Not, of course, that there can be any doubt of the author's sincere conviction that Christ is all and in all and that his insistence was demanded by the empty deceit of a rival philosophy.

After the prayer it is Pauline custom to begin dealing with the matters in hand (see p. 8), but in Colossians the writer turns back to unit 4 and offers an elaborate thanksgiving for God's action in producing the Colossian community (1:12–23). Verses 15–20, however, are notably different from the rest of the paragraph. They make no mention of the community, they are written in a formal and rhythmical style that resembles a hymn or creed, they are rather loosely attached at the beginning and the end, and they display features uncharacteristic of the epistle in general (Lohse, Schweizer and Bruce on Col 1:15–20). Hence, before considering their significance at this place in the epistle, it is instructive first to examine verses 12–14 and 21–23, which, by themselves, form an intelligible unit.

Reconciliation

The first part, using traditional baptismal language (Lohse), refers to the transfer of people from one authority to another, from the power of darkness to the power of light. This metaphor is used to indicate not a future eschatological hope, as in 1 Thes 5:5, but a present condition of life as in hellenistic Judaism (e.g., in *Joseph and Asenath* 8:9, 12:1–2, 15:12). The possible Exodus imagery is not more than an exhausted metaphor[5] and makes little contribution to what is being said, namely, that a gospel of knowledge and wisdom can be thought of as illumination and can be expressed by familiar baptismal phrases, including 'son of his love', which echoes 'my son, the beloved' in Mk 1:11.[6] In him they have their liberation *(apolytrōsis),*[7] which is identified as 'the forgiveness of sins'. The phrase is absent from Paul and indeed is uncommon in the New Testament, where it is associated with proclamation of the faith and baptism.[8] It is inherited from Jewish piety and is an answer to one of two questions that might be asked about wrongful actions, namely, What have those actions done to God? and What have they done to the sinner? To the first question: they have offended God, who must be induced to forgive them, put them out of mind, and restore his favour; the sinner who repents and puts his trust in the Lord Jesus Christ has a sponsor who will secure forgiveness for his wrongful actions. To the second question which would be more natural for Paul: they have become enslaved, ruining themselves as well as harming their victims, so that God must be persuaded to come to their aid and find ways of liberating them from their slavery and alienation. If one is *hopelessly* enslaved, nothing would help; if enslaved by ignorance or inexperience, forgiveness would be unnecessary, for knowledge and wisdom would set him free. But if one is enslaved by a dark power, more than knowledge is needed—as the writer makes apparent when he moves from the traditional language of verses 12–14 to the case of the Colossian community in verses 21–23.

By the word *reconciled* these four verses are, of course, linked with *reconciled* in verse 20 (*apokatallassō* in both places, also Eph 2:16—an extremely rare word, known only in Christian writers). But for the moment we are considering them as the application of verses 12–14. The Colossian Christians had formerly been estranged and hostile in mind—that is, estranged from the truth (verses 5–6) and hostile to it—and had shown their condition by discreditable behaviour. But their estrangement and hostility had been removed, they had been reconciled to the truth; they could now be presented as fit members of God's people: holy, blameless and irreproachable,[9] provided they maintained their adherence to the hope of the gospel. How was this reconciliation achieved?—'In his body of flesh by his death' (Col 1:22). The phrase is clumsy, perhaps an afterthought (Martin, p. 57); but the 'body of flesh' reappears in Col 2:11 and has parallels in Qumran, Sirach and 1 Enoch (Lohse). Because elsewhere in the epistle *body* is an important metaphor, it is necessary here to explain that the reconciling death took place in a physical, human body. It is not immediately obvious why the physical death of Christ should reconcile people to the truth; nor is it evident that the writer is drawing upon Rm 5:10–11 and 2 Cor 5:18–20 (where Paul used *katallassō*). But Pauline hints can be put together from the deception myth of 1 Cor 2:6–8 (see p. 26) and from the triumphal procession image of 2 Cor 2:14 (reflected in Col 2:15), so that we are again provided with an explanatory myth. It is as if God were a king surrounded by enemies in his own royal court, who wished to draw his people from their support of his enemies and reconcile them to himself. So he set a trap for his enemies, allowing them to kill his Son and thus, by displaying their vicious nature, destroy their own power. When the Son is raised from the dead, he triumphantly mocks the discredited enemies, and their formerly deluded victims are reconciled to God.[10]

The myth gives no factual information about the existence and activities of supernatural beings. Only those who have forgotten or never realised how myths operate in society would suppose they do. Indeed, the myths may seldom appear except in fragmentary form and conventional language; but at a deep level of awareness they preserve an understanding of the kind of situation in which we find ourselves. According to this myth, therefore, we are not required (shall we say?) to appease an offended God but to respond to his initiative in sending Christ to our rescue.

The Hymn of Creation and Victory

Now we must return to the distinctively structured section, Col 1:15-20. For the purposes of this enquiry it is not necessary to enter into detailed exegesis of the whole, but use will be made of the strongly supported conclusion of recent study that the verses contain, probably in modified form, a previously existing composition.[11]

> Who is the image of the invisible God,
> the first-born of all creation;
>
> for in him all things were created,
> in heaven and on earth,
> visible and invisible,
> whether thrones or dominions,
> or principalities or authorities—
> all were created through him and for him.
>
> And he is before all things
> and in him all things hold together.
>
> And he is the head of the body *the church,*
> who is the beginning,
> *the first-born from the dead,*
> that in everything he might be pre-eminent,
>
> For in him all the fulness was pleased to dwell,
>
> *and through him to reconcile to himself all things,*
> *having made peace by the blood of his cross*
> *through him*
> whether things on earth or things in heaven.[12]

This was originally a hellenistic Jewish hymn to the divine Wisdom. It was adapted by the writer of the epistle to a Wisdom Christology, and it is commonly agreed that there are two stanzas, one describing the preeminence of Christ in creation (the first eleven lines) and one describing his preeminence in redemption. If two phrases, "The church" and "by the blood of his cross" are removed (one in each stanza), the hymn no longer carries evidence of Christian composition. Hence, the phrases are reasonably attributed to the writer of the epistle. In the former he used a Pauline convention to change the reference of *body* from the world to the Church, as in Col 1:24. In the latter he introduced the Pauline theme of Christ's death (though admittedly in non-Pauline phrasing) to qualify the triumph of the Redeemer.[13] With that in mind, the writer's theological intention can be distinguished from the sense of the material he was using.

So far as it goes, that may be accepted; and there is no need to play about with the wording in attempting to reconstruct two rigidly parallel stanzas. Nevertheless it is unsatisfactory for two reasons. First, remaining phrases in the second stanza raise questions. Why does 'the first-born from the dead' appear when death has not yet been mentioned? Why is reconciliation introduced without any previous hint of hostility or disorder? If the first stanza establishes the total dependence of all things on, say, the divine Wisdom (and that emphatically includes all human and nonhuman powers capable of hostility), why is a Redeemer needed in the second stanza to reconcile them? It can be replied that everyone knows how much the world needs reconciliation and knew it in the first century. It can be argued that a hymn has no need to supply information but addresses God, praising him both for creation and redemption (Schweizer, pp. 84–85). True enough; but even complacent congregations may sometimes notice the deficiency of a creation that needs to be reconciled and may think the reconciliation as precarious as the creation. And it is evident from the rest of the epistle that the Colossian congregation were far from complacent. Second, if the second stanza is given to redemption, what is "For in him all the fulness was pleased to dwell" doing there? Indeed, *have* two stanzas been properly defined, and their content recognized? I think not and therefore propose a different division. Stanza 1 is concerned entirely with *creation* and comprises the first *ten* lines. Stanza 2 is concerned with rule or lordship and comprises the rest except the words in italics. The hymn is thus a typical hellenistic evocation of the invisible God, whose two perceptible functions are creation and lordship, or ruling power. When the hymn is used in a Christian congregation, it provides an expression of the Wisdom or Logos Christology: Christ is God's agent not only in the total process of creation but also in maintaining the created order in being and in controlling its existence.

Hence, what the writer of the epistle has done is fused the creation myth with the victory myth. He has introduced ideas from death-and-resurrection theology (sometimes using existing hymn phrases as a model, so that the thirteenth line reflects the second) and placed them in the stanza of the divine lordship, thus extending rather than overturning its meaning. In particular, he has used Paul's teaching about reconciliation and given it a second verb, namely, peace making, either from Paul's frequent references to 'the God of peace' or from his stock greeting formula, 'Grace to you and peace'. There are important consequences of this bold action.

Death-and-resurrection theology is firmly moved from eschatology to protology, that is, from discussion of the end of this age to discussion of the beginning of all ages. It is not necessarily a radical shift, since God's relation to the world can be considered either as regards his original intentions or his final purposes. Nor is the dependent and precarious existence of the world overlooked. But the imagery is changed. In eschatology this age or this world must pass away, and a new one must be created. In protology this world passes, but a new world already exists in heaven. In the first stanza, the Son of God is the prototype of all created being, whether on earth or in heaven. Note that that stanza is untouched by modifications.

The modifications are introduced in stanza 2, dealing with Christ's preeminence, his ruling and sustaining function. Elsewhere it is commonplace that his lordship derives from the resurrection; but here his death, too, belongs to his lordship. How that can be so is shown by analogy in the victory myth. The divine lord is surrounded and nominally served by disaffected powers: the visible earthly powers called thrones and authorities and the invisible heavenly powers called dominions and principalities (Houlden, p. 163). By the death of the Son, a trap is set for these powers, who overreach themselves and are caught by their own hostility. Hence, they can be restored to their proper service of the first-born from the dead (Bruce [p. 76] calls it 'pacification'); mankind is no longer manipulated by the powers and so is reconciled to the truth. The purpose of this placing of the myth is to indicate that mankind is threatened not so much by individual wrongdoing but by semiautonomous accumulations of power, in society and the environment.

Christ is the beginning of a new ruling authority, which, of course, already exists in heaven. As the first-born of the dead, he is the head of the body but in a new phase of existence. Hence the original world-body is redesignated as the *ekklēsia*. The Christian now lives in two worlds. In one the divine mystery has been hidden but is now disclosed, and the gospel is bearing fruit and growing (Col 1:6, 26). The elemental spirits of the universe are no longer a threat or an imposition; at the most they are a shadow of what is to come (Col 2:8, 17, 20). The soma belongs to Christ. *Sōma* ('body') is usually translated 'substance' (RSV; it is 'solid reality' in NEB) by contrast with 'shadow' in this passage; but the link with "For in him the fulness was pleased to dwell" must not be lost: 'holding fast to the Head, from whom the whole body . . . grows with a growth that is from God'. For the body is the hope laid up in heaven, the things that are above, the reality where Christian existence is concealed with Christ in God (Col 1:6, 3:1– 3). From this perception the moral instructions flow, and it must be said that they are of a reconciling kind—just as reconciliation is the theme of verses 21– 23 already considered.

The closing words of verse 23 make a smooth transition to the next surprising section, 1:24–2:5, which begins, 'I rejoice in my sufferings [*pathēmata*] for your sake, and in my flesh I complete what is lacking in Christ's afflictions [*thlipseis*] for the sake of his body, that is, the church' (verse 24). The history of exegesis shows what theological anxieties this incautious statement has raised (Schweizer). Who could possibly think that the afflictions Christ suffered were insufficient to complete his redeeming work? Can the anxiety be allayed by noticing that, elsewhere in Paul *pathēmata* are the sufferings of Christ's passion that Paul and others can share, whereas *thlipseis* are common hardships that Christians must endure (so that verse 24 reverses the usage in 2 Cor 1:4–5)? If so, 'Christ's afflictions' would mean 'afflictions we must bear as Christians'. But that will not remove the problem; for if 'the blood of his cross' has reconciled everything on earth and in heaven, why must Christians be afflicted? The problem is not that Christ's afflictions are insufficient but that his reconciliation is not properly effective. At this point it is customary to mention the stock apocalyptic view that God requires a predetermined amount of hardship before the old age gives way to the new (Lohse); but no expression of that view places the reconciliation of all things before the transitional hardships. Nor is Colossians a conventionally apocalyptic writing. The writer does not imagine a new age succeeding this present age; he supposes that a new, reconciled world already exists in heaven, 'where Christ is, seated at the right hand of God'. Christ is head of the body, namely the Church. In due course it will be disclosed what *he* is and what *we* are (Col 3:1–4). At present the heavenly reconciliation is available in the Church, which knows the hidden mystery (namely, 'Christ in you, the hope of glory') and its rich possibilities among the Gentiles. God has devised a strategy (*oikonomia*, curiously rendered 'divine office' in RSV) for making his intention fully known. It requires Paul to be the supreme mystagogue, not despite his suffering but because he suffers disproportionately, taking on himself the burden of the Church. Even if heavenly opponents have been reconciled, earthly enemies must still be pacified. The death of Christ shows the only guaranteed way for overcoming the earthly enemies of God. So Paul rejoices in his sufferings on the community's behalf, and, in his own physical body, completes for the sake of Christ[14] whatever hardships are required by the divine strategy for so extending the boundaries of the body that everyone can be presented mature in Christ.[15]

The Fulness of Deity

In this section, which deals mainly with the apostolic hardships, warning hints are being given to the community in case they are deluded with beguiling speech (Col 2:4). Moving onwards, verses 6–7 encourage them to live in Christ Jesus the Lord—of which the obvious continuation is Col 3:1–4, 'If then you have been raised with Christ. . . .' But that sequence is interrupted by a repudiation of the beguiling speech that threatens the community—the philosophy and empty deceit of Col 2:8. Even that promising beginning is not followed up at once—it has to wait until verses 16–23—but gives place to another section on Christ, verses 9–15. It contains important references to the death of Christ, notable for the variety and prodigality of the images employed.

There are no less than six images: the first two and the last one are dependent on *in him* phrases, two others on *with him* phrases.

Fulness (Verses 9–10). *In him* dwells the fulness of deity (see 1:19). Therefore, insofar as the community is *in him,* they are brought to fulness, either in the sense that they are fully possessed by deity or, more probably, in the sense that no room is left for any other allegiance or cult. The statement that Christ is 'the head of all rule and authority' picks up 'head of the body' and 'principalities or authorities' (1:18[a], 16[e]), thus neatly drawing on the Christology both of creation and lordship.

Circumcision (Verse 11).   *In him* they had been circumcised, though in a non-literal sense: not removal of the foreskin but removal of the body of flesh (see 1:18ᵃ). That sounds like the language of the mystery cults, but it is probably no more than a literary convention in the manner of Philo, e.g., 'The soul that loves God, having disrobed itself of the body and the objects dear to the body . . . gains a fixed and assured settlement in the perfect ordinances of virtue'.[16] It is something more than grave moral language, but not *much* more; and it becomes meaningful in Col 3:9–11, where the community are said to have put off the old identity and put on the new. At this point, however, it probably arises from this train of thought: Christ in his body of flesh (Col 1:22) achieved the reconciliation and then rose with a heavenly body, to which Christians now belong. Christians are therefore concerned with heavenly existence, and need not be terrified by the problems of earthly existence.[17]

Baptism (Verse 12).   *With him*[18] they had been buried in baptism, in which[19] also they were raised *with him* through faith in the working of God who raised him from the dead. This has been borrowed from Rm 6:4 and adapted to suggest that rising with Christ, which is prospective in Romans, is immediate in Colossians. This is made clear in Col 3:1: 'If then you have been raised with Christ', though the resurrection life is still concealed with Christ in God and has still to be disclosed. Their hope is laid up for them in heaven (Col 1:5). Hence, they have nothing more to achieve but must simply hold fast to what is theirs until it is disclosed.

Death and Resurrection (Verse 13).   This new metaphor is marked by a change in construction (*you* ceases to be subject and becomes object) and by a novel understanding of the word *dead*. The previous verse has 'You were raised *with him*', and the present verse has 'He [presumably God] brought you to life *with him*'. But being buried with Christ in baptism implies having died with him (as in Col 2:20, 3:3), whereas here they are dead in trespasses[20] and the uncircumcision of their flesh. Death means the condition in which they were beyond recognition by God because they behaved wickedly and were not God's people. As it is, however, God has brought them to life with Christ and graciously disregarded their trespasses.[21] The most apt illustration is the father in the parable of the Prodigal Son who welcomes back his erring but repentant son with the words, 'This my son was dead, and is alive again; he was lost, and is found' (Lk 15:24, 32). Under this image, all the emphasis falls on the saving initiative of God in Christ's resurrection.

The Bond (Verse 14).   This comes as a complete surprise to us, though not, perhaps, to the Colossians. Even the Greek is rather perplexing:

> *Exaleipsas to kath hemōn cheirographon*
> *tois dogmasin*
> *ho ēn hypenantion hēmin,*

kai auto ērken ek tou mesou
prosēlosas auto tō staurō.

He blotted out the bond that was against us,
with its requirements,
which was contrary to us,
and took it from our midst
having nailed it to the cross.

That inelegant clause bears signs of injudicious tampering. It is overloaded by its double insistence that the bond was unfavourable to us and hampered by the awkward dative in the second line. Moreover, by the vividness of the metaphor, it prompts the question, If the bond had been blotted out, why had it to be removed from the midst and, of all things, nailed to a cross? If any sense is to be made of it, it may perhaps be along the following lines. We begin with the first three lines. A *cheirographon* was a bond knowingly entered into by the signatory. Something like that was the case with anyone in the Colossian community who had adopted the 'philosophy and empty deceit' put forward to trap them (Col 2:8). With their eyes open they had, as it were, signed a bond greatly to their disadvantage and were caught up in its inevitable requirements. But in the manner of God's promise in Is 43:25 ('I am He who blots out your transgressions for my own sake, and I will not remember your sins'), God had blotted out the bond. If it is argued that only the power with whom the bond had been made could blot it out, the answer is twofold: if the power had acted deceitfully in encouraging the bond to be made, a higher power could properly cancel the bond; and in any case God had an absolute right to relieve the disability of his people, however foolish they had been. It is not to be supposed that God is hampered by commercial morality and legal contracts when they are destructive of mankind. In that case, however, in blotting out the bond God has done nothing more than he did in verse 13, when he graciously disregarded their trespasses. Therefore in addition (last two lines) he does what at first sight seems superfluous: he removes the bond from the midst (of the beguiled signatories) and nails it to the cross, where it replaces—or is added to—the accusation brought against Christ. Thus, Christ bears their indebtedness; though in fact the bond has been cancelled, and the bondholders in their turn are deceived and trapped.[22]

The Victory Procession (Verse 15). God has stripped the principalities and powers (of their ability to cause harm), has publicly disgraced them, and has triumphed over them in him (namely Christ).[23] The image seems to have been adopted from 2 Cor 2:14. The principalities and powers are more pacified than reconciled (see p. 136): these powerful forces are reduced to their proper subservient role.

It must be obvious that these six images have been collected together to make a rather incoherent paragraph. The writer has presumably fired off some familiar images on the principle that if you shoot enough arrows into the air some will probably find an appropriate target. He is preparing his attack on the 'philoso-phy' that requires Christians to undertake additional rituals if they are to be sure of surviving in this world and the other. So the *fulness* of deity resides in Christ, who is *victor* over, and now head of, possibly hostile forces on earth and in heaven. Christians are concerned not with progress in this earthy life of *flesh* and *bondage* to elemental forces of earthly existence but with heavenly existence. By *baptism* they share Christ's *death and resurrection* and must therefore know that it is not the performance of rituals but solely the divine initiative tht assures them of a hope laid up in heaven.

That summary, however, assumes that the victory procession image, which certainly refers to the principalities and powers (*archai* and *exousiai;* see Col 1:16ᶜ, where RSV translates the same words as 'principalities or authorities'), is different from the bond image where the bond is due to someone unnamed. Now, verse 8 warns the community against a deceitful philosophy that derives from human tradition bearing on 'the elemental spirits of the universe' and is not in accordance with Christ. Later, verse 20 takes for granted that they have died with Christ (and so have been set free) from 'the elemental spirits of the universe' (*ta stoicheia tou kosmou;* see Gal 4:3, 9). Is it not, therefore, likely that the unnamed bondholders of verse 14, which has the one explicit reference to Christ's death in the whole paragraph, are the *stoicheia?* If so, what are they? Almost all commentators now take the view that they are part of the angelic or demonic inhabitants of the spirit world of antiquity (hence the RSV translation). They are much the same as the thrones, dominions, principalities and authorities of Col 1:16—or such part of them as are supernatural. But that view is unsatis-factory and perhaps even misleading, for three reasons: (1) If the *stoicheia* are no more than run-of-the-mill angelic powers, why are they named specifically? (2) If they are supernatural powers, why are they said to be according to human tra-dition? and (3) Why is there no evidence as early as the New Testament for *sto-icheia* meaning angelic powers? What is needed is an interpretation that includes the *stoicheia* in a kind of *philosophia,* a popular man-made system of belief (Col 2:8, 22) with corresponding rituals that include the observation of seasons and tabus. It has been well said that a Greek contemporary of Paul would take *sto-icheia tou kosmou* to mean the basic constituents of everything in the universe (and so of mankind), namely, earth, air, fire and water.[24] Schweizer (p. 132)[25] finds the distinctive themes of Col 2:16—23 in a Pythagorean text of the first century B.C.E. and it is certainly possible that the neo-Pythagorean revival, dem-onstrated vividly in the contemporary Apollonius of Tyana, was proving attrac-tive to the Colossian community. It must also be remembered that Josephus *Ant.* 15.371 described the Essenes as 'a group which follows a way of life taught to the Greeks by Pythagoreans'. All authorities are agreed that the Essenes had not borrowed their religious philosophy from the Pythagoreans, though their life-style was sufficiently similar for contemporaries to notice the affinity. What is more, the determination of the Qumran sectarians, with their special calendar, 'to live according to the structure of the universe' is an example of Jewish par-ticipation in an ambition widespread in antiquity.[26] Even nearer home is a pop-ular Jewish writing saying that God 'gave me unerring knowledge of what exists, to know the structure of the *kosmos* and the activity of the elements *(stoicheia);* the beginning and end and middle of times, the alternatives of the solstices and the changes of the seasons, the cycles of the year and the constellations of the stars. . . . I learned both what is secret and what is manifest, for wisdom, the fashioner of all things taught me' (Wis 7:17—22). That is a kind of practical wisdom. At least in some circles, to practise the rituals appropriate to the seasons would maintain the continuity of the *kosmos;* to ignore them might destroy it. And the writer of Col 2:23 disdainfully says of the rituals he is opposing that they 'have indeed an appearance of wisdom'.

If, then, the *stoicheia* are component elements of the world that must be maintained in their proper balance and harmony by the necessary rituals, the following suggestion can be made. In the Colossian community it could be accepted that Christ's lordship was dominant over the spiritual powers; but it could not so confidently be agreed that he had mastery over the structural elements of the world. Hence it was necessary to rewrite Christology in the form of a creation hymn and to place both resurrection and death in the stanza dealing with his rule over every area of creation. Suffering and death suggest that the elements of the world are in severe disorder; but if you have died with Christ, you are set free from the elements (Col 2:20). You no longer have the anxious responsibility of maintaining the structure of the world, for you are already risen with Christ.

Risen with Christ

At least, in a sense you are. The section Col 3:1-4 (which both concludes the problem of the deceptive philosophy and introduces practical wisdom for Christians) is explicit—with qualifications. Christ has died and has been raised to lordship at the right hand of God (in the hard-worked image of Ps 110:1). His lordship, though genuine, is at present concealed in God; but Christ in his lordship will be disclosed in glory, which has the standard overtones of majesty and power (see Col 1:11, 27). Correspondingly, Christians have died (with Christ, Col 2:20) and been raised with Christ. That must imply that they are protected, governed and fulfilled under his lordship (insofar as they hold fast to the head of the body, Col 2:19); though the fulfillment is still concealed with Christ in God. Its disclosure waits upon his disclosure. Meanwhile, they are to seek and give consideration to the things that are above not in order to withdraw from the world but to learn endurance and patience, to receive instruction in the divine will and to experience fruitful growth (Col 1:9-11).

That this modified resurrection teaching does not recommend a flight from earth to heaven is shown by the next section of the epistle, 3:5-4:5, which contains the imperatives for a cautiously regulated life within existing society. They are to conduct themselves wisely towards outsiders (those not in possession of the mystery) and to make the most of opportunities (Col 4:5). The divine instructions are clear enough: wives are to be submissive and children and slaves to be obedient, while husbands, parents and masters are to be considerate (Col 3:18-4:1). The community as a whole is to rid itself of bad qualities and adopt good ones—

advice that is not entirely commonplace and is marked by imperative formulas. *Put to death* is matched by *put away* (verses 5, 8), which suggests that *put to death* is only a rhetorical variant. Both are replaced by *putoff* and *put on* (verses 9–10), which is baptismal imagery and signifies the renunciation (though not, as in Rm 6:6, the crucifixion) of the old identity and the adoption of the new, undifferentiated mankind under the universal lordship of Christ (though not quite so clearly formulated as in Gal 3:26–29). Finally, the clothing metaphor is reinvigorated to introduce the good qualities and activities of Christians—without any reference to the Spirit.[27] Thus, it is evident that when the writer of the epistle comes to the ethical consequences of his Christology, when he sets out admirable guidance in opposition to the demands of the 'philosophy', the death of Christ is no longer in his mind. The resurrection has secured the Christian's hidden life with God but has only indirectly shaped his continuing life in the world.

Summary

It is evident that Colossians is controlled by the resurrection of Christ, which gave him triumphant lordship over the threatening supernatural powers and earthly authorities and abolished the tyranny of the constitutive elements of the world. Believers are under Christ's lordship; they belong to his risen body, so their life is concealed with him in God. They benefit from the death of Christ, which has both reduced the hostile powers to their proper subordinate function and reconciled alienated humanity to God. Yet they still live in this world, not in heaven. By baptism they share in the death, as well as in the resurrection, of Christ. From the sufferings of the apostle they understand the divine strategy of making the mystery fully known and accept patience and endurance as necessary consequences of being under the lordship of him in whom the fulness of deity is present bodily. In gratitude to God (Col 3:17) a great change in behaviour is required and is indeed possible; but actual instruction on standards of behaviour is judiciously drawn from the more peaceable kinds of hellenistic morality. It is possible to say that the death of Christ keeps this kind of Christianity within bounds by keeping the actual, historical death of Jesus in sight, by linking it with the church's experience of lordship, and by reaffirming Christ's destruction of ritual demands for our appropriation of the fulness of God.

EPHESIANS

Scholars are sharply divided on almost all questions about Ephesians: whether it was written by Paul or someone else, whether it is a letter or not, what situation caused it to be written, whether it does or does not contain liturgical components, whether it is chiefly infuenced by Gnostic or by hellenistic Jewish thought, and how it is to be interpreted.[28] If it were necessary to resolve these disagreements before considering what Ephesians says about the death of Christ, progress would be slow. In order to move onward, it is necessary to learn from the debate and to single out the contributions that most advance this enquiry.

The foremost question is not Who wrote the Epistle? but What kind of writing is it? The following matters must be taken into account.

If we read through the Pauline epistles, we know that real people are being addressed, argued with, rebuked and encouraged and their problems dealt with. In Ephesians the readers are cardboard figures. They are all-purpose Gentiles who once led morally disreputable lives, had been made fellow citizens with the saints, but still needed encouragement not to fall back into the bad habits of surrounding society. Hence this epistle is not a letter of the normal Pauline kind.

It has the appearance of a letter because it begins and ends appropriately. Paul (1) names himself in 1:1, and then (2) names the recipients (though the syntax is awkward and the text very doubtful). He greets them in verse 2. The ending in 6:23–24 is curt and lacks personal greetings. It is immediately preceded (in verses 21–22) by a commendation of Tychicus, lifted almost word for word from Col 4:7–8. These, however, are minor oddities. The body of the epistle is more unusual. After the three opening units, we expect (4) a thanksgiving or blessing of God. Pauline thanksgivings always thank God for the recipients; the one blessing (in 2 Cor 1:3–11) blesses God for benefits received by Paul and shared with his readers. Eph 1:3–14 follows the blessing pattern (presumably because no particular community of Christians was being addressed) and, with high-flown language in a loosely structured, single, long sentence blesses God for benefits conferred on 'us.' It continues in verses 15–16[a] with the standard introduction of a thanksgiving unit, which immediately, however, is turned into (5) the prayer for the recipients. It is, in fact, a didactic prayer, setting out the benefits that are offered; and it concludes dramatically with Paul on his knees and a final doxology in Eph 3:14–21. The construction of this prayer unit is a remarkable artifice, quite unlike anything else in Paul's letters. After the prayer unit, we expect Paul to deal with the matters in hand; but in this eipstle there are none—no questions to be answered, no mistaken views to be argued against. The writer moves at once to advice about conduct: how they should walk (*peripatein,* promised in 2:2, 10 and repeated in 4:1, 17 and 5:2, 8, 15). The pattern of Eph 4:1–6:20 is similar to the paraenetic section of Colossians with an important preface on 'the unity of the Spirit' and a rhetorical conclusion on the Christian soldier. The epistle is therefore intended as a basic text to be used by Tychicus (or those whom Tychicus represents) when visiting Christian communities.[29]

Ephesians is closely related to Colossians: characteristic words and phrases in one are used (though not always in the same context or with the same meaning) in the other. The relation is so close that in principle it is possible to argue the priority of either (though what must be said of Colossians—that it is so obviously Pauline and yet non-Pauline—cannot be said of Ephesians). It can therefore be deduced that something in the circumstances that called forth Colossians must have prompted the writing of Ephesians. To this must be added the fact that Ephesians contains proportionately more reminiscences of the Pauline corpus than Philippians (an epistle of similar length). So it can further be deduced that

Ephesians was written when it was helpful to draw upon a collection of Pauline material. Moreover, the peculiar linguistic style of Ephesians is surprisingly akin to the sentence structure and word pairing of the Qumran writings. These writings can also be used to illuminate the distinctive vocabulary and themes of the epistle but so can the familiar language and to some extent the characteristic assumptions of Gnostic writings. It is clear that Ephesians is not written from within a Qumranic or a Gnostic community; but it certainly makes use of the style and language of syncretistic forms of hellenistic Judaism. Even though Paul may have ventured into that area, it is difficult to believe that *he* was in firm control of the language of this epistle.

The first three chapters of the epistle, despite their ornate language and elaborate use of metaphor, irresistibly press home a few important convictions: the absolute authority of the divine will, the enormous generosity of the divine grace, the magnitude of the divine power, and the total claim of the divine glory. This divine being is the God and Father of Our Lord Jesus Christ, and receiving the benefits of knowing him is dependent on understanding the mystery of his will, namely, his plan to unite all things in Christ, things in heaven and things on earth. In particular the mystery requires the Gentiles to be members of the same body as the Jews in order that the Church may thus make known the divine wisdom to the principalities and powers in the heavenly places. What is said in Romans 11 to warn Gentile Christians against despising Jews and in Romans 15 to encourage Jewish and Gentile Christians to live joyfully and peaceably in the same community is here transformed into a basic principle of the gospel: that Gentile Christians can have access to God only if they enter into the Jewish heritage by faith. The epistle is a sustained assertion to the effect that Gentiles can benefit from the overwhelming power of God only on these terms.

In the epistle, Paul has a formal prominence. It goes without saying that the vigorous character of the apostle is stamped all over the Pauline Epistles; but it is hard to find it in the baroque language of Ephesians, and the explicit appearances of Paul seem contrived. There is the notorious break in construction after 3:1 (not formally resumed until 3:14), when the writer inserts an elaborate description of Paul's apostolic ministry and his insight into the mystery of Christ, now revealed to his holy apostles and prophets by the Spirit. Paul's imprisonment and suffering on behalf of his Gentile readers (though they are scarcely acquainted) are emphasised, and he is called the very least of all the saints (3:1, 8, 13; 4:1; 6:19–20). That combination of abasement and privilege can properly be explained as a signal expression of divine grace; but it was not necessarily seen in that fashion by Gentile Christians. If, they may have thought, Paul was the true mystagogue, how could he be subject to hardship and humiliation? Ephesians may therefore have been composed in order to deal with some consequences of Paul's withdrawal (for whatever cause) from the control and expansion of his missionary empire. It is being asserted that even if Paul himself is no longer an active force in the creation of the Gentile churches, faithfulness to his unique perception of the mystery is necessary if Gentile Christianity is not to lose the very purpose of its existence in the age-long divine intention. There is one body and one Spirit and a structure of apostles, prophets and the rest to continue what Paul could no longer maintain.[30]

The Praise of God

With these considerations in mind, the text may now be examined. The 'blessing unit' 1:3–14 is regarded by some as a previously composed hymn; but the evidence is not persuasive. A better guide to the structure is the observation of four smaller units:

1. "Paul" blesses God for the blessing he has conveyed to *us* in the heavenly places *(en tois epouraniois)* in Christ, namely, in the Beloved (verses 3–6),
2. in whom *we* have the redemption *(apolytrōsis)* through his blood, the forgiveness of *our* trespasses, according to the riches of his grace that he lavished on *us* in all wisdom and insight, having made known to *us* the mystery of his will, namely, the plan to unite all things in Christ, things in heaven and things on earth in him (verses 7–10),
3. in whom indeed *we* have been chosen . . ., *we* who first hoped in Christ (verses 11–12),
4. in whom *you* also, having heard the word of truth, the gospel of *your* salvation—in whom indeed having believed, *you* have been sealed with the promised Holy Spirit, which is the guarantee of *our* inheritance, for the sake of the redemption *(apolytrōsis)* of God's property to the praise of his glory (verses 13–14).[31]

Subunit (1) praises God for his special benefits, which are 'in the heavenly places in Christ'. Since Christ was raised by God to sit at his right hand in the heavenly places, and we are raised to sit there with him (Eph 1:20, 2:6), this opening blessing is intended to give the first indication that a Christian's competence in the transcendent realm inhabited by good and bad supernatural powers (Eph 3:10, 6:12) is based on Christ's resurrection. The remaining subunits specify the benefits. In subunit 2 the writer speaks for *us,* presumably for 'us Christians', but at the end of subunit 3 suddenly specifies 'us who first hoped in Christ', either the first generation of Christians, or more precisely, the first Jewish Christians. These are contrasted in subunit 4 with *you*—presumably Gentile Christians, though the writer slips back to *our* before ending his enormous sentence. It seems to have been his intention to set up a kind of dialogue between Paul with other founding members of the Gentile mission on one side and the Gentile converts on the other.

The resurrection of Christ is implied but not mentioned; his death is added to a clause taken from Col 1:14 (see pp. 132–33) 'in whom we have the redemption *through his blood,* the forgiveness of our trespasses' (Eph 1:7). Both redemption *(apolytrōsis)* and forgiveness *(aphesis)* are uncommon words in the New Testament (both may mean 'release'), and they are seldom associated with the shedding of blood.[32] Further, it must be noted that there is another redemption in verse 14: the setting free of God's property, his own true people (if that is what the Greek means). In Eph 4:30 (as in Rm 8:23) redemption is eschatological: the final liberation. Therefore, it must be suspected that Christ's death creates the possibility of the first redemption (namely forgiveness), and that his resurrection begins the process that leads to the second redemption (namely, the more-inclusive liberation).

Resurrection and the Church

Death and resurrection play a part in the next section of the epistle. After the nominal thanksgiving in 1:15–16[a], the standard prayer begins, with the writer sometimes remembering to say *you* and sometimes saying *us*. It is a prayer for wisdom, revelation, knowledge and enlightenment as regards the Christian hope and the divine power. In verses 20–23, the divine power is explicitly the power that raised Christ from the dead, seated him at God's right hand in the heavenly places above all supernatural powers in the present and the coming ages, and put everything under his feet (Ps 8:6). This exalted person, by God's gift, is head over all things for the church that is his body, the fulness of him who fills all in all (or whatever the right translation is).[33] This is a narrative myth that tells a story about the divine power: it is as if a mighty Being from another realm used his magical power to bring the dead lord back to life to confer upon him the Being's active sovereignty over the totality of all powers in that realm and gave him authority in our earthly realm over a community of people that can be called his body. What does that story imply for those who are members of his body (Eph 5:30)? In the course of chapter 2 the significance of the story is explored by an astonishing number and variety of images, with apparently random variations between *us* and *you*.

In 2:1–9 the death-resurrection image is immediately applied to those who were dead (i.e., beyond the reach of God) in trespasses and sins; who, under the domination of evil powers, were giving way to damaging physical and mental pressures; and who therefore were abandoned to the wrath (as in Rm 1:24–31). But God, entirely of his own gracious initiative, raised them with Christ and gave them a share in the lordship of Christ in the heavenly places.[34] The image originates in Col 2:13, and stresses resurrection, not Christ's death.

As a pendant, 2:10 introduces the new creation image, possibly from Col 3:10–11 and 2 Cor 5:17, Gal 6:15. The image ingeniously combines stress on the divine initiative with the necessity for Christians to behave well (perhaps from baptismal instruction, Schnackenburg).

2:11–13[35] changes to circumcision imagery. The Gentile readers were uncircumcised and separated from Christ, alien to the Jewish community and promises and without hope and God. But now they have been brought near *in the blood of Christ;* that is to say, they are still uncircumcised but faith in the blood of Christ has replaced the blood of circumcision.

Ephesians 2:14–18

These verses set down a string of reconciliation images whereby peace is established in place of hostility.

Jews and Gentiles are made one by destruction of the wall between them (verses 14–15ᵃ). The Greek calls it *to mesotoichon tou phragmou*. A *mesotoichon* is a partition wall; and a *phragmos* is fence, wall, or hedge. Its obvious metaphorical meaning is the (rabbinic) fence round the Torah which was intended to prevent Jews from straying into Gentile habitual ways and correspondingly to exclude Gentiles unless they accepted the obligations of Torah.[36] That meaning is suggested by mention of the law in verse 15ᵃ, but the phrasing is by no means clear. The Greek runs:

> *kai to mesotoichon tou phragmou lysas,*
> *tēn echthran en tē sarki autou,*
> *ton nomon tōn entolōn en dogmasin katargēsas.*

A prosaic rendering would be:

> Having broken down the partition wall [consisting] of a [protective] fence,
> the enmity in his flesh,
>
> having annulled the law [composed] of commandments [expressed] in requirements.

The first and last lines fit easily together: Christ broke down the barrier between Jews and Gentiles by annulling the legislative Torah. In the middle line *the enmity* indicates that *wall* is a metaphor for hostility. *In his flesh* is usually taken as qualifying one of the participles (though awkwardly separated from both): either he broke down the wall, or annulled the law, by an action undertaken in his flesh, namely, his crucifixion (rather than, say, by providing knowledge) or by acting as God's agent of flesh and blood (rather than as a divine being in human form). That may all be true, but it is not obviously intended by the brief wording. Is it not possible that 'enmity in his flesh' means exactly what it says? Käsemann long ago suggested that the flesh is the cosmic barrier that separates God and man. The Gnostic redeemer destroys that barrier—gathers his own into one new man, which is called his body.[37] But apart from general criticisms of the Gnostic interpretation, at this point in the catena of images the writer is not yet talking about the barrier between God and man. He is talking about the notorious enmity that a Jew carried in his flesh and blood, specifically by the sign of circumcision.[38]

The image of the new identity (verse 15ᵇ) sums up Col 3:10–11, where renewal in the image of the creator eliminates the distinctions between Greek and Jew, circumcised and uncircumcised, and so on. No doubt, if pressed, this image could associate happily with Gnostic new men; but it stands comfortably on its own (Bruce).

In verse 16ᵃ *peace* is replaced by *reconciliation* (*apokatallassō* from Col 1:20, 22) and the image of 'the one body' appears in a phrase that must be compared with Colossians:

> and might reconcile us both to God in one body through the cross (Eph 2:16ᵃ)

> . . . to reconcile to himself all things, . . . making peace by the blood of his cross. And you . . . he has now reconciled in his body of flesh by death (Col 1:20–22; pp. 132–33).

The formulation in Ephesians omits the qualification *of flesh* and therefore allows *body* to mean the resurrection body—no longer the presumed cosmic body of the original hymn in Colossians, because it has been subjected to death and resurrection, but nevertheless the resurrection body of which Christ is head (Eph 1:23). Hence, in the presentation of this image, the words *through the cross* need mean no more than the death that makes resurrection possible or the shocking death that makes resurrection necessary. The glory, power and extent of the resurrection body are in inverse proportion to the shame, weakness and particularity of the crucifixion.[39]

The presentation, however, runs on in verse 16ᵇ with the phrase 'having killed the enmity in it' (i.e., the cross, unless *autō* means 'himself,' as it does in verse 15). The word *enmity* was probably suggested by the adjective *hostile* in Col 1:21–22 (*echthra* and *echthroi*), and no doubt that same explanatory myth lies behind (though more remotely behind) verse 16ᵇ as it lies behind the Colossians passage.

Verse 17 presents a completely new image: the welcome envoy who announces peace, which is good news both to those who are far off and those who are near. This was a familiar image in antiquity but is here expressed in words provided by deutero-Isaiah (Is 52:7, 57:19)—as in verse 13.

The last of the reconciliation images appears in verse 18 and provides a neat transition to what follows. It is the image of access to the royal presence of the Divine Being (Eph 3:12, Rm 5:2); and the two parties have the privilege of joint audience because both possess the Spirit of wisdom and revelation (Eph 1:17, 3:5), which guarantees their present and prospective membership of God's special people (Eph 1:13, 4:30).

By a perceptible shift, the reconciliation of hostile groups is replaced in verse 19 by the naturalization image. Strangers become fellow citizens, and resident aliens become household members.

Finally this section of the epistle is rounded off in verses 20–22 by the elaborate image of a building that is in process of construction and is turning out to be a temple. The readers are component parts of this building, for which apostles and Christian prophets provide the foundation (on earth) and Jesus Christ comprises the cornerstone (in heaven).[40]

In all this, what is the writer saying to Gentile readers about their membership in the Church that is the (resurrection) body of Christ? He is certainly not putting before them a coherent argument or a systematic explanation. All he can manage is a catena of images collected from several sources because he has not made up his mind about the status of Gentiles before they became Christian or even about the nature of their conversion. In 2:1–10 they were like the rest of mankind, separated from God in their wickedness and therefore dead. By the direct love of God they were made alive with Christ and given the benefits of Christ's session in the heavenly places. By God's initiative they were newly created in Christ Jesus to be active agents in doing good. In that respect they are viewed in relation to mankind; but everywhere else they are viewed in relation to the Jews. For one thing, Jews and Gentiles are *different*. Therefore they are given a new, common identity in Christ, 'one new man in place of the two'. To both, Christ gave access to the Father by the Spirit of wisdom. But Jews and Gentiles are also *hostile,* separated by the wall of the law, which Christ broke down. He killed the hostility by his cross and reconciled the hostile parties in one resurrection body, made possible and necessary by the cross. Finally Gentiles are at a *disadvantage* when compared with Jews: whereas Jews are near to God and his Temple, Gentiles are far from him.[41] Therefore Christ proclaimed peace to 'you who were far off and peace to those who were near'. Those who once were far off were brought near no longer by circumcision but by the blood of Christ. Thus Gentiles were admitted to full citizenship with Jewish Christians and to the household of God. In the process of constructing a new dwelling place of God in the Spirit, 'you [Gentiles] also' were included.

That orderly scheme, however, was scarcely intended by the writer. He was putting together a number of images—without attempting consistency—to express the significant range of resurrection consequences for believers: exaltation with Christ, membership in his resurrection body, a new creation, peace with God and access to him, a new social identity, membership of God's household and his chosen people, participation in the New Temple which is the place of the Divine Presence. Only in some parts of this range was it necessary to mention the *death* of Christ. The cross is the (perhaps) necessary precondition of the resurrection body, the means by which enmity between Jews and Greeks was destroyed; and faith in Christ's blood, rather than the practice of circumcision, admits one to membership among the holy people. The death of Christ is therefore preliminary to the major action of God by his grace, love and wisdom.

What follows in Ephesians 3 is mainly devoted to the mystery specially revealed to Paul, namely, that 'the Gentiles are fellow-heirs, members of the same body, and partakers of the promise in Christ Jesus through the gospel' (verse 6). This is a revision of the mystery disclosed in Rm 11:25–32, where Paul, speaking as a hellenistic Christian, addressed himself to the Jews. Here Paul as a Jewish Christian addresses himself to the Gentiles. The passage is devoid of all reference to the cross and of any explicit reference to the resurrection.[42]

The paraenetic second half of the epistle has two explicit references (Eph 5:2, 25); but rather more significant are the passages from which expected references are absent.

The Unity of the Spirit

The general purpose is to encourage appropriate behaviour for Christians. The opening verses commend modest and considerate behaviour within the community and then single out maintaining 'the unity of the Spirit in the bond of peace' (Eph 4:3). Behind that phrase, which to our ears sounds encouragingly ecumenical, lies not merely a warning but a threat. The writer says, in effect, 'You Gentiles cannot have the varied benefits of the Spirit unless you maintain unity with the Pauline Jewish tradition; and you are bound to be at peace with Jewish Christians, not in conflict. (*Bond* [*syndesmos*] comes from Col 2:19, 3:14.) The variety of endowments is not denied nor the possibility that they will produce energetic confusion; but they ought to be coordinated harmoniously in the growing body (Eph 4:14–16). Hence the curt insistence on oneness:

> One body and one Spirit,
> as indeed you were called [by God] in one hope when he called you;
> one Lord, one faith, one baptism,
> one God and Father of all
> who is above all and through all and in all. (Eph 4:4–6)[43]

Whether this is an already-familiar credal formula or a creation of the writer, it is clearly intended as a summary of fundamental realities and securities by which a Christian community exists. It ignores the death and resurrection of Christ. It may be argued that *body* means the resurrection body and that *Spirit* is the Spirit of him who raised Jesus from the dead (Rm 8:11). It may perhaps be thought that the death of Christ is presumed by the one baptism, though baptism is not elsewhere mentioned in the epistle. It may be suggested that enough has already been said about the death and resurrection of Christ for them now to be taken for granted; and the formula can obviously be read in a fully Pauline way. Yet a formulation that takes the death of Christ for granted is not being serious about it. The thought cannot be evaded that the writer has borrowed from something like a Jewish Gnostic source. That suspicion is strengthened by what follows, namely a midrash on Ps 68:18 in a form akin to the Aramaic Targum: 'When he ascended on high he led a host of captives, and he gave gifts to men'. The interpretative words of verses 9–10 probably mean that he first ascended with his captives, then descended with his gifts. Possibly the Gnostic scheme is in mind whereby a redeemer descends to claim his own and before returning to the Fulness provides them with necessary gifts to maintain and develop their new self-awareness. But whichever it is, the exposition is developed in terms of ascension, not resurrection; and the descent has no hint of crucifixion. Consequently, what is said about the unity and ministry of the church is not directly exposed to teaching about the death of Christ.

Christ and the Church

This must not be taken to imply that the writer has abandoned a Christocentric manner of thought. When at Eph 4:17 he again turns to direct moral instruction and speaks against customary Gentile morality, he assumes that his readers had heard about Christ and had been taught in him, as the truth is in Jesus (4:20–21). They must therefore put off the old identity and put on the new (baptismal imagery as in Col 3:9–10; see p. 142). They are to be kind to one another, tender-hearted, generously overlooking one another's faults as God in Christ had overlooked theirs (Eph 4:32; the verb is *charizomai* as in Col 2:13; see Chap. 3, n. 21). Just as children who deserve their father's love follow his example, so they must behave lovingly.

> [just as Christ loved us
> and gave himself up for us,
> a gift and sacrifice to God
> as a fragrant offering] (Eph 5:1–2)
>
> *kathōs kai ho Christos ēgapēsen hēmas*
> *kai paredōken heauton hyper hēmon,*
> *prosphoran kai thysian tō theō*
> *eis osmēn euōdias.*

The first two lines are modelled in Gal 2:20, and are used again at Eph 5:25: 'just as Christ loved the church and gave himself up for her'. *Prosphora* occurs again in Paul only at Rm 15:16, where the Gentile Christian community is Paul's gift to God, acting as apostolic *leitourgos*. The 'fragrant offering' comes from Phil 4:18, namely, the acceptable service performed by Epaphroditus. Just as Epaphroditus, risking his life to serve Paul, had pleased God, so Jesus in greater measure had pleased God by giving himself for our sake. Thus, the death of Christ for the benefit of others is a proper example for Christians. Throughout Ephesians, God is presented as working in Christ—and Christians must be imitators of God.

With that conviction in mind, in Eph 5:22–23[44] the writer of the epistle, for reasons unknown to us, decides to rewrite the simple subjection formula of Col 3:18–19, which runs 'Wives be subject to your husbands, as is fitting in the Lord. Husbands, love your wives, and do not be harsh with them'. The instruction to wives is relatively simple: they are to be subject to their husbands, as to the Lord. Since Christ is the head of the church and the church is therefore subject to Christ, so the husband is the head of the wife and therefore wives are to be subject in everything to their husbands. Assuming that the parallel is valid, that is, that the husband corresponds to Christ and the wife to the church, there is only one difficulty: in verse 23 after 'Christ is head of the church' the Greek adds 'He is saviour of the body' (not clearly represented in RSV). Thus, the instruction ought properly to say (and perhaps implies) that wives are to be subject in everything to their husbands provided the husbands act as their wives' guardians and protectors. The instruction to husbands is more elaborate, even embarrassed. The writer wants to say 'Husbands, love your wives' and to support it with pleading and argument. In fact he sets out the case for his plea but fails to press it home and turns to argument instead. Here is the beginning of his plea:

> Christ loved the church
> and gave himself up for her,
> that he might consecrate her
> having cleansed her by the washing of water with the word,
> that *he* might present to himself the church in splendour
> without spot or wrinkle or any such thing
> but that she might be holy and without blemish. (Eph 5:55^{b}-27)

At that point the writer could be expected to say something like, 'Since Christ loved the church in that way, husbands ought to love their wives by ensuring their virtue, maintaining their honour—if necessary at the cost of their own lives'. It could not be called an unheard-of demand and perhaps we are to draw that conclusion. But instead, the writer moves away to his argument, though he sets it out back to front: the basic justification of marriage in Gn 2:24 says that man and wife become 'one flesh', the word *flesh* can be read as meaning both 'kin' and 'body'; therefore a man's wife is his own body; since no one ever hated his own body (a statement open to reservations), but provides and cares for it, so a man should love his wife; and that indeed is how Christ treats the church, of which we (men and women?) are living parts. It is an odd inducement to love and care; and it need concern us no more.

But the statement of verses 25^{b}-27 needs further attention. Clearly Christ's actions on behalf of the church—including his death if 'gave himself up' has that meaning—are exemplary, though not in a simplistic fashion. The writer does not say, Christ acted in such and such a manner; in similar circumstances we must follow his example. His argument is more profound and less individualistic: since Christ acted in such and such a manner to bring into being the body of which we are living parts, we must live within that body in a way that corresponds to his creative activity. Since he displayed lordship and self-giving love, the relation of husbands and wives must reflect both. The wife's subjection to her husband displays his lordship; the husband's care for his wife displays the self-giving love due to her. That principle—less defensible today than the mutual subjection of verse 21—at least hints at the possible subtlety of exemplary views of Christ's death.

It must also be said that the story in verses 25^{b}-27 scarcely began as a marriage illustration, though it draws on ancient Near Eastern marriage imagery. It is, of course, an old folk story doing duty as an explanatory myth. There is the marvellous prince seeking a bride for himself not from among the shining ones of the kingdom but from among the neglected, the despised, even the immoral. Having, at cost to himself, found his bride, he makes known her destiny, cleanses her from present impurity, and then (in the great transformation scene) presents her to himself in splendour, without a suspicion of spot or wrinkle, a sacred and unblemished person. The marvellous prince and his splendid bride are depicted in Ps 45. The discovery of the filthy and neglected girl, betrothed by God when ready for marriage, cleansed with water, who 'grew exceedingly beautiful and came to regal estate' is described in Ez 16:8-14. Even the prince's death and miraculous return to life may have been present in the 'myth and ritual pattern' of Near Eastern religion, and it certainly appears in the marriage of the Lamb in Rv 19:7–9; 21:2, 9; 22:17.[45]

It is thus possible to find the origins of this story in Old Testament and related sources. But what of its use? The image of the church as a bride cannot easily be transferred to individual Christians. Nor can the image of the bridal bath easily represent the baptism of individual converts. When we ask what impurity is removed by the cleansing, it is not obviously the defilement that results from disobeying God's commands. Is it perhaps the impurity of being immersed in the material world? Does the Redeemer come to seek and find his bride who does not know she is his bride because she is ill treated, corrupted and ignorant? When he finds her in his self-giving love, he removes her corruption by washing her in the bridal bath, and her ignorance by his word disclosing her real identity. Hence, it is possible and probably suitable to interpret the myth in a mildly Gnostic manner—in line with the epistle's firm, though modest, interest in wisdom and knowledge. It is well known that wedding imagery featured prominently in second-century Valentinian Gnosis, and the early third century Acts of Thomas has a particularly splendid example.[46] The Epistle to the Ephesians does no more than suggest a way of thinking that could bear that elaboration. Instead of describing Christ's self-giving as performed for all sinful people, we are allowed to think of it as an act of supreme love on behalf of those who are truly God's people, but are deprived of the benefits and knowledge of their true nature. If those people are limited in number, the thought is Gnostic; if unlimited, the thought may be catholic.

Illumination

There remains one final reference in Eph 5:14 that can also be regarded as Gnostic.[47] The section Eph 5:3–20 takes up again, at first somewhat repetitively, the task of renouncing vices and commending virtues (verses 3–4). There are threats against those who continue in vice, including withdrawal of association (verses 5–7). These are justified with the imagery of light and darkness (verses 8–14), followed by a call for prudence (verses 15–17), the replacement of drunkenness by spiritual exaltation and thanksgiving (verses 18–20). The religious use of light-and-darkness imagery was, of course, commonplace in antiquity and, as everyone now knows, was much favoured at Qumran. Here it would scarcely call for comment were it not accompanied by a proverb and a hymnic quotation:

> When anything is exposed by the light it becomes visible,
> for anything that becomes visible is light.
> Therefore it is said,
> Awake, O sleeper, and arise from the dead,
> and Christ shall give you light. (Eph 5:13–14)

In view of 'dead through our trespasses' in Eph 2:1, 5 it is tempting to suppose that the unknown quotation calls on the sleeper to awake from the death of sin;

and the whole context suggests a choice between acting wrongly and acting rightly. But the theme of illumination in fact changes the focus to acting in ignorance and acting with knowledge. The sleeper must awake from a stupor and receive illumination from Christ. The preceding proverb is indeed somewhat cryptic but seems to suggest the purifying and therapeutic qualities of light. The verb in the last line *(epiphauskō)* is not used elsewhere in the New Testament, but *phōtizō* (of similar meaning) appears at Eph 1:18 and 3:9 of the illumination provided by knowledge of the Christian mystery. The quotation may well come from a baptismal hymn in which the resurrection of Christ is applied to the baptized and explained as the gift of illumination and knowledge.[48]

Summary

The epistle begins with the glad acknowledgment that God 'has blessed us in Christ with every spiritual blessing in the heavenly places'. The phrase *en tois epouraniois* is not only peculiar to Ephesians (six times) but strongly influences all its teaching. Christ's resurrection caused him to be seated at the right hand of Divinity in the heavenly places where the world-controlling powers, whether favourable or malignant, have their basis. We too have been brought to life with Christ, have been raised with him and caused to sit with him in the heavenly places. Those who were dead in sins, dominated by evil powers, and handed over to the wrath have, by God's own initiative, been raised with Christ. In their exaltation with Christ they are members of his (resurrection) body, a new creation, at peace with God and granted access to him, possessing a new social identity as members of God's household, his chosen people, his new Temple. With this wealth of heavenly conviction, it is not surprising that the imagery of ascent can replace the language of resurrection and that direct death-and-resurrection teaching can disappear from a credal formulation. It is equally understandable if what may be baptismal teaching stresses illumination and if an explanatory myth describes the discovery to the Bride of her true identity, her purification and exalted betrothal. In general, resurrection is foremost; Christ's death is its necessary preliminary.

It is not to be supposed, however, that a safe seat in the heavenly places diverts interest from this earthly existence. It is clearly perceived that spiritual blessings are required for life within the community of people of which Christ is head, as well in this age as in the age to come. Even if the death of Christ is the necessary preliminary to the spiritual blessings, it is still present within the community as the precondition of reconciliation between Jew and Gentile and of a sound family life for Christians. To the writer of Ephesians it is plain that Gentiles can become members of the Church and join Christ in the heavenly places only if they accept (his understanding of) the Jewish heritage. That was—and is—made possible, no longer by the blood of circumcision but by the blood of Christ. Once they are members of the community the death and resurrection of Christ is exemplary, in the sense that his self-giving and lordship should be expressed in the relation of husbands and wives. To modern ears, family relations sounds very prosaic and rather unprofessional for a writer obsessed with the heavenly places; but he thought it a profound mystery concerning Christ and the Church. When he used the old language of liberation *(apolytrōsis)*, he knew that it began with the death of Christ and the forgiveness of sins; but it was not completed until all that properly belonged to God shared in the liberation to the praise of his glory.

4

The Pauline Writings (3)

THE PASTORAL EPISTLES

Study of these Epistles has long been dominated by the question of authorship. In my judgement it is now more profitable to begin by asking what function they were expected to fulfil in the Pauline communities.[1]

2 Timothy

Let us begin with 2 Timothy, which lacks the formal structure of the other two and contains much more personal Pauline material. There are two centres of dramatic interest, Rome and Ephesus.

In Rome, Paul is a prisoner, wearing fetters like a common criminal, and greatly embarrassed by his situation (2 Tm 1:8, 16; 2:9). He takes the gloomiest view of his future: he is about to be sacrificed, the time of his departure has come, he has finished the race (2 Tm 4:6–7); yet he urges Timothy to join him before winter and to bring Mark as a useful assistant, as well as his cloak, books and parchments (2 Tm 4:11, 13). He is aggrieved by everyone's failure to support him at his first defence; he has been deserted by Demas and harmed by Alexander, and others have left him so that he is alone except for Luke (2 Tm 4:10–16). Not quite alone, however, because the epistle ends with greetings from four named persons and all the brethren. Paul seems scarcely to know whether he is coming or going. No doubt his sufferings, and remembrance of earlier hardships (2 Tm 3:11), and his disappointments have taken their toll.

So, too, perhaps has the situation in Ephesus. All who are in Asia have turned away from him, including Phygelus and Hermogenes—though the Ephesian Christian Onesiphorus was a welcome exception when he sought out Paul in Rome (2 Tm 1:15–18). The communities are involved in ruinous disputes, godless chatter, senseless controversies and quarrels. Hymenaeus and Philetus have swerved from the truth by holding that the resurrection is past already (2 Tm

2:14–26). There is at least the risk that the communities will disintegrate from the self-will and immoral pressures of people with corrupt minds and counterfeit faith (2 Tm 3:1–9). The time is coming when Christians will not accept wholesome teaching but will choose teachers to their own liking, preferring the myths to the truth (2 Tm 4:1–4).

That is an alarming picture. No doubt a good deal of it is conventional, but the convention is clearly being used by the writer to describe the collapse of Paul's missionary work in Asia. Why was that collapse taking place and when? Perhaps it happened when Paul's presence was removed and his leadership was put in doubt by his arrest. Perhaps faith wavered and disputes began when the messages from Paul became confused, and it was thinkable to write that he was breaking up (*analysis,* 2 Tm 4:6). Perhaps Hymenaeus and Philetus, who held that our resurrection had already taken place, thought Paul a failed leader for not displaying resurrection life. And this presumably took place while it was still plausible to circulate letters in the name of Paul and still possible to hope that something might be saved from the wreckage. Therefore the writer of 2 Timothy attempted a complex task. He could not conceal the serious condition of Paul and his dependent communities. In fact, it was not in his interest to do so. Using Paul's own teaching, he gave an encouraging theological interpretation of Paul's condition; but he withdrew from some of Paul's more exposed theological positions, and dictated a programme for bringing the disturbed communities into line with the model of sound teaching that could be derived from Paul.

After a brief form of the standard epistolary beginning in 2 Tm 1:1–2, the apostle begins to put his support and encouragement behind Timothy in 1:3–12. This unit includes what appears to be an outline of the gospel, speaking of God

> who saved us
> and called us with a holy calling,
> not in virtue of our activities
> but in virtue of his own purpose and grace
> which he gave us in Christ Jesus ages ago,
>
> and now has manifested
> through the appearing of our saviour Jesus Christ,
> who abolished death
> and brought life and immortality to light through the gospel.
> (2 Tm 1:9–10)

This hymn, liturgical unit, or collection of stereotyped catechetical phrases (Kelly) answers the preceding reference to suffering by stressing God's purpose and grace, rather than *our* activities. Its theme is our salvation from darkness and death. Death has been abolished by the appearing *(epiphaneia)* of our Saviour, which also brought to light life and incorruption *(aphtharsia).* This is phrased in the language of hellenistic religion. It may *imply* that Christ substituted life for death by his cross and resurrection, but it actually speaks of his epiphany—a good hellenistic term for the appearing of a saviour god—used here uniquely for his earthly life rather than his eschatological coming.[2]

Timothy's Responsibilities.   In 2 Tim 1:13–4:5, Timothy's responsibilities are set out by a series of imperatives. He is (1) to use the Pauline model of sound teaching and guard the spiritual property entrusted to him, as some were refusing to do (1:13–18); (2) to be strong, to convey Paul's teaching to others, and hence to share suffering (2:1–7); (3) to keep in mind, according to Paul's gospel,

> Jesus Christ, risen from the dead,
> descended from David, (2 Tm 2:8)

to be matched with the 'reliable statement' *(pistos logos)* that

> If we have died with him
> we shall also live with him;
>
> if we endure,
> we shall also reign with him;
> if we deny him,
> he also will deny us;
>
> if we are faithless,
>
> he remains faithful—
> for he cannot deny himself. (2 Tm 2:11–13)

These two formula quotations justify Paul's humiliating sufferings on behalf of the elect, by which they too will obtain their salvation in Christ. Hence, dying with Christ implies the martyrdom theme, although the *pistos logos* probably originated as a baptismal formula—as is suggested by the past tense *we have died (synapethanomen)* and by a comparison with Rm 6:3–4.[3] The formula encourages endurance and thus turns away criticism of the apostle, who is still enduring, not yet reigning with Christ. The earlier formula on the resurrection shows that the hellenistic epiphany formula is not by itself satisfactory when the Christian has to experience persecution and death. To that extent these are therefore polemical quotations, and it may be (in view of the reference to 'Jewish myths' in Ti 1:14) that *descended from David* is slipped in for similar reasons. No doubt it was part of a resurrection formula, similar to Rm 1:3–4, here included with the implication that the Risen Lord and his kingdom (verse 12[b]) comprehends all proper Jewish expectations.

Timothy's fourth responsibility is to avoid controversies (2 Tm 2:14–26) such as those provoked by the assertion that 'our resurrection has already taken place' (verse 18). Such a teaching may have been encouraged by Rm 6:3–11 (but it ignores Paul's care in placing our resurrection with the Lord's return) and may well have been developed from Col 3:1 ('If then you have been raised with Christ'; compare 'You have passed from death to life' in Jn 5:24 1 Jn 3:14). It could be interpreted in various ways: as a Greek protest against bodily resurrection, as the beginning of a double resurrection theory as found in Revelation 20 or as a conviction that dying and rising with Christ in baptism is the total promise of the gospel. Such ideas certainly appear in early Christian writings of an eccentric or Gnostic character; but to me it seems unlikely that the waywardness of Hymenaeus and Philetus had taken them in those directions. What is said in verse 18 scarcely implies that they had attacked a fundamental Christian conviction or had unduly emphasised the baptismal background of verse 11. What they were saying was, 'Christians are those who reign with Christ', like the enthusiasts at Corinth. This verse tells us nothing about an Ephesian heresy, though it indicates the prominence of resurrection teaching in this Christian environment.[4]

Timothy's remaining responsibilities are to be aware of what causes a community to disintegrate (2 Tm 3:1–9), to rely on what he has learnt from Paul and from scripture (3:10–17), and to give sound teaching and to endure suffering as an evangelist (4:1–5). He is solemnly charged

> in the presence of God and of Christ Jesus
> who is to judge the living and the dead,
>   and by his appearing [*epiphaneia*] and his kingdom.

Lordship is attributed to Christ, rather than to God (so Brox); and it can perhaps be presumed that lordship derives from resurrection. His epiphany is to establish his kingdom, not, as in 2 Tm 1:10, to initiate his earthly life. The wording is probably traditional and formulaic; and it is used rather to add solemnity to the apostle's words than to make a theological point.

All possible passages have now been mentioned where some reference, explicit or implicit, to the death and resurrection of Christ may be detected. All are formulaic, which implies that the writer either thinks it unfitting to use his own words or is not accustomed to expound his religion in terms of death and resurrection. When, at the beginning, he comes nearest to describing the gospel he uses hellenistic epiphany language. The other three formulations are used more for apologetic or polemic or solemnity than for theology; and it must be suspected that the writer was not entirely clear about the meaning of his terms (hence the variation of *epiphaneia*) or their original force.

Titus

It seems that 2 Timothy sketches a general policy for replacing Paul's direction of his churches and that Titus and 1 Timothy give attention to the details of church management. Their aim is to establish wholesome teaching, that is, sensible, modest behaviour that encourages people to renounce irreligion and worldly passions and to live sober, upright and godly lives in this world (Ti 2:12). This is offered to all people and so is opposed both to Jewish sectarianism and to Gnostic elitism. But when the epistles oppose contrary attitudes they rely more on what is sensible than on what is theologically important. Nevertheless, from time to time they introduce traditional theological statements that indicate the doctrinal background of their sober, unspeculative but caring religion.

Thus Ti 2:13–14 speaks of the community waiting for 'the epiphany of the glory of our great God and Saviour Jesus Christ

> who gave himself for us
> that he might redeem us from all iniquity

> and might purify for himself a people of his own
> who are zealous for good deeds.'

The very exalted language, which leads to a modestly moralizing conclusion,[5] is compounded of common hellenistic phrases, an echo of *gave himself* in Gal 1:4 and 2:20 and more remotely in Mk 10:45, and themes from the Septuagint (the second line is close to Ps 130:8; the third to Ez 37:23 and Ex 19:5). In Pauline writings we expect the self-giving of Christ to refer to his death, but there is no explicit reference to it in this epistle. The presumption may be that his self-giving was his epiphany. The appearance among us of so great a God and Saviour was itself an act of condescension, in the course of which he used his power to release us from the consequences of every prohibited action[6] and 'by the washing of regeneration and renewal in the Holy Spirit' (Ti 3:5) purified for himself a people of his own. The model statements of the Septuagint—that God will redeem Israel from all its wrongful acts, that God will cleanse them so that they are his people, a special possession above all other people—demonstrate that God can do all this without any intermediary and human bloodshed. So long as it is a matter of dealing with the results of prohibited actions and of encouraging good deeds, the problem is not difficult. But when the power of wickedness and the tragedy of human existence are present to the conscience, it becomes meaningful to speak about the death of Christ.

Titus contains a *pistos logos* in 3:4–8.[7] It is to be used by Titus as an encouragement to good behaviour, in contrast to their highly discreditable behaviour (verse 3) before becoming Christians. It insists strongly on the sole initiative of God our Saviour who by his mercy

> saved us by the washing of regeneration
> and renewal in the Holy Spirit
> which he poured upon us richly
> through Jesus Christ our Saviour

so that we became heirs in hope of eternal life. This is doubtless a reference to baptism (see Eph 5:26), but nothing is said (as for example in Rm 6:3–4) to relate this saving rite to the death of Christ. Regeneration and renewal are made possible through the Holy Spirit, which we receive through Christ. The verb *poured upon us* recalls Acts 2:33, which relates the pouring out of the Spirit to the resurrection of Christ. That may be the implication here, but it is not explicit. Indeed, very little is explicit. The *pistos logos* is mostly a compound of the popular words of hellenistic religion (*goodness, philanthrōpia, saviour, epiphany, regeneration,* and *renewal*) and some memories of Paul's teaching on righteousness and the inheritance of eternal life.

1 Timothy

1 Timothy is more explicit but also more perplexing. After the address and greeting, the writer of the epistle at once addresses the problem of certain persons who,

wishing to be teachers of the law, are putting out variant speculative doctrines. He indicates proper uses of the law, and ends with Timothy's authorization (from Paul and prophecy) to engage in conflict for the faith and to reject Hymenaeus and Alexander (1 Tm 1:3–20). Inserted into this strongly worded instruction and using the form of a thanksgiving unit is an elevated passage on Paul as the primary example and prototype of conversion (verses 12–17). Formerly Paul had treated Christ with abuse, persecution and outrage (NEB); but he had been treated mercifully because he had acted in ignorance. So indeed he had, for in persecuting the church he supposed himself to be doing the will of God. Therefore, the grace of God in Christ had saved him from ignorance, provided rather painful enlightenment, and appointed Paul to the service of Christ. This account—oversimple by the standard of Romans 7—is supported by the *pistos logos* that 'Christ came into the world to save sinners' (verse 15). Nothing is said about the means by which the saving actions were performed (perhaps by his epiphany?) and the statement is thrown in to give dramatic force to the apostle's self-presentation.

The epistle turns to the prayers of the community and urges that all men should be prayed for since God desires all men to be saved and to come to knowledge of the truth. The request is supported by a catechetical or liturgical formula:

> There is one God,
> And there is one mediator between God and men,
> the man Christ Jesus,
> who gave himself
> as an *antilytron* for all,
> the testimony for the proper time. (1 Tm 2:5–6)

It is not entirely easy to distinguish the original intention of this formula from the use that is made of it here. The argument may go like this: there is one mediator (perhaps as against numerous Gnostic mediators) between the one God and the multiplicity of mankind, and the one mediator gave himself to be an effective *antilytron* for all (as against Gnostic elitism); therefore, prayers may properly be made for all men. The formula is introduced for the word *all* in 6[b]. If *gave himself* refers to Christ's death, *antilytron* describes the effect of his death—though the statement only reinforces an argument for the scope of intercessions. The saying is constructed in the tradition of Mk 10:45 (see Appendix A) and says little more than Ti 2:14. In the hellenistic religious environment of the Pastoral Epistles, it cannot be taken for granted that *gavehimself* means 'gave himself in death'. A mediator represents A to B and B to A. In giving himself as an *antilytron,* Jesus Christ represents all mankind to God; in being the testimony at the proper time (whatever that obscure phrase means), he represents God to all mankind. Hence, again it is possible that his self-giving is his epiphany and that *antilytron* signifies not a mediator offering himself as a ransom but a mediator dispelling the ignorance of mankind and disclosing the One God.

Somewhat later, the writer describes 'the mystery of our religion' before defending it against (possibly Gnostic) perversions:

A   Who was manifested in the flesh,
B   vindicated in the Spirit,
B   seen by angels,
A   preached among the nations,
A   believed on in the world,
B   taken up in glory. (1 Tm 3:16)

How these carefully structured, rhythmically arranged lines are to be understood is a matter of discussion. Following Hanson and Houlden, I have used A to indicate the earthly world and B the heavenly. Hence, the general thrust of the formula may be that the Christian 'mystery' has full heavenly authority while operative in the earthly world (so that the prohibition of marriage and certain foods, mentioned in 1 Tm 4:3, is misguided). But for this investigation a decision about the structure of this formula is not very important. It is clear that the *mystery* runs from the epiphany to the exaltation without any mention of the death of Christ.

Finally the apostle encourages Timothy, as the man of God, to keep the commandments unstained and free from reproach until the appearing of our Lord Jesus Christ (1 Tm 6:13–14). He speaks with great solemnity 'in the presence of God who gives life to all things, and of Christ Jesus who in his testimony before Pontius Pilate made the good confession'—as indeed Timothy had made the good confession *(tēn kalēn homologian)* before many witnesses (verse 12). In that case Timothy is being reminded of his baptismal profession (or possibly his ordination) by reference to Christ's testimony before Pilate (silence in the synoptic accounts, religious argument in the Johannine) or possibly *in the days* of Pontius Pilate (Kelly). Here at last in the Pastoral Epistles is a traditional formula that preserves a historical memory of the death of Christ, though, even so, it does not actually say that he died, still less that he was crucified. The traditional formula, which makes an uncertain comparison and belongs to a paragraph that sits oddly between the encouragement to moderation in verses 6–10 and advice about dealing with the rich in verses 17–19, is used to encourage heroic endurance. The death of Christ is exemplary, whatever theological consequences may be implied elsewhere.

Summary

It is a notable feature of the Pastoral Epistles that statements about God and Christ stand out from their contexts by their formulaic character. The writer uses them for apologetic reasons, to advance his polemic, to add dramatic force and moral earnestness to his advice, to induce a proper solemnity. It is often effective, sometimes extravagant. The writer makes use of theological statements without giving the impression that he is capable of thinking theologically. He has a powerful sense of the majesty of the One God, of his saving purpose and grace. The man Jesus Christ is the sole mediator between this majestic Being and the multiplicity of ignorantly wicked human beings who are the declared objects of God's mercy. Indeed, Christ came into the world to save sinners, so that God's grace has its epiphany in him, and he will judge the living and the dead at his future epiphany and realization of his kingdom. In this kind of theology the word *epiphany* is more than a hellenistic cliché. It means not only disclosure of the majestic Being to those who can perceive him in the earthly world but also confirmation of that disclosure in the heavenly world. It involved the self-giving of Christ to release us from wrongdoing and to be the means of release for all mankind. Christ in fact died in the time of Pontius Pilate (showing us how to make a good confession) and rose from the dead, thus bringing life and immortality to light. If in baptism we have died with him, we shall also live with him. That is a sufficient structure of wholesome doctrine, and Christians can devote their energies to sobriety and modest behaviour.

5

The Synoptic Gospels and Acts

MARK AND THE OTHER SYNOPTICS

When, in an enquiry about the death and resurrection of Christ, we turn to Mark's Gospel, it is necessary to distinguish the kinds of material we may expect to find.

There may be material of use to a historian, for example, the date of Christ's death, the customary practice of crucifixion, the administrative powers of the Roman governor and the Jewish Sanhedrin, the plausibility of the trial narratives, and the possible reasons for Christ's condemnation and execution. It would have some bearing on our theological interpretation (ours and Mark's) if it could be decided that Christ died before or after Passover or could be shown that the Jews had no hand in his condemnation. But such historical matters are best examined in a comparative study of the Gospels, and it can scarcely be said that historical questions have any bearing on Mark's treatment of the resurrection.

There may be statements—attributed to Jesus or not—that can be extracted from Mark's narrative and used to throw light on the subject; but if so, there are difficulties.

Apart from a possible reference in Mk 2:20 ('when the bridegroom is taken away'), there is no saying of Jesus in the first half of the Gospel relevant to his death. Thereafter there are three predictions of the death and resurrection, which teach about the cross of Jesus' disciples and his experience, which they will share (described as cup and baptism); the service of the Son of Man and his life as a ransom; the parable of the wicked husbandmen; the prior anointing of Jesus' body for burial; predictions of betrayal and denial and falling away; a promise that he will go ahead of his followers to Galilee; the designation of bread and wine as his body and blood; the prayer in Gethsemane; and the apparently despairing word from the cross. These quite numerous sayings make it plain that the death of Jesus and (with less emphasis) his resurrection were God's intention; but why they were necessary and what they were expected to achieve is indicated but seldom—

for instance, in the ransom saying and the designation of his body and blood, perhaps also in the vineyard parable. The character and distribution of these sayings is somewhat perplexing. They cannot sufficiently be accounted for on the assumption that Mark gave verbatim records of the words of Jesus and plausibly displayed a decisive change of attitude on his part from confident proclamation of the kingdom in the former half of the Gospel to ominous announcement of his death in the latter half.

Apart from the actual narrative of the passion, Mark's own contribution is formal: in the former half of the Gospel a statement that the Pharisees made plans with the Herodians to kill Jesus (3:6), the betrayal by Judas (3:19) and the Baptist's death (6:14–16); in the latter half that Pharisees and Herodians tried to trap him in debate (12:13) and that chief priests and scribes sought a way to destroy him and made plans to arrest him by stealth and kill him (11:18, 14:1–2). These have the appearance of narrative links; if so, they should not be extracted from Mark's narrative as if they were independent items of information.

That comment, however, leads on to the further observation that not even the sayings attributed to Jesus should be extracted from the Markan narrative. Their significance and force, at least in Mark's judgement, must be discovered from the composition in which he has set them. No doubt Mark used traditional units, preserved either singly or in small groups, and fitted them together into a continuous narrative, which he called a Gospel. The included sayings of Jesus play their part in his conception of the Gospel, and that determines where we must first look for their significance. But since the traditional units were previously preserved to meet the needs of early Christian communities, the sayings of Jesus may have given a somewhat different impression at that earlier stage; and even that impression may not have coincided with the intention of Jesus when he uttered them. However that may be, it is sensible to begin by attempting to discover Mark's intentions and his theological understanding.[1]

Thus, we are required to consider the structure of the Gospel, to discern if we can the principles on which Mark arranged his material, and to notice recurrent themes that convey his intentions. This is a relatively new approach. At the end of last century, E. P. Gould's commentary simply bestowed titles on successive episodes (following indeed the tradition of Codex A)[2] without suggesting that they were arranged in any significant order. More recently, on the grounds that very little can be discovered about Mark's scheme if he had one, Haenchen did the same.[3] But in the first half of the present century, when it was firmly believed that the Gospel gave a chronological though incomplete history of Jesus,[4] it became commonplace to mark the progress of the narrative by such indicators as the following: introduction, the Galilean ministry, the height of the Galilean ministry, the ministry beyond Galilee, Caesarea Philippi and the journey to Jerusalem, the ministry in Jerusalem, and the passion and resurrection.[5] That is at least reasonably convenient. It is undeniable that Mark often names places (though it is sometimes difficult to make topographical sense of them) and adds vague notes of time. The thematic character of his Galilean narrative changes notably after the episode at Caesarea Philippi. It can be argued that Mark regards

Galilee as a place of good omen and Jerusalem as a place of bad omen—though whether that conveys theological symbolism or provincial rivalry is another matter. It is worth noting that Galilee appears nowhere in the New Testament except in the Gospels and Acts, and that Mark's presentation (in which Jesus is active mainly in Galilee, with only one significant visit to Jerusalem) conflicts with John's presentation (in which Jesus is mainly active in Judaea and makes only occasional visits to Galilee). Hence, there is a temptation to read Mark's topography as partly symbolic and to make the best one can of the rather inconsequential sequence of many episodes. Even some exegetes who are strongly persuaded that Mark's intention was theological, not historical, preserve relics of a topographical framework; and those who prefer a theological framework neither agree with one another nor offer a persuasive description of Mark's intentions.[6]

Hence, it is reasonable to ask whether the evangelist has not left structural indications that have partly escaped notice because his modern interpreters expect something different. After all, it is well known that Mark used stylistic devices (such as the sandwich narrative enclosing one episode between the separated parts of another) and made extensive use of duality.[7] The neglected indicators may perhaps be provided by certain features that have puzzled or irritated exegetes: for example, the grotesque narrative of the death of John Baptist and the two narratives of miraculous feedings. It can scarcely be supposed that the Baptist's death was included for trivial reasons; and Mark himself makes it plain in 8:18–21 that the presence of two feedings was deliberate. Hence the suggestion may be made that Mark shaped his narrative by means of paired units, using the method of *inclusio*. The narrative is thus built up from five main groups of material. It is not implied that the narrative can be cut into separable sections; clearly, the narrative as a whole has a continuity provided by the passage of time, the movement from place to place, by connecting links and cross references. But the five groupings provide significant indications of the manner in which Mark views the material.

Mark begins with John Baptist in 1:1–8, defines the beginning of Jesus' activity in 1:14 as 'after John had been handed over', refers to John's disciples in 2:18, and describes his end in 6:16–29. That comprises the *first* group. Links with subsequent groups are made by 8:28, which reproduces the conjectures about John, already reported in 6:14–15; and by Jesus' polemical use of the Baptist's reputation when discussing his authority with Jewish leaders in 11:27–33. The *second* group 6:30–8:21 begins and ends with miraculous feedings, and most of the material is related to feeding. The description of Christ's blessing and distribution of the bread seems to be echoed in the later narrative of the Last Supper. The *third* group is clearly indicated by the restoration of sight to the blind man of Bethsaida in 8:22–26 and to blind Bartimaeus in 10:46–52. These are the last two healings recorded in the Gospel, only here is blindness mentioned, and nowhere else are there *two* examples of a particular type of cure. The healing at Bethsaida has cross-references to the cure of the deaf stammerer in 7:31–37 (removal from public view, and the therapeutic use of spittle and touch); and the deaf stammerer is rather artificially linked with the exorcism of the dumb and deaf spirit in 9:14–29, the final exorcism reported in the Gospel. This third group also contains the three passion predictions, and a good deal of teaching on rather controversial subjects. The *fourth* group begins with Jesus giving instructions for two disciples to find a colt by prearrangement, and his consequent entry into Jerusalem and the Temple. It probably ends with Jesus giving instructions to two disciples to prepare the Passover in a prearranged room and his consequent meal with the disciples and disclosure of his intentions. The *fifth* group begins with a prediction that the twelve will fall away and a promise that Jesus will go before them to Galilee (14:28) and ends with a message for the disciples and Peter that Jesus is indeed going before them to Galilee (16:7).

Hence, the structure of the Gospel may be displayed thus:

1. 1:1–6:29   In the Shadow of John Baptist
2. 6:30–8:21   The Problems of Feeding
3. 8:22–10:52   The Problems of Perceiving
4. 11:1–14:25   The Temple and the Kingdom of God
5. 14:26–16:8   Death and Resurrection

In the Shadow of John Baptist

In this first part of the Gospel, Mark sets the Galilean scene. Jesus briskly sets about proclaiming the presence of God in his kingly power. That is his response to the handing over *(paradothēnai)* of John Baptist. The verb could properly be translated 'taken into custody' or something of the kind and usually is; but it plays so important a part in the Passion narrative and other references to the death of Christ that it may well imply that John was handed over by God to those who would destroy him. When that happened 'Jesus went into Galilee proclaiming the gospel of God and saying, "The time is fulfilled and *ēngiken hē basileia tou theou* [conventionally translated 'the Kingdom of God is at hand']; repent and believe the gospel'" (Mk 1:14–15). The conventional translation is spectacularly misleading if 'Kingdom of God' is understood to be a transformed and reconstructed social life for God's people. When allowance is made for reverential Jewish idiom, *basileia tou theou* means God in his kingly power.[8] The one thing that Jesus has to proclaim is that the divine power is to hand. In a sense, nothing more need be said. It is often remarked that Mark, at least when compared with Matthew and Luke, gives very little of the teaching of Jesus, and with reservations that is true. But what more needs to be said? The prophetic herald of the Lord's coming (Mk 1:3) has been sent to his death; whereupon the authentic representative and agent of the Lord—God's beloved Son endowed with Holy Spirit (Mk 1:10–11)—declares in Galilee the immediacy of the divine power.

As Mark tells the story, it does indeed seem to be so. There are typical examples of exorcisms and healings, summary accounts of great crowds coming from far and near, and four narratives of exceptional acts of power. These miraculous episodes, moreover, are intended to be *instructive*. The exorcism of an unclean spirit on sabbath (1:23–26), like the restoration of a withered hand on sabbath (3:1–6), presumably has a bearing on the Christian view of the Jewish holy day—

so also the reduction of a fever in 1:29-31. The exorcism is placed between the astonishment of his sabbath hearers when he taught with authority, not as the scribes, in the Capernaum synagogue and their subsequent perception that this new teaching could command unclean spirits and make them obedient (1:21-22, 27-28). The cleansing of a leper gave instruction for fulfilling the Mosaic requirement (1:40-44); the restoration of the paralysed man had a bearing on the forgiveness of sins (2:1-12). The four extraordinary acts of power in 4:35-5:43 present in narrative form the main characteristics of the divine kingship as they appear in the Psalter: that God ruling as King controls the raging of the sea and the raging of the heathen and that to him belong the issues of life and death.[9]

Yet all these activities are contentious. In 3:20-35 they are opposed on two levels: by his family, who think that Jesus is beside himself, and by Jerusalem scribes, who regard him as possessed by a demon. In reply, Jesus for the first time makes use of parabolic sayings as a method of defence, threatens his accusers with God's unrelenting hostility, and in effect repudiates his own family. Yet despite this robust self-defence, it is clear that the healing and exorcising activity was something of a problem, perhaps for Jesus himself, certainly for the early Christian communities. Signs and wonders were expected to happen, and they did. It was inevitable that they should become known and should put a particular stamp on the life of the communities. But it was one thing for the communities to take them as they came and refuse to make them the centre of attention; it was another thing for the communities to exploit the persons healed and the methods of healing as a way of gaining influence. In doing that, they would properly lay themselves open to charges of magic and madness. Hence the notorious Markan instructions for secrecy to be kept, at least on some occasions. In fact there are three kinds of situation: (1) sometimes knowledge of the healing caused no embarrassment (Peter's mother-in-law in 1:29-31, the woman with the issue of blood in 5:25-34) or was indeed encouraged (the paralysed man in 2:1-12, the withered hand in 3:1-5, the legion demoniac in 5:1-20); (2) sometimes knowledge of the healing could not remain unknown (the leper in 1:40-45, Jairus's daughter in 5:35-43), but the command of silence, even if ignored, was an attempt to safeguard the interests of the restored sufferer; but (3) when exorcisms were performed, silence was always commanded (the unclean spirit in 1:23-26, the multiple exorcisms in 1:32-34 and 3:11-12, the muzzling of the storm demon in 4:35-41). It is presumed that these are manic demons: they are noisy and talkative, and the recognized way of handling them is to compel silence. (When, later in the Gospel, a depressive demon is exorcised, there is no command of silence, 9:14-29.) The demons, of course, resist and try to blunt the attack on them by claiming to know what is happening and who their opponent is: 'What have you to do with us, Jesus of Nazareth?'[10] Have you come to destroy us? I know who you are, the Holy One of God' (1:24). Indeed the legion demoniac is more explicit: 'What have you to do with me, Jesus, son of the Most High God? I use the power of God to stop you tormenting me'; whereupon Jesus forces the demon to disclose its own name (i.e., its demonic character) and then commands it at will. (The silencing is present, but differently handled.) In the first Markan summary the demons are silenced because they know Jesus; in the second summary they fall down before him and shout out that he is the Son of God. It is true that God's agent (and Jesus is at least God's Son in that sense) subdues the demons agent (and Jesus is at least God's Son in that sense) subdues the demons because when he utters a command it is God's command. The silencing of demons because when he utters a command it is God's command. The silencing of demons has two grounds: it belongs to the recognized technique for exorcising manic demons, and it is unfitting that Christian confession of Jesus as Son of God should be spoken on the authority of demons. Perhaps, indeed, it would be useless to acknowledge the testimony of demons; for whereas Jesus could overcome the malice of demons, he could not contend with the malice of his fellow men.[11]

Patterns of Hostility.   That remark is prompted by the patterns of hostility that Mark sets up in the first part of the Gospel. There is the hostility of unclean spirits and demons (Mark seems to use the two terms indifferently), which may sometimes be described as the hostility of Satan, though *Satan* is old-fashioned and formal.[12]

Disapproval of Jesus and hostility towards him is shown by scribes and Pharisees. Mark may not distinguish them (perhaps correctly for his own day and Pharisees. Mark may not distinguish them (perhaps correctly for his own day and even for the time of Jesus),[13] and he finds no need to explain who they were (contrast 7:3–4). In the synagogue at Capernaum Jesus' authoritative teaching is preferred to that of the scribes (1:22), who think that Jesus' pronouncement of forgiveness is blasphemous (2:6–7). The scribes of the Pharisees (i.e., who belonged to that religious movement) object to his eating with tax collectors and sinners (2:16); and it was noted that Jesus' disciples did not fast, as did John's disciples as well as the (disciples of the) Pharisees (2:18). The Pharisees ask why the disciples of Jesus improperly pluck corn on sabbath (2:24), observe that he himself restores a withered hand on sabbath, and then promptly collude with the Herodians in a plan to destroy Jesus (3:6).[14] Already the first indication is given that Jesus' successful reputation as an agent of divine power *(exousia)* involves behaviour that some Jewish leaders find abhorrent. Since any religious community maintains itself by relation to its tradition, Jesus must obviously have appeared as an enemy. Only rarely (and so far not at all) does Jesus himself identify scribes and Pharisees as his enemies, an indication that it was Mark who set the scene, no doubt in relation to issues that concerned the churches when he was composing his Gospel.

The story units in 2:1–3:6 constitute a section of five controversies, either because Mark incorporated an already-existing collection (though it is not easy to see what purpose such a collection would have served) or because Mark devised the structure to demonstrate a theological intention. It has been argued that the section forms 'a tightly-constructed literary unit, predominantly chiastic in principle: the first two stories having to do with sin; the last two with sabbath law; the first and last stories deal with resurrection-type healings; the second and fourth with eating; and the middle one with fasting and crucifixion. . . . Thus, Mark employed the controversy stories theologically to place Jesus' life in the context of his death, and he used them in his narrative construction to show how

Jesus' death historically was to come about'.[15] To say, however, that the middle story has to do with fasting and crucifixion is to make a judgement about a very complicated discussion.[16] The episode runs thus:

18   John's disciples and the Pharisees were fasting;
      and people came and said to him,
      'Why do John's disciples and the disciples of the Pharisees fast,
      but your disciples do not fast?'

19ᵃ   And Jesus said to them,
      'Can the wedding guests fast while the bridegroom is with them?
  ᵇ   As long as they have the bridegroom with them,
      they cannot fast.

20   The days will come,
      when the bridegroom is taken away from them,
      and then they will fast in that day'.

The following exegetical suggestions may be made.

1. The original purpose of the pronouncement in 19ᵃ was a defence of eating with sinners, as in 2:16.
2. The bridegroom is God himself (as not uncommonly in the Old Testament, e.g. Is 62:5) coming to renew his marriage relation with an unfaithful people; and Jesus is the agent empowered to offer the marriage contract. In other words, this is a picturesque way of saying that 'the Kingdom of God is at hand'. It is a time for expecting, and rejoicing in, the divine generosity.
3. That being so, fasting is inappropriate, even perverse. For fasting is a means of inducing the deity to confer benefits he would otherwise withhold.[17] To fast when God is present in his gracious sovereignty is to misread the *kairos*.
4. It is so misread by John's disciples (Mt 11:17//Lk 7:32 ' "We piped to you and you did not dance; we wailed, and you did not mourn". For John came neither eating nor drinking, and they say, "He has a demon"; the Son of man came eating and drinking, and they say, "Behold, a glutton and a drunkard, a friend of tax collectors and sinners" '); and by Pharisaic piety.
5. While the bridegroom is with them, fasting is wholly misconceived. But— an ominous corollary is added—the time will come when the bridegroom is taken away. No matter that this does not conform to normal wedding procedure, for it speaks of God's withdrawing his intention of making the people his bride.[18] And then they may well return to fasting in the vain hope of recovering what they had lost.
6. No doubt Mark expected his Christian readers to relate that warning to the death of Christ, but it is scarcely a dominant intention of this passage. For it reverts at once to the *reductio ad absurdum* use of parables: it is ridiculous to fast at a wedding, equally ridiculous to use brand-new cloth (for the wedding garmet?) to patch an old cloak, or to store new wine (for

the wedding feast?) in old wine skins. Hence, the note of warning appears only briefly and uneasily; but to the reflective reader it may imply that if the divine bridegroom withdraws his offer, his agent may suffer in consequence.

It is possible that Mk 3:21, 31–35 is intended to display hostility towards Jesus on the part of his family, matched by Jesus' repudiation of their claim on him. But it is perhaps more likely that his family's anxiety is reported simply as a pretext for Jesus to make an impressive statement about the nature of his extended family. The same point is really made in the reverse way when the villagers of Nazareth refuse to regard Jesus as anything more than one member of a local family (6:1–6). This scarcely looks like destructive opposition.[19]

So much for patterns of hostility. In part 1 Mark provides Jesus not only with opponents but also with supporters. On entering Galilee after John's arrest, the first action of Jesus is to recruit four followers (1:16–20). Correspondingly, the last episode before the story of John's death is the sending out of the twelve, two by two (6:7–13); and the transition from part 1 to part 2 is neatly done by a contrast between the disciples of the Baptist burying their master and the disciples of Jesus reporting to him on their mission. As we would expect, the disciples are insiders, not outsiders (4:11): they witness Jesus' actions, hear his interpretations, and make discoveries about him. Sometimes they seem to cause trouble for him, sometimes they seem slow-witted; but Mark, in familiar haggadic manner, is simply using their shortcomings to provide Jesus with an opportunity of teaching. Neither Jesus nor Mark shows hostility to the disciples—with one exception: the appointment of the twelve ends with 'Judas Iscariot who betrayed him' (3:19).

So in part 1 Jesus, supported by his disciples, successfully overcomes demonic hostility and draws a great following. He stands firm against human hostility with only one or two hints that all may not be well. But the story that begins with John Baptist's proclamation ends with John's death at the hands of Herod, in grotesque and humiliating fashion. This unseemly episode (abbreviated by Matthew, omitted by Luke) says, in effect, 'This is what can happen to a man sent by God'; and there can be little doubt that Mark meant it to say that.

The Problems of Feeding

It is clear that the interrogation of the disciples in 8:14–21 is intended to bring together the two feeding narratives and to draw conclusions from them.[20] Those conclusions may apply generally but more likely apply to the intervening material. They are in principle quite simple: once the children have been fed there is plenty left over for others. Whatever the circumstances, however great the need and the numbers involved, what Jesus provided—from almost nothing—was enough for Jew and Gentile. Since it is a basic principle that 'man does not live by bread alone but by everything that proceeds out of the mouth of the Lord' (Dt 8:3), feeding miracles are symbolic of the Lord's teaching. Therefore Jesus' abolition of the purity rules for food (7:1–23) is nourishing teaching, persuasively extended to Gentiles (7:24–30) and perhaps acceptable to Jews if the deaf can be made to hear and the dumb allowed to speak (7:31–37).[21]

If something like that is the implication of the two feedings, Mark is not directing attention to the formulas of blessing, namely, 'He looked up to heaven, said the blessing [of God], broke the loaves and gave them to the disciples' (6:41) and 'He took the seven loaves, said the thanksgiving [*eucharistēsas*], broke them and gave them to the disciples' (8:6). These formulas are obviously parallel to the words of Jesus at the Last Supper (14:22), and it is possible that the actions at the supper are illuminated by the miraculous feedings. But it is unlikely that the feeding stories are intended by Mark to point forward, even in a subsidiary way, to the self-giving of Jesus. It is perhaps possible that the banquet given by Jesus (as Fowler suggests)[22] is to be set against the banquet given by Herod, but it seems far-fetched. Hence, it may be concluded that this part of the Gospel does not refer to the death of Christ. But part 3 certainly does.

The Problems of Perceiving

This part of the Gospel begins with the restoration of a man's sight, requiring two stages before he sees plainly; and it ends with the immediate restoration of sight to a blind beggar. We already know, from the first exorcism in the Gospel, that these acts of healing are both beneficial and instructive. What we learn from this part of the Gospel is that it is easier to cure blindness than to improve perception. It is worth displaying the contents of part 3:

8:22–26	The blind man of Bethsaida
8:27–33	Passion prediction (1) to disciples
8:34–9:1	The cross of disciples (*basileia*)
9:2–10	Transfiguration
9:11–13	Elijah
9:14–29	Exorcism of a dumb and deaf spirit
9:30–32	Passion prediction (2) to disciples
9:33–41	Valid and invalid claims to recognition
9:42–48	Sacrifice to enter the *basileia*
9:49–50	Salt
10:1–31	Questions of divorce, children (*basileia*), possessions, and wealth (*basileia*)
10:32–34	Passion prediction (3) to the Twelve
10:35–45	Position, service and ransom
10:46–52	Blind Bartimaeus

If the two blind men demarcate this part of the Gospel, the three Passion predictions give it a structure. The *first* prediction is accompanied by sayings that transfer the language of crucifixion to those who follow Jesus. The *second* prediction is developed by describing some consequences for followers: even the chosen Twelve have no claim to prominence except as servants of all (9:33–35); even a child received in Christ's name is effectively the representative of Christ (9:36–

37); even a rival exorcist making use of his name has his approval (9:38–40); even the anonymous person who does a kindness to the followers because they are Christ's is recognized and rewarded by God (9:41). If such teaching seems to deprive followers of all status and recognition, it leaves them with decisive influence. For they constitute the locus where Christ is encountered. By their own sacrifice they present the choice between falling into gehenna and entering into life (or the kingdom of God, 9:42–48).[23] When in 10:1 the crowds reappear, the teaching turns from the inner life of a community of followers to matters that relate them to other social groups: divorce, children, possessions and wealth (10:2–31). In these matters, too, there is a disturbing suggestion that established positions must be surrendered by those who desire the benefits of God's kingdom: traditional assumptions about divorce, the growing assumption that followers are no longer children but instructed initiates and the social assumption that possessions and wealth are the result of piety and provide the occasion for it. Taking up the cross has such implications. The *third* prediction leads into the request by James and John for special participation in the forthcoming glory of Jesus and his response in terms of service and ransom. Thus, only at the very end of part 3 does Jesus speak explicitly about the intention of his dying, and that intention arises from a particular conflict within the community of his followers. But it is also true, throughout part 3, that Mark's choice and arrangement of material have indicated that the community's problems have to be solved in the context of death and resurrection. When they really understand that, their blindness will be removed.

Passion Predictions.   Some of these episodes require detailed examination, first of all the three Passion predictions, which can be taken together.[24] They have a common basic structure of three clauses, with variations of wording:

> (1) The Son of man must suffer many things and be rejected by the elders, and the chief priests and the scribes and (2) be killed and (3) after three days rise again. (8:31)

> (1) The Son of man is delivered into the hands of men, and (2) they will kill him; and (3) when he is killed, after three days he will rise. (9:31)

> (1) The Son of man will be delivered to the chief priests and the scribes, and (2) they will condemn him to death, and deliver him to the Gentiles, and they will mock him, and spit upon him, and scourge him, and kill him; and (3) after three days he will rise. (10:33–34)

Clause 1 presents the Son of Man as he falls into the power of his opponents; clause 2 foresees his death at their hands (though opponents are not mentioned in that clause in 8:31); and clause 3 promises his recovery from death (without further reference to opponents). What then is the formula about? If it is about the Son of man's relation to his opponents, why are they absent from clause 2 in 8:31 and from clause 3? Mk 8:31 can best be understood if it began as a threefold statement about the Son of man:

> The Son of man must suffer many things
> and be killed,
> and after three days rise again.

It would then stand alongside 9:31 as a distinct variant of the tradition. Mk 8:31 as it stands has a supplement to clause 1; 10:33–34, with a more elaborate supplement, belongs to the type of 9:31. It cannot be taken for granted that the prospects of the Son of man (who must be considered later) are described identically in the three formulas.

Clause 1.   Clause 1 in 8:31 is reproduced almost exactly by Matthew and Luke except that Matthew substitutes *from* for *and be rejected by,* thus identifying the sufferings as those inflicted by the Jewish authorities. That may well be Mark's intention also, but he introduces the official intervention by the word *apodokimasthēnai* ('rejected'), which appears once more in the Gospel at 12:10–11 in a quotation of Ps 118:22–23:

> The very stone which the builders rejected
> has become the head of the corner;
> this was the Lord's doing,
> and it is marvellous in our eyes.

The builders so referred to are doubtless the vineyard workers of the immediately preceding parable, and they in turn stand for the chief priests and the scribes and the elders who had questioned Jesus about his authority for creating a disturbance in the temple (11:27–28). Mk 9:12 (with a corresponding reference to suffering, though not a close parallel, in Mt 17:12[b]) must be compared with clause 1 in 8:31:

> How is it written of the Son of man,
> that he should suffer many things
> and be treated with contempt [*exoudenēthē*]?

The Greek verb, which comes nowhere else in Mark, is equivalent in meaning to the verb used in clause 1 in 8:31. Further comparison must also be made with Lk 17:25, a surprising intrusion into a collection of Q sayings on the expected coming of the Son of man:

> 24   As the lightening flashes and lights up the sky
> from one side to the other,
> so will the Son of man be in his day.
>
> 25   But first he must suffer many things
> and be rejected [*apodokimasthēnai*] by this generation.

Verse 24 has a fairly close parallel in Mt 24:27; verse 25 has none and is presumably Lukan. Indeed, suffering is a not-unfamiliar theme in Luke: at the Last Supper, Jesus tells the apostles, 'I have earnestly desired to eat this passover with you before I suffer' (Lk 22:15). In postresurrection encounters, he argued from scripture that it was necessary *(edei)* for the Christ to suffer and to enter into his glory (Lk 24:26), and interpreted the scriptures, 'thus it is written, that the Christ should suffer and on the third day rise from the dead' (Lk 24:46). *Suffering* is used of Christ's death in Acts 1:3, 3:18 (foretold by the prophets), 17:3 (proved from scripture). It is so used once by Paul, and appears twice in Hebrews and four times in 1 Peter. So the verb *paschō* ('suffer')[25] represents the death of Christ in important passages, but it is not a Markan word. Hence, 'The Son of man must suffer many things' is not a Markan creation. Nor can it have arisen from a customary Jewish manner of speaking about hardships endured in God's service because *paschō* lacks a recognized Hebrew equivalent and has no significant place in the Greek translation of Hebrew scripture. But it is used three times in 2 Mc and four or five times in 4 Mc in referring to the sufferings of the martyrs.[26] Hence, hellenistic Jewish writing about martyrdom may have given currency to the verb *paschō* for describing both the sufferings and the death heroic men endured at the hands of God's enemies. In Maccabean times those enemies were alien Seleucid rulers and influential urban Jews who conformed to the newly popular Greek style of life. The corresponding enemies in the time of Jesus would be the Roman overlords and their uneasy collaborators in the Jerusalem Sanhedrin.

The Son of man's suffering (which will be explored on pp. 178, 181–84) is linked to the word *dei*: 'The Son of man *must* suffer many things'. In the hellenistic world *dei* often expresses a neutral supernatural necessity, and it is therefore uncommon in Jewish writings.[27] In Mk 9:11 the disciples ask Jesus, 'Why do the scribes say that first Elijah must [*dei*] come?' with obvious reference to God's promise in Mal 4:5 (LXX 3:23) to send Elijah before the great and terrible day of the Lord. That passage neatly combines the rival interpretations of Mk 8:31,[28] namely, (1) Elijah must first come because God has promised in *scripture* that he will, and (2) the promise is part of the sequence of events in the *eschatological drama* God has devised.

Those who rely on the fulfilment of *scripture* point to the word *rejected* (and its possible origin in Ps 118) and to Mk 9:12, as well as to Lukan support (see above), and some think that the suffering servant of Is 53 must be in the background even though direct quotation, allusion and verbal identity are missing.[29] Instead, reference is often made to the numerous complaints in the Psalms about the godly man's unmerited sufferings at the hands of God's enemies. Wis 2:12–20 is a dramatic hellenistic portrayal of conspiracy against the despised godly man; and 5:1–7 shows his escape and triumph, though not, of course, his resurrection. Along similar lines, reference may also be made to the cruel fate of the prophets, which seems to be intended by the Markan allegory of the vineyard (12:1–11) and certainly finds a place in the collection of Jesus' sayings (Mt 23:34–37 = Lk 11:49–51, 13:34).[30]

None of these references can be ignored. But are they sufficient to account for the conviction not that all God's faithful emissaries are bound to suffer but that the 'Son of man must suffer many things' with immediate consequences for his followers? Hence, other interpreters properly call attention to the Jewish *eschatological* scheme (known, for instance, at Qumran), which attributed the sufferings of the godly to the character of this present age, which God had, for a limited period, handed over to the Evil One. But an end time had been set when

God's people would be saved and rewarded for their sufferings. There are traces of a fixed sequence perhaps in Mk 13:7: wars and rumours of wars must *(dei)* take place, but the end is not yet; and in 13:10 the gospel must *(dei)* first be preached to all nations. Hence, 'the sufferings of Jesus were a crucial part of the eschatological drama. So also were the sufferings of John before him and, as chapter xiii should make clear, so are the times of persecution and hardship through which Mark's community was passing'.[31] It may indeed be so. But if Mk 8:31 was spoken by Jesus, does he otherwise give the impression of keeping to a fixed timetable, and deciding that the moment has come for the suffering phase? Or if 8:31 was a community invention, was the community contemplating the eschatological sufferings when it wrote down instructions about divorce, children, possessions and wealth? It is not plausible.

Without doubt, *dei* indicates the intention of God but not necessarily his intention to bring about something already decided. It is perfectly possible to say, A must take place to make B happen. In 1 Cor 11:19 Paul says, 'There must [*dei*] be factions among you in order that those who are genuine among you may be recognized.' Thus, Jesus says what the Son of man must do in order to overcome the opposition to God's intention. This is made clear by the additional words, 'and be rejected by the elders, and the chief priests and the scribes'. The words constitute a denunciation of the three groups as the enemies of God, in much the same way as Hebrew prophets had denounced the religious leaders of God's people. Before the feeding of the five thousand, Mark (6:34) describes them as 'like sheep without a shepherd' in the manner of Ez 34:5, that is, they were without instruction and compassion from their leaders. Since this denunciation provoked or strengthened the hostility of the members of the Sanhedrin, it was predictable that they would act against Jesus. Hence, his words look like a prediction—with one odd feature: the elders are placed *first,* before chief priests and scribes. Everywhere else in Mark the chief priests come first, with the elders second or third or not mentioned at all.[32] Hence, the priority of elders at 8:31 is not due to Mark but perhaps to a clear memory of what Jesus had said. If so, he expected opposition to be led by influential lay members of the Sanhedrin (normally allied with the Sadducean priesthood),[33] perhaps in response to the kind of things that Jesus said about the rich: 'How hard it will be for those who have riches to enter the Kingdom of God!' (Mk 10:24–25). However that may be, the purpose of 8:31 becomes clearer: it is like a prophetic utterance that not only foresees the course of events but—more important—sets them in train. That is why it is twice repeated, with appropriate variations, on the principle that a curse (or a blessing) becomes activated by a threefold incantation.[34] Mark uses the same principle to show the inevitability of Peter's denial ('Before the cock crows twice, you will deny me three times' [Mk 14:30, 72] and the complete rejection of Jesus' plea in Gethsemane ('He came the third time'). So then the words of Mk 8:31 both identify the enemies of God and initiate a plan for dealing with them: namely the suffering, rejection, death and (after *three* days) resurrection of the Son of man. Why that plan should be effective is not immediately indicated, but there is a hint in the word *rejected* and its derivation from Ps 118: the builders used their undoubted power and rejected the stone; when by the Lord's doing the rejected stone was made the head of the corner, the power of the builders was overruled and indeed broken. Thus, the vulnerability of Jesus is designed to entice the religious leaders to come out from their defences and to exercise their power in causing suffering and death to the agent of God. By destroying him, they destroy their own power.[35]

Clause 1 in 9:31 is purely Hebraic. There is a word play on *Son of man/men* (in Aramaic, *bar *naša/bᵉnē *naša* [Jeremias, *New Testament Theology*, p. 260]). *Is handed over* (*paradidotai*,[36] namely, by God, which corresponds to *dei* in 8:31) and *into the hands of* are familiar Old Testament expressions for putting people into the power of others, whether for benefit or harm.[37] Matthew and Luke follow Mark, except in their writing, 'is to be delivered' *(mellei paradidosthai)*, thus reading Mark's present tense as a futuristic present. They understand the statement as a prediction about forthcoming events, whereas Mark (I take it) records what has already happened: the Son of man has left the protection of God's power and is now in the power of men.

The narrative moment at which it becomes clear that God's intention is moving towards its climax is indicated in Mk 14:41–42:

> He came the third time, and said to them,
> 'Are you still sleeping and taking your rest?
> He [God] is far off.[38]
> The hour has come.
> The Son of man is *betrayed* into the hands of sinners.
> Rise, let us b going; see my *betrayer* is at hand.'

In both sentences the verb is *paradidōmi*. In the Old Testament a victory for Israel is often described by saying that God delivered enemies into their hands, a defeat by saying that he delivered Israel into the hands of their enemies. The verb can be used for handing over to, for example, punishment or slavery or death. Thus, in Job:

> *paredōken gar me ho kyrios eis cheiras adikou*
> *epi de asebesin erripsen me*
>
> [The Lord has *delivered* me into the hands of the
>   wicked
> and has thrown me to the ungodly.] (Jb 16:11)

The psalmist relies on God not to *deliver* him into the hands of his enemies (27:12, 41:2), and in Ps 118:18 he says,

> The Lord has chastened me sorely,
> but he has not *given me over* to death [*kai tō thanatō ou paredōken*
>   *me*].

Thus, the words of Jesus echo rather Job than the psalmist. But are they even closer to the suffering servant? In Isaiah 53 the verb *paradidōmi* is used three times in the Septuagint:

All we like sheep have gone astray;
a man has gone astray in his way
and the Lord has handed him over [*paredōken*] for our sins.
· · · · · · · · · · · ·

Therefore he will inherit many
and he will divide the spoils of the strong,
because his soul was handed over [*paredothē*] to death,
and he was numbered with the transgressors;
yet he bore the sins of many
and was handed over [*paredothē*] on account of their sins.

In Isaiah 53 the prophet is writing about the Servant of the Lord, not the Son of man; he is explicitly concerned with sins admitted by the speakers (absent from Mk 9:31); and he does not use the words *into the hands of*. Nobody reading the Septuagint (still less the Hebrew) of Isaiah 53 would easily have constructed Mk 9:31; it is difficult to imagine that Mk 9:31 was intended to recall the Servant. It is much more likely that Mk 9:31 was an original saying expressed in scriptural language and prompted by the situation of Jesus.

The elders have disappeared from clause 1 of 10:33–34 possibly because when the formula was composed, some of them were less hostile.[39] We are left with the chief priests (the Sadducean leaders) and scribes (the mainly Pharisaic professionals). Thus, 10:33–34 is a hybrid from 9:31 (though *is delivered* becomes *will be delivered*) and the denunciatory addition to 8:31—though it is no longer a denunciation but a prediction. Matthew follows Mark; but Luke substitutes fulfilment of everything written of the Son of man by the prophets.

Clause 2.   Mark (unlike the other evangelists) uses the verb *kill (apokteinō)* almost exclusively of the death of Jesus and the parabolic foreshadowing of it.[40] It is conceivable that clause 2 in 8:31 and 9:31 suggested this restricted use. In clause 2 in 8:31 both Matthew and Luke follow Mark, but in clause 2 in 9:31 only Matthew does so; as often, Luke saves some of his material for Acts.

Clause 2 in Mk 10:33–34 sets out a programme of events. It is followed by Matthew with two changes: the spitting is omitted, and the 'killing' is changed to 'crucifixion'. Luke reports only the handing over to Gentiles, and adds 'shameful treatment' to Mark's list. Mark appears to say that the chief priests and scribes will condemn Jesus to death and hand him over to the Gentiles and that they (presumably the Gentiles) will (1) mock him (the soldier's mockery, Mk 15:17–20), (2) spit on him (the soldiers, Mk 15:19), (3) scourge him (*mastigōsousin*), not in fact described, though in Mk 15:15 Pilate orders scourging *(phragellōsas)*, and (4) kill him (he is crucified at 15:24). But that is not all. At 14:65 members of the Sanhedrin spit on Jesus, treat him roughly, and mock him, and the Sanhedrin officers receive him with blows. Later, at 15:29–32, onlookers mock Jesus, and so do the chief priests and the scribes. Thus clause 2 in 10:33–34 has a double fulfilment—at the hands of the Roman soldiers and of the Jewish people and their leaders.

Two consequences follow. First, clause 2 in 10:33–34 may have been, as is widely supposed, not a prediction of future events but a summary of an existing narrative; but it is more likely that it provided the material out of which the narrative was later constructed. Second, the purpose of the narrative, as sketched in clause 2 in 10:33–34 becomes an open question. On the face of it, it might seem that the indignities and cruelties are mentioned in order to increase the pathos of the story or perhaps to conform the narrative to a familiar pattern of the suffering but innocent victim. But mockery and spitting are not mindless activities: they are well-known repudiation rituals.[41] In Mark's passion narrative, when the Sanhedrin had accepted the high priest's ruling that Jesus had uttered blasphemy, all condemned him to death—and by spitting, ill-treatment and mockery they cast him off, thus reinforcing the high priest's tearing of his cloak (Mk 14:63–64). If the mockery was repeated—even by the authorities—when Jesus was on the cross, it is understandable. The man had been a notorious worker of magic, and it was essential to maintain the curse that had been placed on him lest he should break the spell and place a countercurse on his enemies. It is equally simple to understand the soldier's mockery: they had caught and were putting to death a tiresome, possibly dangerous, Jewish disturber. The mockery discredits him and his cause and, conversely, strengthens confidence in the army. But spitting is rather surprising, for it is not a stock insult. Soldiers spit on captives when they have captured people who previously terrified them. It is perhaps unlikely that Jesus had been such a terrifying figure to the soldiery. However that may be, it was probably the intention of the Passion narrative and of the formula that provided some of the wording to insist that Jesus had been repudiated, discredited and rendered helpless by both authorities, Jewish and Roman, with whom he had to do.

Clause 3. In Mark clause 3 is 'After three days [he will] rise again' (with the verb *anistēmi*).[42] Matthew's text says, 'on the third day' and '[He will] be raised' with the passive of the verb *egeirō*.[43] Lk 9:22 has, 'On the third day he will be raised';[44] but Lk 18:33 says 'On the third day he will rise' *(anastēsetai)*. The variations have been much discussed; but it is likely that they are more a matter of habit and fashion than anything else. The Jewish habit of including part days as whole days and the convention of counting three days as a short but sufficient interval modify the apparent contradiction between 'after three days' and 'on the third day'. Leaving aside textual variants, Mark says 'after three days', and Matthew and Luke say 'on the third day'. They may have wished to conform their wording to the timing in the passion narrative, or common tradition may have remembered the expectation in Hos 6:2 that God, having punished his people, would restore them if they returned penitently to him:

> He will heal us after two days,
> on the third day we shall rise again [*tē tritē hēmera anastēsometha*]
> and we shall live before him.[45]

Not that a repentance oracle would be appropriate to the resurrection of Christ, though it might be appropriate to the 'resurrection' of believers in a transferred sense. Incidentally, in that Septuagint quotation *anastēsometha* corresponds to

'God will raise us up' in Hebrew,[46] thus disposing of the suggestion that in Mark's wording the Son of man rises by his inherent energy, whereas in Matthew he is raised by God. All in all, the variation in wording between the evangelists (and indeed between the textual witnesses) is not significant. All agree that the Son of man will be raised up by God shortly after his humiliating death at the hands of God's enemies. So much then for the content of these three logia, but what of their context? In particular, why are they attached to the Son of man, who is mentioned in Mk 8:31 for the first time since the rather isolated references in Mk 2:10, 28?

Towards the end of part 1 of the Gospel, Mark recorded what was being said about Jesus: that John Baptist was risen from the dead, that he was Elijah, that he was a prophet like those of the prophetic age. Herod responded by saying in effect, 'Having beheaded John, now I have the same trouble all over again' (Mk 6:14–16). After the interlude of part 2, Mark begins part 3 by using an artifice to recall those popular conjectures:

> He asked his disciples, 'Who do men say that I am?'
> And they told him, 'John the Baptist; and others say Elijah; and others one of the prophets.'
> And he asked them, 'But who do you say that I am?'
> Peter answered him, 'You are the Christ'.
> And he rebuked them in order that they should not tell anyone about him. (Mk 8:27–30)

The difficulties of that account have long been felt. If Peter's nomination merited a kind of rebuke (about which Mark is not explicit), was Jesus refusing the title or displaying dismay at its disclosure or hastening to modify its impression? Why did he change the subject and begin a statement about the sufferings and triumph not of the Christ but of the hitherto seldom-mentioned Son of man? And why, as a result, did Peter and Jesus have such a sharp exchange of rebukes?[47]

No satisfactory answer is forthcoming if the conversation is taken simply as a dispassionate report. But it can be helpfully regarded as the reflection of a later controversy within Mark's community: Mark asks for a decision on the matter by presenting it in terms of a dispute between Jesus and the disciples. How was Jesus to be regarded? In well-disposed Jewish circles (even perhaps in the 'Q community' if it ever existed) he was a prophetic figure: a successor to John Baptist or the expected Elijah or the disciple of one of the other prophets. His death could be accounted for by the popular view of the prophets' tragic fate. But how did Mark's community regard him? Clearly, Mark would like them to answer, 'He is the Son of God', for his Gospel is directed to that acknowledgement.[48] But in fact they said, 'He is the Christ'. (Of course, Matthew saw Mark's point and made them say, 'He is Christ and Son of the living God' (Mt 16:15); but that is another story.) When Mark was composing his Gospel *Christos* was firmly lodged in the tradition, at least as a proper name (Mk 1:1).[49] Without repudiating it, therefore, Mark used the episode at Caesarea Philippi to bring great pressure to bear with the purpose of returning *Christos* to private use. His motive can perhaps be detected in 13:21–23, where Christians are excitedly looking here and there for the Christ and are warned against false Christs and false prophets. What they meant by *Christ* is not easily decided, because of the variety of Jewish messianic expectation; but there are indications. According to Mk 15:32 the crucified Jesus was mocked as 'the Christ, the King of Israel', and the Roman legal process, from accusation to inscription, depends on a claim to kingship (Mk 15:2–32)—supported, no doubt, by the enthusiastic slogan in Mk 11:9–10: 'Blessed be the coming kingdom of our father David'. Nevertheless, when Jesus is addressed by Bartimaeus as 'son of David', he is being approached as a healer (which is no part of the messianic expectation); and Jesus' controversial use of Psalm 110 looks like an ingenious attempt to dissociate Messiah from son of David (Mk 10:47–48, 12:35–37). Moreover, Jesus' response to Peter's statement and to the high priest's question about *Christos* is at the best ambiguous, even discouraging (Mk 8:29–30, 14:2). Finally, it is possible that Mark regarded the anointing at Bethany as an overt messianic gesture that Jesus indeed accepted— as an omen of his death (Mk 14:3–9).

When, therefore, Peter gives Jesus the title *Christos* and is sharply told to keep it private, in its place comes *Son of man*, which is probably *not* a title at all.

Son of Man.   So at last we enter the most disputed area of Gospel studies, though not in the first instance to ask what Jesus meant by *Son of man*—if, of course, he indeed used it.[50] What did Mark expect his Greek readers to make of *the son of the man*, which in Greek is an odd expression? He would scarcely expect them to be familiar with the idiom of Palestinian Aramaic or to have read a hasty Greek translation of the latest pamphlet from the Enoch scriptorium (supposing the Similitudes to have been available by that time). They may well have found the phrase in the Greek oral tradition, as the Q sayings suggest. Mark has a glancing contact with that tradition (8:38) but otherwise does not draw upon it for Son of man sayings. Even so, how would they understand them? Readers familiar with the Septuagint would perhaps remember that God commonly addressed the prophet Ezekiel as *Son of man,* but in the tradition Jesus is never so addressed. If readers knew the poetic sections of scripture, they would occasionally come across *Son of man* (never *the son of the man*), sometimes in parallel to *man;*[51] and in Daniel 7 they would encounter a visionary figure 'like a son of man'.[52] If they were attuned to the Hebrew manner of speaking, they might draw the right conclusion that words for *man* implied mankind in general and that *a son of man* might indicate a particular human being. Beyond that they would depend on Mark's manner of introducing the phrase and the associations attached to it.

It is well known that Mark introduces *Son of man* abruptly and without explanation in 2:10, 28 and then not again until the first passion/resurrection prediction in 8:31, after which there are eleven more Son of man sayings. It is therefore plain that Mark finds *Son of man* useful, indeed essential, for his treatment of the Passion and triumph of Jesus. It is not difficult to show that the two early and apparently isolated sayings were deliberately placed by Mark (in the shadow of John Baptist) to indicate the sense in which he would use the phrase.

Both are introduced into conflict episodes. In the former (2:1–12) Jesus heals a paralysed man but first declares his sins forgiven, thus prompting the objection that only God can forgive sins. Verse 10 presents a notorious difficulty unless the words are not spoken by Jesus to the scribes but written by Mark for his readers (Cranfield):

> But that you may know that the Son of man has authority
> on earth to forgive sins. (Mk 2:10)

Thus, the first episode is used to define *Son of man* as God's agent on earth to forgive sins and to heal. The second episode (2:23–28) begins with an apparent breach of the sabbath rule, which Jesus seeks to justify by the haggadic example of David. He then gives the ruling that 'the sabbath was made for man, not man for the sabbath' with the additional comment (which must be the presupposition of the ruling, not its consequence) that 'the Son of man is lord even of the sabbath'. Verse 28 is best taken as Mark's addition to the pericope in order to give further definition to the Son of man's authority. It should be noted that nowhere else throughout the Gospels is it suggested that the 'Son of man' has authority to perform cultic actions (like forgiving sins) or to give legal rulings (as on sabbath activities). Consequently, these two sayings cannot be taken into account if we wish to know what *Jesus* said about the Son of man; but they must certainly be taken into account when considering what *Mark* wanted to say about the Son of man. Why he wanted to ascribe precisely these two kinds of authority to him is another question,[53] but he clearly intended to establish the conviction that the Son of man who suffered and triumphed was God's authentic agent on earth. Moreover, by speaking in 2:27–28 of 'man and the son of the man' (i.e., mankind and this particular man) he provides an unmistakable clue for understanding the strange phrase.

The characteristic feature of almost all the other Son of man references is present distress and future glory. That is obviously true of Mk 8:31 and 10:33. In Mk 8:38, which is akin to a Q saying, the contrast is equally present:

> Whoever is ashamed of me and my words
> in this adulterous and sinful generation,
> of him will the Son of man also be ashamed,
> when he comes in the glory of his father with the holy angels.

That logion, however, has been developed from a simpler form. Only here is *Son of man* brought together with *Son of God* (implied by *his father*). When the last line is compared with the Q saying in Lk 12:9,[54] it is most likely that the simpler form read, 'When he comes in glory with the holy angels'. The addition of *his father* corresponds to Mark's preference for *Son of God*; and the simpler form is easily related to Ps 8:5–6 (LXX), where a son of man, made somewhat lower than angels, is crowned with glory and honour.

Further, in the first line it is surprising to find *and my words*, a phrase absent from the Q saying and not required by the episode that Mark is describing.[55] It must be taken as Mark's device for insisting that there is direct authority in the words of Jesus for regarding him who proceeds through shame to glory and thus proves to be Son of God as the 'son of the man' (rather than the 'Christ').

In the first line Jesus refers to himself in the first person and in the third line to 'the son of the man' in the third person. Since Mark has already established 'the Son of the man' as a self-designation of Jesus (in 2:10, 28) and has used the third-person reference for the Passion/resurrection of Jesus, it is unreasonable to suppose that the third line is referring to some person other than Jesus. Later, in the Gethsemane narrative, something similar is found:

> Behold, the Son of the man is betrayed into the hands of sinners.
> Rise, let us be going; see, my betrayer is at hand. (Mk 14:41–42)

Thus, Judas will betray not simply the man Jesus but God's agent on earth. Similarly in 8:38 Jesus endures the shameful rejection of this adulterous and sinful generation but, when 'the son of the man' is crowned with glory and honour, they will discover that in rejecting him they have rejected God's agent on earth. So they, too, will be rejected by God.[56]

Henceforth, I shall take it for granted that the third-person style refers to Jesus in some aspect of his work. So in Mk 9:9–13, a somewhat contrived composition by the evangelist, the witnesses of the transfiguration are instructed to tell no one what they had seen 'until the Son of the man should have risen from the dead' (verse 9), for it is written that the Son of the man 'should suffer many things and be treated with contempt' (verse 12). In Mk 10:45 the disciples are told that the Son of the man 'came not to be served but to serve, and to give his life a ransom for many' (which they are told because the sons of Zebedee had requested special honours when Jesus came in his glory [10:37]). Chapter 13 is mostly given over to apocalyptic distress and disaster, trouble 'far worse than any the world has ever known from the very beginning when God created the world until the present time' (Mk 13:19 with a reminiscence of Dn 12:1). In the confusion there will be false Christs and false prophets who may lead the elect astray (cf. Mark's dissatisfaction with 'prophet' and 'Christ' in 8:27–30) but the Son of the man will come in clouds with power and great glory, and send out angels to gather the elect (Mk 13:26). Distress is replaced by divine protection and the Son of the man is at the centre of the change. When Jesus eats his final meal with the Twelve, he foresees that one of them will betray him, 'for the Son of the man goes as it is written of him, but woe to that man by whom the Son of the man is betrayed' (Mk 14:21); yet he speaks of again drinking the fruit of the vine new in the Kingdom of God (Mk 14:25). Finally, before the Sanhedrin Jesus is interrogated, condemned, spat on, mocked and ill-treated—and promises them that they will see the Son of the man seated at the right hand of Power, and coming with the clouds of heaven (Mk 14:62). Thus Mark uses Son of the man to display the varied significance of contrast between present distress and future glory— with one exception: in Gethsemane there is only the betrayal of the Son of the man, unrelieved by any memory or expectation of glory (Mk 14:41–42).

In this collection of sayings the Son of the man is despised, rejected, betrayed, made to suffer and put to death; his troubled life is intended to serve others, and his death is their ransom. He will shortly rise from the dead and be seen in the presence of God, restored with power and glory. Without doubt the Son of the man means Jesus, who also speaks in the first person of being rejected and betrayed, of drinking the cup and undergoing the baptism. Why then does Mark present so many sayings as statements about the Son of the man? Because Jesus wishes to speak not of what he himself will do but of what God will do through him. When Jesus says, 'People who are ashamed of me now will find that the Son of the man is ashamed of them when he comes in glory with the holy angels', he does not mean by *Son of the man* 'the well-known and long expected figure of the last days' but simply 'the person concerned'. Vermes has said that 'in Galilean Aramaic the *son of man* occurs as a circumlocutional reference to the self. . . . It is employed in a context where humiliation or death are mentioned, but there are also instances in which the avoidance of the first person is motivated by reserve and modesty'.[57] In the Markan Greek renderings of that idiom, reserve and modesty are displayed in regard to both present distress and future glory. Jesus does not represent the significance of his death and resurrection by referring to traditional agents of God (such as the prophet) or to expected agents (such as the Messiah): he simply uses the general anonymity of the Son of the man, which at one and the same time gives his statements the widest application (and often their general quality) and ascribes the saving consequences to God—in accord with the psalmist's plea,

> Let thy hand be upon [the] man of thy right hand,
> [the] son of man whom thou hast made strong for thyself. (Ps 80:17)

Mark retained this peculiar but significant idiom and found a way to express it in Greek: *the son of the man*—which I represented above by 'the person concerned'. It makes sense when Jesus uses it of himself but not if someone else tries to use it of him. To be meaningful it needs the context of Jesus speaking. Thus, in the Gospels it is confined to Jesus (except for the puzzled enquiry in Jn 12:34) and has no application in other Christian writings. To provide an appropriate manner of referring to Jesus, Mark proposes *Son of God*.

Bearing the Cross. After that long examination of the three predictions it is now necessary to direct attention to what Mark writes after the first of them. In 8:34–9:1 he cleverly uses and adapts sayings from the tradition to give the death and resurrection of Jesus an analogue in the expectations of disciples. What Jesus foresaw and what Christians afterwards preserved in credal form could be read as an indication of their own prospects. The section begins with taking up the cross and ends with the arrival of the Kingdom of God with power, and it combines as many as six logia. It begins with Jesus calling to himself 'the multitude with his disciples'. A crowd at Caesarea Philippi is unexpected (Matthew omits it), but Mark is extending the teaching to the expanding mission of his community. Indeed the passage is infused with mission language. *Come after me* recalls the summons to the first fishers of men in Mk 1:17, 20. *Follow me* recalls the earliest followers in 1:18, 2:14–15 and prepares the way for the rebuke in

9:38–40 when disciples inhibited an exorcist because he was not *following* them; for the instruction in 10:21 to the rich man that he should sell all and follow Jesus; and for the promise to Peter in 10:28–31 (who had said, 'We have left everything and *followed* you') that they would receive a greatly extended family in this present time (with persecutions) and in the age to come eternal life. In that very passage 'for my sake and the gospel' picks up the words of 8:35 (*gospel* also in 1:1, 14, 13:10, 14:9—otherwise strongly Pauline). 'The whole world' in 8:36 comes again in 14:9, 'wherever the gospel is preached in the whole world' (*kosmos* a strongly Pauline word, especially in 1 Corinthians). The verb *to gain* in 8:36 can mean to gain profit by trading—which would be possible but not obviously relevant here; but it can also mean to gain converts, as in 1 Cor 9:19–22 and 1 Pt 3:1, and that sense should be considered. All in all, the passage is not so much concerned with the personal lives of Christians as with the Christian community going about its mission of preaching the gospel.

8:34. The first saying has its corresponding parallels in Matthew and Luke (in the latter with a significant addition), and there is also something very similar in Q.

> If any man would come after me,
> let him deny himself
> and take up his cross [Lk 9:23 adds *daily*]
> and follow me. (Mk 8:34ᵇ)

The Q saying has the form not of an instruction but of a ruling:

> He who does not take [bear] his [own]
>   cross
> and follow [come] after me
> is not worthy of me [cannot be my disciple].
> (Mt 10:38, Lk 14:27)

In Matthew it follows a warning against loving parents and children more than Jesus; in Luke it is preceded by the statement that anyone who does not hate his own parents and family cannot be a disciple.[58] Hence, *discipleship* means the risk of being reckoned undutiful sons, bad husbands, dangerous agitators; and he who becomes a disciple accepts these risks with his eyes open, and is prepared to accept all the consequences that may follow.[59] According to Hengel, among the Romans, crucifixion 'was inflicted above all on the lower classes, i.e. slaves, violent criminals, and the unruly elements in rebellious provinces, not least in Judaea.'[60] The stock example of ruthless official action is provided by Varus, who in 4 B.C.E. put down disorder that followed the death of Herod with two thousand crucifixions.[61] But could Jesus really have thought that the family disturbances of his Jewish followers would become so threatening as to bring on them the final and most terrible Roman punishment of crucifixion? Or, returning to Mk 8:34, is it likely that Jesus was inciting his followers to attack Roman administrators at the risk of terrible vengeance—though in fact none of his followers was crucified? The answer must be *no*. If *cross bearing* is to retain a literal, historical meaning, the sayings must derive from a period after the resurrection when Christians were expecting the outbreak of savage persecution, hinted at in Mk 10:30 (quoted on p. 185) and made explicit in Mk 13:20: 'If the Lord had not shortened the days, no human being would be saved' (Haenchen, p. 297). In that case the section Mk 8:34–38 is an impassioned call to martyrdom *for every Christian,* even though 9:1 says that some will not need to die. But if so, what follows? Nothing to encourage Christians keyed up for martyrdom. The vision of Jesus transfigured has no comforting words about the fate of the persecutors, and subsequent episodes deal with epilepsy, communal rivalry, divorce and wealth. Even the two succeeding passion/resurrection predictions show no urgency about martyrdom; instead they lead up to the statement that the Son of man will give his life as a ransom for many, that is, he will die and they will not.

Perhaps, then, *cross bearing* is metaphorical. In the ancient world crucifixion was known as the slave's punishment, intended to punish, to humiliate and to deter. Thus, 'taking up his cross' could perhaps imply accepting extreme humiliation, rejection by his family and community, saying *no* to himself, even (in the phrasing of Lk 14:26) hating his own life. That Luke's parallel to Mark speaks of taking up the cross *daily* does not allow us to conclude that Luke has converted cross bearing into the endurance of commonplace hardships. He is simply aware that self-abnegation is not so much a decisive renunciation as a daily necessity. The model is Jesus who leaves on one side the title *Christos* and adopts for himself the anonymity of 'the person concerned'. In proper Jewish manner someone who goes after and follows a teacher learns from both his words and his practice.

8:35

> For whoever would save his psyche will lose it;
> and whoever loses his psyche
>   for my sake and the gospel's [Mt and Lk omit 'and the gospel's']
> will save it.

The Q parallels (in Mt 10:39 immediately linked with the preceding saying, in Lk 17:33 separated from it) have 'find . . . lose' and 'seek to gain . . . preserve'. These, however, are translation variants since Mark uses *save (sōzō)* in the sense of 'preserve'. In typical Semitic fashion *psychē* means 'life and reputation'—not simply the physical life, not specially the interior life ('soul' in a modern sense) but the full personal existence of human beings as they see themselves and are seen by others to be (something like our modern sense of 'identity'). In effect the logion says, 'If you guard your standing you will lose it; if you put it on the line, you will preserve it'.[62] The version in Q^{lk} is as simple as that; the version in Q^{Mt} adds 'for my sake'; so does Mark, who also adds 'and the gospel'. Putting your psyche to the test for Christ's sake means 'for the sake of copying his example of humiliation'; and 'for the gospel' means 'for the sake of spreading the gospel and gaining converts.'

8:36. That thought is carried on in the next saying:

> For what does it profit a man
> to gain the whole world
> and forfeit his psyche?

The word *kosmos* occurs only twice in Mark: here and 14:9, where the gospel is proclaimed in the whole world. So, here, to gain the whole world means to gain converts throughout the whole of society. Is it worth doing *that,* says Jesus, if you discredit your reputation? In other circumstances the saying may mean, 'Why make a large fortune at the risk of your life'; but Mark applies it to the mission of the church and its manner of operation. The church can destroy the very life it is trying to spread, and its own reputation, by the manner of its evangelism.

8:37. Remove a community's reputation and its life is destroyed. Destroy a man's reputation, and what is left?

> For what can a man give in return for his psychē?
> [copied by Matthew, not by Luke]
> So much for the growing emphasis on 'life and
> reputation'.

8:38. Now comes the denouement in terms of 'shame' (not 'denial', which would be appropriate to martyrdom, as in Q). The saying in verse 38 significantly reintroduces the Son of the man (it has already been quoted on p. 182) and says that anyone who is ashamed of Jesus and his words (about suffering at the hands of a sinful and adulterous generation) will get no approval from 'the person concerned' when he comes in glory. *They* are ashamed of his humiliating slave's death; *he* is ashamed of their self-confident judgement on him. It must be noticed that the logia sequence begins with death, that is, taking up the cross, and ends not with resurrection but with the Parousia of God. Mark shows some reserve in dealing with resurrection.

8:39. Finally, there are the enigmatic words of Mk 9:1, introduced by 'And he said to them' which are a common Markan device for presenting the punchline:[63]

> Amen I say to you,
> there are some standing here who will not taste death
> before they see the Kingdom of God come with power.

The initial *Amen* marks this as a saying of special authority. It concerns the divine rule and power—not, as Matthew confusingly makes it, the coming reign of the Son of the man (Mt 16:28). Its intention is not to give (mistaken, or at least misleading) information about the proximate arrival of the Kingdom of God but to set the divine power in opposition to the experience that causes some Christians to be ashamed of Jesus. They are ashamed not in despising Jesus for what he has done but in taking it for granted that his shameful death frees them from any corresponding experience. By his death and resurrection they have entered at once, it may be, into the resurrection condition. The clue lies in the words 'will not taste death' which is a Jewish way of saying 'will join the immortals' (2[4] Esd 6:26). Some early Christians certainly believed that Jesus tasted death for every one (Heb 2:9) and others cherished the promise that anyone who kept Jesus' word should never see or taste death (Jn 8:52–53).[64] No doubt it was intended in some spiritual sense, but it was apparently enough to dispense some Christians from applying to themselves the spiritual consequences of crucifixion. But not tasting death in this saying is opposed to seeing the Kingdom of God come in power. In the episode that follows (six days later according to Mark) Jesus is seen transfigured in the company of the 'immortals' Moses and Elijah (9:2–8). Peter uncomprehendingly proposes to construct three booths, no doubt to detain the three presences on earth, but is told by a heavenly voice, 'This is my beloved Son; listen to him'; and immediately Peter, James and John are alone with Jesus. Mark's point has been made. On the most impressive authority the Christian's model is not (yet) the transfigured person but the Jesus who instructs followers to take up the cross.

Mark, however, is still not satisfied that he has dealt sufficiently with the problem, so he pursues it in a conversation between Jesus and the Three, as they descend the mountain (Mk 9:9–13). This passage is notorious for its difficulties (e.g., Haenchen) but at least some of them can be solved if implicit assumptions of the interchange are made explicit.

What the Three have seen is not to be disclosed until the Son of the man (i.e., the person concerned) has risen from the dead (9:9).

This command becomes a matter of discussion. They are, of course, familiar with teaching about resurrection. But what is here implied by 'the Son of the man'? What is required of Christians in the Markan community? Must they conceal the transfigured majesty until they themselves are 'risen from the dead'? (9:10).

In the earlier conversation (Mk 8:27–28, repeating 6:14–15) it had appeared that some regarded Jesus as (the successor to) John Baptist, and others regarded him as (the promised) Elijah—according to Mal 3:1 the messenger who would prepare the way for the coming of the Lord God. Scribal teaching confirmed that expectation. If Jesus was Elijah (or even the successor of the Baptist as Elijah), surely God had come: Christians could disclose the transfigured Christ and live the risen life (9:11).

Jesus agrees that Elijah comes first (but he moves beyond Mal. 3:1 to Mal 4:5–6) 'to restore all things'. Against that intention must be placed scriptural indications that the Son of the man (the person concerned) 'should suffer many things and be treated with contempt' (9:12).[65]

So it must be perceived that Jesus himself had come to fulfil the role of Elijah and restore all things; but people 'did to him whatever they pleased, as it is written of him' in the scriptural indications just mentioned. Restoration had not yet been made, and Christians must still take up the cross (9:13).[66]

To me that seems the probable meaning of an obscurely difficult passage, but who can be sure what was intended? Mark is not at his best in that style of debate; he is far better when he writes out a story, as he does successfully in 9:14–29. It is linked not merely with the descent from the mountain but also with the previous logia on discipleship and the expansion of the community in 8:34–9:1.[67] The scribal arguments cannot help an epileptic boy, but neither can the disciples. Their ineffective presence simply encourages uncertainty. So Jesus rebukes the deaf-and-dumb spirit, and the boy seems to be *dead* from a convulsion. Jesus lifts him up *(ēgeiren)* and he arises *(anestē):* the dead child is raised.[68] It is beside the point to argue that this was not the return to life of a genuinely dead person. Mark describes an episode to which the death/resurrection paradigm can be applied in order to confront ineffective disciples who might be tempted to believe that they are already living the resurrection life. In an appended comment, when the disciples wonder why their attempted exorcism failed, they are told, 'This kind cannot be driven out by anything but prayer'. What is needed is not an exorcistic technique provided by their invulnerable condition but prayer that asks God to exercise his power in casting out a deaf-and-dumb spirit.

Servant and Ransom.   Then follows the second passion/resurrection prediction, which the disciples neither understand nor question. It is they who are deaf and dumb (Mk 9:32). I have already suggested (p. 184) that the episodic material separating the second from the third prediction illustrates from the affairs of the community, the taking up of the cross. It begins with a dispute about prominence and so provides the context for the request of James and John in Mk 10:35–36. The Twelve are instructed that anyone wishing to be first must be servant of all *(diakonos,* Mk 9:35), thus introducing the theme echoed in 10:43.[69] The commendation of children and little ones in 9:36–37, 42 and 10:13–16 prepares for the self-abnegation described in 10:42–44 and realised in Jesus' acceptance both of the designation Son of man and his destiny.

Mk 10:35–45 is clearly a Markan construction (Pesch, Gnilka) very effectively composed from earlier units. The occasion is a request from James and John, the sons of Zebedee, for a special relation to Jesus in his glory, which (in Mark's terms) means 'when he comes in the glory of his Father with the holy angels' (8:38, cf. 13:26—the only other uses of *glory* in Mark). Two results follow: the pericope is concerned with a community dispute, and at least some members of the community assume that the resurrection of Jesus *is* his glory (as Mt 20:21 substitutes *in thy Kingdom* for *in thy glory)*[70] so that he can and does confer immediate honours. What James and John request is not entirely clear. Reference is often made to the Q saying (Mt 19:28‖Lk 22:30) that promises the Twelve that they will eat and drink at Christ's table in his kingdom and sit on thrones judging the twelve tribes of Israel, a saying Luke, indeed, connects with his account of a dispute at the Last Supper and with Jesus' teaching about service (parallel to Mk 10:42–45). Perhaps so; but that promise may not be determinative for Mark. Thought should be given to the actual request: 'one at your right hand and one at your left'. That is formal cultic language, used of the structure and furnishings of the temple, of the ceremonial array of guards for the holy place, of the divine court, of the formal assembly for the reading of the law, of the two anointed ones who stand by the Lord of the whole earth; and of the protectors of the fugitive David.[71] It is not impossible that James and John are represented as encouraging Jesus (after his third, explicit prediction of suffering and death) by claiming the honour of supporting and guarding him. Moreover, attention must be drawn to the wording of their initial request: 'We want you to do for us whatever we ask of you' (*ho ean aitēsōmen se*, verse 35). Two points must be made. First, Christian tradition, especially in the Johannine communities,[72] preserved a conviction that Christians would receive whatever they asked in the name of Jesus. Mk 11:24 contains a peculiar form of that conviction (the parallel in Mt 21:22 is easier): 'Whatever you ask in prayer [*panta hosa ... aiteisthe*], believe that you have received it, and it will be yours'. So James and John may simply be applying the rule, as some in the community no doubt thought they could. Second, Mark has already told the shameful story of the Baptist's death, a consequence of Herod's rash promise to the daughter of Herodias: 'Whatever you ask me, I will give you' (*ho ean aitēsēs dōsō soi*, Mk 6:23). A careful reader or hearer of Mark's Gospel would be very slow if he missed this parallel.[73] Its meaning is plain: even though the request of James and John is wholly different in motive and intention, the consequence is equally disastrous.

The first reply to their request is elusive and even contradictory (verses 38–40)—not surprisingly, because they do not understand what they are asking for. Two images are presented: the cup and the baptism. *Cup* was a familiar Jewish image for good or bad fortune; *baptized* in popular Greek could mean 'overwhelmed by calamities'; and of course both images were known in early Christian tradition from eucharist and baptism.[74] Within Mark's Gospel, is it the cup of thanksgiving or the cup of terrible destiny (14:23, 36)? Elsewhere in Mark, baptism is associated exclusively with John Baptist: in 1:4, 5, 9 offering a baptism of repentance for the forgiveness of sins (and Jesus was of course baptized); in 11:30 in Jerusalem recalling a baptism that may have been from heaven or from men. The parallel in Mt 20:22 omits 'the baptism with which I am baptized'; but in a context of community dispute Luke has the striking saying, 'I came to cast fire upon the earth; and would that it were already kindled! I have a baptism to be baptized with; and how I am constrained until it is accomplished! (Lk 12:49–50). The indications are that Jesus himself experiences the ambiguity of his situation. It is not that he is passively willing to accept martyrdom but that he must force himself to say and do what will be resisted and so bring suffering upon himself—and on his disciples. That is why he tells James and John that they *shall* drink his cup and undergo his baptism. The statement cannot exclusively imply their martyrdom even if it is regarded as a *vaticinium ex eventu* (despite historical uncertainties); for Mark has already interpreted *taking up the cross* by a variety of deprivations and hardships. So they will in some manner share the sufferings of which they are in part the cause, but whether they will do so in public defiance or private humiliation is not for Jesus to decide. In verse 40, Jesus says, 'To sit at my right hand or at my left is not mine to grant: but [that is for those] for whom it is prepared' ('by my Father', as Matthew rightly adds). Since readers of the Gospel already know the crucifixion story (as Best frequently insists in his *Mark*),

they will not miss the ironical reference to Mk 15:27: 'They crucified two robbers, one on his right hand and one on his left'. Nor should they fail to understand that for Jesus as much as for James and John, the outcome is entirely in the
hands of God.

Jesus' second response is simpler: a contrast between patterns of domination
in Gentile society and patterns of relationship within communities of his disciples. A disciple who wants to be important must be the community's servant
(diakonos); if he wants to be prominent he must be the community's slave *(doulos).*[75] It is a consciously anti-Gentile pronouncement and best suits a community
living among Gentiles. To these verses there is a Lukan parallel with wording
rather closer to language of the early church; and Luke's context for it is a dispute at the Last Supper, thus recalling in some respects the Pauline dispute at
Corinth and in other respects the Johannine foot washing. That observation
becomes significant when we compare the sayings in Mark and Luke that conclude this episode:

> For the Son of man also came not to be served but to serve,
> and to give his life as a ransom for many. (Mk 10:45)

> For which is the greater,
> one who sits at table, or one who serves?
> Is it not the one who sits at table?
> But I am among you as one who serves. (Lk 22:27)

Both present Jesus as a servant *(diakonos)* with the intention of changing the
disposition of his hearers. But Luke relies on argument in the manner of a wisdom teacher who is himself the example of what he teaches; whereas Mark relies
on prophetic assertion by one who is committed to acting on behalf of his hearers.
In Luke the saying has the more formal structure and applies to behaviour in
general; in Mark the saying contains a Christological formula, a direct reference
to death, and the technical word for 'ransom'. For those reasons it is often
thought that Luke provides an earlier (even the original) form of the saying. But
it is unsound to argue that a persuasion towards moral action must be prior to a
statement about theology; it is at least equally possible that a theological statement has been moralised (especially by Luke, who favours simple Jewish piety).
And why should Mark or his tradition spoil the structure of the Lukan saying
and introduce the ransom word *(lytron),* which nowhere else appears in Mark?
The only evangelist to use *lytron*-type words is Luke (see Appendix A), always
in the sense of liberation or redemption and never connected with the death of
Jesus. Thus, the Markan saying has a good claim to be interpreted as it stands.

The interpretation of that saying has been much hindered by two presuppositions. First, it has been taken for granted that *Son of man* was a familiar *title,*
corresponding to an expectation in early Judaism that a heavenly figure, so
named, would be sent by God at the end of the age to rescue the Jewish people.
It is now clear that whatever the Jews thought about heavenly figures, the phrase
Son of man in Aramaic was not a title but a self-designation to be used by a
speaker who wished to be modestly self-effacing[76] when referring to distress or

preeminence. Jesus was 'Son of man' in order to refer everything done through his agency to the will of the divine king. It is entirely God's will that his agent should not require others to serve him ('If I were hungry, I would not tell you; for the world and all that is in it is mine', Ps 50:12) but that he should serve them. This saying was perhaps framed to counteract the feeling, on the part of Peter (Mk 8:32–33), James and John—and even the accompanying women (Mk 15:41)—that Jesus needed protection.[77]

Second, it has also been supposed that *ransom* is a surprising cultic term that consorts badly with the language of service and must therefore be attributed to later hellenistic influence or must be defended as Jewish by deriving it from Isaiah's poem of the suffering servant of the Lord (Is 52:13–53:12). In fact, the saying looks decidedly Semitic when compared with the hellenised form in 1 Tm 2:6 (see Appendix A), and *lytron* ('ransom') is scarcely a Pauline intrusion, for he never uses the word, preferring *apolytrōsis* ('redemption' or 'liberation'). A *lytron* is a recognized or negotiated compensation to someone who is prepared either to surrender what he rightly possesses or not to demand full enforcement of his rights. Here it is not a monetary payment (as it often is) but the offering of one man's life instead of the forfeiture of many lives. There is some analogy (though it is not a parallel) in Nm 3:12 (LXX), where God says, 'Behold, I have taken the Levites from among the people of Israel instead of [*anti*] every firstborn that opens the womb among the people of Israel; they shall be their ransom [*lytra*] and the Levites shall belong to me.' The words of Jesus presume that God's people are suffering disastrously at the hands of the Temple hierarchy (as indicated in the passion predictions) and that their sufferings will continue till God brings them to an end. Unless God shortens the days, no one will be saved (Mk 13:20); but it is possible that God will accept the freely offered life of his agent and not exact the full measure of lives that are forfeit. There springs to mind the tortured and dying Eleazar in 4 Mc 6:28, who prays 'Take my life as a ransom [*antipsychon*] for theirs', and the holy martyrs of 17:19–22 who became 'a ransom [*antipsychon*] for the sin of our nation'.[78] Jewish martyrdom theology—adapted to an internal, not an external, enemy—is behind the words of Jesus.[79]

The Suffering Servant

Nevertheless, it has been said that the saying cannot be understood without the background of Is 53:10–12 (Gnilka, p. 104). Something must therefore be said about the fourth Servant poem. It is by no means easy to understand: sometimes the Hebrew scribe was at fault (see the Isaiah scroll from Qumran), sometimes the Greek translator could not fathom his meaning or wrote something different. The structure of the poem is uncertain, the formal identity of the speakers a matter of guesswork, and the theologically important words have a rather wide range of meaning. If the four poems are read together, it is clear that (1) Yahweh has commissioned his *Servant* (which, in oriental fashion, is a title of honour, not of humiliation), so that (2) Israel is restored to God by undertaking the new task of bringing light to the heathen Gentiles; (3) though the Servant is not aggressive and his work is unrewarding, he is despised by the Gentiles and tormented, yet

(4) he will convince by his very suffering and will finally be vindicated. This is a subtle development of the old argument that unless God bestirs himself for Israel he will lose credibility with the heathen (Ex 32:11 14, Is 48:9–11, Ez 20:9, 14, 22). Even if readers in antiquity did not consider the four poems as a whole—indeed, even if they did not regard the fourth poem as a special unit—surely they could not overlook the constant presence of the nations, the kings and princes and the rulers of the earth throughout deutero-Isaiah. Hence, for them the Servant was Israel or represented Israel. In the fourth poem Yahweh promises that his Servant will be greatly exalted, so that heathen nations and kings will be astonished and put to silence at the transformation of one whose appearance was so disfigured as to be scarcely human (Is 52:13–15). Then the nations respond: they had seen nothing attractive in the Servant: he was humiliated by sickness, God had turned his face from him, and they despised him (53:1–3). But now they knew that Yahweh had not rejected him but laid on him the consequences of their wrongdoing and the possibility of their well-being (53:4–6). Then the image changes from sickness to violence, denial of justice and criminal disgrace—though whether the nations are speaking or Yahweh resumes cannot be decided (53:7–9). Finally, Yahweh announces a constructive outcome, though some of the Hebrew is impenetrable and much of the Greek is divergent. The most that can be said with confidence is that three statements make the suffering of the Servant effective for others. Verse 10 may be a covert plea to Yahweh, and verses 11–12 his response:

10   Yet Yahweh was pleased to crush him, he made him sick.
     If you make his soul a compensation [*ašam*]
     he will see his offspring and extend his days.
     and Yahweh's pleasure shall prosper in his hand.

11   From the trouble of his soul he shall see [light]
     and be satisfied by his understanding.
     My servant shall vindicate himself
     and show himself vindicated to the many,
     for *their* wrong-doings it was that he bore.

12   Therefore I will allot him a portion with the many,
     and he will share the spoil with the mighty,
     because he poured out his soul to death
     and was numbered among the rebellious,
     for it was he who bore the sin of many
     and made intercession for their rebellions.[80]

Thus the unparalleled sufferings of God's people, painfully displayed in the sight of the Gentiles, need no longer be a reproach to God, for two things have happened. The sufferings have changed the Gentiles' attitude towards Israel and have made them repent of their own wickedness; and the people who once were despised are now glorified and given their share of power among the mighty. The Servant poems are not a novel solution for the general problem of forgiveness but a way of restoring credibility to God and hope to Israel.

What, then, is the relation between the fourth Servant poem and the words of Jesus in Mk 10:42–45? It is at the most indirect. In the Servant poem, the speakers are God and 'we'—presumably the Gentile rulers; in Mark the only speaker is Jesus. In the Servant poem the actors are God, the many (nations, kings, men), and the Servant; in Mark the actors are Gentile rulers, disciples, and Jesus. The Servant poem welcomes the favour of the Gentiles and a Jewish share in their power; but Jesus derides the Gentile rulers and warns his disciples against imitating them. It is true that the *'ebed Yahweh* (called *pais* in Greek, a word that does not appear in Mark) suffers and gives his life *(psychē)* because of them and perhaps for their healing (Is 53:5), but his work is not described as a *diakonia,* as it is in Mark. The Servant makes his soul a compensation, which is a possible meaning of *ašam;* though in Greek it is called a sin offering *(peri hamartias),* not, as in Mark, a *lytron.* It is said repeatedly that the Servant suffers because of their sins: he bears them, is wounded and hurt by them, is classed with lawless men, and is handed over to death for that reason (Is 53:4–6, 8, 11, 12). But it does not say that he died in place of many *(anti pollōn)*[81]—as Mark says, without ever mentioning any word for sin. Indeed it almost appears in Mark that the problem of sins had been sufficiently dealt with by the disclosure that the Son of man had power on earth to forgive sins and had come to call sinners (Mk 2:10, 17; see chap. 5, n. 77). Nothing in Mark permits the conjecture that Jesus, having read Isaiah's poem about the grossly disfigured, sick, and violently ill-treated Servant, decided that he must fulfil that prophecy in order to gain Gentile approval and power for his people. Something simpler may have been possible: if the acute suffering of Israel might do so much to transform the attitude of the Gentiles and the position of Israel, how much more might the willing suffering of 'the person concerned' do, in the hands of God, to transform Israel and his own position! Even so, it would tell us rather little about Mark's text.

Matthew and Luke

Before considering the Jerusalem section of the Gospel, it is convenient to notice how Matthew and Luke have treated the Markan material so far considered.

Matthew accepts the boundaries of Mark part 1, with a little extra emphasis on the Baptist's death by the addition that John's disciples 'went and told Jesus' (Mt 14:12). He adds new material from Q and M, brings in a few sayings from Mark 13 (in Mt 10:17–22), and makes an orderly arrangement of three discourses and one miracle collection. The added material contains a few sayings that may refer obliquely to the death of Jesus: the temptation for Jesus to throw himself from the pinnacle of the temple (Mt 4:5–7), the observation that the Son of man was a fugitive (Mt 8:20) and the admission that Jesus came to bring not peace but a sword (Mt 10:34), and the sign of Jonah's three days and three nights in the belly of the whale as a symbol of the Son of man's three days and three nights in the heart of the earth (Mt 12:40)—all from Q. More significant, however, is Matthew's stress on the hardship of disciples, prepared for at the end of the Beatitudes (disciples will be reviled and ill-treated for Jesus' sake, so repeating the experience of the prophets, Mt 5:11–12) and amply developed in Mt 10:16–

39: sent as sheep among wolves (Q), ill-treated by tribunals, their families disrupted (from Mark 13), hunted from town to town until the Son of man comes (M), discovering that a disciple is not above his teacher (Q + M), assured of God's protection if he acknowledges Jesus (Q, cf. Mark 8), called to take up the cross, to find his life by losing it (Q). The suffering and death of Jesus is a model for Christians, given the times in which they live: 'From the days of John the Baptist until now the kingdom of heaven has suffered violence, and men of violence take it by force' (Mt 11:12, Matthew's version of a difficult Q logion).

Mt 14:13–16:12 follows Mark part 2, beginning and ending with the feedings. Mt 16:13–20:34 corresponds to Mark part 3 except that Matthew omits the blind man of Bethsaida, so damaging Mark's symbolic use of a miracle of sight at the beginning and the end.[82] But Matthew's aim is different from Mark's. Having displayed the perilous situation of the church in Chapter 10, Matthew is not inclined to embarrass Peter with a questionable acclamation (Peter gains praise for his forthright 'son of the living God') or to concede that apostolic perception is poor. Jesus announces that Simon Bar-Jonah is Petros, that he will build his *ekklēsia* on this *petra,* and that the gates of Hades (through which Jesus will pass when he dies) will not destroy the *ekklēsia.* Until the Son of man returns in his kingdom (Mt 16:28, changed from Mk 9:1), Peter is given the keys of the kingdom of heaven and so the power of binding and loosing. Hence, the community rules and perhaps also the parables that Matthew adds to the Markan material in this part of his Gospel; but beyond that he has nothing to add about the death of Jesus.

Luke follows the general outline of Mark but is less interested in preserving the Markan composition. He makes an emphatic beginning with John Baptist, having given him a major role in the birth narratives, by adding extra material from Q and L; but then he dismisses him from the scene in Luke 3:19–20. The narrative of John's death, which ends Mark part 1 is simply omitted. Luke does not use the Baptist's death as an ominous indication of Jesus' fate. Like Matthew, he includes Jesus' temptation to throw himself from the pinnacle of the temple (Lk 4:9–12) though in Luke it is the third temptation (not, as in Matthew, the second). At that point the temptations were ended—ruled out by scripture, as Luke would suppose—and the devil went off, biding his time.[83] Thereafter, Jesus appears as the spirit-filled prophet in the Nazareth synagogue (Lk 4:16–30, with the consequent omission of Mk 6:1–6) to announce the acceptable year of the Lord not only for Jews but for Gentiles too if Jews will not listen—whereupon his hearers would have lynched him if they could, but Jesus evaded them (verses 29–30). Already Luke is suggesting that the death of Jesus was prompted by opposition to teaching about God's universal grace, matched perhaps with the complementary suggestion of the miraculous fishing in Lk 5:1–11 (replacing Mk 1:16–20 and akin to Jn 21:1–11), that the risen Christ would produce fishers of men. Luke uses Mark's healing of the withered hand but changes its ending: instead of a conspiracy between Pharisees and Herodians to kill Jesus, the scribes and Pharisees are filled with fury and discuss what they might do to Jesus (Lk 6:11). No doubt Luke was cautious about the Herodian entourage (the wife of Chuza, Herod's steward, had been cured by Jesus and was one of his supporters,

Lk 8:3). He is willing to put blame on Herod himself (Lk 3:19–20) and to report
that Herod wanted to kill Jesus (Lk 13:31), though in the end even Herod could
find nothing against him (Lk 23:7–15). Hence, as Luke writes his Gospel, the
death of Jesus occasionally comes to mind but is not prominent, even when (as
in Matthew) the disciples must expect to be ill-treated like the ancient prophets
(Lk 6:22–23). It is rather more characteristic of Luke's mood that he tells an
explicit story of a raising from the dead (Lk 7:11–17).

Mark part 2 is almost entirely discarded, apart from the first feeding narrative. Mark part 3 is the scene of Luke's most extensive and least systematic alterations. Like Matthew, he omits the blind man of Bethsaida. To the transfiguration narrative he adds the statement that Moses and Elijah appeared in glory and
spoke of Jesus' *exodus,* which he was to accomplish in Jerusalem (Lk 9:31). Two
of the (Jewish) immortals have an impressive conference with Jesus, not about
death but about something more dignified: *exodus.* Luke has added a touch of
pomposity, as Josephus did in writing Moses' farewell speech (*Ant.* 4, 189). The
long, unsystematic insertion Lk 9:51–18:14, coming soon after the second Passion prediction, contains teaching and episodes nominally assigned to the journey
to Jerusalem. It begins with high-sounding words: 'Now it came to pass while
the days preceding his assumption [*analēmpsis*] were being completed that he set
his face to go to Jerusalem' (9:51). It is clear that Jesus, who sent messengers
ahead of him, was making a stately, deliberate progress—teaching, occasionally
healing, and making his way to Jerusalem (13:22). When it is reported that
Herod wants to kill Jesus, he replies,

> Go and tell that fox,
> Behold, I cast out demons and perform cures today and tomorrow,
> and the third day *teleioumai.*
> Nevertheless I must go on my way today and tomorrow and the
>   day following;
> for it cannot be that a prophet should perish away from Jerusalem.
>   (Lk 13:32–33)

Despite critical and exegetical problems (see Marshall, *Gospel*), the broad impression is clear: (1) Jesus goes about his work according to a plan that is not determined by threats from Herod or by the malice of his informants; (2) in due
course—as determined by God—he will experience the hostility of the Jerusalem
establishment and share the well-known fate of the prophets; (3) and so he will
finish his course, reach his goal, and 'be perfected' (reminiscent of *teleiotēs* in
Hebrews)—however the verb is translated. Once again, Luke uses a dignified
euphemism—not because he is squeamish but because for him the death of Jesus
is a solemn and moving component of a predetermined progress.

Jesus' intention of suffering in Jerusalem is complemented by his dirge over
the city that kills the prophets and stones those who are sent to her (Lk 13:34
[Q]), to be compared with his rebuke to the lawyers who build the tombs of the
prophets, whom their fathers killed. The blood of all the prophets, shed from the
foundation of the world, is to be required of this generation, 'from the blood of
Abel to the blood of Zechariah, who perished between the altar and the sanctu-

ary' (Lk 11:47–51). This disturbed situation has consequences for the disciples, but when they are under threat they can expect God's protection if they acknowledge Jesus (Lk 12:2–12 [Q]), who himself is a cause of disturbance, not simply a hapless victim of malice and conflict:

> I came to cast fire upon the earth;
> and how I wish it were already kindled!
> I have a baptism to be baptized with;
> and how I am constrained until it is accomplished! (Lk 12:49–50)

These are entirely Lukan sayings, though the third line is a close parallel to Mk 10:38, part of the challenge made by Jesus to James and John (omitted by Matthew and more extensively by Luke); and Luke 12:51 is Luke's version of a Q saying, 'I did not come to bring peace but a sword [or division]' (cf. Mt 10:34). Thus, 'fire' is a symbol of strife and division. For a century, the second line has been translated as above because it is thought to be 'the required meaning',[84] on the pretext that it is a Semitism (for it cannot be got out of the Greek). But what is wrong with the train of thought, 'I came to set the world on fire, and what more can I wish if it is already alight? I have an ordeal to undergo, and I am so restricted until it is finished'? As Luke writes these words, he implies that the disturbance caused by Jesus is no miscalculation but the beginning of Jesus' necessary submission to the forces of social dissolution and death. So, in teaching on discipleship, Luke includes the Q sayings on bearing one's own cross and following Jesus (Lk 14:27‖Mt 10:38) and immediately follows it by two parables on counting the cost. Towards the end of the great teaching section he manages to insert into a collection of mostly Q sayings on the day of the Son of man the qualifying remark, 'First he must suffer many things and be rejected by this generation' (Lk 17:25). And then, having rejoined the Markan narrative, he rewords the beginning of the third Passion/resurrection prediction: 'Behold, we are going up to Jerusalem, and everything that is written of the Son of man by the prophets will be accomplished' (Lk 18:31—which becomes a stock theme in the Lukan Passion narrative. Finally, Luke omits Mk 10:35–40 (the request of James and John and Jesus' reply) and replaces Mk 10:41–45 by similar teaching in response to a dispute after the Last Supper, though totally ignoring the Markan saying that the Son of man gives his life a ransom for many (Lk 22:24–27).

Thus, John Baptist and Jesus appear as prophets and must expect to suffer a prophet's fate. John's death is removed from prominence, and Jesus' death—as fitting for one who could raise the dead—is described in dignified language. All that must happen is planned by God and laid down in scripture, and everything that Jesus does is carried out according to timetable. He comes to announce the acceptable year of the Lord, and if his fellow countrymen are hostile, Jesus knows that it was his task to provoke them. His death is a necessary event that must take place in Jerusalem before the Son of man can bring the lightning of God's power to expose a world that is in as bad a case as in the days of Noah and Lot (Lk 17:22–37). But *why* his death is necessary Luke does not indicate—except that he prefers not to describe it as a ransom.

The Q Material

It must now be remarked that we have exhausted all possible Q references to the death of Christ. Even if Q material may be found in the Jerusalem phase of the Gospels, it contributes nothing further to this enquiry. For convenience I will summarise what Q contains:

1. One indirect mention: Mt 10:38‖Lk 14:27, namely, taking up the cross, a variant of Mk 8:34–35 and parallels.
2. Statements about persecution of the prophets: Mt 5:11–12‖Lk 6:23–24, where disciples can expect insults, persecution and false calumny because of me/hatred, separation, insults, and their name cast out as evil because of the Son of man—for so the prophets were persecuted; Mt 8:20‖Lk 9:58, where the Son of man has nowhere to lay his head; Mt 23:37–39‖ Lk 13:34–35, Jerusalem that kills the prophets; Mt 23:29–30‖Lk 11:47, You build tombs for the prophets, your fathers killed them. Cf. 1 Thes 2:15.
3. Mt 12:40‖(Lk 11:30), the Son of man in the heart of the earth three days and three nights; Lk 17:25, the Son of man must first suffer and be rejected (inserted into a Q context).

If Q had separate existence as a single or multiple source of sayings of the Lord, it is clear that it contained no direct statement about his death and little that was even obliquely related.

The Temple and the Kingdom of God

Mark part 4 belongs exclusively to Jerusalem. It describes Jesus' decisive visit (not necessarily his only one), when for the last time he was able to act on his own initiative. What he did was to force a confrontation with the temple authorities, settle scores with his opponents, provide guidance to the disciples for reading the signs, and prepare them for his own destruction though confidently expecting the Kingdom of God. In composing this part, Mark uses a diversity of material, much of it used to display disenchantment with the temple and hostility towards the hierarchy.

He begins and ends with prearranged episodes. In Mk 11:1–10 two disciples are sent to fetch a colt, which, presumably by prearrangement, has been made available. Then Jesus, riding on the colt and surrounded by disciples and supporters, makes a noisy but joyful and devout approach (chanting from the festival psalm, Psalm 118) to the city. He ends the day by inspecting the temple and departing. But next day the temple is under attack: symbolically, its ritual function is brought to an end. An objection could no doubt have been made to the wealth of the temple, but in fact Mark identifies the protest by two prophetic fragments: Isaiah's oracle admitting foreigners to the temple cult (56:6–8) and Jeremiah's castigation of Jews who use the temple as a base for violence (Jer 7:9–11), which is followed by the threat that God will destroy the temple in Jerusalem as he had already destroyed the temple in Shiloh. Mark reports that chief priests and scribes understood the threat and sought a way to destroy Jesus. Here, in Mark's view, was the cause of Jesus' death,[85] He had accused the hierarchy of using their position and the inviolable sanctuary of the temple to harbour violence against the Gentiles. Whether rightly or wrongly, Jesus used his prophetic power to set in train the forces that would destroy the temple. Whether, therefore, he was a prophet sent from God or an accomplice of Satan, he had to be stopped. So, according to Mark, thought the hierarchy—and Jesus too. The disturbance in the temple is surrounded (in Mark's familiar fashion) by the destruction of the fig tree, the only miracle performed by Jesus in Jerusalem according to this Gospel (and omitted by Luke).[86] It is perhaps the only destructive miracle, and the protests of the liberal conscience are right: it was a dreadful thing to do. But it was meant to be. As everyone realises, the fig tree is a symbol of fruitless Israel, specifically of its temple. It is unthinkable that the temple should be destroyed, but Jesus has set in motion the forces of destruction. What is unthinkable will be accomplished by faith (verses 22–24); but when we pray for God to use his power, we must pray not angrily but forgivingly—for *we* shall need forgiveness for having injured those whom God loves (verse 25). Mark appends these assorted sayings on faith, prayer and forgiveness (which appear elsewhere in the tradition) in order to bring out the necessary but ambiguous quality of what Jesus did. Of course, Mark said to his own community that their opposition to the temple must not allow faith to make them insensitive to temple worshippers.[87]

So much for the beginning of part 4. In the ending, which we must consider later, two disciples are sent off to make arrangements for Passover in a room that, again by prearrangement, has been made available. This time only Jesus and the Twelve are concerned: their mood, as they eat the meal is (unsuitably for Passover) sorrowful, not joyful. Jesus accuses the Twelve of treachery and pronounces a woe on his actual betrayer: 'It were better he had not been born'. Then he takes bread and wine, which they consume, and, surprisingly, designates them as his body and blood—accompanied by a vow that he will not drink wine again until he drinks it new in the Kingdom of God. He is vowed to die that God may bring in his Kingdom. Jesus, having called on God to destroy the temple, now calls on God to destroy him. So, at least, it seems, is the intention of the Markan story.

There is, however, a great deal of debate and prophetic instruction between the attack on the temple and the Last Supper. In Mk 11:27–12:44 Jesus is shown settling accounts with his opponents and leaving himself singularly few friends:

1. with chief priests, scribes and elders on the authority for his attack on the temple—which he refuses to disclose unless they will rule (significantly) on John Baptist (11:27–33)
2. with the same audience, using an allegory of God's vineyard (12:1–12)
3. with Pharisees and Herodians on a political question about tribute to Caesar (12:13–17)
4. with Sadducees on belief in resurrection (12:18–27)
5. with a scribe on the chief commandment, for once in agreement (12:28–34)

6. with the scribes on their ruling that the Messiah is David's son (12:35–37ª)
7. with the scribes and their despicable practices (12:37ᵇ–40).

This suggests not that Mark is using up polemical material (though Matthew seizes the opportunity to introduce more sayings of that kind), but is intent on representing Jesus as carrying the attack to his opponents. He wants to make them come out fighting.

This is evident in item 2, the allegory of God's vineyard.[88] 'Jesus began to speak to them in parables', that is, in the manner of a Hebrew *mashal* (Lagrange). It was one of the most ancient functions of a *mashal* to describe a scene of prosperity or humiliation in order to create the reality of good or bad fortune (e.g., when Balaam took up his *mashal* in favour of Israel [Nm 24:3] or Isaiah his *mashal* against Babylon [Is 14:4]); and that indeed must have been the intention of this parable. From the long history of God's relations with his people it picks out those leaders who refused God his due, who dealt violently with God's agents (once more the rejected prophets), and were determined to kill his son, thus depriving God of his rights. Whether we now read this as a plausible or implausible story is beside the point. As it stands in the Gospel, it is Jesus' denunciation of the hierarchy that will force them to do what it describes. It is not a piece of narrative theology about the status of Jesus or a reflexion on the course of history (Drury) but a forceful *mashal* which will leave the hierarchy no option but to kill him. No other parable contains any reference to the death of Jesus:[89] it was not his practice to use the teaching parable to explain the necessity or the meaning of his death. The allegory of the vineyard is not persuasion but condemnation: 'He will come and destroy the tenants, and give the vineyard to others'. It was a terrible thing to say, not unlike the condemnation of the fig tree; and if it were not countermanded, it might come true (cf. 2 Chr 36:15–16). In terms of the parable, God is sending not one final prophet but his beloved son, that is, not a servant but the one person who can undoubtedly assert God's authority over his property (Pesch). If that idea, with its hostility to the hierarchy, was taken up by the crowd, the consequences would be fearful; so the authorities sought to arrest Jesus.

But there is more: as Mark presents the denunciatory *mashal*, it is complemented by a quotation from Psalm 118, which has already appeared at the 'triumphal entry' (Mk 11:9–10). It is a liturgical psalm (belonging, according to Jewish tradition, to Tabernacles) which assigns speaking parts to Israel, the priests and proselytes but gives the main ritual thanksgiving to an individual, probably the king. He celebrates the Lord's help in giving victory over his enemies even though it was a bare victory:

> I shall not die, but I shall live,
> and recount the deeds of the Lord.
> The Lord has chastened me sorely,
> but he has not given me over to death. (Ps 118:17–18)

He then passes through the temple gate and confesses the saving power of God. On that cue the congregation begin their rejoicing with the words,

> The stone which the builders rejected
> has become the head of the corner.
> This is the Lord's doing;
> it is marvellous in our eyes. (Ps 118:22–23)

Mark quotes these words in their Septuagint form. That modern perception of the psalm[90] may correspond to ancient liturgical practice and may perhaps have survived in the time of Jesus by a kind of cultic folk memory. But it seems unlikely—just as unlikely as the suggestion that makes it refer to the king-Messiah (because the Jewish messiah is not humiliated by God) or the conventional opinion that it is a cryptic symbol for the resurrection of Jesus (for which verses 17–18 of the psalm were to hand).[91] It is much more likely that it had become a proverbial saying, used by Jesus to say, 'The establishment rejects the outsider, but the outsider overthrows the establishment'. It is Jesus' second attack: first the temple invasion justified by Isaiah 56 and Jeremiah 7, then the vineyard *mashal* derived from Isaiah 5 and justified by Psalm 118. If, indeed, Jesus provokes his own death, he is prompted by his reading of scripture.[92]

The provocation may also be seen in item 4, where Jesus answers Sadducean derision of (Pharisaic) resurrection beliefs (Mk 12:18–27), an episode that has received too little consideration. It can scarcely be the result of Christian invention, since the primitive church related the *manner* of resurrection to the risen Lord (not to angels) and established the *fact* of resurrection not by an argument from Torah but by Christ's resurrection. We must allow the probability that it was indeed Jesus who disputed with some Sadducees[93] and if so, not about a peripheral matter but about convictions at the centre of his message. Why did the Sadducees (on whom see Lane) persist, against popular feeling, in denying resurrection? Because the only kind of 'eternal life' known to the normative tradition of ancient Israel was present in male semen and therefore in children, grandchildren, and so on. If, however, eternal life could be given after death to individual people, the safeguarding of semen and the obligation of marriage would no longer have any proper function.[94] Therefore, resurrection was felt to be destructive of family continuity, even of family existence; and the old-fashioned, conservative Sadducees understandably raise the question by means of a story about the continuity of a man's identity in his offspring. It is not necessary, for my purpose, to deal fully with the implications of this debate, but some things need to be said. The Sadducean position was this: the continuity of human existence and hence the fulfilment of God's promise to mankind depends on simple, linear continuity. The Pharisaic position opens up another possibility: that existence and promise are secured by transformation, namely, death and resurrection. Jesus asserts the second possibility for himself not because he is Christ or Son of God but because he is 'the Son of the man' (in the three Passion/resurrection predictions)—not a unique process for a unique person but the common process for mankind in general and Israel in particular. So, against the incredulity of the Sadducees, Jesus sets the power of God to do what seems to them impossible and the reputation of the patriarchs to commend what seems undesirable.

One more episode requires attention, namely, item 5 (Mk 12:35–37ᵃ) on the scribal ruling that the Messiah is David's son. Yet 'David himself, inspired by the Holy Spirit, declared,

> The Lord said to my Lord,
> sit at my right hand,
> till I put thy enemies under thy feet. (Ps 110:1)

David himself calls him Lord; in what way then can he be his son?' This is not so much a rejection of a son of David Christology as a riddling development of one passage of scripture in terms of other passages. According to Mk 14:62, Jesus told the high priest that he would see 'the Son of man seated at the right hand of Power, and coming with the clouds of heaven'. If the second clause takes up Dn 7:13, the first clause recalls Ps 110:1, beyond which lies Psalm 8 (where God puts all things under the feet of the Son of man) and also Ps 80:17: 'Let thy hand be upon the man of thy right hand, the Son of man whom thou hast made strong for thyself'. Thus, the expected son of David (as in Ps 89:19–37) need not model himself on King David or even act according to the modified image of Ps Sol. 17. He could do what the son of David needed to do; but since he was Lord in his own destiny, he could do it as Son of man, who would die and rise again. If that—or something like it—is the intended train of argument, Mark's statement of it is the barest hint; but any reader might grasp the thought that Jesus leaves everything to God, confident that God held him at his right hand and God himself would subdue the enemies.

The death of Christ comes to the centre of attention in Mk 14:1–2. Before then the episode of the widow's mite (which illustrates popular devotion to the temple, Mk 12:41–44) introduces a prediction of the temple's destruction, with the disciple's enquiry when it would happen and what would be the sign (Mk 13:1–4). The questions are answered in reverse order: 13:5–23 refers to indications that can easily mislead but in fact are not signs of the final events, which are instead symbolically indicated by celestial imagery from the prophets (13:24–27); and 13:28–37 refers to the nearness of the time known only to the Father. In all this no doubt Mark is recording traditional eschatological sayings, but he makes it serve the particular needs (as he sees them) of his readers. First, although the discourse is unlike anything else in the Gospel and seems to interrupt the narrative flow, Mark ties it to the Passion narrative by numerous links.[95] The implication is that the only necessary sign for the fulfilment of God's promises is the death of Jesus. Second, Mark has no doubt that Jesus is risen from the dead, but he describes no resurrection appearances and discountenances the belief that Christians can live the resurrection life. Yet the possibility of that life is real—despite the persecution of Christians, the outbreak of general violence, and the misleading claims of messianic pretenders—and it will soon be displayed when the Son of man sends out his angels to gather his chosen from the four winds, from the ends of the earth to the ends of the heaven (Mk 13:27). Third, the phenomenon of Jesus is not the sole property of the Markan community, as the words just quoted indicate. Mark's readers may be punished by Jewish councils and synagogues or by Gentile governors and kings, and the gospel must first be preached to all nations (13:9–10). It was not by chance that both Jews and Romans were implicated in the death of Jesus. The celestial imagery of 13:24–25 implies even a cosmic significance; that is to say, the death of Jesus and his coming as the Son of man has consequences for mankind, for the world in which mankind lives, and for the forces that sustain it.

Before we turn to the final part of Mark part 4 it is worth noticing how Matthew and Luke treat Mk 11:1–13:37 by way of omissions, insertions and rewritings. Matthew omits very little but inserts a great deal of polemical and eschatological teaching and does little rewriting. Luke omits rather more, makes few additions but undertakes some significant rewriting and rearranging. None of this has any bearing so far on the death of Jesus, but the case is different when we turn to Mk 14:1–25.

The Last Supper. All the Gospels associate the Passion with Passover (Mk 14:1 ‖Jn 11:55, etc.). That must be a firm historical item, all the more so because the evangelists (apart from John) make very little of it. Mark and Matthew confine further direct Passover references to the story of preparing a room for the meal; Luke, however, in addition to saying that the parents of Jesus went annually to Passover (2:41), reports Jesus' earnest but unfulfilled desire to eat this particular Passover (22:15). No one reading the accounts of the Last Supper would guess that it was a Passover meal (though of course it may have been) and Jesus makes no use of specifically Passover imagery to explain his intentions or the purpose of his death. Paul's account of primitive tradition (1 Cor 11:23–26) neither refers to Passover nor recalls his own earlier description of Christ as 'our passover' (1 Cor 5:7). By contrast John places several stories at Passover and says that Jesus was crucified about the sixth hour on the day of preparation for the Passover (Jn 19:14). Since at that time preparations were made for sacrificing the Passover lambs, John may have intended his readers (if they knew the usual ritual arrangements in Jerusalem) to conclude that Jesus took the place of the Passover lamb. But the indications are oblique, and it is by no means clear what would follow from concluding that Jesus replaced the paschal offering.

Shortly before the Passover, then, the hierarchy were seeking how to arrest Jesus by stealth and kill him (perhaps phrased so as to recall the peril of the godly man in the psalms, for instance, Ps 37:32–33 'The wicked watches the righteous and seeks to slay him' and Ps 71:9–11 'My enemies speak concerning me, those who watch for my life consult together, and say, "God has forsaken him; pursue and seize him, for there is none to deliver him"'). According to Mt 26:1–5 Jesus knows perfectly well what will happen ('After two days the passover is coming, and the Son of man will be delivered up to be crucified'), and the hierarchy were gathered in the palace of the high priest to make their decisions. Since Passover celebrated Israel's revolt against Egyptian tyranny, the authorities might well feel sensitive about popular response to Jesus, who had announced his presence at the festival by a disturbance in the temple. Their hope of dealing with him 'not during the feast, lest there be a tumult of the people' is understandable. But Mark's expression of it troubles commentators, and their perplexity is not much reduced when they (quite properly) replace *festival* by *festival crowd*. How can the death of Jesus be separated from the festival or concealed from the festival crowd if (as Mark presumes) Jesus was killed *after* Passover *during* the festival of unleavened bread? The perplexity arises from taking Mark's statements as direct information about events rather than as oblique comments on intentions. No doubt the hierarchy intended to keep control of the festival and to remove a threat to their authority (see Mk 11:32, 'They were afraid of the people'); but in fact they provoked a public disturbance (Mt 27:24 sees the point) and prompted Jesus to designate a ritual that would rival and, for some, replace Passover.

The intentions of the hierarchy are sharply contrasted with the action of an unnamed woman who interrupted a meal at Bethany and poured ointment— 'pure nard, very costly'—on the head of Jesus (Mk 14:3–9). This anointing story is copied by Matthew, omitted by Luke (who uses a different form of it in Lk 7:36–50), and provided with a parallel in Jn 12:1–8; and a variety of applications can be detected. In Jewish custom washing and anointing are standard routines,[96] washing for cleanliness, anointing for physical well-being. Both are forbidden at times of fasting and mourning, when it is customary to appear distraught and unkempt—though (acc. to Mt 6:17) Jesus told his hearers when they fasted to anoint their heads and wash their faces. According to Ber. Rabbah 74 Jacob so behaved when secretly mourning Joseph, and God declared that because he concealed his sorrow, he would manifest it to the world. At least that was the effect of this woman's extraordinary anointing of Jesus. That guests at a meal should anoint their heads—no doubt with a few drops of perfumed oil—was a commonplace courtesy from a considerate host (contrast Lk 7:44–46), but that a woman from outside the company should single out one guest and waste a pot of most expensive ointment was surprising beyond measure. But the massive waste is the essence of the story—corresponding to the attack on the temple, the destruction of the fig tree and the parabolic ruin of God's vineyard. Jesus accepts the woman's impulsive action as an omen: she has performed a good deed, valid according to Torah. When the company protest that such a costly gift could easily have both honoured a guest and given support to the poor, Jesus invokes the principle that burial of the dead takes precedence of all other duties. The woman's extravagant gift had not replaced sorrow by joy but had been a proleptic anointing of a dead body—and that, in symbolic form, is the gospel that is preached throughout the world.[97] By another sharp contrast the betrayal by Judas is next presented (see Appendix D), and then the conclusion of Mark part 4 begins with two disciples being sent off to prepare for Passover. Whether Mark's information actually suggests a Passover depends on his account of the meal, which falls into two parts: (1) verses 17–21 and (2) verses 22–25.

The first part is considered in Appendix D. In Mark's narrative it has the function of placing the reader alongside Judas: we know (as he knows) who the betrayer is, something Jesus and the others do not know. They will find out in Gethsemane. Meanwhile, they are shocked and sorrowful and apprehensive as they eat their meal, which can certainly not be the normally joyful Passover.

The second part runs as follows:[98]

> And as they were eating
> he took bread, said the blessing,
> broke it, and gave it to them
> and said,
> 'You take it;
> this is my body'.
>
> And he took a cup, said the thanksgiving,
> gave it to them, and they all drank from it.
>
> And he said to them,
> 'This is my blood of the covenant,
> which is poured out for many.
>
> Amen I say to you
> I will not drink again of the fruit of the vine
> until that day when I drink it new in the Kingdom of God.'

Matthew follows that text closely, with a few changes: the fifth line becomes 'You take and eat it'; the statement of the eighth line becomes a command 'Drink of it, all of you'; the outpouring in the eleventh line is 'for the forgiveness of sins'; the *amen* is dropped from the twelfth line, and the phrasing is delicately emended: '*From now on* I will not drink from *this* fruit of the vine until that day when *with you* I drink it new in the Kingdom *of my Father.*' These modest changes do no more than conform the Markan wording to the cultic practice and devotional tradition of Matthew's community. Elsewhere in Matthew the forgiveness of sins is not a theme characteristic of the Gospel, and that addition must be regarded as an important though unusual definition of the statement about the cup.

Luke's account, however, presents three major divergences. First, before Mark's account begins Luke 22 has the following:

> 15 And he said to them,
> 'I have earnestly desired to eat this passover with you
> before I suffer;
>
> 16 for *I tell you* I shall not eat it
> until it is fulfilled *in the Kingdom of God.*'
>
> 17 *And* he took *a cup, said the thanksgiving*
> and said, '*You take this,*
> and divide it among yourselves;
>
> 18 for *I tell you* from now on *I shall not drink of the*
> *fruit of the vine*
> until *the Kingdom of God* comes'. (Lk 22:15–18)

I have italicized the phrases common to these verses and Mark's account. Whether Luke obtained them from an independent source or composed them from Mark is disputed. It is clear that they take their cue from Mk 14:25 and insist that Jesus declared his intention of not consuming either food or drink of the festival before the arrival of the Kingdom of God. The phrasing is ambiguous: it is not clear whether Jesus' intention was to partake of this festival and then to abstain or to abstain forthwith. In verse 16 a well-supported but probably unacceptable variant reading has 'I shall not *again* eat it'.

Second, the next Lukan verses, 19–20, which are formally parallel to Mk 14:22–24, are subject to severe textual uncertainty:

> 19 *And he took bread,* said the <u>thanksgiving,</u>
>
>   <u>broke it,</u> and gave it to them,
>
>   saying,
>
>   ' *This is my body* which is given <u>for you,</u>
>
>   <u>Do this in remembrance of me</u>'.
>
> 20 *And* likewise the *cup* after supper,
>
>   <u>saying,</u>
>
>   ' *This* <u>cup</u> is the new *covenant* in *my blood,*
>
>   *which is poured out for* you'.

I have again italicized the parallels with Mark and underlined the parallels with 1 Cor 11:24–25. Luke does not exactly reproduce either Mark or Paul. His text (as I have quoted it) may be a conflation of the two or a variant of the Pauline tradition or derived from an independent source that may or may not be earlier than either Mark or Paul. This text is given by all Greek manuscripts except D (Codex Bezae) and by most of the ancient versions; but D and the Old Latin manuscripts omit everything after *my body*. Moreover, two Old Latin manuscripts rearrange the verses in the order 19ᵃ, 17, 18; and three forms of the Syriac version make their own selection and arrangement. Obviously, Luke's divergent account conflicted with primitive Christian liturgical tradition and attempts were made to bring it into line. Either an originally shorter text was filled out from Mark and Paul by almost the entirety of Greek Christianity; or an originally longer text was abbreviated and sometimes rearranged by Latin and Syriac Christianity in an attempt to cure its oddity. It is now widely agreed that the longer text is to be preferred (Metzger).[99]

Third, with Mk 14:25 the supper has ended, but not in Luke. He adds (1) verses 21–23, his rewriting of Jesus' prophecy of betrayal, which Mark placed earlier; (2) verses 24–30, a dispute about greatness, rebuked by Jesus' teaching about service and his own presence among them as one who serves, with a promise that the Twelve will eat and drink at his table in his kingdom and sit on thrones judging the twelve tribes of Israel (material parallel to Mk 10:42–45 and Mt 19:28); (3) verses 31–34, sayings directed at Simon Peter predicting his denial (placed later in Mark); and (4) verses 35–38, in which Jesus, surprisingly, tells the disciples to arm themselves because what is written of him is coming to its appointed end and he will be 'reckoned with transgressors' (quoting Is 53:12). This additional material, which reflects the hopes and internal conflicts of the

Twelve, firmly binds the problems of early Christian leadership to the words and actions of the Last Supper. Whereas in Mark Jesus offers his body and blood in expectation of the coming Kingdom of God, in Luke his self-abnegation is instructive for church leaders as they await the kingdom.

In addition to variations between the synoptic accounts, thought must be given to Paul's account and to the peculiarities of the Johannine tradition. The result of intense study is that we cannot be certain whether the meal was a regular or irregular Passover or a special meal at Passover season; we cannot discern which account was most reliable, the original source of the others, nor can we confidently reconstruct such an original; we cannot be certain what Jesus would actually have said in Aramaic (possibly Hebrew), for there are problems in translating back from the Greek; and it is by no means clear what the theological scope of the two sayings about the bread and wine was and whether he had any intention of instituting a continuing liturgy. This admission of failure, however, is no cause for dismay. It may have no further implication than that the direct pursuit of what Jesus actually said is unrewarding. It may also imply that a determination to find support for the Church's subsequent eucharistic practice is likely to obscure the very information that the Gospel texts can provide; for they certainly have meanings of their own even if *our* uncertainties go unresolved. The place to begin is the Greek text of Mark as it stands and its possible meaning to readers.

The giving of bread with appropriate ceremony was nothing unusual. Mark has shown it happening twice before, when Jesus *took* the five loaves, said the *blessing, broke* the loaves, and *gave* them to the disciples (6:41) and when he *took* the seven loaves, said the *thanksgiving, broke* and *gave* to the disciples (8:6, where the blessing is said for the fish—so *eulogeō* and *eucharisteō* are equivalent words for Mark).

But wine was very different. For one thing it is uncommon in Mark (the parable of the wineskins in 2:22, and the wine mingled with myrrh in 15:23) and is here mentioned in a formal, old-fashioned way, *fruit of the vine* (so M. Berakoth 6.1). Moreover, only here does Mark use *covenant* and *pour out;* only once is there a parallel to *for many (hyper pollōn),* namely, 10:45 *(anti pollōn).* It looks as if we have pre-Markan tradition that has been carefully fitted into the Passion narrative. The words *cup* and *drink* are obviously ordinary words (though not common in Mark), but they have a special resonance: on one side, 'Are you able to *drink* the *cup* that I *drink?*' and on the other side, 'Remove this *cup* from me' (Mk 10:38–39, 14:36).

The words *body* and *blood* are also uncommon in Mark. They occur together in 5:25, 29 where a woman touched Jesus' clothing, the flow of her *blood* ceased, and she knew in her *body* that she was healed. In 14:8 the woman at Bethany anointed the *body* of Jesus beforehand for burying; and at 15:43 Joseph of Arimathaea requested the dead *body* of Jesus. It is not to the point to debate the proper Aramaic equivalent of *my body* (*bisri* or *guphi*), and argue that the latter could mean 'myself'. No doubt it could, but Mark's readers would be thinking of an actual physical body that would bleed and die.

Jesus performs significant actions, intended to set in train the events they indicate. Without himself partaking, he gives them bread and wine, which is an invitation to eat and drink communally with the well-known range of benefits of such a fellowship. But he does more: he designates them as his body and his blood. His abstinence and this designation imply quite plainly that they will eat and drink at the cost of his life. Since he explains his abstinence by reference to the near arrival of the Kingdom of God, he intends to sacrifice his life to make possible that arrival so that his companions can eat and drink in the Kingdom of God. (Lk 22:30 is not far astray.)

The bread is simply designated 'my body', and nothing more is said; but the statement about 'my blood' is longer, a little awkwardly phrased[100] and more complex. If something like the more symmetrical Pauline form were known to Mark's community, it might be conjectured that Mark had simplified the statement about Christ's body in order to concentrate most attention on his blood—perhaps to restore an emphasis that was in danger of fading. True or not, that is where the emphasis lies—on 'my blood of the covenant, which is poured out for many'. The question is whether those interpretative words carry their own meaning, or whether (as is usually supposed) they point to a large store of partly hidden information. One thing can at once be said: the emphasis on blood rules out any meaning from Passover. The head of each household slaughtered his own offering; the priest caught the blood in a basin and tossed it against the base of the altar. It has no further part to play and is not even mentioned in M. Pesachim 10 or the modern Haggadah.[101] None of the traditional four Passover cups of wine would have suggested blood. Passover, of course, is a great celebration of God's saving power: the mood of the worshippers moves from shame at the beginning to praise and glory at the end. But the liturgy makes nothing whatsoever of the atoning and expiatory power of blood.

That is not surprising, since 'the blood of the covenant' is not part of Passover. The phrase is rare: it appears as 'the blood of the [eternal] covenant' in the rhetorical language of Heb 10:29 and 13:20, as the ground of God's promise to set captives free in Zec 9:11, and as the means by which the Sinai covenant between Yahweh and Israel is ratified in Nm 24:8. That remote historic episode would scarcely be foremost in the minds of Mark's readers;[102] and even if they knew it, they would not be much enlightened. For one thing, the covenant is ratified in two diverse manners: in Ex 24:1–2, 9–11 Moses, Aaron, Nadab, Abihu and seventy elders ascended the mountain, saw the God of Israel without injury, and shared a meal; in Ex 24:3–8, at the foot of the mountain Moses arranged for the whole people to present burnt offerings and sacrifice peace offerings. Half the blood he threw on the altar (thus securing Yahweh's sponsorship), and the rest (once they had promised obedience to Yahweh's commands) he threw on the people. Targum Pseudo-Jonathan on Ex 24:8 (parts of which may be pre-Christian) says, 'Moses took blood . . . and sprinkled it on the altar *to make atonement for the people,* and he said: Behold the blood of the covenant'. The words in italics are a targumic addition, but they merely provide the standard explanation for the use of blood: when people claim divine sponsorship, they are wise to have its potent protection. It is no indication that the covenant sacrifice is being assimilated to rituals for expiating sins. But could Mark presume that his readers, on the basis of one phrase, would understand Jesus to be asserting that a cup of wine signified his blood, that human blood was replacing animal blood, part of which was to be applied to God and part to his supper companions, who agreed to keep a code of divine law? Even the author of Hebrews, who exploits sacrificial imagery, and twice uses *blood of the covenant* is muddled about the covenant sacrifice when he refers to it in Heb 9:18–21 (see p. 268). It may be reasonable to argue that Mark's covenant implies Paul's (and Jeremiah's) *new* covenant; but Jeremiah says nothing about blood or sacrifice. It is all done by the implantation of Torah, a quite different covenant from that made with the fathers (Jer 31:31–34). All in all, it must be said that the reference back to Exodus 24 leads nowhere, and the phrase 'my blood of the covenant' must be allowed to carry its own meaning. Everyone agrees that the phrase *to haima mou tēs diathēkēs* needs supplementing, though not everyone realises how it is tacitly expanded. Familiar exegesis requires 'my blood, which corresponds to the blood used in ratifying the covenant that God made with Moses'. It is much simpler to read 'my blood required by the covenant that God made with me'. Only very rarely in Mark does Jesus give any intimation of his direct relation with God: the three Passion/resurrection predictions are of such a kind and they dominate the early part of Mark part 4. The most explicit intimation brings Mark part 4 to an end. It is the divine pleasure that if Jesus gives his life, the kingdom will come; and he says to the supper companions (using the words of Lk 22:29), 'As the father has covenanted to me, so I covenant to you, a kingdom'.[103]

The blood is 'poured out for many'. The very (which in Mark occurs only here) is commonplace and has no particular theological significance. When Matthew and Luke tell the parable of the wine skins, wine is poured out when the skins burst; but Mark makes nothing of it. In the Q rebuke of the scribes and Pharisees (Mt 23:34–35 = Lk 11:49–50) the verb describes the suffering of the prophets:

> Therefore the Wisdom of God said,
> 'I will send them prophets and apostles,
> some of whom they will kill and persecute,'
> that the blood of all the prophets,
> *poured out* from the foundation of the world,
> may be required from this generation,
> from the blood of Abel to the blood of Zechariah,
> who perished between the altar and the sanctuary.
> (Lk 11:49–50 = Mt 23:34–35)

That, of course, is from Q, not Mk; but we have no better indication of what Mark thought. The Suffering Servant may be described as the supreme example of martyrdom, but Isaiah's poem does not help, for it makes no mention of blood. Even when the Hebrew says 'He poured out his soul to death', the Greek does not. It is true that the Servant bears the sins of many (Is 53:11, 12—presumably the 'many nations' of 52:15); but the Greek has nothing like Mark's *hyper pollōn* (on behalf of many), and Mark has nothing at all about sins. It cannot be supposed that Mark's readers, having been fruitlessly persuaded to absorb Passover and covenant sacrifices, should now take up—on the hint of the one word *many*—the perplexing fate of the Suffering Servant. The Markan *many* is much simpler. It corresponds to 'They *all* drink of it' in verse 13. Because all drank, all were under obligation. If all would eat and drink in the coming kingdom, all came within the scope of his death. This is Mark's familiar theme: no Christians yet live the risen life, all live by means of death and resurrection.

But what of Matthew? With one exception, his minor alterations of Mark convert narrative statements into rubrics. It is reasonable to suppose that the same liturgical source supplied the addition 'poured out for many *eis aphesin hamartiōn* [for the forgiveness of sins].' Matthew knows that the Son of man announced the forgiveness of sins and that God would forgive all sins except one (opposition to the Holy Spirit). He knows that God can be asked to forgive our sins, regarded as debts, if we have forgiven our debtors; and he has a procedure for forgiveness within the community.[104] In the Matthean birth story there is a double divine operation: in Mary by the conception from the Holy Spirit and in Joseph by the naming commanded by the angel of the Lord. He is to be named Jesus, which means that 'he will save his people from their sins.'[105] Jesus is shown to be God by the salvation he accomplishes (Ps 130:8). When, later, the people flock to John Baptist, they confess their sins as they are baptized; but as Matthew tells the story, he avoids saying that John offers a baptism *for the forgiveness of sins*—which is presumably reserved for Jesus. But nothing in Matthew prepares us for the statement that the blood is poured out for the forgiveness of sins. The evangelist has picked up the thought from the secondary stage of early Christianity[106] and has included it without asking whether it matches his standard teaching that forgiveness is available to those who repent and forgive others.

Many exegetes think that Matthew's addition admirably gives the intention of Mark's statement. For example, Pesch (p. 359) says, 'When the statement about the significance of the cup speaks of "my blood", that has meaning only if the atoning action of blood is in mind, the forgiveness of sins because of Jesus' death.' But he gives no reason, apart from the implication that he cannot think of anything else. Why should forgiveness require the shedding of human blood? Where in Jewish teaching is that found? The new collection of *The Old Testament Apocrypha*, edited by J. H. Charlesworth, includes sixty-three Jewish documents probably from the seventh century B.C.E. to the ninth century C.E. Under *vicarious atonement* the comprehensive index provides reference only to 4 Mc from the first century C.E. The introduction to that book (p. 539) can add only 2 Mc Test Benj. 3.8, 'The sinless one will die for impious men' and passages in 1QS where the community think they are making atonement for all Israel. H. Anderson calls this 'a doctrine that already enjoyed at least a limited currency in Judaism'. Rabbinic teaching is displayed by E. P. Sanders in *Paul and Palestinian Judaism* (pp. 157–82) and summarised thus: 'God has appointed means of atonement for every transgression, except the intention to reject God and his covenant' (p. 157). Rabbi Ishmael taught that repentance, cultic rites, death and suffering effect atonement, that is, God forgives when sinners make appropriate offering of one or more of these four. But death 'does not avail in the case of one who has denied God, thrown off the yoke of the covenant and remained defiant to the end' (p. 176). Since Jesus was condemned by the Jewish hierarchy and—as they would have seen it—went unrepentant to his end, how could his death atone either for himself or for others? And why should others need his atoning death when the means of atonement were readily available? There is one possibility, suggested in the description of Jesus as both sponsor and expiation for our sins in 1 Jn 2:1–2, which epistle also has a quite-different view in 1:9 that sins can be atoned for by repentance (see pp. 280–81). Granted that Matthew likes to compare sins to debts, he may perhaps have supposed that God would remit our debts the more readily if we pleaded the name of one who had died to please God.

When we turn to Luke[107] we find that the scene looks rather different. Luke does not avoid, and does not wish to avoid, the mention of blood (if the longer text is what he wrote); but he accommodates it to a more optimistic view, less fraught with danger. It is possible that the first references to eating and drinking in Lk 22:15–18 are part of a longer account of a Passover meal, with a lamb on the table and the traditional four cups of wine. But it is unlikely, because Luke, like Mark, makes nothing of the Passover associations. What he does, however, is bring into the forefront a vigorous expectation that the Kingdom of God soon will come. Jesus' earnest desire to eat the Passover (Lukan narrative colouring) is not gratified, not is his wish to drink with the Twelve, because by self-abnegation he will hasten the kingdom, which is vividly present to his mind in the conversation after supper. Having made that point in verses 15–18, he delicately repeats it in verses 19–20: the body 'which is given for *you*', and the blood 'which is poured out for *you*'. Moreover, the bread is a symbol of his body which is indeed *given* for them (clearly the language of sacrifice, though, by the addition of the rubric, it is at once defined as a *memorial* sacrifice). Likewise, the blood sacrificially poured out for them is represented symbolically by the cup that stands for Jeremiah's *new* covenant, which is effected by the implantation of Torah, not by the effusion of blood. What might sound like a literal human sacrifice is transformed, without diminution of force and with great increase in intelligibility, into remembering what Jesus has done for the Twelve and into defining what is henceforward the constituent principle of their common life (the Torah of their new covenant), namely self-giving on behalf of others.

The proof of that intention comes in the subsequent conversation. Judas serves as a dreadful warning to Christians who presume on the tolerance of Jesus; but for the rest, Jesus is among them as one who serves their needs (as in Acts the *name* of Jesus performs necessary miracles) rather than as the Son of man who gives his life a ransom for many (Mk 10:45). He is the Lord who rewards his supporters. By undergoing hardships and trials he claims the kingship covenanted to him by his Father. In turn, he shares his sovereignty and his table with those who have remained faithful to him in trial and appoints them to judge the twelve tribes of Israel. Not that all will be plain sailing: Peter himself will scarcely manage to remain loyal and yet will be needed to strengthen the loyalty of his brothers, for the community is moving from a period of security into a period of risk. Formerly the seventy had gone out 'as lambs in the midst of wolves' (Lk

10:3–4) but had required no protection; but now the missionaries would need a sword, though not much more. When the Twelve say, 'Look, Lord, here are two swords', Jesus replies 'It is enough', that is, there is danger, but it is not very great.[108] The danger comes from something foreseen in scripture, namely, that Jesus would be 'reckoned with transgressors' (Is 53:12). Whenever Luke has to admit that the death of Jesus was regrettable, he takes comfort if he can find it foreseen in scripture. What is written about Jesus now comes to its fulfilment, and for once Luke finds an apt passage in Isaiah to excuse (as it were) Jesus' forthcoming crucifixion at the hands of the Romans. But it is notable that he avoids Isaiah's words about vicarious atonement as carefully as he avoided Mark's saying about the ransom for many. This is by no means Luke's final estimate of the death of Jesus, but it is characteristic of his approach to it.

Death and Resurrection

When at last we come to Mark part 5, we know how the story will end. The three passion/resurrection predictions have warned us that Jesus will fall into the hands of the Jewish hierarchy and the Roman authorities, will be put to death, and after three days rise again. The promise in 14:28 that after his resurrection he will go before them to Galilee and the announcement in 16:7 that he has already gone there contain all that Mark intends to say about the resurrection.[109] In that case, the present narrative is not so much a historical outline of the course of events (though it certainly contains historical information) as an account of the relentlessly increasing isolation of Jesus. When finally he utters a loud cry and dies, he can expect nothing from his disciples—from Peter at the head of the Twelve or from Judas at the tail—nothing from the Jewish hierarchy or the Jerusalem crowd, nothing from the Roman governor or his soldiers, nothing from God or Elijah his prophet. The proclaimer of the Kingdom of God, who by the self-designation *Son of man* modestly hid himself from prominence, now has the prominence of total alienation. That is the central theme of Mark's Passion narrative, not the discreditable behavior of the opponents of Jesus or the disappointing behavior of his supporters (though of course they were stupid and ignoble) but, as Luke would say, 'the determinate counsel and foreknowledge of God' that brought Jesus to this pass.

The first indication comes in the reference to scripture:

> I will strike the shepherd,
> and [so that] the sheep will be scattered, (Mk 14:27)

which changes the imperative of Zec 13:7 (reproduced in the Damascus Rule)[110] into a future (see Jn 16:32), thus more explicitly indicating a divine intention. But why should that scriptural passage have been chosen? If their hymn singing before leaving the upper room comprised the remainder of the Hallel (and even if it was not a Passover meal, the Hallel psalms would be much in mind at that season) they would perhaps have sung Pss 114–18 according to Shammai, or 115–18 according to Hillel (M. Pesachim 10:6–7). Leaving aside Ps 114, which is obviously apt for Passover ('When Israel went forth from Egypt'), there are statements that point rather directly at the situation of Jesus:

> The dead do not praise the Lord,
> nor do any that go down into silence. (Ps 115:17)

> I love the Lord,
> because he has heard my voice and my supplications.
> Because he inclined his ear to me,
> therefore I will call on him as long as I live.
> The snares of death encompassed me;
> the pangs of Sheol laid hold on me;
> I suffered distress and anguish.

> . . . .

> For thou has delivered my soul from death.

> . . . .

> I walk before the Lord
> in the land of the living.

> . . . .

> I will lift up the cup of salvation
> and call on the name of the Lord. (Ps 116:1–3, 8–9, 13)
> Out of distress I called on the Lord;
> the Lord answered me and set me free.

> . . . .

> I shall not die but live,
> and recount the deeds of the Lord.
> The Lord has chastened me sorely,
> but he has not given me over to death.

> . . . .

> The stone which the builders rejected
> has become the head of the corner.
> This is the Lord's doing;
> it is marvellous in our eyes. (Ps 118:5, 17–18, 22–23)

These psalms raise the question, What could the godly man expect when caused to suffer and brought within the grip of death? Will he go down into Sheol where God can no longer be known? Or will God answer his plea and surprisingly restore him? According to the words of Zechariah (not elsewhere explicitly quoted in Mark or identified as scripture), God himself will smite the shepherd and separate him from his flock. They too will be caught in God's trap. The words of the Revised Standard Version, 'You will all fall away' represent *pantes skandalisthēsontai*. A *skandalon* is a trap; and the passive verb indicates being caught in a trap, whether for good or bad reasons. Of itself the verb does not carry moralizing implications of disgraceful failure or shocked surprise, for an effective trap can make use of the victim's strengths as well as weaknesses. In the broadest terms, it indicates circumstances that allow nothing but a particular course of action—though in this case, Mark would say, not 'circumstances', but 'God'.

The prediction that Peter will repudiate Jesus—not once but three times, and therefore totally—is shown to be true in 14:66–72. So Peter moves from complete support in verses 29, 31 to complete separation in verses 66—72. In between we are shown why.

Gethsemane

The Gethsemane story is told in the manner of a storyteller who knows private conversations and discerns inner motives and feelings (Mk 14:32–42). This emotive occasion was certainly impressed on the early tradition, but its details had to be indicated by scriptural words (Pss 42–43, Jon 4:9 LXX, and Sir 37:2 for 'My soul is very sorrowful, even to death'; Is 51:17–23 for the cup of God's wrath that must be drunk or can be taken away) and seemingly by phrases of the Lord's Prayer (which Mark does not otherwise record). 'Abba, father' recalls its first words (especially in the Lukan form), 'not what I will, but what thou wilt' recalls 'Thy will be done', and 'Watch and pray that you may not enter into temptation' echoes the final petition (Mt 6:9, 10, 13). From that perception it should be clear that Mark is expressing two convictions. First, on this very stressful occasion, Jesus approaches God as he taught his disciples to, as indeed the early Christians did. *Abba father* is evidenced in the hellenistic Pauline communities (Gal 4:6, Rom 8:15), and the vocative use of *Father* is established in the writings associated with John, Peter and James. Thus, Jesus addresses God with full submission to the divine will but with the intimacy of a son who is the agent of that will.''' Moreover, he recognises that the hour has come for a decisive trial of strength *(peirasmos)* against God's enemies. Second, Jesus, like Jonah and the psalmist, objects to God's treatment of him: 'I say to God my rock: "Why hast thou forgotten me? . . . Vindicate me, O God, and defend my cause against an ungodly people. . . . For thou art the God in whom I take refuge; why hast thou cast me off?"' (Ps 42:9, 43:1–2). So Jesus prays God to take the cup from him, and he has three witnesses to support his appeal ('at the mouth of two or three witnesses'). But he discovers that his witnesses have been silenced. Mark uses the familiar threefold pattern—a little awkwardly it is true—to show Jesus three times making his appeal, three times rallying his witnesses, and then drawing the conclusion that his appeal will not be heard. If it is asked *what* Jesus wanted God to take from him, what acutely distressing imposition should be removed, it is safest to avoid speculation and to hold firmly to what Mark says. He gives no indication whatever that Jesus is unwilling to suffer or die, least of all that Jesus begs God to spare him death and yet is willing to be fobbed off with three fractious visits to somnolent disciples. If we avoid romantic elaboration, Mark's intention is tolerably clear. Jesus has previously said that if the shepherd is smitten, the sheep will be scattered or—to put it plainly—if he dies, his followers will be disbanded. What Jesus is therefore seeking in Gethsemane is an assurance from God that when he dies his name and work will be preserved by his followers. But no assurance is given. The followers are already beyond his reach. So the scene ends ironically with chiastic comments about the followers and God:

> Sleep on and take your rest.
> He [God] is far away[112]
> behold the Son of man is surrendered [by God] to the hands of sinners.
> Rise, let us be going;
> behold my betrayer is at hand. (Mk 14:41–42)

Jesus is separated from God, from disciples—and now from Judas.

Matthew's Gethsemane is essentially Mark's, though written a little more fully. But Luke's is very different. It does not name Gethsemane or distinguish the group of three. The emotion of Jesus is not recorded: he prays only once (in somewhat different words) and visits the sleeping disciples only once. The final chiastic comments disappear, but Lk 22:43–44 (which are textually unreliable)[113] add a strengthening angel and the bloody sweat. Those fervid verses are quite unsuited to the rest of Luke's prosaic narrative; indeed, they were probably added to give it a lift. As it is, Luke begins and ends by telling disciples not to enter into *peirasmos* (verses 40 and 46). What Luke means by *peirasmos* can be seen by comparing 'in time of *peirasmos*' in Luke's interpretation of the parable of the sower with Mark's corresponding words, 'when hardship or persecution occurs' (Lk 8:13, Mk 4:17): it becomes difficult to be a Christian and the temptation is to give up. Thus, Luke has cleared away much of the familiar story in order to emphasise the need of his contemporaries. The remedy is prayer, not to ease the situation but to keep Christians loyal. Jesus is not Mark's distraught prophet, but Luke's impressive teacher and healer who attracts by his goodness and nobility. He goes to the Mount of Olives, as his custom was, and is entirely self-possessed. He does not throw himself on the ground (as in Mark) but simply kneels, and his prayer is wholly lucid. Then with composure (if verses 43–44 are omitted) he rejoins his disciples, finds them sleeping (from sorrow, says Luke, thus excusing them) and warns them against the temptation of disloyalty. In Luke's Gethsemane Jesus does not appeal to God or look for an omen or display extreme agitation: he betrays knowledge of God and a controlling personal goodness, both directed to the benefit of his followers.

Arrest and Trial

After the Gethsemane episode, Mark relates how Judas arrives with an armed band from the hierarchy. Judas embraces Jesus to identify him (Matthew and Luke variously give Jesus words of reply); there is a scuffle, and someone injures the high priest's servant. (Matthew and Luke attribute the aggression to supporters of Jesus. Luke says that one of them asked permission to fight but that Jesus accepted the situation and, in line with his characteristic generosity, healed the injured ear. In Matthew, Jesus moralizes by saying that violence begets violence and claims the [magical?] power to call up legions of angels.) Jesus protests against the armed attack but says, 'Let scripture be fulfilled'; whereupon all left him and fled (omitted by Luke), and a youth lost his linen cloth and fled naked (omitted by Matthew and Luke). In Mark, Jesus rebuts the charge of being a terrorist *(lēstēs)* and insists that he is a religious teacher. His followers betray him, use unacceptable violence, and shamefully run away. The incident of the young man who flees naked is symbolic: whereas Jesus goes off to captivity, suffering and death, an unnamed follower (perhaps representing a section of Mark's readers) gains his freedom at the price of his own shame. All three evangelists underline the nonaggressive character of Jesus and his submission to the divine will, features that have some bearing on his death. And Jesus is now literally separated from his disciples and in the hands of people who will reject him.

In the next section of the Gospel Jesus is first in the hands of the Jewish the next section of the Gospel Jesus is first in the hands of the Jewish hierarchy and then of the Roman authorities. Mark draws on memories of the process that led to the condemnation of Jesus. To the modern critical reader, his account is full of historical difficulties. If Mark is treated as an amateur law report, it is defective and inconsistent with what we know (or can presume) about ancient Jewish and Roman judicial procedures. For example, when we remember the enormous confusion and delay in dealing with Paul's case, is it conceivable that the Jewish and Roman authorities could have arrested, tried and executed Jesus in a couple of days or that they should have been so desperate to be rid of him? If, however, Mark is treated as a simple recorder of early Christian tradition, how can we make out a reasonable story from these fragments of rumour and popular reporting? Only John's Gospel makes a dramatically coherent narrative of the confrontation between Jesus, Pilate and the hierarchy and pointedly omits anything like a trial before the Sanhedrin. If Luke rewrote Mark in order to produce a more persuasive story, he was scarcely successful and obscured Mark's intention. For Mark is neither an amateur law report nor a simple record of tradition but a theological argument for readers who already know the tradition.

The structure is announced in Mk 14:53–54: verse 53 brings Jesus before the high priest and his colleagues, and is continued in verses 55–65 where Jesus makes his confession, and verse 54 brings Peter into the company of the high priest's servants and is continued in verses 66–72, where he makes his denial. Peter's story once again uses the three-times motif and ends with him denying— with God as his witness—that he knows the man and calling down a curse, possibly upon himself but probably rather upon Jesus.[114] Thus, under severe pressure—not simply from understandable human weakness—Peter totally repudiates Jesus. He repeats outside the judgement hall what is going on inside.

Inside, in the presence of the hierarchy, proceedings take place in two phases. In the *former* an attempt is made (as Mark tells the story) to find Jesus guilty of a capital offence on the grounds that he intended to destroy the temple and use magical powers to rebuild it. (Matthew tones down the implication of magic, and Luke omits the whole episode.) But, not surprisingly, the witnesses do not agree, and although the hierarchy had proper grounds of complaint in Jesus' attack on the temple, they sensibly decide not to make fools of themselves by pressing this charge. Jesus, of course, had no witnesses in his favour, but he needed none. Nor did he need to defend himself against the natural suspicion that anyone under arrest must be guilty of something. So the first phase ends negatively. Jesus' attack on the temple has obviously been successful in arousing the hostility of the hierarchy, but on that ground they are not ready to fight.

In the *second* phase, therefore, Mark turns the enquiry to the identity of Jesus, both for historical and theological reasons. Historically, it is necessary to explain why Jesus was crucified as king of the Jews; theologically, it is necessary to identify him as Son of God. When the high priest, in effect, makes what looks like a proper confessional statement, he has in the word *Christ* what allows him to denounce Jesus to Pilate, and in the phrase *Son of the Blessed* what suits Mark's theology. So the high priest knows the truth and regards it as blasphemy, whereas Peter knows the truth and can only deny it. But Mark's subtlety perplexed the other synoptic evangelists. According to Mk 14:61–62,

> The high priest asked him,
> 'Are you the Christ, the Son of the Blessed?'
>
> And Jesus said,
> 'I am;
> and you will see the Son of man
> seated at the right hand of Power [see Ps 110:1],
> and coming with the clouds of heaven [see Dn 7:13]'.

In the words 'I adjure you by the living God', Mt 26:63 shows that we are reading about a cultic trial of strength rather than a legal trial. By substituting 'You have said so' *(su eipas)* for 'I am' he allows Jesus to evade the question and to substitute his own statement of what they will see 'from now on' *(ap arti)*. Jesus speaks as a prophet who points to scripture and warns about what is to come. Luke, on the other hand, separates the Sanhedrin's question about 'the Christ' from another question about 'Son of God,' which they have deduced from Jesus' reference to scripture (Ps 110:1 only). To both Jesus replies in an assured, dismissive manner—the appropriate response from a noble-spirited person to a tribunal whose servants (acc. to Lk 22:63–65) had already treated the accused shamefully. But Mark's intention was different. The indications are already present in Peter's confession (Mk 8:29–32), in the controversy over the Christ as David's son (Mk 12:35–37), and in the apocalyptic discourse of Mk 13:26. When Jesus is asked to declare his identity as the Christ *or* as the Son of the Blessed (for it is still unlikely that the titles are equivalent),[115] he (1) answers chiastically that he is the Son of the Blessed and (2) avoids the title Christ by referring to himself as the Son of man.

Since Christians certainly addressed Jesus as *Christos*, why does Mark show this caution about the title? Partly for reasons internal to his own community (as suggested on p. 180) but also because *Messiah* had strongly defined associations among Pharisees and Essenes, to judge by the Psalms of Solomon and the Dead Sea Scrolls. He is an entirely human leader of God's people fighting against their enemies, particularly against Gentiles.[116] If Jesus had confirmed a claim to be Messiah, it might have been religiously comic or politically risky; but no one would have thought it blasphemous. Both Jesus and the high priest avoid the

name of God and hence blasphemy as (later ?) technically defined (M. Sanh. 7.5).
But there is such a thing as constructive blasphemy, namely, statements that
(whether intentionally or not) imply and encourage contempt for the divine
being. If Jesus, pointing to scripture, claimed his position at the right hand of the
(divine) Power and prophesied that he would return with the overwhelming
authority of that Power, it might indeed sound like blasphemy to a body of men
already persuaded that Jesus was a dangerous person. And if he answered that
he was Son of God, a commonplace designation among the Greeks but used of
individual persons only with great hesitation by Jews, it might sound both blasphemous and alien. But on the lips of Jesus these are not defiant or provocative
words. They are his essential testimony to faith in God and his power to save,
even though God has handed Jesus over to those who wish him dead. This is
easily understood if we are no longer misled by the fable of a supernatural being
called 'the Son of the man'. As elsewhere in Mark, Jesus refers to himself with
reserve as the Son of man when his authority or his suffering (in this case both)
are in question. He is the Son of man (acc. to the Passion/resurrection predictions) already handed over to the hierarchy, about to be handed over to the Gentiles; he will be given the Son's position at the right hand of God and (taking a
lead from Daniel 7) will receive an everlasting kingdom. Thus, Mark permits us
to interpret the 'rising after three days' of the three Passion/resurrection predictions as the exaltation of Jesus and his subsequent Parousia. But to the hierarchy
his words are shocking and horrifying: the high priest performs the alienation
ritual of tearing his clothes, and the whole company display their repudiation
with blows and spitting. Comparison with the servant of God in Is 50:6, who
'hid not his face from shame and spitting' illustrates the process of shaming an
accused prisoner but does not undermine Mark's account of this trial of strength
between the religious establishment and the isolated prophet from Nazareth. The
blindfold mockery of his prophetic pretentions—still part of the shaming process
to neutralise his (as they think) evil power—incidentally suggests the likely
grounds on which a legal charge could have been brought: he was a false prophet
leading Israel astray (Dt 18:20, M. Sanh.11.5).

Pilate and the Crucifixion

The next section of the Gospel, which recounts Jesus' appearance before Pilate,
bears contradictory indications (Mk 15:1–21). Read as the account of a Roman
trial, it is defective and implausible (Nineham, Anderson). Something can be made
of it by borrowings from the other Gospels and by informed guesses (Lane) and
it cannot reasonably be doubted that Jesus was sent to the cross by Pilate, that
Jesus and the Jewish hierarchy appeared before the governor, and that Barabbas
was released. But it should be clear that Mark was not providing a coherent
account of the course of events. It should be equally clear that this section was
not perfunctory. There are structural features that hold the narrative together
and give it a distinctive character. The title King of the Jews—neither proposed
by Pilate nor accepted by the hierarchy—appears for the first time and becomes
the leading matter of dispute (Mk 15: 2, 9, 12, 18, 26, 32). It is matched in that

respect by 'crucifixion', introduced by the crowd, assented to by Pilate, carried out by the soldiers, and mocked by others (Mk 15:13–15, 20, 24, 25, 27, 32). The actual crucifixion and death of Jesus are set down in terms that recall Psalm 22 (verse 18 for the division of his clothing in 15:24; verse 7 for the mockery in 15:29; and verse 1 for the loud cry in 15:34). Thus, considerable care has been taken in displaying the participants and their significant parts in the action. The *Jewish hierarchy* hand Jesus over to Pilate, they stir up the crowd to demand Barabbas, and join in the (precautionary) mockery as described above (pp. 179, 218). Yet they frame their mockery in words that admirably express a Christian perception of the death of Jesus: 'He saved others, himself he could not save' (Mk 15:31, though Lk 23:35 may have overlooked the point). The Jerusalem *crowd* demands the release of Barabbas, who had committed murder in some uprising, and the crucifixion of Jesus. They mock him when crucified but try to keep him alive to see if Elijah will come to his rescue. Even the two *lēstai* crucified with him ('bandits', probably Barabbas' people) revile him. So Jewish hostility to Jesus is plainly shown, and perhaps Mark's hostility to the Jews. That is partly confirmed and partly modified by Mark's presentation of the *Romans,* chiefly (but not only) Pilate. He first associates *King of the Jews* with Jesus, who, however, pointedly declines to relieve Pilate of responsibility for using the title. In the end of course it must be Pilate who endorses the title by the inscription of the charge against Jesus; but before he condemns him to scourging and crucifixion, he seems impressed by Jesus' silence and (knowing the motives of the hierarchy) unconvinced of his guilt. In this uncertainty, he is even willing to release him. As Mark tells his story, unbearable pressure from the Jews decides Jesus' fate: a political murderer goes free, and an innocent man is executed—with political and moral consequences for both parties. Pilate is by no means excused, though he is shown in a more favourable light than the hierarchy; yet it is Pilate who makes sure that Jesus is proclaimed King of the Jews and dies a political death. When the *soldiers* carry out an elaborate mockery—with parodies of the imperial purple, the radiate crown,[117] and the royal acclaim—they are saying derisively what in fact is true: that Jesus, who died as a Jew and for the Jewish people, will prove to be the universal imperator. When derision is renewed at the cross by the hierarchy, they too are ensnared by the necessity to speak the truth out of anger and disbelief: 'Let the Christ, the King of Israel, come down now from the cross, that we may see and believe' (Mk 15:32). And finally there is *Simon* of Cyrene, who is clearly known to Mark's community ('the father of Alexander and Rufus', Mk 15:21) and, in a sense, represents Christian participation in the Passion. He carried the cross for Jesus (and each Christian must take up his cross and follow Jesus, Mk 8:34). Just as it is possible for the truth to be spoken out of hostility, so it is possible for the right response to be made out of compulsion. It is rightly argued that what Mark wrote was 'of great significance for Gentile churches living in a Roman environment' (Anderson, p. 334); but something begins to emerge of general theological importance: of how God absorbs the powerful hostile attack and brings good out of evil.

In the corresponding section of Matthew, the evangelist somewhat obscures Mark's subtlety by introducing anecdotes that place more blame on the Jews. In

Mt 27:3–10, Judas repents and kills himself—thus presumably securing himself a place in the world to come—and the hierarchy find a religiously acceptable way of spending the blood money. In Mt. 27:19, Pilate's wife—warned by the gods in a dream—urges her husband to have 'nothing to with that innocent man'; and Pilate washes his hands of the matter, disclaiming responsibility. Whereupon, 'all the people answered, "His blood be on us and on our children"' (verses 24–25). That bears all the marks of hostile judgements presented anecdotally. If (Matthew is saying) even Judas can repent, if pagan Romans can see that Jesus is innocent, Jews who condemn him must bear responsibility for his death. He judges, of course, in anguish, not in hatred; for he is deeply concerned for Israel and Israel's faith.

Luke is quite different. In the main he follows Mark, but with many omissions, additions and rearrangements. He is dissatisfied with the impression left by Mark's narrative and, either from the storytelling tradition in his own community or from his own imaginative historical gift, he redirects the reader's attention. Jesus is charged with creating political disturbance and urging non-payment of tribute throughout Galilee and Judaea. He is examined by the rulers of both territories. Herod of Galilee finds him disappointly unimportant, and Pilate three times finds him not guilty as charged. But finally he gives in to Jewish demands and Jesus is taken away to be crucified. As he goes, he speaks with solemn foreboding to the women mourners, customary on such an occasion. His cruel fate is an omen of the more severe fate that will fall on them and their children. It is the same Lukan Jesus who earlier spoke of the Galileans whose blood Pilate mingled with their sacrifices and of the inhabitants of Jerusalem on whom the tower in Siloam fell (Lk 13:1–5), and prophesied that Jerusalem itself would be trodden down by the Gentiles (Lk 21:24). In going to his death, Jesus is concerned not about himself but about the fate of others; but in Luke he has not been flogged (Lk 23:25 omits the Markan *phragellōsas,* despite the prediction in Lk 18:33), nor has his kingship been derided by the soldiers (Luke omits Mk 15:16–20). Jesus goes like a great prince to the place called The Skull, with a countryman to carry the cross behind him.

In the last few hours of Jesus' life, there is no such dignity in Mark's account. It begins with the offer of wine mixed with myrrh (which Jesus refuses, no doubt because he had vowed not again to drink wine until he drank it new in the Kingdom of God, Mk 14:25) and ends with the spongeful of sour wine that he cannot refuse (Mk 15:22–37). His clothing becomes the property of his executioners, and he is like the man in Luke's parable who fell among bandits who stripped him, beat him, and left him half dead (Lk 10:30). In fact, he is crucified between two bandits and so is associated with men of violence—hence the inscription of the charge against him: 'The King of the Jews'. He dies for a political offence, as a deluded nationalist rebel against the Roman government. He is subject to jeers from the spectators, who revive the temple accusation and tell him to save himself, for they will do nothing for him. Members of the hierarchy deride him—both his reputation as a healer who cannot save himself and his claim to kingship that cannot escape the Roman condemnation. And he is reviled by the bandits with whom he is crucified. When finally he gives a loud cry and calls on God, his prayer is grotesquely (or perversely) misunderstood as if he were calling for Elijah. So he dies with a final, inarticulate, loud cry: helpless to save himself or to preserve his vow; despised by his fellow sufferers, the hierarchy, and the people of Jerusalem; and misunderstood as regards his work and his message: 'Jesus uttered a loud cry, and breathed his last'.

That is followed by three short episodes that have great symbolic force. First is the tearing of the temple curtain, presumably that which separates the holy place from the most holy place (Ex 26:33). Its symbolic meaning is debatable (Gnilka)—either that access to the most holy place is no longer confined to the high priest but is open to all, including Gentiles (but Christians would scarcely be interested in that) or that the deity had abandoned the most holy place (which agrees with Mark's dislike of the temple).

Second, there is the centurion's verdict 'Truly this man was [a] Son of God'. Even if that is not a properly Christian confession (*was* referring to the dead Jesus, not the risen Christ and *son of God* suggesting nothing more than a familiar hellenistic category), it clearly implies that the death of Jesus, with all its suffering, humiliation and defeat, brings a Gentile within striking distance of the truth. The centurion stands symbolically for the Gentile constituency of Christian missionaries. In contrast to the soldiers' mockery previously recorded, when 'he stood facing Jesus and saw that he thus breathed his last', he gave his verdict. Thus, Mark implies that testimony to the Gentile world does not require extraordinary and supernatural signs but (however embarrassing) confession of the death of Jesus. Presumably that needed to be said to Mark's church. Matthew, however, thought differently. He links the tearing of the temple curtain with an earthquake when rocks were split and tombs were opened: 'Many bodies of the saints who had fallen asleep were raised and coming out of the tombs after his resurrection they went into the holy city and appeared to many' (Mt 27:51–53). The addition is inept, for the astonishing revival of dead saints anticipates his later reference to the resurrection of Jesus; but it is clearly deliberate. It was when the centurion and the other soldiers saw the earthquake and the other consequences that they were overcome wih religious dread and recognized that Jesus had been Son of God. Matthew is not averse to framing his revision of the Gospel by means of prodigies (see the star in his birth story), and no doubt his earthquake has symbolic meaning: the death of Jesus is a key to the final events that include the resurrection of the godly.[18] Of that Mark says nothing.

Third, there are the women who watch from a distance—three are named, many others are mentioned. They had been his followers in Galilee and had looked after him there, and they had accompanied him to Jerusalem. They have a symbolic significance just as much as the woman at Bethany who identified Jesus as the one who should be anointed (Mk 14:3–9), and the serving woman in the courtyard who identified Peter as a companion of Jesus (Mk 14:67–69). Shortly, the women witnesses of the crucifixion will identify the place of burial and will then discover the empty tomb. It must be suspected that Mark is saying to his church that the way to the resurrection life begins in discipleship and service and depends on a clear view of the crucifixion and teaches that lesson by the presence of the women.

Granted that the crucifixion demonstrates the complete alienation of Jesus from all helpers, Mark has adopted two devices to give it a theological context. In the first place the final day is divided arbitrarily into four three-hour periods,[119] suggesting divine control of human actions and the physical world. Secondly, the procedures of crucifixion are related to the words of Psalm 22 (see p. 219), and the final words of Jesus repeat the godly sufferer's approach to his God. At the very least Mark implies that Jesus dies as other godly sufferers have died, that Jesus is not alienated from Jewish tradition, and that Jewish understanding of God is capable of containing what is here described. Matthew and Luke do not dissent, for they allow echoes of Psalm 22 to appear rather more plainly—[120]even though Luke suppresses the so-called cry of dereliction.

The Words From the Cross. All the evangelists agree that Jesus spoke words from the cross. The synoptic evangelists agree that he uttered a loud cry just before he died, though Luke alone gives the words of that cry. Matthew and Mark agree that Jesus uttered an earlier loud cry and give the words; but Luke omits both the cry and the words. In addition, Luke has one earlier word from the cross, or possibly two, not found in Mark and Matthew. John reproduces none of the synoptic words but has three of his own. Even without further enquiry it looks as if the 'words' come from interpretative traditions, in some measure reflecting the evangelists' theological intentions.

Matthew knew and used Mark, and he knew no other tradition about the words from the cross, despite his use of some information (independent of Mark) about the death of Jesus. He reported the one saying of Mark in a revised form.[121] Luke used Mark with rewritings, replacements and excisions. His Passion narrative may draw on an independent source, though it includes some direct borrowings from Mark. Luke either knew the Markan word from the cross and excluded it because it offended his interpretation of the death of Jesus or, knowing Mark, enquired about the surviving tradition (which may be the implication of Lk 1:1–4), found no memory of Mark's word, and decided to exclude it or substitute other more instructive sayings, or at least not intrude it into a rival Passion narrative. The implication would be that Mark was wrong or misleading. John may have known Mark and even Luke, or at least comparable traditions. He betrays no evidence of Mark's word, because he either found it in Mark but chose not to repeat it or failed to find it in the comparable tradition. Hence, either Mark's word was not part of the early tradition, or Luke and John suppressed it in favour of their own totally divergent sayings. They may have suppressed it as unedifying and *therefore* untrue. If so, they were capable of believing that a saying was true because it was edifying; that is, the sayings of Jesus from the cross may be "true" not because he said them but because they *truly* indicate a possible interpretation of his death.

But let us examine the Markan word (15:34).

> At the ninth hour Jesus cried with a loud voice,
> 'Eloi, Eloi, lema sabachtani?'
> which means, 'My God, my God, why hast thou forsaken me?'

The words are first given in Aramaic, then in translation Greek. In Mt 27:46 there are small variations: 'Eli, Eli', which may be a Hebrew form (adopted into Aramaic) to make the subsequent confusion with *Elijah* more plausible; and a Greek translation slightly nearer the Septuagint.[122] To a Greek reader the words no doubt would be striking and unusual, but to a Jewish reader they would be familiar. They introduce a psalm of complaint; and if the personal laments of the psalter are examined, similar expressions can easily be found. God is begged not to forsake the petitioner or cast him off utterly in Pss 27:9; 38:21; 119:8; and 71:9, 18 (where in verse 11 enemies say 'God has forsaken him', which is almost as bad as if God had). Further, he begs God not to forget him (Ps 13:1, 74:19) but in Ps 42:9 says "Why hast thou forgotten me?" In Ps 43:2 he says, "Why has thou cast me off?" and in Ps 88:14–16, "Why dost thou cast me off? Why dost thou hide thy face from me? . . . Thy wrath has swept over me; thy dread assaults destroy me." In Ps 89:38, "Now thou hast cast off and rejected, thou art full of wrath against thy anointed" (see Ps 27:9). And in Ps 77:7–9,

> Will the Lord spurn for ever,
> and never again be favourable?
> Has his steadfast love for ever ceased?
> Are his promises at an end for all time?
> Has God forgotten to be gracious?
> Has he in anger shut up his compassion?

All those complaints, all that pleading come from the experience of the godly sufferer. Almost always the sufferings are attributed to 'enemies' (as in Psalm 22), very seldom to wrongdoing. When the psalmist says that God has forsaken and forgotten him, cast him off and exposed him to wrath, it is not because he has sinned. Indeed, these complaints are not to be read naively as statements of theological fact; they are a form of pleading, intended to provoke a reassuring response from God: 'I certainly have not forsaken you and I remember you constantly; you are not cast off but only chastened as a beloved son.' When, therefore, Jesus uttered the loud cry from Ps 22:1 he was following a central tradition that affirms not that God is absent from the sufferer but that he is present and open to entreaty. It is not required that sin should be expiated or that God should be propitiated—simply that God should be stirred to action by the plight of his faithful saint. As Mark tells the story, particularly the story of Gethsemane, Jesus knows that his sufferings are not to be attributed to the inactivity of God and the malice of enemies but to the activity of God.

Why should Jesus break his silence to utter those words? Not, presumably, to give his hearers a theological indication of the mystery of his suffering. Who, indeed, were his hearers, and who remembered these words? Was a crucified person, after three hours, capable of speaking coherently in a loud voice? Was it indeed not Mark himself but the early Aramaic-speaking Christians who reflected on the loud cry of verse 37 and found an interpretation of it in Psalm 22? It is said that 'the cry from the cross can hardly have been invented, because of the offence caused by its implications for christology'.[123] But that is perhaps an anachronistic misunderstanding. We must remember that Jesus died under a curse uttered by the high priest: 'For thirty-six transgressions is extirpation prescribed in the law . . . if a man blasphemes' (M. Kerithoth 1.1); 'The blasphemer's judges stand upon their feet and rend their garments' (M. Sanh. 7.5). Thereupon they spit on him and mock him (Mk 14:65; 15:20, 31), both methods of reinforcing a curse.[124] And plainly the curse is effective: his chief disciple repudiates him: 'He saved others, himself he cannot save,' that is, his potency has left him. The best protection against a curse is to acquire a blessing, and Psalm 22 is a vigorous plea for the removal of a curse and the restoration of blessing. Hence Jewish Christians would not have been embarrassed by these words on the lips of Jesus: knowing that he was the helpless victim of a curse, Jesus turned to God for a blessing.

Did God grant a blessing and remove the curse? He at least fulfilled the Jewish maxim, 'He who utters a curse, the curse returns to him':[125] God turns the high priest's curse back on the temple, and the temple curtain was torn in two from top to bottom. By contrast the centurion says, 'Truly this man was son of God'. The rest of Mark's story is the disposal of the body, the belated preparation of unguents, the discovery of an empty tomb, and the announcement that he was risen—news that the women received with shock, fear and silence. It is characteristic of Mark that he should interpret the loud cry as the request of a desperate man for a blessing. Whether his readers think such a blessing is available depends on their inclination to turn the centurion's *was* into their own *is*.

Luke's account conveys a quite different feeling. In addition to minor changes,[126] the Lukan words from the cross transform the narrative.

Two criminals (not bandits or terrorists as in Mark) were to be put to death with Jesus. 'When they came to the place which is called the Skull, there they crucified him, and the criminals, one on the right and one on the left [23:33]. And Jesus said, "Father forgive them; for they know not what they do" [verse 34ª]. And they cast lots to divide his garments [verse 34ᵇ].' Clearly, the prayer is intrusive between 'They crucified him' in verse 33 and 'They cast lots' in verse 34; and it is textually most uncertain. All the main early traditions have some witnesses that omit the words, though the prayer was known in the second century to Marcion and Irenaeus in the West. But the manuscript tradition is divided: in Alexandria the prayer was known to Origen in the third century but omitted from Papyrus 75. The Old Syriac is divided, suggesting that manuscripts including and excluding the prayer were known at an early stage. Hence, it was added or deleted at an early stage of the textual tradition. If it is not a genuine part of Luke's text, it could have been inserted to show Jesus putting his own teaching (Lk 6:27–28) to the final test, and to suggest that Jesus was not less magnanimous than Stephen (Acts 7:59–60). If it was a genuine part of Luke's text, it may have been omitted because Jesus ought not to appear forgiving to the Jews or because his prayer went unanswered in the events of 70 C.E. In my opinion, the balance of the evidence is against this prayer;[127] but its existence is at least evidence for early reflection on the death of Christ. What the prayer *means* is another question. It is commonly supposed that the ignorance of the executioners may be the ground of their forgiveness, but it is more likely that ignorance is the sin that needs forgiveness. Luke has a dozen passages where people are said *not to know;* in every one a rebuke is implied or given, or a dangerous situation is exposed: for example, when Jesus wept over the city and predicted its dreadful destruction 'because you did not know the time of your visitation' (Lk 19:41–44). Jesus shows his magnanimity by asking forgiveness for people who ought to know what wickedness they are doing.

In Mark and Matthew the two brigands crucified with Jesus revile him, but in Luke it is different:

> One of the criminals who were hanged railed at him,
>   saying,
> 'Are you not the Christ? Save yourself and us!'
> But the other rebuked him, saying,
> 'Do you not fear God, since you are under the same
>   sentence of condemnation? And we indeed justly; for we
>   are receiving the due reward of our deeds; but this man
>   has done nothing wrong.'
> And he said, 'Jesus, remember me when you come into
>   your Kingdom.'
> And he said to him,
> 'Truly I say to you, today you will be with me in Paradise'.
>   (Lk 23:39–43)

The reproach of the first criminal is modelled on the gibe of the spectators (omitted by Luke) and the gibe of the 'rulers' in verse 35. The reply of the second criminal strengthens Jesus' claim to innocence (a favourite Lukan theme) and his request suits Lukan theology (Lk 24:26). Jesus' response promises an immediate experience of paradise. In the Septuagint *paradeisos* is used of the garden of Eden (Gn 2:8, etc.), to which condition the wilderness is to be restored (Is 51:3, 2[4] Esd 8:52–54, Rv 2:7) and which was seen by Paul in an anticipatory vision (2 Cor 12:3). Luke is fond of dreams and visions (in 10:18 he reports a visionary experience of Jesus) and no doubt he correctly represents a genuine impulse behind the teaching of Jesus: the restoration of paradisal conditions. Hence, Jesus, with oriental munificence, promises a place in the new world to a criminal who shows penitence by accepting his punishment. It is straightforward Jewish teaching: penitence and death atones for all.

Luke omits the opening words of Psalm 22 and instead says that Jesus died after saying in a loud voice the words from Ps 31:5, 'Father, into thy hands I commit my spirit' (Lk 23:46). It was the night prayer of devout Jews,[128] and Jesus could not be supposed to have died less confidently than Stephen (Acts 7:59). When Luke's community sought an understanding of the death of Jesus they turned not to Psalm 22 but to Psalm 31—equally a psalm of ill-treatment but expressing untroubled confidence in God: 'Let me never be put to shame' (Ps 31:1).

Thus, Jesus magnanimously forgives his executioners (either Lukan or in the Lukan spirit). In princely fashion he promises paradise to a criminal and at last defiantly commends himself to the Father. The exalted mood overcomes the death, which becomes the means by which Jesus displays his nobility. In fact Jesus dies as a hero—as nobly as Socrates, though in more distressing circumstances; with at least the gravity of a Roman nobleman; and not without a touch of the Eastern monarch. He was, as the centurion said, a good man (Lk 23:47).

Mark's account is equally self-conscious writing, but it done with irony, especially his account of the burial and the empty tomb (Mk 15:42–16:8). He decides to give credit where credit was due: Joseph of Arimathea, a respected member of the council,[129] who was commendably looking forward to the arrival of the Kingdom of God (and so a sympathiser with Jesus but scarcely a disciple)[130] persuades Pilate to permit the burial of Jesus, who had died unexpectedly quickly. So he recovers the body from the cross, wraps it in a linen shroud, and lays it decently in a secure rock tomb. Pilate has confirmed the death, and Joseph the burial. Sabbath ended, the women buy spices (which are not needed), and approach the tomb worrying about entering it (but the stone is rolled away), expecting to perform the last rites for a corpse (but they are shown an empty place):[131]

> Do not be amazed; you seek Jesus of Nazareth, who was crucified. He has risen, he is not here; see the place where they laid him. But go, tell his disciples and Peter that he is going before you to Galilee; there you will see him as he told you. (Mk 16:6–7)

Joseph and Pilate and the women did what was right as best they could—and it was all ironically inappropriate. The final irony is that the women were overcome with trembling and astonishment 'and they said nothing to anyone, for they were afraid'. Just as it is inappropriate to treat Jesus as one of the respected dead, so (when you discover that he is not dead) it is unsuitable to speak confidently of encounters and visions. If the life-giving power of God has done this tremendous thing, trembling, astonishment and awe measure the reality of the experience.

The words of the angelic messenger are concise in expression, subtle in implications. First, the central figure, though identified by the centurion as Son of God, is still Jesus of Nazareth *(ton Nazarēnon),* which is in Galilee where Jesus came from and where he was first greeted as the holy one of God (Mk 1:9, 24). Hence, he has returned to Galilee, where the women and the disciples will see him. It would be a little simple-minded to discuss whether Mark located resurrection appearances or the Parousia in Galilee: for Mark is again talking about *seeing,* a question earlier introduced and concluded by the giving of sight to blind men. In effect, Mark says to his readers, 'Go back to the Galilee part of my Gospel, and you will now see him'.

They will see him who was crucified but is risen. The resurrection does not cancel the crucifixion. Everything that was said about the shadow of John Baptist's death, the constraining presence of the Passion/resurrection predictions, is now full of significance. It is, of course, also true that preoccupation with the death of Jesus does not cancel the resurrection. *Ēgerthē*—'He is risen'. But Mark thinks it wrong to expect the resurrection life until the Kingdom of God comes with power.

Meanwhile, they must go to Galilee, as he had told them. The stress on Jesus' instruction is notable, intended, no doubt, to counteract some other intention. Equally notable is the phrase, 'Tell his disciples and Peter', where Peter is singled out. That probably corresponds to the prominence of Peter in Mk 8:29–33, where he rebukes Jesus for predicting death and resurrection and is in turn rebuked as Satan or Satan's agent. When Peter, after confidently asserting, 'If I must die with you, I will not deny you', actually denies him, breaks down, and weeps, he has died a death and a very painful kind of death. So, according to Mark (and despite 1 Pt 5:1), he cannot be 'a witness of Christ's sufferings' or the first to hear that 'he has risen'. The women have that alarming privilege. But if with the others he will go to Galilee, there he will see him and be raised to the foremost position in the Christian confession: 'He appeared to Cephas, then to the twelve'. That is a consequence of, and a form of, resurrection.

Matthew makes major (as well as some minor) changes to Mark. A great earthquake takes place, a shining angel descends to remove the stone and terrify the guards who are present with Pilate's consent and are later bribed to tell an improbable lie (Mt 27:62–66; 28:2–4, 11–15). The earthquake and the angel cause no difficulty: they are conventional symbols for indicating an earthshaking event initiated by God. But the story of the guard is directed against a Jewish explanation of the empty tomb, that disciples had removed the body. Matthew simply shows how difficult it is to cancel the death of Jesus by a physical resurrection; but of course he does not stop there. For him the women, experiencing awe and great joy, run to tell the disciples, though what they say is not recorded. Instead, Matthew has their banal encounter with Jesus, who needlessly repeats the angel's message and the story of how the guard was bribed. Matthew rounds off his story by taking the eleven disciples[132] to a mountain in Galilee where Jesus claims full authority from God and authorises the eleven to make disciples by baptizing in the triune name and instructing the converts to obey his teaching. Matthew ends where his interest really is, not with death and resurrection but with the teaching of Jesus.

Luke's interest is certainly in the resurrection, less in the death of Jesus. His two dazzling angelic witnesses reproach the women for seeking the living among the dead, which implies that too much interest in the death of Jesus distracts attention from his living presence. They announce his resurrection[133] and (having with some effrontery evaded their Markan colleague's command to go to Galilee) recite a form of the standard prediction of Passion and resurrection,[134] whereupon the women tell their story to the eleven and others and are disbelieved, though Peter runs to see the empty tomb and is puzzled.[135] So much for Lk 24:1–12. The story naturally continues in verse 34 with the baffling, off-hand mention of an appearance to Peter,[136] and with the arrival of Jesus himself, which frightens them until he identifies himself and proves himself not a spirit by eating a piece of fish.

Before that happy resolution of their doubts, there is the elaborate story of the journey to Emmaus (Lk 24:13–32). The two travellers confess their hope that Jesus would have redeemed Israel, for he was a prophet mighty in deed and word before God and all the people; but their chief priests and rulers delivered him up to be condemned to death, and they crucified him. That was three days ago, and now some women of their company had found no body in the tomb but had seen a vision of angels who said he was alive. Others of the company confirmed that report but did not see Jesus; whereupon their unrecognized companion rebukes them for slowness in grasping and accepting all that the prophets had said, namely, that of necessity 'the Christ should suffer these things and enter into his glory'. And he remedies their ignorance by interpreting to them, beginning with Moses and all the prophets, the passages in scripture that concern him. They still do not perceive who he is and recognition comes only with the breaking of the bread. The story seems to imply that recognition of the risen Lord comes from the conjunction of scriptural understanding and the Church's meal—at least as far as the redemption of Israel is concerned.

There is more to be said when Jesus shows himself to the larger company, namely, 'that repentance and forgiveness of sins should be preached in his name to all nations, beginning from Jerusalem'. That is included in his teaching about scripture. He first repeats what he had previously taught them (the reference is to the Lukan form of the third Passion prediction, Lk 18:31) that 'everything written about me in the law of Moses and the prophets and the psalms must be fulfilled'. Then he opened their minds to understand the scriptures and said to them, 'Thus it is written, that the Christ should suffer and on the third day rise from the dead, and that repentance ...' (Lk 24:44–47). In one sense, Luke implies that the whole scriptural tradition of Israel finds its fulfilment in the death and resurrection of Jesus, which should lead to the salvation of Israel and the conversion of the Gentiles. No doubt he could give some colour to that conviction by references in the Passion narrative to Pss 22 and 69; but if Jesus really gave a comprehensive scriptural justification of his death and resurrection, his hearers omitted to record it. As far as *we* are concerned it is not possible to find in scripture the expectation that *Messiah* would suffer and rise on the third day. As far as Luke is concerned, it has to be said that he gives no thought to any reason why Messiah should suffer except that it is in scripture and must be willed by God. Nor does he explain why Messiah should suffer and rise from the dead to promote repentance and forgiveness (which are freely available to Jews) among the Gentiles.

There is, of course, more ample information about Luke's view of death and resurrection in Acts. In composing his Gospel, he was much dependent on Mark and doubtless on familiar traditions, but he succeeded in giving a distinctive presentation. In writing Acts, he was more at liberty to draw on tradition and use his narrative gift according to his own judgement and so to produce an account that was not untrue to the earliest apostles and not irrelevant to Christians in his own day.

Acts

References to the death and resurrection of Christ, sometimes brief, sometimes extended, are found in every major division of the book.[137] Not surprisingly, they mostly occur in speeches, largely to Jewish audiences. Though the wording is often more credal than narrative,[138] the following schema can be recognized.

Preceded by John with his baptism of repentance for all Israel (13:24), Jesus of Nazareth (from the family of David, 13:23) was anointed by God with the Holy Spirit and with power (4:27, 10:38). God attested him with mighty works and wonders and signs (2:22), and he went about doing good and healing all that were oppressed by the devil (10:38). But God foretold by the prophets that his Christ should suffer (3:18, 17:3, 26:22–23); and so he did, by hanging on a tree (5:30, 10:39, 13:29—with Dt 21:22–23 in mind). Jesus was delivered up according to the definite plan and foreknowledge of God (2:23, 4:28); nevertheless, it was the Jewish people, behaving like their fathers, who had persecuted and killed the prophets (7:52, 13:27–28), that crucified and killed him by the hands of lawless men (2:23, 36; 3:14–15; 4:10). In Jerusalem, Herod and Pontius Pilate, Gentiles and the people of Israel were arrayed against him (4:27–28); and Judas, one of the Twelve, was guide to those who arrested him (1:16). He was laid in a tomb (13:29); but God raised him from the dead (2:24, 32; 3:15; 4:10; 5:30; 13:30, 33, 34; 17:31) on the third day (10:40), as had been promised in scripture (2:25–31, 4:11, 13:33–37, 17:3, 26:22–23). He showed himself alive to many witnesses for forty days and indeed ate and drank with them (1:3, 10: 41, 13:31). He was thereby made Lord and Christ, Leader and Saviour (2:36, 5:31), and exalted to the right hand of God (2:33, 5:31), where indeed he was seen by Stephen and Paul (7:56, 59; 9:5; 22:8; 26:15). Christ would be in heaven until the appointed time, when he would return (like the prophet promised by Moses) to be judge of the living and the dead (3:20–25, 10:42, 17:31), replacing life in this crooked generation (2:40) with times of refreshing from the Lord and the restoration of all things (3:19–21). Until then, he was active in the world by means of his NAME, producing healing and salvation (3:16; 4:10, 12, 30), forgiveness of sins and the Holy Spirit (2:38, 5:31, 10:43, 13:38).

From the number of references, it is at once obvious that Christ's death and resurrection are the most important features. And of the two, resurrection more than death. Even the mention of the mighty works, wonders and signs of Jesus is no more than a preparation for the marvels performed by the NAME of the living Christ at the hands of the apostles and their associates (e.g., 4:30). Clearly, the risen and exalted Christ is the giver of the Holy Spirit (2:33), so that it is possible to say 'Repent, and be baptized . . . in the NAME of Jesus Christ for the forgiveness of your sins; and you shall receive the gift of the Holy Spirit (2:38). For 'God exalted him at his right hand as leader and saviour, to give repentance to Israel and forgiveness of sins' (5:31). Consequently, 'there is salvation in no one else, for there is no other NAME under heaven given among men by which we must be saved' (4:12). These benefits come from the resurrection of Christ, and it is not surprising that Luke represents the apostles as 'teaching the people and proclaiming in Jesus the resurrection of the dead' (4:2). At the other end of the story, he respresents Paul as claiming that his troubles are entirely caused by his loyalty to the Pharisaic belief in resurrection (23:6; 24:15, 21; 25:19). In the presence of Festus and Agrippa, he is more candid: 'Saying nothing but what the prophets and Moses said would come to pass: that the Christ must suffer, and that, by being the first to rise from the dead, he would proclaim light both to the People and to the Gentiles' (26:22–23). But neither here nor elsewhere does Luke indicate the scriptural passages that require the Christ to suffer.

No doubt 'Christian interpretation applied to Jesus all passages in the Psalms and Isaiah which refer to suffering' (Foakes Jackson and Lake, vol. 4, p.37), but Luke fails to quote any of them. On the resurrection and exaltation, he is more forthcoming; and perhaps he takes it for granted that at least Jewish audiences will know the suffering passages. In Peter's speech at Pentecost and Paul's speech at Pisidian Antioch use is made of Psalm 16, which expresses unbounded confidence in God and therefore excludes the thought of suffering. As quoted by Peter in Acts 2:25–28, it runs thus:

> I saw the Lord always before me,
> for he is at my right hand that I may not be shaken;
> therefore my heart was glad, and my tongue rejoiced;
> moreover my flesh will dwell in hope.
>
> For thou wilt not abandon my soul to Hades,
> nor let thy Holy One see corruption.
>
> Thou has made known to me the ways of life;
> thou wilt make me full of gladness with thy presence. (Ps 16:8–11)

It is argued that David, in composing the psalm, was speaking prophetically about God's guarantee that a descendent should sit on his throne. Since David, in fact, died and was buried, he could not be speaking of himself. Therefore, he must have spoken of the Christ who was not abandoned by God to death, nor did his flesh undergo disintegration—hence he spoke of Christ's resurrection. It is an ingenious argument and no doubt persuaded Luke, though it has certain consequences. It required Luke to exclude 'My God, why has thou forsaken me?' (Mk 15:34), for the Holy One of the psalm sees the Lord *always* before him and knows that he will not be abandoned (the verb is *enkataleipō*, as in Mark), and also prompts Luke to demonstrate the incorruption of Christ's risen flesh by the eating of broiled fish (Lk 24:42–43).[139] Paul uses a similar argument and more modest quotation of the psalm in Acts 13:35–37 but prepares for it by citing (in a characteristic example of rabbinic free association), 'I will give you the holy things of David which are sure' from Is 55:3. To promote Christ's exaltation, Paul uses Ps 2:7, 'Thou art my Son, today I have begotten thee', a psalm for the King's enthronement and his triumph over enemies (sung to good purpose by the exultant Jerusalem community in Acts 4:23–31). Peter uses Ps 110:1, 'Sit at my right hand', and Ps 118:22 'The stone . . . become the head of the corner' (Acts 2:34–35, 4:11)—all of which is, in its way, instructive. It should not be denied that passages from Psalms and Isaiah can properly be used to put the resurrection of Christ into context; but the striking discovery is that Luke does it for the resurrection and exaltation, not for the death. It often appears as if the death is no more than a precondition for resurrection.

Speeches to Gentiles. It has already been noted that most of the speeches in Acts are delivered to Jewish audiences; but the remainder also merit attention.

In 8:26–40 an Ethiopian court official is converted by reading Isaiah. Though he had been to worship in Jerusalem, it is left unclear whether he was a proselyte or not (Haenchen). The evangelist Philip explained to him the good news of Jesus from Is 53:7–8:

> As a sheep led to the slaughter
> or a lamb before its shearers is dumb,
> so he opens not his mouth.
>
> In his humiliation his *krisis* was taken away;
> who will tell his *genea?*
> for his life is taken up from the earth. (8:32–33)

Luke's Greek (which corresponds to the Septuagint) is as obscure as the Hebrew but not identical with it; and that can be seen by the numerous attempts to produce a sensible modern translation. But we have to ask how Philip could produce the good news of Jesus from the words in the last three lines. On the face of it, the meaning is that the deplorable condition of the accused denied him a fair trial, his life on earth was ended, and he lacked descendants to preserve his memory. But that clearly will not do. If *humiliation* implies death, *krisis'* ('punishment's') being removed implies resurrection. As a consequence, who will tell [the number of] his descendants, for his life was taken up from earth [into heaven] (Haenchen, Conzelmann), so death, resurrection, exaltation and Church are surprisingly present. It may be so. After Qumran pesher exegesis, all things are possible to him that believes. But the most significant fact is that the quotation stops short: the conclusion of the passage quoted would be, 'For the wickednesses of my people he was led away to death'. The striking feature of Is 53 is that the servant of the Lord suffers because of the sins of others and bears those sins. Nothing of that appears in Luke.[140]

Peter's speech to the Gentile Cornelius is not addressed to a Gentile at all but to fellow Galileans ('You know the word which he sent to Israel . . . beginning from Galilee') and contains some Galilean animus against the Jews in Jerusalem who put Jesus to death. When the spirit phenomenon happened, it was 'believers from among the circumcised' who were amazed and, apparently, convinced (10:34–48). In any case, Cornelius was 'a devout man who feared God' (10:2). Luke therefore provides the standard kerygma for Jews.

When at last Luke produces a speech for Gentiles at Athens (17:22–31), it is at first sight very different. But reflexion on the formal structure suggests a basic similarity. The Areopagus speech begins, like one form of hellenistic Jewish propaganda, with Greek religious tradition about the unknown deity who is nevertheless the source of our being; then it moves to resurrection, repentance and judgement. Compare that pattern with Paul's speech to men of Israel and those that fear God at Pisidian Antioch (13:16–40): it begins with the familiar but obligatory account of Israel's religious tradition and shows that even to Jews God was unknown 'because they did not recognize him or understand the utterances of the prophets'. So they had Jesus killed. But then it turns to resurrection, forgiveness of sins and judgement. It should now be clear that Luke regards the death of Jesus as the final proof that the Jews need salvation, and regards the gospel as beginning with resurrection.

Paul's farewell speech at Miletus is a remarkable composition (20:17–38), an apologia combined with instructions to the community.[41] Its leaders are told 'to care for the church of God which he obtained with the blood of his own Son': thus the Revised Standard Version, though there is strong manuscript support for 'the church of the Lord'; and the word *Son* is not present in the Greek, which reads 'the blood of his own' or even 'his own blood' (see Metzger). The awkward phrasing accounts for the textual variants and is often thought to be a none-too-successful recording or imitation of Pauline language. The thought is more pastoral than theological. Paul knows that the community is in danger, but he is innocent of their blood (i.e., their injury and death) because he has told them God's full intention and himself goes to suffering and death. They are leaders of a community that God has obtained for himself by the blood of his own—his Son's of course, very soon his apostle's and perhaps therefore that of the community leaders as well.

Apart from that isolated passage, Luke ascribes no positive benefit to the death of Christ. He dies in order to be raised from the dead; but that will not do. If Jesus was attested by God with mighty works and wonders and signs, his death contradicts the attestation. Therefore, two more things must be said. First, the death took place at human hands: the Jews were guilty, the Romans were only instruments. They 'denied the Holy and Righteous one . . . and killed the Author of life', though they acted in ignorance (3:17), for did not recognize Jesus or understand the utterances of the prophets (13:27; compare the Gentile 'times of ignorance' 17:30). It should be remembered that the excuse of ignorance may justify a plea for mercy but does not remove guilt and blame. Second, the death was according to 'the definite plan and foreknowledge of God'. Herod, Pilate, Gentiles and Jews were gathered together against Jesus to do whatever God's hand and plan had predestined should take place. And God foretold by the mouth of all the prophets that his Christ should suffer.

For Luke the death of Christ seems to be ambiguous. He treats it in the spirit of Aboth 3.16, 'All is foreseen, but freedom of choice is given'. Jesus died in obedience to the will of God, but Luke does not describe it as an atoning death.

Summary

Mark. The work of Jesus begins in Galilee and, after the resurrection, is resumed there. In prophetic manner Jesus announces that the divine power is at hand and by miraculous actions apparently demonstrates God's power over nature, the heathen, demons, disease and death. The demonstration, though drawing great crowds, meets with hostility; and it becomes clear that miracles can be misread, that God himself may withdraw his benefits and may hand his prophet over to death (as with John Baptist). Thus, Galilee is indeed the scene of God's power, but also of hostility, misunderstanding, and risk.

Clearly, Jesus is by common consent a prophet, but the prophet's fate has often been rejection and death. Therefore, Peter wishes to replace *prophet* with *Christ,* that is, the triumphant protector of God's people and the destroyer of their enemies. Jesus, however, replaces *prophet* with *Son of man,* that is, the authoritative agent for what God will do, in a context of present distress and future glory. This is embedded in a pre-Markan formula: that the Son of man *must* suffer many things ('must' in that it is written in scripture and so is God's will) or that the Son of man is delivered (namely, by God) into the hands of men and will be killed and after three days rise again. The formula is filled out by identifying those who cause the suffering: the Jewish hierarchy as originators, the Romans as instruments. The threefold use of the formula not only stresses its significance but must also be regarded as a prophetic utterance that both announces the course of events and also sets them in train.

The *first* use of the formula leads to the cross of Christ's disciples, which stands for self-abnegation, disclaiming access to the resurrection life until the arrival of the kingdom in power—even though there are paradigms of death *and* resurrection in healings and exorcisms. The *second* use of the formula leads to instruction about personal problems in the community, problems that can be solved only by surrendering established status. And the *third* use of the formula leads to the fundamental teaching that Jesus is not a passive martyr but a prophetic person who must say and do what will be resisted and so will bring suffering upon himself—and on his disciples—in order to serve others. Nevertheless, the Son of man—the authoritative agent of God, not demanding service but required to serve—offers his life by an agreement with God that at the cost of his life many other lives should not be forfeit.

Once that is established, Jesus begins his attack on the temple hierarchy and the scribal establishment. He makes prophetic use of symbolic action, cursing and parable and enters into scriptural disputation with opponents, thus leaving himself almost no support but his closest disciples. What he has prophetically set in train will have consequences for Jewish society and for himself. The so-called apocalyptic chapter, Mark 13, is in fact a preview of consequences for the disciples. Most of it is taken up with telling them what is *not* to be the end of their troubles: not endemic violence or natural disaster, not civic disabilities or family dissension, not the capture of the temple by an alien force, not religious pretenders and imposters. Only when the basic structure of their world is shaken will they see the Son of man coming in power and glory—but only God knows the time, and they must watch. And that must be Mark's way (for he has no resurrection appearances) of expanding the last clause of the formula, 'after three days rise again'.

The consequences for Jesus himself occupy the rest of the Gospel. At supper he again speaks and acts prophetically to set events in train. He invites the disciples to eat and drink. If he himself partakes, he joins them in communion with him. If, however, he surprisingly abstains (as I think likely), they benefit while he does not. He designates bread as his body and wine as his blood, thus indicating that they benefit at the expense of his life. In particular, blood indicates the sacrifice of his life in accordance with a covenant made with God, namely, that if he gives his life the kingdom will come and all disciples will eat and drink with him in that kingdom.

Since Jesus acted wholly within the Jewish tradition, he could properly model himself on the godly sufferer of the Psalms, who pleads with God and expects God to rescue and restore him. But Jesus also drew upon the alarming prophecy of Zechariah that God himself would strike the shepherd and scatter the sheep. They are all caught in the trap that God has set. So, from Gethsemane to the crucifixion, Mark's account is controlled by the psalms of complaint wherein God is pictured as present and open to an appeal. At the same time, all is within the control of God. Jesus submits himself wholly to God's will; therefore God becomes entirely responsible for what happens, namely, that Jesus becomes more and more isolated from his disciples and successors, from the Jewish hierarchy, from the Roman administration and from those who rebel against it. Finally, he is isolated even from God and any emissary of God. By that isolation Jesus dominates the reader's attention. Correspondingly, what is at issue ceases to be Jesus' attack on Jewish institutions and authorities and becomes instead the identity of Jesus. It is presented both by the denials of Peter (representing disciples who want to make him *Christos*) and by the questioning and derision of the hierarchy, who certainly reject him as *Christos*. Before the Sanhedrin, on one interpretation Jesus accepts both designations *Christos* and Son of the Blessed but rephrases them as Son of man at the right hand of Power and coming with the clouds of heaven; on another interpretation (which I prefer), he accepts the designation Son of the Blessed but replaces *Christos* by the exalted Son of man. So the designation *Son of man* appears for the last time, with the proper implication of present humiliation and future glory. Readers of the Gospels have become familiar with the words 'Son of man', and regard them as a recognised though perhaps elusive title that Jesus applied to himself. There are good reasons for thinking, however, that 'the Son of the man' (as the words appear idiomatically in the Greek of the Gospels) was not a title at all. Jesus did not use it to identify himself with some expected emissary from God; he used it in accordance with Semitic idiom as a modest, oblique self-reference: 'the person concerned'. The expression could imply that the words and actions of the person concerned were of ultimate importance, whether in present suffering or future glory—while leaving all credit and responsibility with God.

Yet even the Romans saddle him with the title King of the Jews, their equivalent of *Christos,* and the soldiers mock him as if he were an imperial pretender. But when he died, the centurion replaced the ridicule with his rueful judgement that this man was Son of God, whether reflecting on his outstanding heroic character or transferring to him the *divi filius* of the oriental emperor cult. Of course, Mark does not present this acclamation as an expression of Christian faith; but surely he is saying that the death of Jesus (not exclusively the resurrection) can put Gentiles on the road to faith. As for disciples, if they return to Galilee, they will find Jesus of Nazareth who was crucified (for resurrection does not cancel crucifixion)—at least until he comes in the glory of the Father with the holy angels.

In following the course of Mark's Gospel and trying to discover his intentions, it seems to many recent students that we are reading not a simple attempt to put earlier traditions into a useful chronological order or to give a plausible historical account of the life and death of Jesus but a carefully judged theological account of the traditions in order to remedy unsatisfactory attitudes within the community for which Mark was writing. For the Gospel to succeed in persuading its readers, it must of course contain recognizable tradition, not simply invention. Its aim was to show that all the experiences of the community from miracles to morals and martyrdom could be read—and should be read—within the context of the crucifixion of Jesus.

Matthew.   Matthew in general reproduces Markan episodes and teaching but chiefly uses the Gospel as a framework for his collections of teaching. They include Q sayings on the sufferings of disciples (see p. 198), no doubt with the general implication that the sufferings of Jesus are a model for Christians, given the times in which they live.

Matthew has no authentic additional information about the Galilean work of Jesus. His additions to the Passion and resurrection narratives are either apologetic anecdotes (Pilate's wife, the guard at the tomb, Judas) or legends—the earthquake, the resurrection of the dead saints, and the mountain in Galilee. One may compare the legends with which the Gospel begins—the magi with their magical star and the flight into Egypt. Only two angels appear in the Gospel: one announces the birth of the Son of God, and the other accomplishes his resurrection. To call these episodes legends needs explanation. A legend is commonly an inauthentic story handed down by tradition and popularly regarded as historical; but it is not, for that reason, to be dismissed. It is necessary to ask about its function. The Latin *legenda* meant 'what is read', and the French *légende* explains the conventions used in mapping; so *legend* in its technical sense indicates how the account is to be read and understood. As a clear example, consider Mt 12:40, where the evangelist expands a Q saying about the scriptural sign of Jonah: 'As Jonah was three days and three nights in the belly of the whale, so will the Son of man be three days and three nights in the heart of the earth'. Thus, the death of Jesus and his escape from death is to be understood in the light of God's command to Israel (symbolised by Jonah) to bring light to the Gentiles (see Bonnard and Ellis on Mt 12:40), a thought that comes to full expression on the mountain in Galilee.

Thus, Matthew's earthquake may usefully move the stone and tear the temple curtain, it may be the first instalment of earthquakes foreseen in Mt 24:7 and may open Sheol for the dead saints to appear. But its overarching significance is that the death and resurrection of Jesus release the most fundamental earthshaking forces (as indeed, according to Mt 2:1–12, the heavens were moved at his birth). The angel of the resurrection in Matthew is in appearance like lightning and his clothing white as snow. Obviously we are dealing with the Lord of heaven and earth—and, indeed, on the mountain Jesus says, 'All authority in heaven and earth has been given to me' (Mt 28:18). But it is primarily the authority of a teacher: 'Go therefore and make disciples of all nations ... teaching them to observe all that *I have commanded you*'. And he gives the assurance that he is with them to the close of the age. In what sense is he with them? That question is already answered in Mt 16:18–19: in order that the powers of death (in Greek, 'the gates of Hades') should not prevail against Christ's church, Peter is given the keys of the Kingdom of Heaven, so that whatever he binds on earth is bound in heaven and whatever he looses on earth is loosed in heaven—that is, Peter is given authority to rule on the proper way of interpreting Christian Torah (there is a complementary saying in Mt 18:19–20). And when the Son of man comes with his angels in the glory of his Father, he will (acc. to Mt 16:27) *repay every man for what he has done.*

For Matthew, God's Torah is the source of all power. That is why Jesus is a teacher of Torah, and why disciples have to teach all the Gentiles what Jesus has taught them. The Kingdom of Heaven is the condition in which it is possible to keep Torah at every time and in full detail, to be free from discrimination, to develop and preserve the gentle virtues, and to find complete satisfaction and fulfilment in this constant practice. But Matthew has nothing to add to Mark's understanding of the death of Christ—if indeed he grasped it.

Luke. Luke draws a striking parallel between the origins of John Baptist and Jesus, in some way connected, perhaps, with the people who knew only John's baptism in Acts 19:1–7. He uses his parallel to establish the prophetic character of Jesus (even if he, like John, is more than a prophet). But, unlike Mark, Luke makes no use of the Baptist's death as a pointer to the fate of Jesus. When Jesus is threatened by religious and political hostility, the danger comes from the unacceptable character of his message; but that is nothing more or less than the occupational risk of a prophet. Luke certainly does not avoid mention of Jesus' death, but he dignifies it. At the transfiguration, which discloses his proper glory, Jesus discusses with Moses and Elijah his forthcoming 'exodus' in Jerusalem. Thereafter, he moves in a deliberate progress towards his 'assumption', taking his time, unhindered by Herod, until he finishes the course and reaches his perfecting. After all, he came to kindle a fire and undergo an ordeal ('I have a baptism ...' Lk 12:50). The Son of man first must suffer many things and be rejected by this generation. Why *must*? Because it is the will of God. In suffering he obeys, but Luke does not say that he atones or expiates.

At the Last Supper, Luke supplements the Markan tradition by making Jesus' self-giving instructive for Church leaders as they await the kingdom. He certainly includes sacrificial language (if the longer text is original): 'my body given for you' and 'my blood poured out for you'. But the sacrifice of his body is a memorial sacrifice, and the blood poured out achieves the new covenant that implants Torah. Thus Luke is not describing a literal human sacrifice but the remembrance of what Jesus had done for the Twelve, and the definitive principle of their future life, namely, self-giving on behalf of others. Jesus is amongst them as one who serves (in Acts he is present in his NAME), rather than giving himself as a ransom for them. By undergoing hardship and trials (a popular storyteller's motif) he claims the kingdom covenanted by the Father, even though he has to be reckoned with transgressors.

That Jesus is not in fact a transgressor is shown by his nobility of character. In Gethsemane he is self-possessed, prays lucidly, and utters grave warnings. When arrested, he displays calm generosity. At the Sanhedrin trial, he speaks with the assured, dismissive manner of a noble-spirited person suffering indignities. At the Roman trial he is repeatedly pronounced not guilty. On his way to death he speaks of others with solemn foreboding, and goes to his execution like a great prince. On the cross he magnanimously prays for his executioners (if the reading is accepted), with oriental munificence promises paradise to the repentant criminal, and dies showing untroubled confidence in God—certainly not shouting that he is forsaken. In fact, death is overcome by nobility of spirit, and the centurion calls him 'a good man'. In its way this is fitting; but it looks as if Luke's centurion is confirmed in the pagan virtues, whereas Mark's centurion is on the way to being a Christian.

The corporeal reality of the risen Christ in Luke proves that he is the identical noble person that died. He explains from scripture that the Christ must move through suffering to his glory. In Acts his death and resurrection are everywhere present, but the emphasis lies on resurrection, which is amply argued from scriptural proof. But not the death. Healings, miracles, forgiveness of sins and the Holy Spirit come from the NAME of the risen and exalted Lord. Opportunities for using the expiatory statements of Isaiah 53 are not taken, even avoided. The church that God obtained as his own possession by the blood of his own—his Son, his apostle, perhaps his Church elders—goes no further than the thought of martyrdom. Some of Luke's statements about the death of Jesus seem ambivalent: it was the dreadful work of human hands and also the carefully decided initiative of God. It is almost enough to say that the death was the proof that Jews needed salvation and that the Gospel began with the resurrection.[142]

6

1 Peter

This epistle has long been the centre of a bold and sometimes imaginative argument about its author, date of composition and its unity and nature as a written document. That kind of discussion is now partly replaced by sociological enquiries,[1] but it has provided the base line for exegetical study. Certain positions are now widely agreed. First, the epistle is a genuine circular letter, written by a single author, though it may combine two originally separate components. Second, it contains a great deal of traditional material, such as Old Testament quotations, fragments of hymns and liturgical phrases, credal statements and codes of conduct. It does not reproduce a baptismal liturgy or paschal homily. Third, all this material is welded together by a writer of very competent Greek, with some pretentions to style, who has produced a clearly indicated literary structure. Fourth, the thought and phraseology of the epistle has some obvious similarities to sections of the Pauline corpus, notably to Romans, Ephesians, Titus and 1 Timothy. But it is not easy to decide whether the similarities suggest the writer's direct knowledge of later writings in the name of Paul or use of oral tradition known in the Pauline circle or both. 1 Peter also betrays non-Pauline features and has some thought and language in common with James, Hebrews and 1 Clement. It is best to think of a semi-independent Petrine circle, not uninfluenced by the Pauline circle. Fifth, it is not easy to date the epistle, but a time after the death of Peter and before the correspondence between Pliny and Trajan seems best suited to internal information. Statements about Christian 'persecutions' or (better) disabilities cannot be identified from our scanty historical knowledge of the period, and are best explained by the socially disagreeable impression made by Christian communities. The simple, undeveloped theology of the epistle suggests an earlier date. The spread of Christian communities into the four Asian provinces mentioned in the address (remembering that Pontus and Bithynia comprised a single province) suggests a rather later date. There is considerable support for the period 65–80, perhaps earlier rather than later. Because Silvanus wrote the letter and Mark sends greetings (1 Pt 5:12–13)—both of them elsewhere associated with Paul—[2] I myself propose the view that 1 Peter was part of an attempt to deal with the disintegration of the former Pauline missionary empire. Sixth, the epistle contains some ten references to the death and resurrection of Christ. It also uses the verb *paschō* twelve times and the noun *pathēma* four times ('suffer' and 'suffering'),[3] far more than any other New Testament writing. These observations already suggest that the writer intends not simply to give instruction on faith and morals but to deal with the distress of the Asian communities by appealing to Christ's death and resurrection.

The structure of the epistle is quite simple. 1 Pt 1:1–2 is the opening address and greeting (very like the Pauline beginning, though not identical with it). 1 Pt 4:7–11 looks like a kind of ending, so that 4:12–5:14 (beginning with 'Beloved') is a distinct section. 1 Pt 1:1 addresses the recipients as 'exiles of the Dispersion'; and 1 Pt 2:11 begins with 'Beloved', addressing them as aliens and exiles. Thus 1 Pt 1:3–2:10 establishes the status of God's people in exile; 2:11–4:11 gives instruction about Christian behaviour in such a situation; and 4:12–5:11 deals with the crisis of the exiles.

It is obvious that the recipients of the epistle knew well enough their own stressful condition. We, however, have to acquaint ourselves with their situation and keep it firmly in mind from the opening verses onwards. This we can conveniently do by beginning with the third section, 1 Pt 4:12–5:11, where the writer deals explicitly with his readers' difficulties. Some interpreters think the difficulties mentioned in this section are so much greater than those mentioned earlier that 4:12–5:11 must belong to a different letter; others deny the necessity of that judgement. For my purposes, it does not greatly matter.

THE FIERY ORDEAL, 4:12–19

The writer turns directly to his readers at 1 Pt 4:12 with a wordplay that may be represented thus:

> Do not feel alienated by the fiery experience in your community
> which is taking place for your testing,
> as if something alien were happening to you.

> But as you share the sufferings [*pathēmata*] of Christ rejoice
> in order that, at the disclosure of his glory too,
> you may rejoice and be glad. (1 Pt 4:12–13)

Adequate understanding of this pastoral advice depends on some exegetical points. The alienation imagery is a variation of the exile theme that opens the first and second parts of the epistle (1 Pt 1:1, 2:11).[4] The readers are consistently treated as resident aliens (not as a minority group seeking to escape) who must expect some unpleasant social disabilities and must, by their conduct, at least deserve social toleration.

The phrase 'fiery experience' is intended as a fairly neutral rendering of *pyrō-sis*. Standard translations such as 'fiery ordeal' or 'painful trial' may suggest overreaction to the word, which Goppelt (from its Septuagint use) renders 'refining fire'. There is a reference back to 'the genuineness of your faith, more precious than gold which though perishable is tested by fire' (1 Pt 1:7). But this stock metaphor is used here less to excuse God's infliction of trials on the faithful than to explain that adherence to Christ requires sharing his fortunes.

The phrase 'sufferings of Christ' *(ta pathēmata tou Christou)* comes again in 1 Pt 5:1 and something very like it in 1 Pt 1:11 *(ta eis Christon pathēmata)*. It has a strongly Pauline flavour from its appearance in 2 Cor 1:5–7, Phil 3:10 (sharing the sufferings of Christ) and Rm. 8:17 (sufferings opposed to glory and the condition 'Provided we suffer with him that we may also be glorified with him'). The Pauline passages refer to the sufferings of those who are members of Christ's body and belong to Christ (see p. 52); but in this epistle the thought is different, as will appear in due course.

The encouragement to rejoice while presently suffering is strengthened by the promised rejoicing when Christ's glory is revealed. The joyful acceptance of suffering is a common theme of ancient Jewish teaching, especially in martyrdom theology (see Goppelt, pp. 299–304). In this epistle, it finds prominence early and late, in expectation of the *apokalypsis* of Jesus Christ (1 Pt 1:6–7, 4:13).

The sufferings that Christians can expect are indicated in the succeeding verses: they are social disabilities. Not, of course, the punishments given to murder, theft and other unsocial activities; but the consequences of hostility towards those who bear the name of Christ[5] and are known as Christians. According to the Gospel tradition, Jesus himself was attacked and derided for the designation Christ by both the authorities and his own people. If, then, the Asian Christians suffered for the name, they were suffering as Christ had suffered.

It is characteristic of 1 Peter that those who thus suffer according to the will of God (i.e., not for some disreputable offence) are to entrust themselves to a faithful Creator by continuing to do good (see 2:20, 'But if when you do good and suffer for it you take it patiently, you have God's approval'). The need to work out a social strategy for the Christian community is very evident, and that is why the death of Christ is important.

Humble Yourselves, 5:1–11

In the next section 1 Pt 5:1–5, the writer addresses older and younger men about their responsibilities in the community. The advice is conventional: elders are not to be domineering, young men are to be submissive. Two features, however, are novel in this context. First, the writer declares himself 'a fellow elder and witness of the sufferings of Christ as well as a partaker of the glory that is to be revealed' (verse 1); second, the community is to humble itself under the mighty hand of God. The former means not necessarily that the writer saw Christ die or that he was a preacher of Christ crucified but that he bore witness to the name and in similar manner shared the disabilities of the Asian Christians. The second feature, not unconnected with the first, establishes humility as a desirable attitude. The Greek sense of the adjective *tapeinos* is but partly revealed by the translation 'humble'. It meant subservient, abject, debased, vile. Hence 'humble yourself' meant 'accept your humiliations' as coming from God who in due course will exalt you; for God resists the proud and favours the humble (from Prv 3:34). That sounds well enough in a Semitic culture but less acceptably in Greek life. When the young men are urged to wrap themselves, as it were, in the clothing of humility, Greek neighbours would regard it as an encouragement to discreditable thoughts or a humiliating attitude.

Thus, in this section of the epistle, the sufferings of Christ (1) are shared and witnessed to by Christians as they experience social disabilities, (2) belong to the grace of God as a necessary prelude to the glory, and (3) give rise to the surprising Christian posture of social humility. Having established these associations by examining the section of the epistle that mentions most directly the affairs of local churches, we can now turn to the earlier sections.

Christians as Exiles, 1:1–2:10

The exiles are initially addressed as chosen *(eklektoi)*

> according to the foreknowledge of God the Father
> by the sanctifying work of the Spirit,
> for obedience to Jesus Christ
> and sprinkling by his blood. (1 Pt 1:2 NIV)[6]
>
> *Kata prognōsin theou patros*
> *en hagiasmō pneumatos*
> *eis hypakoēn*
> *kai rhantismon haimatos Iesou Christou*

This triadic formula can be regarded as expressing the ground, means and purpose of their election (according to, by, for) or, more probably, the structure of the main argument. Thus, the foreknowledge of God lies behind 1 Pt 1:3–12; the sanctifying work of the Spirit behind the description of the holy community in 1:13–2:10; and the obedience of Christians who share Christ's sufferings in 2:11–4:11. In any case, the sprinkling with Christ's blood is not a reference to ransom or sacrificial atonement but to the establishment of a covenant between God and the people in Ex 24:3–8 (Hort). Moses threw half the sacrificial blood against the altar; then he read the book of the covenant; and when the people promised obedience, he threw the remaining blood on them. First, the promise of obedience, then the sprinkling with blood—so in 1 Pt 1:2, where the words are metaphorically intended. There was no blood-sprinkling ritual. But in reality Christians, in their obedience to Christ, might expect to share his disabilities and perhaps shed their blood for his name.[7]

The first main section of the epistle 1:3–2:10 discusses the status of God's people as a group of exiles, first in relation to God's foreknowledge (1:3–12) and then in relation to his holiness (1:13–2:10). There are four explicit references to the death and resurrection of Christ (1:3, 11, 19, 21) and also an impressive concentration of equivalent images: (1) rebirth (1:3, 23; 2:2), (2) the life of the seed (1:23–24), (3) the stone rejected and then prized (2:7), (4) the transition from darkness to light (2:9), and (5) the nonpeople who become the people of God (2:10). This prodigality may imply either that the resurrection has been assimilated to a common expectation of rescue and renewal or that Christ's resurrection is so greatly effective that it gathers up and surpasses other images. The latter seems to be right. The resurrection has regenerated hope, which therefore becomes, in this epistle, the equivalent of faith and the distinguishing name of Christian existence (1 Pt 1:21, 3:15). It is a living and active expectation that the Christians' inheritance—imperishable, undefiled and unfading—is protected by God's power and due to be revealed soon (see 4:7, 'The end of all things is at hand') despite the predictable[8] likelihood of present discriminatory treatment. Thus, the resurrection of Christ is firmly linked to his *apokalypsis* and therefore to the readers' (eschatological) salvation, which in turn is previewed by the prophets (1 Pt 1:9–12).

The prophets appear at this point in order to show that suffering for believers has been foretold by God. Indeed to the prophets who searched and enquired about the time or circumstances (NEB; 'person or time' RSV) of such suffering, it was disclosed that the writer's generation of Christian believers were first to suffer and then to experience the glory. Since the writing prophets were little concerned with such enquiries, it is commonly held that the apocalyptic writers are in mind (Schrage, Goppelt). The object of their enquiries (much in the style of Qumran) was 'the sufferings in store for Christ' (NEB, *ta eis Christon pathēmata;* see 'the grace of God awaiting you', *tēs eis hymas charitos* in the previous verse);[9] and 'the spirit of Christ in them' indicated a fulfilment not for themselves but for the readers of the epistle. Whether the writer's intention can be discovered in his two references to Christ is not clear. What could he mean by saying (only here in the New Testament) that the spirit of Christ was in the prophets when they conducted their search for time and circumstances? It would perhaps be possible to translate *Christos* as Messiah and explain the reference by means of Is 61:1: 'The spirit of the Lord is upon me, because he has anointed me' (see Is 11:1–2, quoted in 1 Pt 4:14). But that would be a complicated and unlikely thought to present to Gentile Christians, who would inevitably read *Christos* as Jesus Christ. Or the words 'spirit of Christ' might mean 'Christ as spirit' and so imply his preincarnate existence (Kelly), but it would be a surprisingly casual introduction of a major Christological assertion. It would therefore be far better to be guided by the practice of Qumranic interpreters, who used the events in their own day to give proper significance to ancient scripture. The writer of the epistle compares 'the spirit of Christ in the prophets' with 'the Holy Spirit sent from heaven in the evangelists who preached in Asia'. Now that they know about Christ and his sufferings—partly by sharing them—they can find it foretold in scripture, where the glory was also foretold. If one prediction came true, so will the other.

In 1 Pt 1:13–21 the writer neatly relates first obedience and then the blood of Christ (which together will be the subject of 1 Pt 2:11–4:11) to the holiness of God. The relation of obedience to holiness (verses 13–16) is sufficiently obvious to any reader of the Old Testament; the relation of Christ's blood to holiness is a little more subtle. It must be studied in verses 17–21, which consist of an introductory conditional clause in verse 17ᵃ, a specific command in verse 17ᵇ and various supporting statements in verses 18–21, which can be written out in rhythmical form and look like fragments of credal or liturgical formulas. It is presupposed that the readers call on God as Father and know that he judges impartially each person's activity. They are not worshipping an ancestral god or a family deity but the universal judge of all. If exodus imagery has left its impression on this passage, it is because they have already moved away from the passions of their former ignorance (verse 14) and been set free from the futile ways inherited from their fathers (verse 18).[10] In that case, it is not surprising that the society from which Christians had withdrawn resented their separation and put pressure on them to abandon their threatening stance. The writer of the epistle therefore invokes a powerful sanction. They are to conduct themselves with fear *(en phobō)* throughout the time of their exile—fear not of social pressure but of God who has set them free from the futilities of ancestral custom by the blood of Christ. They have not been set free by the payment of a ransom but by the power of shed blood (Beare). It is as if a lamb, unblemished and spotless, had been sacrificed on their behalf and its blood had become the guarantee that God would liberate and protect his people.[11] All readers in antiquity, Jewish or Greek, would understand the power of sacrificial blood and would realize the need to regard it with extreme caution (the meaning of *fear* in this context).[12] If they were to worship the universal judge of all and to benefit from his liberation (*elytrothēte;* see Appendix A) they must behave with *phobos,* all the more so because they had been set free not by the blood of a lamb but by the blood of Christ, which is worth far more even than precious metals—silver or gold. The 'precious blood' (verse 19) is justified by two traditional fragments: verse 20 has the primeval time–end time formula: 'destined before the foundation of the world but made manifest at the end of the times for your sake'; and verse 21 has the death-resurrection formula: 'God raised him from the dead and gave him glory'. *Lytron* words can be linked with the end of the age (Rm 8:23, Eph 4:30), with the death of Christ (Rm 3:24, Eph 1:14), and with the purification of a people (Ti 2:14, cf. 1 Tm 2:6), themes included in this passage of 1 Peter. They imply deliverance from prohibited actions (Ti 2:14); from folly and weakness (1 Cor 1:30), which are somewhat akin to the ancestral futility of 1 Pt 1:18, though 1 Peter does not describe this as forgiveness of sins (as in Col 1:14, Eph 1:7). Thus, 1 Peter's solitary reference to *lytrōsis* reflects general New Testament usage but provides it with a far more substantial basis in the original creative purpose of God and its consummation in the new age, focussed on the decisive divine act in raising Christ from the dead and giving him glory. Hence, the sacrificial blood of Christ is regarded as freeing human beings from the seemingly unalterable pattern of ancient society and then making it difficult (or at least dangerous) for them to sin. The precious blood is not so much an atonement for past sins as an inhibition of sinning, as believers wait for their exile to cease and the age of glory to begin

(1 Pt 1:11; 4:13–14; 5:1, 4, 10). Nor are they merely kept in order by a religious threat but are fed on pure spiritual milk so that they may grow up to salvation—meaning that their life is nourished by the abundant images of regeneration.

The Model of Submission, 2:11–4:11

The remaining section 1 Pt 2:11–4:11 gives practical advice about the conduct of Christian aliens. Though they are God's true people, by circumstance they are rootless strangers. Therefore, they must abstain from the natural human desires, which may damage their proper interests. Even as it is, they are being spoken against as criminals because (as appears shortly) they stir up trouble and make people disrespectful to authority and ready to answer back. Therefore, they must be models of good behaviour to refute such charges and to persuade their Gentile neighbours to honour God (1 Pt 2:11–12). That optimistic hope is then supported by a strategy of submission. They are to be subject for the Lord's sake to civil authorities (1 Pt 2:13–17); slaves are to be submissive to their masters (1 Pt 2:18–25), wives to their husbands (and husbands are to be considerate to their wives, 1 Pt 3:1–7). A similar code of submission is found in Col 3:18–4:1 and Eph 5:21–6:9, and there may well be fragments elsewhere. The instructions it sets out go back to Aristotle's teaching 'concerning household management', and there is plentiful evidence that they were well known and accepted in the Roman Empire: 'Many minority religious communities, including Christians, had to relate to this Greek and Roman household ethic. Devotees of the god Dionysus and of the goddess Isis, as well as Jews and Christians, were forced to come to terms with these household customs'. As far as Christians were concerned, 'persons in Roman society were alienated and threatened by some of their slaves and wives who had converted to the new, despised religion, so they were accusing the converts of impiety, immorality and insubordination'.[13] Therefore, the writer of the epistle tells wives and slaves to be as submissive as society requires them to be.

To the modern liberal mind that is a shocking, even wicked, thought. But we must make an effort to understand why it was not shocking to Roman authority or the writer of the epistle or even to ourselves in certain situations. I leave aside the question of wives (see pp. 151–52) and mention only house servants _(oiketai)_. To use a modern parable, let us imagine ourselves on a package holiday in a tourist hotel. To our dismay we discover that our rooms are not cleaned, that the telephone is not answered and that service in the dining room is erratic and surly. We tell the tour agent that hotel staff should do the work for which they are hired and take orders from the manager, otherwise the package business will not work. So, by analogy, in late antiquity it was taken for granted that civil and political benefits would be lost unless slaves were submissive to their masters. In this enquiry we are not compelled to debate that assumption, but it is certainly necessary to consider how the death of Christ is used to give it support.

First it is useful to compare the structures of the two paragraphs on slaves and on wives.[14] They are remarkably similar:

(1) House servants are to be submissive in all fear *(en panti phobō)* to their masters, (2) not only to the kind and gentle but also to the harsh. (3) They may have to endure pain unjustly inflicted, to suffer patiently for doing what is right, (4) for that is approved by God. (5) They were called to follow the example of Christ's suffering (2:18–25).

(1) Wives are to be submissive to their husbands, (2) some of whom are not Christian, but may perhaps be won by their chaste behaviour in fear *(en phobō)*. (3) They are to replace outward adornment by a gentle and quiet spirit, (4) which is precious to God. (5) Their example is Sarah's behaviour to Abraham. (6) Husbands are to live considerately with their wives (3:1–7).

Thus, both groups must be submissive *(hypotassomenoi)* for reasons of fear—not simply, I think, fear of masters or husbands or even fear of God (as in 1:17) but also fear that the social system may fall to pieces. Since it is easy to be submissive to gentle masters or Christian husbands, the case of harsh master or non-Christian husband is singled out. Wives may perhaps influence their husbands, slaves will scarcely influence their masters. Slaves must be prepared for unjust suffering, wives must cultivate a gentle and quiet spirit. The example for slaves is the suffering of Christ, for women the obedience of Sarah. Husbands (presumably Christian husbands) are to be considerate to their wives, but nothing is said about Christian masters (contrast Col 4:1, Eph 6:9). From this comparison it seems clear that (1) the position of slaves was likely to be more unpleasant than the position of wives, (2) the references to Christ and Sarah are to provide examples of proper conduct, and (3) both paragraphs are intended to encourage conformity, for whatever reason, to current social expectations.

Slaves and the Suffering of Christ, 2:18–25

The submission of slaves is doubly justified in view of their relation to God and to Christ. The relation with God begins *touto gar charis* and ends *touto charis para theō* (verses 19–20). This is not quite the normal Pauline *charis* (grace) but means 'what brings God's favour or blessing', much the same as *kleos* in the fifth line below (good report, what produces favourable consideration):

> For this is what brings God's blessing
> when, in awareness of God,
> someone endures pain
> suffering unjustly.
>
> For what favourable consideration [can you expect]
> if you do wrong
> and are beaten?
> But if you do right
> and suffer for it
> and you take it patiently,
> that is a claim on God's blessing.

The explanation in terms of 'blessing' is justified by 1 Pt 3:9, 'To this you have been called [see 2:21], that you may obtain a blessing'. And you obtain it by not returning evil for evil or reviling for reviling. When a person is abused and suffering severe pain, he may shout abuse at his tormenters and curse them. Perhaps the curse may make them desist, or punish them if they persist. But the more effective method is to call on God for a blessing, which is more powerful than the malice of tormenters, and that indeed is the counsel of this paragraph. It does not imply that God takes pleasure in the patient endurance of unjust suffering but that such endurance in one who is 'mindful of God' calls forth his blessing.

That conviction helps to interpret the verses relating the slave to Christ (21–25):

21 For to this *you* have been called,
because Christ also suffered for *you*,
leaving *you* an example,
that *you* should follow in his steps,

22 who 'committed no ⟨sin⟩' [Is 53:9]
'nor was deceit found in his mouth',

23 who being reviled did not revile in return, [see Is 53:7]
suffering, he did not threaten,
but trusted to him who judges justly;

24 who 'himself bore *our* sins' [Is 53:4, 12]
in his body on the tree
that, having finished with sins,
we should live for righteousness,
with whose 'wound *you* have been healed'. [Is 53:5]

25 For *you* like 'sheep were going astray', [Is 53:6]
but have now returned to the Shepherd
and Guardian of *your* lives.

This piece of rhythmical prose[15] is divisible into three parts: lines 1–4, verse 21 which speaks to *you*; lines 5–13, which have four clauses introduced by *who* or *whose*, and in verse 24 speak of *our* and *we*; and lines 14–17 which revert to *you*. Moreover, verses 22–24 are largely constructed of phrases taken from Isaiah 53. If verses 21 and 25 are put together they make an intelligible sequence: the thought of verse 25 follows that of verse 21. Thus the function of verses 22–24 is to draw out the implications of the traditional formula, 'Christ suffered for our benefit' (see pp. 14–15), with *our* changed to *your* when specifically directed to house servants. At the end of verse 24, the writer neatly linked the *we*-phrases of the verse to his *you*-style address by taking an additional image from Is 53:5 and saying in effect, 'It is true for *us* all, and so *you* too have been healed by his wound.' A final image, again in *you*-style, leads the section to its reassuring conclusion. So 'Christ died' is developed in verses 22–23; and 'for you' in verse 24 (Goppelt).

Thus, the instruction is given that a house servant was called (by God)[16] for this purpose, namely, to do what was right and to suffer for it. To fulfil his calling he has been provided, as it were, with a master copy *(hypogrammos)* and a line of footprints to follow,[17] namely, Christ's suffering on behalf of others.[18] In verses 22–23 that suffering is interpreted by the scriptural model of the sufferer in Isaiah 53, or at least by selective references to that chapter in its Septuagint version. It has long been held (though of course not in antiquity) that Isaiah 52:13–53:12 comprises the last of four poems about the servant of the Lord,[19] which in Semitic culture is a title of honour. The Hebrew word for servant *('ebed)* can be rendered in Greek by the standard word for slave *(doulos)* or by *pais* (child, servant). In the second poem, both are used (Is 49:3, 5, 6); in the fourth, only *pais* (Is 52:13). The Hebrew text of the fourth poem is unsatisfactory (therefore amended by modern translators) and difficult to understand. The Septuagint diverges from it and, in places, is more obscure. Even though the text is not wholly intelligible, it is eminently quotable, especially as the fourth poem alone brings together sin and suffering. In the language of Is 53:9, 'Christ committed no sin, nor was deceit found in his mouth', that is, he had not offended God, he has not been found a cheat and liar by his neighbours; yet he suffered. When abused, he did not answer with abuse; when he suffered, he did not threaten those who caused his suffering (these statements possibly prompted by Is 53:7). Instead, 'he trusted to him who judges justly', which can scarcely be extracted from Isaiah's poem, for it portrays the sufferer as wholly passive.[20] The effect of these statements is to present Christ as innocent, suffering unjustly but without retaliation and firmly relying on God. Correspondingly, the house servant must hope to be beyond reproach, to accept unjust punishment without resenting it and to hold firmly to God.

In turning now from the manner to the purpose of the suffering, it will be convenient to refer to the Greek of verse 24:

> 24   *hos* tas hamartias hēmōn autos anēnenken
>   en tō sōmati epi to xylon,
>   hina tais hamartiais apogenomenoi
>   tē dikaiosynē zēsōmen
>   *hou* tō mōlōpi iathēte. (translation above)

The words in roman in the first line look like a conflation of Is 53:4[a], *houtos tas hamartias hēmōn pherei* ('He bears our sins') and Is 53:12[e] *autos hamartias pollōn anēnenken* ('He bore the sins of many'). Since *anēnenken* is the aorist form of *anapherō*, there may be little difference in meaning between the two verbs. As far as this passage is concerned, *anapherō* means 'to take up' (e.g. sacrificial objects to an altar), hence 'to offer up spiritual sacrifices' (as in 1 Pt 2:5). Since a cross is not an altar (and even if it were, putting sins there would defile it), the sacrificial meaning is excluded.[21] Hence, the wider meaning must be intended: 'to take upon oneself'. So, 'He took [the consequences of] our sins upon himself when he suffered bodily on the cross'.[22] Since he suffered innocently, his fate was the result of the wrong that others did. But more than that is meant. Christ takes up the consequences of our sins in such a way as to inhibit sinning (see the 'precious blood', p. 243). We have finished with sins[23] and their consequences and now should live to do what is right (which is the meaning of righteousness, here and 1 Pt 3:14). Thus, the result of Christ's death (including the reasons for which he died and the manner of his dying) changed the possibilities for sinners. Not that he innocently suffered the punishments due to sinners but that he without cause or complaining suffered the damage that sinners can do and by his injury promoted their healing. In Isaiah and elsewhere in the Septuagint, 'healing' is a metaphor for God's work of restoration, which includes forgiveness. In a complementary metaphor, these Gentile slaves once were straying sheep but now have been returned to the shepherd who is indeed responsible for guarding their lives. That is God's blessing, and it is called forth by the suffering of Christ. He therefore is the representative sufferer, and those who are in the same case likewise receive the blessing.

If that interpretation is satisfactory, it operates well within the boundaries of hellinistic Jewish teaching. It includes the conviction that innocent suffering can compensate for wickedness: that God, who is moved by the innocent sufferer to confer his blessing, is willing to extend the blessing to those who belong to the sufferer and for whom he suffers. Moreover, the blessing is not chiefly an individual benefit, but a powerful means of 'healing' the social situation in which the individuals must live. It seems likely that the suffering of Christ is indeed exemplary and provides a test of belonging to him but is more than exemplary in calling forth God's blessing. To some extent the suffering of slaves must be said to call forth that blessing even though it cannot confidently be said that slaves were in the same case as Christ. Had Christ no choice but suffer, or did he freely choose? Is the use of Christ's suffering to encourage the subordination of slaves morally objectionable (in pacifying the sufferers and ignoring the oppressors), or are the lessons taught to slaves extensible to all Christians?

The Imitation of Christ (1), 3:18–4:11

Such an extension indeed takes place in 1 Pt 3:8–22, where the same themes are developed for the whole community[24] without the use of Isaiah 53 but relying on recognizable catechetical or liturgical excerpts. The community is reminded of the conduct that corresponds to God's blessing, and is assured that no one will harm them if they eagerly do what is right. Even if they did suffer from abuse for their Christian good behaviour, it would be better to suffer from doing good (if that were God's will) than for doing wrong.[25] And the writer continues:

18   For Christ also suffered for sins once for all,
      the righteous for the unrighteous,
      that he might bring you to God,
      put to death in the flesh,
      but made alive in the spirit;

      . . . .

21   by the resurrection of Jesus Christ,

22 who is at the right hand of God,
having gone into heaven—
angels, authorities and powers made subject to him. (1 Pt 3:18,
21^d–22)

These verses, like previous sections of the epistle, have a structured form appropriate to a catechetical or liturgical unit. They do not fit perfectly (in the absence of verses 19–21^c), and it may well be that an existing formula has been adapted by the writer. If so, attempts to define its limits or to restore its original form fail not from lack of ingenuity but from lack of controlling evidence.[26] However that may be, the formula deals with familiar themes, which may briefly be noted.

Christ's death for sins[27] *(peri hamartiōn)* was a pre-Pauline formula (see pp. 14–15) and is used here to justify the writer's proper disapproval of sins, where the word *hamartia* has the straightforward ('non-Pauline') sense of 'wrong actions, especially those that displease God'. What was said about *hamartia* as regards slaves in 1 Pt 2:22, 24 is here strengthened and will be confirmed in 4:1—with an antidote suggested in 4:8.

His death was 'once for all' *(hapax)*. This is not the insistent *hapax* or *ephapax* of Hebrews (applied to Christ's death for sins in Heb 9:12, 26, 28; 10:10), which contrasts the unique sacrificial act of Christ with repeated sacrificial acts in the cult; but is perhaps reminiscent of Rm 6:10, 'The death he died he died to sin, once for all'. But whereas Paul was positing a wholly decisive act in the (mythically presented) conflict between Christ and the power of Sin, the Petrine writer implies that Christ's innocent suffering established the decisive pattern for dealing with sins.

The righteous *(dikaios)* suffers for the unrighteous *(adikos)*. In this context, what do those words mean? Since the writer contends that Jesus suffered unjustly, *dikaios* may mean 'innocent' (see Appendix B), and the phrase may thus mean, 'the innocent suffering for the unjust', giving a somewhat legal flavour. But the previous paragraph quotes from Psalm 34, 'The eyes of the Lord are upon the *dikaioi* and his ears are open to their prayer, but his face is set against those who act wickedly' (1 Pt 3:12). Further, in 1 Pt 4:18 there is a quotation from Proverbs 11: 'If the *dikaios* scarcely manages to be saved, what will happen to the godless and sinful?' Thus, the *dikaios* is someone who attracts God's favour, the *adikos* his disfavour. The formula is not so much making a point about justice and injustice as asserting that he who knows God's favour suffers for him who knows his disfavour.

By this self-sacrificing action Christ gives us access to God (see p. 95 on Rm 5:2, where the noun *prosagōgē* is used, corresponding to the verb *prosagō* here). Precisely what that implies is not clear from the phrase itself. Does it mean that Christ overcomes God's displeasure and induces him to receive sinners or that he brings sinners into encounter with God so that they understand his pleasure? Probably something simpler, in an image familiar from Greek private religion: Christ is the mystagogue who takes his followers into the divine presence, where they are accepted because they are his (Beare; and compare Christ as *parakletos* in 1 Jn 2:1).

The phrase 'put to death in the flesh but made alive in the spirit' is no doubt equally untechnical, no more than a simple scheme for making death and resurrection thinkable. He was put to death and therefore terminated the existence called flesh; he was brought to life, and so began the other existence called spirit. The same contrast is present in the perplexing statement of 1 Pt 4:6, 'Though judged in human fashion in the flesh, they might live in divine fashion in the spirit'. In these two passages, at least, flesh and spirit are not regarded as components of each personal existence but as different ways of existing personally.

'Made alive in the spirit' is traditionally represented by the resurrection of Christ, his departure to heaven (the realm of the divine spirit existence), and his presence at the right hand of God (implying his share in the divine ruling authority). Thus the benefits of Christ's death are not achieved simply by his dying and the manner of his death, but by the transfer from flesh to spirit and his reception by God. It is both death and resurrection that provide the pattern for Christian life.

The subjection to Christ of angels, authorities and powers is a piece of conventional imagery, current in the Pauline circle, too firmly embedded in the catechetical or liturgical tradition to be left out. For the Petrine author they were doubtless hostile powers in the realm of spirit, but they had no significance for the social or political threats with which Christians had to live. If he mentioned them deliberately, it was to imply that Christ could deal with them; sufficient for Christians to keep sane and sober and become good stewards of God's varied grace (1 Pt 4:7–10).[28]

The Spirits in Prison, 3:19–21ᶜ

Unfortunately that is not the whole story. Into the rhythmical phrases of this catechetical or liturgical piece has been pushed a very clumsy prose insertion, verses 19–21ᶜ. Syntax, vocabulary and interpretation are all disputed.[29] Even to write down a translation of the passage requires decisions about its meaning, and they in turn depend on guessing the motive for including the passage. Let me first offer a translation:

19 in which spirit he went and proclaimed [the nearness of judgement] to those imprisoned spirits
20 who formerly disobeyed [God's will] when the patience of God was eagerly expectant in the days of Noah, as an ark was being prepared. Only a few people, eight persons in all, were saved by entering that ark from the [divine judgement effected by] water.
21 So also water correspondingly saves you[when used in] baptism, not as a removal of dirt from the body but as a pledge to God arising from a good conscience [see verse 16].

What could have prompted so normally lucid a writer to include that laboured paragraph? Surely not the hope of encouraging his readers to stand firm under provocation or the desire to enlighten them about baptism (Kelly) but rather the need to rebut damaging objections to the advice he was giving. That advice is to testify to their faith with the evidence of good deeds, to silence the ignorance of foolish men by doing what is right (1 Pt 2:12, 15). Slaves are not expected to influence their masters, though husbands may be won—without a word being spoken—when they observe the reverent and chaste behaviour of their wives (3:1). Christians are allowed to give an account of the hope that is in them, though with gentleness and reverence, if someone demands it (3:15); but the implication throughout is that Christians do not undertake propaganda for the faith. They have entered into the ark of salvation, as Noah's family did of old and have no direct responsibility for those who will perish in the coming judgement. The Noah story seemed an appropriate foreshadowing of the present situation, for in those days the wickedness of mankind had been so great that God had decided to destroy his creation and had spared only the eight members of Noah's family. Now in these days God was again preparing to destroy mankind except for the minority family of Christians. They have entered their ark through the waters of baptism which is not merely a bodily purification. When supported by a good conscience (see verse 16) it is a pledge directed to God: Christians entrust themselves to the waters and trust God to bring them safely through, as he did in raising Christ from the dead.

Such teaching, which arose perhaps more from prudence than fanaticism, must have met with objections, especially from those who wished to follow Paul in preaching the gospel whatever the cost. The Petrine writer therefore attempted to remove some of the offence by relating an explanatory myth. It was as if a king had long ago captured his hostile subjects and confined them in prison until he saw fit to consider their misdeeds and pass judgement. But the prisoners were not forgotten. When, in recent days, the king himself had rescued an obedient and faithful lord from unjust imprisonment, he despatched the lord to tell the prisoners that their cases would now be decided. The moral is that Christ, who made his proclamation to those who disobeyed in Noah's day, will equally take responsibility for the besotted pagans of the present day. In those days, out of all mankind, only a few people—eight in all—were saved. In these days also Christians are only a small minority. But their duty is (if need be) to accept suffering, to be careful that it is innocent suffering, to rely on their baptismal assurance, and to leave the heathen to God.

The Imitation of Christ (2), 4:1

In 1 Pt 4:1 the writer returns to the *imitatio Christi:*

> Since therefore Christ suffered in the flesh,
> arm yourselves with the same thought,
> for [or that] he who suffered in the flesh
> has ceased from sin.[30]

We know from 1 Pt 3:18 that he refers to Christ's innocent suffering for the guilty, to his death in the flesh and his coming to life in the spirit. In the flesh he suffered the consequences of sin; in the spirit he ceased from (encountering the

damage and frustration of) sin.[31] Christians must therefore have the same intention as Christ's when he suffered in the flesh, and they must dissociate themselves
from the morals of their social environment (verses 2–4). Why not? The end of
everything was at hand, and soon their hostile neighbours must give account of
themselves 'to him who is ready to judge the living and the dead' (verse 5). This
traditional phrase (Acts 10:42, 2 Tm 4:1, see Rm 14:9–10) turns the writer's
mind back once more to the imprisoned spirits of verse 19;[32] and he gives the
reason for that evangelizing foray. It was to bring the dead before the judgement
seat, those who had already received human judgement in their existence called
flesh but were now to receive divine judgement in the existence called spirit. The
whole argument takes for granted that Christians were under testing, and could
expect to be. If their suffering was innocent they had nothing to fear and much
to look forward to. Their hostile neighbours were *God's* concern, whether to
punish or to save. All necessary preparations had been made for the great transformation whereby one world was ending and a new world was coming into
existence.

Summary

The most striking feature of 1 Peter is the wealth of traditional material, its
bearing on the death and resurrection of Christ, and its single-minded application
to the unpleasant and precarious situation of the communities for which it was
intended. They are represented as aliens and exiles who, having withdrawn from
the manners of society, have not attempted to escape from it because they were
expecting the end of all things and their own special inheritance. Not surprisingly, this withdrawal had produced a hostile response, always testing and on
some occasions even severe. The writer, in attempting to reassure his readers,
tends to discount the likelihood of trouble provided they adopt his strategy of
public humility and overt good behaviour. Nevertheless, they may endure social
disabilities for being known as Christians. But there is no going back. They have
been liberated from the futile manner of life of their forefathers by the blood of
Christ; they are no longer worshipping their ancestral gods but God, who is judge
of all. This liberation can be compared metaphorically to the ransoming of prisoners (though in fact there was no money to be handed over and no one to whom
it could be paid); or it can be put alongside the rescue of the Israelites from slavery
in Egypt (though the Asian Christians were not captive Israelites, nor had they
escaped unharmed from the devastation of their masters). The image of 'precious
blood . . . like that of a lamb without blemish or spot' is in fact more ancient: it
refers to sacrificial blood that makes a bond between the worshipper and the
deity. On one side it guarantees God's acknowledgement and protection; on the
other side it requires the worshipper's submission and renewal. Since the blood
was *not* provided by a sacrificial lamb but by Christ, the required submission was
in sharing his sufferings and the expected renewal was anticipated in his
resurrection.

Sharing Christ's sufferings, in this epistle, refers to the clear expectation that
in being obedient to Christ the communities might share his disabilities and per-

haps shed their blood for bearing his name. The picture of Christ, sketched with the help of Isaiah's servant, shows him innocent, suffering unjustly without retaliation and firmly relying on God. This picture provides an example of proper conduct for slaves and supports their submissiveness. In his example they have a master copy and a line of footprints to follow. Yet not only slaves but the whole community must regard Christ's innocent death on behalf of others as establishing the decisive pattern for dealing with sins. Indeed the writer is at pains to commend this pattern, and not only on grounds of prudence: first the suffering, then the glory as foretold in scripture and embodied in the Passover liturgy. Suffering, which belongs to the grace of God, is a necessary prelude to the glory. The Christian hope, built on the death and resurrection of Christ, includes within itself a range of other experiences—rebirth, natural death and fertility, rejection and acceptance, and the transition from darkness to light and from communal hostility to communal acceptance. In addition, it is also expected that there will be immediate benefits not only from Christ's resurrection but from his blood, his death for sins, his bearing our sins in his body on the tree (1 Pt 1:2, 19; 2:24; 3:18). At this point it is tempting to find refuge in the word *sacrifice,* to assume that atonement is indicated even though its method is lost in the cultic world of antiquity. It is indeed proper, and sometimes helpful, to speak of sacrifice; and *atonement* has so wide a meaning that it must certainly have a bearing on the affairs of human beings and God. But it is important to examine any indications the writer may have provided about his intentions in using sacrificial language. 'Sprinkling with Christ's blood' is a forceful covenant image to those who know Exodus and suggests a robust willingness to accept social disabilities as the price of being Christian (1 Pt 1:2; see above, pp. 241–43). Christ's blood was shed not only to liberate from former patterns of life but also to inhibit sinning (1 Pt 1:19; see above, pp. 243–44). Christ's innocent death, when he took our sins upon himself, was exemplary and representative. It called forth God's blessing both on himself and on those he represented. It changed the possibilities for sinners by healing their social situation (1 Pt 2:24; see above, p. 245). He who knew God's favour suffered for those who knew God's disfavour, who were taken into the divine presence and there accepted because they bore Christ's name, that is, they shared the same intention as Christ when he suffered in the flesh (1 Pt 3:18, 4:1; see above, pp. 251–52). With Christ alive in the spirit, at the right hand of God, these benefits provide a confident Christian hope, which they must carefully safeguard. The disobedient and the deceased can safely be left to God.

7

Hebrews

Chester Beatty papyrus 46, dated about 200 C.E., includes 'To Hebrews' *(pros Hebraious)* immediately after Romans.[1] That unique placing reflects the opinion of Alexandrian Christians that the writing, though not carrying Paul's name, was somehow Pauline. Eusebius *(H.E.* 6.14, 25) reports that it was discussed by Pantaenus (who suggested that Paul had withheld his name out of modesty) and by Clement of Alexandria, who thought that Luke translated what Paul had written in the Hebrew tongue and that Paul withheld his name to avoid a hostile reception from prejudiced Jewish readers. Origen more sensibly said that the style could not be Paul's but that the thought might be. Some held that Luke was the writer, some Clement of Rome (who in fact quoted Hebrews 1 or used a parallel tradition); but only God knew the truth. About the same time, in North Africa Tertullian followed a tradition that ascribed *ad Hebraos* to Barnabas, and said that the writer learnt his rejection of a second repentance (Heb 6:4) from apostles and taught it with apostles *(De pudic.* 20). From this it must be concluded that no reliable tradition was available in the second century about the authorship of Hebrews.

Paul's direct authorship is excluded by the Greek style and type of scriptural exegesis (Bruce, p. xiii). Paul's indirect authorship might be supported by the common features shared by Hebrews and the Pauline corpus, but it is overthrown by the more striking divergences.[2] And it is excluded by the statement in Heb 2:3 that salvation 'was declared at first by the Lord, and it was attested to *us* by those who heard him'. Yet the reference to Timothy in Heb 13:23 brings Hebrews within the Pauline orbit; and the similarities between Hebrews and the Lukan writings[3] at least keep *pros Hebraious* on the periphery of Paul's universe. To some extent the evidence pulls in opposite directions; and that is a characteristic discovery in almost every enquiry about this writing.

The evidence is equally perplexing when we ask whether the title 'To Hebrews' gives a satisfactory indication of the intended readers. The writing gives extensive quotations from the Greek Old Testament, is much concerned with the literary account of the Jewish cultus, and insists on Christ's high priestly superiority to the Levitical priesthood. But the intended readers are clearly Christian believers (e.g., Heb 5:11–6:4, 10:29), who were familiar with a kind of scriptural exegesis best known to us from Philo of Alexandria. Perhaps therefore they were hellenistic Jewish Christians, a view that finds some support in the unquestioning use of the superscription *pros Hebraious* by the Alexandrian fathers. If so, their antecedents and associations would be with nonconformist Judaism (of which the Qumran sect was only one example) rather than with the mainstream.[4] Even so, there are reservations. The writer nowhere suggests that he is encouraging Jewish Christians to resist and oppose actual Jews. His discussion of the sacrificial system is based entirely on regulations for the wilderness Tabernacle (which sometimes he misreads) and not on the Jerusalem cultus. And he seems to presume that the elementary instruction of his readers had included 'repentance from dead works and faith towards God' (Heb 6:1, see 11:6), which sounds more appropriate to Gentiles than Jews.[5]

It is not easy to accept confident conclusions either on this matter or in regard to dating. It is tempting to read descriptions of priestly activity as implying that they were still offering sacrifices in the Jerusalem Temple; in that case Hebrews must have been written before 70 C.E. But the present tenses in Heb 9:6–10 need carry no such implication, since Josephus, Clement of Rome (end of the century) and the Epistle to Diognetus (second or third century) also write as if the Mosaic ritual were still in existence (Moffatt, p. xxii). It has been said that the argument of Heb 10:1–2 (that continually repeated sacrifices cannot make the worshippers perfect, otherwise they would have ceased to be offered) would have been greatly strengthened if the writer could have said 'and indeed the Temple worship has now ceased'. But that might be irrelevant to the argument, which is contrasting the purposes of a sacrificial system with the result of a once-for-all human death. In former days the community had suffered hardship and public abuse (Heb 10:32–34) but had 'not yet resisted to the point of shedding blood' (Heb 12:4). Does that place the letter before Nero's savagery in 64 C.E. or within the age of Domitian (81–96 C.E.)? The clear indication in Heb 2:3 that writer and readers were not first-generation Christians, the mention of Timothy in Heb 13:23 and the links between Hebrews and Luke-Acts permit—but do not demand—a date towards the end of the century, probably before the writing of 1 Clement.[6]

The beginning of *pros Hebraious* gives no indication of the place it was sent to, though the ending (ambiguously) does: 'They of Italy send you greetings' *(hoi apo tēs Italias)*. It may imply that Hebrews was written in Italy and that Italian Christians sent greetings to the recipients; or that it was written to Italy from a place where Italian residents sent greetings to Christians at home. The latter is usually preferred, but we cannot be certain. We cannot even be quite sure that the thirteen chapters were composed as a whole. It is not impossible that the first twelve chapters were originally a written-out speech, intended for one group of readers and then adapted for another group by the addition of chapter 13.

What cannot be doubted is that Hebrews was not a general theological dissertation but was written for a particular community. That is demonstrated not so much by the rather general advice in chapter 13 as by the rebukes and encouragement of the previous paraenetic sections. The readers must attend more closely to what they have heard if they wish to escape retribution (Heb 2:1–4). In the community's confession *(homologia)* Jesus is not only the pioneer of salvation (Heb 2:10) but also God's emissary *(apostolos)* and high priest. Therefore they must hold fast their confidence and not fall away from the living God through an evil, unbelieving heart. Life is not all hardship and trial; but God, whose word is sharper than any two-edged sword, could easily refuse admission to his promised rest (Heb 3:1–4:13). As it is, they are no longer listening to his word. They are like children needing milk, not like adults eating solid food. And some are falling back into their condition before being enlightened, and so crucify the Son of God on their own account and hold him up to contempt (Heb 5:11–6:12). Genuine Christian conviction knows a new and living way to enter the divine presence. Therefore, Christians should be whole-hearted, holding fast their confession, prompting one another to love and good deeds, not neglecting the common assembly but encouraging one another as the Day draws near. But if they sin deliberately there is no hope of salvation if, that is, they have trampled on the Son of God and profaned the blood of the covenant and outraged the Spirit of grace. But it need not come to that if they practise endurance until the promise is attained; and they are supplied with many scriptural examples of the heroic virtue required by endurance and loyalty. They are to look to Jesus the pioneer and perfecter of faith, who, for the joy that was set before him endured the cross, despising the shame, and is seated at the right hand of the throne of God. So they are encouraged to take part in the contest of faith, to meet hostility without becoming faint-hearted, to treat hardship as a kind of training, to try harder and to reject bitterness (Heb 10:19–12:17). If they reject him who warns from heaven they will learn that God is a consuming fire (Heb 12:25–29).

Warnings about Christ's Death

When the author is giving biblical exposition he writes with cultured ease, but in these paraenetic sections he becomes anxious and overprotective. The courteous *we* becomes a pointed *any of you*. He fears they are drifting away, becoming hard-hearted, rebellious, disobedient disbelievers. He berates them for making so little progress, warns against falling away after being enlightened, and threatens them with the penalties of deliberate sinning. He sounds like an intellectually able teacher who is determined to deliver a prepared lecture even if the audience is restive and inattentive. His remedy for their faults is difficult to assess because it is not clear (at least to us) what their faults really were. At two or three points, however, his voice becomes exceptionally shrill; and here he may be touching on his deepest anxiety. First in Heb 6:4–6:

> 4   It is impossible
>   for those once *(hapax)* enlightened
>   who have tasted the heavenly gift
>   and have become partakers of the Holy Spirit,

5   and have tasted the goodness of God's word,
     the powers of the coming age

6   and have fallen away,
     to renew their repentance,
     thereby [re]crucifying the Son of God for themselves
     and making [his death] a public mockery.[7]

Behind that formulation lies a sequence: repentance, the once-for-all death of Christ (as in Heb 7:27; 9:12, 26–28; 10:10), and once-for-all enlightenment (compare the once-for-all cleansing in 10:2). It is clear that the benefits of enlightenment—experiencing the heavenly gift by participating in the Holy Spirit's miraculous activities (Heb 2:4) and scriptural illuminations (Heb 3:7, 9:8, 10:15), experiencing the goodness of God's word by anticipation of the energies promised for the coming age—all these proceed from the death of Christ. That is their exclusive source, effective once only for each believer. Once accepted, they may perhaps be rejected, but they cannot then be taken up again. Those who fall away cannot (in a bitter phrase) expect Christ to be crucified again for their convenience or to tolerate a public mockery of his death.

The author returns to the same subject in Heb 10:26–29. If Christians deliberately persist in sinning after receiving knowledge of the truth, there is no longer available to them a sacrifice to deal with sin (which takes up the theme of his main exposition) but only the expectation of terrible judgement. What he means by deliberate sinning is persistent refusal to enter the divine presence by the blood of Jesus (as he sets it out in 10:19–25, the closing summary of his main theme). This refusal he calls trampling on the Son of God, regarding the covenant blood (by which a Christian is consecrated) as unholy, and insulting the gracious Spirit. This severe and emotional language must be directed against those who regard Jesus with little respect, give no sacred meaning to his death, and (perhaps from a perverse desire for the Spirit)[8] actually insult the Spirit by which God's gracious truth is disclosed. It may be on their account that the author describes Jesus as 'the pioneer and perfecter of our faith, who for the joy that was set before him endured the cross, *despising the shame,* and is seated at the right hand of the throne of God (Heb 12:2). At least it can be said that such a reading of the paraenetic sections justifies the two major expository sections, the first of which is christological (working out the proper status of the Son of God against his detractors), and the second soteriological (offering an elaborate cultic argument for the necessity and sacredness of his death).

It now becomes necessary to say something about the structure of Hebrews. Clearly it is not an epistle in the style of the Pauline corpus, but a thematic writing on carefully limited christological and soteriological subjects. It proceeds by alternating expository and paraenetic sections in the following manner:[9]

Christological

1:1–14	The Son of God above the Angels
2:1–4	First Appeal
2:5–18	The Son of God below the Angels

The Son of God above the Angels

It has long been recognized that Hebrews begins with a Wisdom or Logos Christology; that is to say, in 1:1–3 the Son of God is described in language that
elsewhere is applied to the divine Wisdom and by Philo to the Logos.[10] Having
thus begun, however, the writer apparently abandons that theme for a prolonged
scriptural proof that the Son is above the spiritual beings called angels. Having
sufficiently made that proof, he interposes a brief, sharp warning (the First
Appeal) against neglecting a spoken tradition so much more powerfully authen-
ticated than the message declared by (mere) angels. Why are angels deliberately
introduced here, kept in sight in the next section as well, and sturdily repudiated?
Perhaps because a rival angel christology had to be rejected.[11] In the writer's view,
such a christology would be unsatisfactory for at least two reasons: it would
ascribe too low a status to the Son and would also (since angels are spiritual
beings) make him incapable of dying a human death. The status of the Son is
assured by elevating him above the angels as the Wisdom or Logos; his death
becomes conceivable if he is made lower than the angels as Man *(anthrōpos)*.

The Son of God below the Angels

The second part of this argument therefore begins with the assertion that God
has not placed the coming world under the control of angels but under the control of mankind (demonstrated from Psalm 8; see Moffatt). Even though all
things are not yet seen under mankind's control, we do 'see Jesus, who for a little
while was made lower than the angels, crowned with glory and honour because
of the suffering of death, so that by the grace of God he might taste death for
every one' (Heb 2:9). Despite some modest exegetical problems,[12] the meaning is
reasonably clear. Although the Son as Wisdom or Logos is far superior to angels,
in the person Jesus he was temporarily made inferior to the angels because God's
intention is to prepare a new world for mankind. For mankind (though not for
angels) death is the insoluble problem. Therefore, Jesus, bearing the divine favour,
becomes representative Man and tastes (i.e., experiences) death not for all of us
together but for every one of us. Therefore, he dies by the most shameful death,
crucifixion, and because of the suffering of death is crowned by God with glory
and honour. For that reason the crucifixion cannot be treated with disrespect;

and it receives its sacred character because it is a representative act accepted by God on behalf of mankind.

In 2:10–18 that consideration is strengthened partly by reformulation in scriptural phrases (which need not detain us) and partly by argument, which (supplied with necessary steps) runs like this:

> In principle God has crowned mankind with glory and honour, though as yet we see only Jesus, representative mankind, so crowned (verses 7–9). But it is always God's intention to bring many sons to glory[13] [a familiar Hebrew description of God's own being; see Heb 1:3]; therefore it was appropriate for God, through sufferings, to equip Jesus completely[14] as the pioneer of mankind's rescue (verse 10). [If human beings are to share God's glory, they must become holy, because God is holy.[15] Therefore it is the task of Jesus to make them holy: he is the consecrator, they the consecrated. On the principle that like responds to like,] the consecrator and the consecrated must belong to one category of being; therefore Jesus is not ashamed to call human beings his brothers (verse 11). . . . Since these children of God share in flesh and blood, he too partook of flesh and blood in order that through death he might destroy him who has the power of death, namely, the devil (verse 14), and might deliver all those who through fear of death were subject to lifelong bondage (verse 15). . . . Therefore he had to be made like his brothers in every respect, that he might become a merciful and faithful high priest in what concerns the being of God so that he could expiate the sins of the people (verse 17). Because he suffered (i.e., died) and so was put to the test,[16] he has power to help those who now face the test (verse 18).

Even that expanded rendering demands some comments, for it makes plentiful use of ideas current in late antiquity but uncommon now.

The phrase 'expiate the sins of the people' (verse 17) is unique in the New Testament. Indeed, the verb *hilaskomai* ('expiate' or, in other contexts, 'propitiate') and its related forms are infrequent, though common in the cultic parts of the Old Testament.[17] It signifies the procedure for removing from persons or objects whatever renders them incompatible with God's purpose and presence. Hence, it belongs to the act of consecration, of making holy, and so is of a piece with verse 11.

The main concern of this section is not sins but death. The writer is not worried about dying but about being dead. For what goes down into death is unholy, that is, incapable of communion with God. Death, according to hellenistic Jewish tradition, was the invention of the Devil: 'God did not make death, and he does not delight in the death of the living. . . . God created man for incorruption, and made him in the image of his own eternity, but through the devil's envy death entered the world' (Wis 1:13, 2:23–24). The fear of death is not the fear of dying but the fearful knowledge that when we die we shall be outcasts in the Devil's power, beyond the reach of God. The Devil's power is not to inflict death but to separate us from God in death.

Jesus shared the mortal fate of all human beings.[18] He was put to the test of death (verse 18). But by his death, instead of entering the realm of the Devil, 'when he had made purification for sins [performed his sacrificial consecrating work] he sat down at the right hand of the Majesty on high' (Heb 1:3). Thereby he was completely equipped as the pioneer of mankind's rescue (verse 10). Human beings still die and suffer the pains of death; but the unclean, unholy realm of death becomes the sacred enclosure of 'a judge who is God of all and the spirits of just men made perfect' (Heb 12:23).

As the consecrator, Jesus acts on behalf of God; as representative mankind he expiates the sins of the people. Therefore, he may be regarded as a faithful and merciful high priest, for such a person has a dual function: speaking on behalf of mankind to God and on behalf of God to mankind. He is an intermediary between the physical and transcendent worlds—a thought that had not escaped the mind of Philo, who sometimes explains the high priest as an image of the Logos (Moffatt, p. xlvii–xlviii). Within Christian circles, the author of Hebrews may have been the first theologian to expound a christology of this kind, but it seems that he found the title in his Church's confession:[19] 'Consider Jesus, the apostle and high priest of our confession' (Heb 3:1), the beginning of the Second Appeal. And when, at the conclusion of that appeal, the christological exposition is resumed, a similar reference appears: 'Since then we have a great high priest who has passed through the heavens, Jesus the Son of God, let us hold fast our confession' (Heb 4:14). With even fuller emphasis, at the beginning of the Final Appeal, Christians are urged to hold fast their confession, which includes 'a great high priest over the house of God' (Heb 10:21–23). The author of Hebrews evidently developed the possibilities he perceived in an existing honorific title in such a way as to solve a christological problem and extend the scope of soteriology.

The Son of God as Heavenly High Priest (Foundation)

In Heb 4:14–5:10 the author begins a statement of the priestly work, at least as far as its foundation is concerned. The mature teaching must wait until his readers have been urged to go beyond 'the first principles of God's word' (Heb 5:12). It may be taken for granted that the Son of God, as great high priest, has passed through the heavens. It may equally be taken for granted that the earth and the heavens are the work of God's hands and that they will perish (Heb 1:10–12, see 12:26, 'Yet once more I will shake not only the earth but also the heaven'). Therefore the Son of God has passed from earth, through the heavens, to his position at the right hand of God (acc. to the quotation of Ps. 110:1, the author's culminating proof of the Son's superiority to angels [Heb 1:13]). That conviction is later confirmed by the spatial imagery of 7:26 and 8:1; but in this section is expressed by completion language (see n. 14). If, as Dey argues, *teleiōsis* ('completion' or 'perfection') means complete and unmediated access to the deity, 'being made perfect' (Heb 5:9 RSV) implies that the Son completed the transition from the impermanence of the earth and the heavens to the unshakable being of God, and has himself been named 'a priest for ever, after the order of Melchizedek' (Heb 5:6, 10). But to make that transition so that the Son could become 'the source of an eternal salvation to all who obey him', it was necessary that he himself should be brought to completion ('made perfect' RSV). The manner of it is expressed in a complicated sentence that contains traditional material:

> Who, in the days of his flesh,
> with both prayers and supplications
> to him who could save him from death,
> with loud cries and tears,
> made his offering and was heard
> for his godly fear;
> though he was Son
> it was obedience he learnt from what he suffered.
>   (Heb 5:7–8)[20]

To the question why the writer of Hebrews mentioned this dramatic episode (usually identified with the synoptic tradition of Gethsemane, or sometimes with the loud cries from the cross) it is commonly replied that he wished to stress the Son's sympathy with the human condition. But since he mentioned the sympathy in Heb 4:15, it is more likely that here he is defending the emotional disturbance. Using a familiar theme from Greek *paideia* ('training', 'discipline')—learning by suffering, or experience—he argued that bringing the Son to completion (perfection) required him to undergo the training that suffering provides. (He uses exactly the same teaching for suffering Christians in Heb 12:3–11.) The loud cries and tears are less evidence of personal distress than of prayerful intensity (Dey, p. 224). Thus, by means of his death and the strength of prayer by which he endured it, the Son underwent the necessary training and reached the condition of total access to God. Here is a version of the redeemed redeemer (Dey, p. 225) and for a reason plainly indicated in Heb 6:19–20: 'We have this as a sure and steadfast anchor of the soul, a hope that enters into the inner shrine behind the curtain, where Jesus has gone as a forerunner [*prodromos*] on our behalf, having become a high priest for ever after the order of Melchizedek'. The same conviction bestows on Jesus the designations 'pioneer and completer of our faith' in Heb 12:2. In himself the Son is 'the radiance of God's glory, and the exact representation of his being' (Heb 1:3 NIV). But when he had been made a little lower than the angels, it was necessary for him to undergo *paideia* by suffering, that he might both be crowned with glory and honour (Heb 2:9) and also become our forerunner, the originator and source of our salvation, the completer of our faith (Heb 2:10, 5:9, 6:20, 12:2)—and therefore a unique high priest.

The Son of God as Melchizedek High Priest (Mature Teaching)

The mature teaching about the Son of God as high priest prefigured by Melchizedek is developed in 6:13–10:18. The writer returns to Abraham (since God's intention was directed to descendants of Abraham, not to angels, Heb 2:16) and establishes the unshakable certainty of God's promise to him (Heb 6:13–20). In that case, why should our hope have anything to do with priesthood? Because in those formative days there was someone greater than Abraham, a mysterious priest capable of exacting tribute from Abraham and of conferring a blessing on him. So in Heb 7:1–25 numerous arguments are advanced for the superiority of Melchizedek's priesthood to the Levitical priesthood.[21] The priesthood of the law made nothing perfect (i.e., gave no direct, permanent and complete access to God); the priesthood prefigured by Melchizedek introduced a better hope 'through which we draw near to God' (Heb 7:19). Jesus, who is the fulfilment of the promise to Melchizedek (Ps 110:4), is therefore the guarantor of a better covenant. Unlike priests who relinquish their priesthood by death, 'he holds his priesthood permanently, because he continues for ever (see Heb 7:16, 'the power of an indestructible life'). Consequently, he is able for all time to save those who draw near to God through him, since he always lives to make intercession for them' (Heb 7:22–25). It is well known that Hebrews makes no mention of the resurrection of Christ (except for the doxology, Heb 13:20–21, which demonstrates his familiarity with the liturgical tradition).[22] Yet clearly, the fact that Christ *lives* and that his life is indestructible is all-important for his argument against the Levitical priesthood and for the security of the Christian hope. But the argument is carried forward by exaltation, not by resurrection. The reason is supplied by the earlier observation that the writer is not worried about dying but about being dead. Resurrection might imply being rescued from the Devil's realm of death while leaving that realm intact. Exaltation implies ascending through the experience of death into the presence of God. Not that death has been abolished or has become an illusion nor that any initiate may exercise his right or use his gnosis and enter the divine presence at will (for access to God is still God's gift, and each suppliant must be interceded for) but that it is necessary to pass through death in the way of our forerunner who, though he lives for ever, knows what it is to die with the prospect of living forever without God.

> It was fitting that we should have such a high priest,
>> holy, blameless, unstained,
>> separated from sinners,
>> exalted above the heavens.
> He has no need, like those high priests, to offer sacrifices daily,
> first for his own sins and then for those of the people;
> he did this once for all when he offered up himself.
> Indeed the law appoints men in their weakness as high priests,
> but the word of the oath, which came later than the law,
> appoints a Son who has been made perfect for ever.
>> (Heb 7:26–28)

Now at last the Son is no longer our forerunner, marking out the path and leading us onward; he is already on the other side of the great divide, beyond the created worlds of earth and heaven, enjoying full and immediate access to God. He is separated from sinners (either from the contamination of their sinfulness or from the hindrance of their opposition) but can still speak for them as the Son of a God whose intention it is to bring many sons to glory (Heb 2:10). And his present status is now to be disclosed in the imagery of his high priesthood, which operates in Hebrews as an explanatory myth does in Pauline writings. In effect, the author is saying that what Christ has done for our salvation can be described *as if* he were a high priest, in some respects like but in other respects unlike the Levitical priesthood of Moses' law. In a literal sense, of course, he was not a priest at all. On this matter the author is explicit and insistent. Our Lord, he says, was descended from Judah, in connection with which tribe Moses said nothing about priests (Heb 7:14). We must dismiss from our minds Western notions about priesthood's involving God's call and the bestowal of necessary gifts and graces. In antiquity priestcraft was a trade, practised exclusively by members of a priestly family (by whatever legal or illegal methods the family had acquired that status). Therefore, Jesus was not, and could not be, a priest in the literal sense: 'If he were on earth, he would not be a priest at all, since there are priests who offer gifts according to the law' (Heb 8:4). If, then, Jesus is described as high priest, he belongs to a class of priests of which he is the only member. Moreover he has only one priestly act to perform, namely, the offering up of himself, and only one occasion on which to perform it, once for all. That unique self-offering is the substance of Heb 9:6–10:18.[23]

The Unique Self-Offering

'To offer up' is cultic terminology. The Greek *anapherō* has various applications of the general meaning 'to bring (or carry) up or back'. In the Septuagint the word is commonly used for presenting the burnt sacrifice on the altar. Three times it is used of human sacrifice: in Gn 22:2 Abraham is instructed to offer his son Isaac as a burnt offering, though in fact a substitute is provided and accepted; in 2 Kgs 3:27 the desperate king of Moab offers his eldest son as a sacrifice and so removes the Israelite threat; in Jer 32:35 the people of Judah offer up their sons and daughters to Molech, an abominable thing to the Lord. Thus, human sacrifice belonged to the unacceptable face of Semitic religion, and later tradition was quite clear that God repudiated it entirely. There is no parallel to the statement that 'he offered up himself', and therefore nothing in the uses of *anapherō* to suggest that God might be better pleased with human self-sacrifice than with human sacrifice (for it must be remembered that the victims of human sacrifice are not always unwilling and that self-sacrifice may arise from discreditable motives). It is true that the Maccabean martyrs could be portrayed as offering themselves, a willing substitute for the sinful nation;[24] but that does not seem an entirely probable source of the cultic imagery of Hebrews. It has been suggested that a more likely source is the double reference to the Servant in Isaiah 53, who 'bore the sins of many' (verse 11, where the verb *anapherō* is used). That is not the same as 'he offered up himself', but it is not far from 'offered once to bear the sins of many' in Heb 9:28. But an echo of the Isaianic phrase there, which may suggest an echo here, need not imply that the whole doctrine of the Servant's vicarious death undergirds the teaching in Hebrews. It is characteristic of the author that he should pick on memorable phrases and use them (if need be) entirely out of context, retaining only the obvious meaning.[25] 'To bear the sins of many' at least means to suffer the consequences of their wicked actions; and that indeed provides a clue to the intention of 'he offered up himself'. In response to the will of God, he bore the shame of the cross (as the author says in Heb 12:2)—not only the public humiliation and the religious revulsion of it but also the sense (which attaches to shame in Jewish culture) of being discredited and repudiated by God. He went to his death (as the Markan Passion narrative insists) with no confidence that God would rescue him or indeed wished to rescue him. It is not difficult to see why a Son who thus died should be crowned with glory and honour; or why he should become our forerunner. Not everyone, it is true, dies with that sense of desolation, but many do; and for their sakes Jesus took the hardest path.

The preliminary presentation of this unique self-offering is not perfectly phrased. Verse 27 runs as follows in the Greek wording:

> He is not daily required
> as are those high priests
> to offer sacrifices,
> first for their own sins,
> then for those of the people.
> For that he did once for all
> when he offered up himself. (Heb 7:27)

The plural 'high priests' is acceptable as a general way of referring to the historical succession of high priests.[26] The statement that *high priests* (not simply priests) were required to offer sacrifices for sins, first for themselves, then for the people applies exclusively to the Day of Atonement, as the author knows (Heb 9:7, 'once a year, not without taking blood which he offers for himself and for the errors of the people'; see Heb 5:3, where the high priest 'is bound to offer sacrifice for his own sins as well as for those of the people'). It is therefore difficult to understand *daily* at the beginning of the verse (possible explanations in Michel, pp. 281–83); difficult, indeed, to avoid the impression that the writer's desire to sharpen the contrast (between the great high priest and 'those high priests') had run away with him. Their weakness was shown by the necessity of their daily cultic activities, his strength by the permanent results of his unrepeatable cultic self-offering. But one difficulty remains: the closing statement of the verse may be taken to imply that the self-offering was made for the sins of Jesus as well as for the sins of the people. Is this another instance of incautious writing, which could be amended by attaching 'that he did' only to the nearest phrase, 'those of the people', and ignoring 'first for their own sins'? How otherwise can the statement be reconciled with the eulogy in verse 26 ('holy, blameless, unstained, separated from sinners') and with 'tempted as we are, yet without sin' in Heb 4:15? But the eulogy belongs to the high priest when exalted above the heavens, and the 'temptation' refers to the test of death, not to temptations of every kind.[27] There must be a strong suspicion that the author deliberately wrote as he did to assure his readers that the self-offering of Jesus was amply sufficient to deal, not only with the sins of the people but also with anything that might ruin his effectiveness as their high priest. Unlike us, his readers were not long familiar with a christological tradition of the sinlessness of Christ. They might dismiss from their minds any suggestion that Jesus had knowingly, carelessly, or ignorantly defied God or that he had played with unworthy temptations before rejecting them. But could they be sure that he had escaped inadvertent defilement, to which mankind is subject and for which the Day of Atonement is provided? That anxiety, says the writer, is finally put to rest by his inclusive self-offering.[28]

The Heavenly Sanctuary

The argument now enters its final and most elaborate phase in Heb 8:1–10:18. The structure is not always clear because three main themes divide and coalesce:[29] but 8:1–5 sets out what they are. We already know that we have 'a high priest, one who is seated at the right hand of the throne of the Majesty in heaven' (see Heb 1:3, 4:14, 7:26). We are now to learn that he is (1) a *minister (leitourgos)* of the sanctuary and of (2) the true *tent* which is set up (3) *not by man but by the Lord*. Since he is a *minister*, like other priests he must have gifts and sacrifices to offer (verse 3). Since he ministers in the true *tent*, his place of ministry differs from the shadowy copy that Moses erected (verses 4–5). Indeed, he exercises a better ministry *(leitourgia)*, since he is mediator of a *better covenant*, enacted on better promises than those attached to the Mosaic covenant (verse 6). This scheme is then developed in the reverse order: 8:7–13 explains from the *new covenant* of Jer 31:31–34 that the first covenant has been superseded. Heb 9:1–5 describes the structure and furnishings of the Mosaic *sanctuary;* and 9:6–10 discusses the *high priest's* ritual duties there performed. The description is oddly restricted. It is confined to the sacred enclosure of the priests, namely, the Holy Place (the outer tent, containing the lampstand and a table for the sacred bread) and the Most Holy Place, the inner tent containing the altar of incense, the ark, and the cherubim overshadowing the place of expiation *(hilastērion)*. According to Ex 30:6 the altar of incense belonged to the *outer* tent, and on this point the author's description is contestable; but on the Day of Atonement incense was taken into the Most Holy Place, so that the cloud of smoke covered the *hilastērion* which also was smeared with blood of the animals sacrificed to atone for the high priest and the people. Everything is concentrated on that particular day. Nothing at all is said about the great court where the people assembled, which contained the laver and the main altar of sacrifice.[30] When the author mentions the gifts and sacrifices of the cult, and treats them, with conventionally dismissive words, as comprising 'food and drink and various ablutions', he ignores the main sacrificial offerings. The enormous slaughter of animals and effusion of blood is kept out of sight,[31] though further references to blood come a little later. At this point, however, the author makes criticisms of the two tents and their rituals. First, the existence of the outer tent indicates that entrance to the inner tent is not (generally) disclosed so long as the outer tent is standing. This is obviously a spiritual interpretation,[32] on popular Platonist lines: earthly worship is but a poor copy of heavenly worship, as is shown by the very limited access to the deity in the Day of Atonement ritual. Second, the offering of gifts and sacrifices cannot fully expose the worshipper to awareness of God (verse 9[b]). That sentence contains two technical words that appear as 'perfect' and 'conscience' in the RSV rendering: 'cannot perfect the conscience of the worshipper'. In Hebrew *to per-fect* means to give complete access to God (see p. 259 n. 14). 'Conscience' is *syneidēsis,* which is certainly not conscience in the modern Western sense but either consciousness/awareness or the inner pain that arises from having done what is morally wrong. In antiquity, to perfect the conscience might conceivably mean to increase its sensitivity to the least suspicion of wrong (which would not suit the argument of Hebrews) or to develop the awareness of God until it became complete. That indeed must be the meaning. The author is convinced that per-petual sacrifices are constant reminders of separation from God, in no sense a route to complete awareness of him.[33]

Thus prepared, the author now sets down a carefully articulated statement of themes of the *minister* and the *tent* under the new covenant:

11 But when Christ appeared[34]
     as a high priest of the good things that have come,[35]
     then through the greater and more perfect tent
     (not made with hands, that is, not of this creation),

12 taking not the blood of goats and calves
     but his own blood,
     he entered once for all into the Holy Place,
     thus securing an eternal redemption.

13 For if the blood of goats and bulls
     with the ashes of a heifer
     sprinkling defiled persons
     sanctifies for the purification of the flesh,

14 how much more shall the blood of Christ,
     who through the eternal Spirit
     offered himself without blemish to God,
     purify your conscience [*syneidēsis*] from dead works
     to serve the living God. (Heb 9:11–14)

This statement draws on the imagery of a high priest, on the Day of Atonement, entering the sacred enclosure and carrying sacrificial blood through the cloud of incense to be smeared on the *hilastērion.* But the high priest presents not the blood of sacrificial animals but his own blood. Since the offering is different, the consequences are different.

We begin with the need for purification, or cleansing, in verse 13. It is agreed that the sacrificial blood of the Day of Atonement 'sanctifies for the purification of the flesh'. That does not mean, as is so commonly said, that Jewish ritual conveyed only bodily, external or ceremonial cleanness.[36] In Hebrew flesh means our existence in the earthly world, in distinction from existence in the heavenly world.[37] The purification sought on the Day of Atonement was the removal from the Jewish community of whatever threatened its corporate existence. They used the imagery of defilement; we, in our age of atomic radiation, might speak of contamination. By such symbolism they intended to say that destructive influences in their society could be removed, or at least held in check, by sacrifices and ritual techniques.

If the sacrificial blood of animals could accomplish so much, how much more could the blood of Christ accomplish when he 'offered himself without blemish to God'. The sacrificial verb is used and the sacrificial condition is fulfilled; but, of course, the body of Jesus was not offered on an actual altar, and his blood was not smeared on a sacred object. When the death of Jesus is regarded *as if* it were a sacrifice, the consequences of it can be displayed and understood. The author says that Jesus offered himself through the eternal Spirit, that is (to put the point a little too simply), it was the spiritual force of the offering that made it effective. In Hebrews the Holy Spirit is one of the heavenly gifts of the age to come (Heb 2:4, 6:4); it is called eternal *(aiōnios)* Spirit because it belongs to the eternal world. In this world it indicates the higher meaning of scripture and so discloses the secrets of the new world. In this instance it transforms a shocking death into a sacrificial act that purifies the *syneidēsis;* that is, it delivers us from the constant awareness that our activities belong to the way of death (Bruce) and sets us free to serve the living God. Thus, Christ's death secures an eternal redemption (verse 12). The word is *lytrōsis* (see Appendix A). It can scarcely mean a ransom so magnificent as to be perpetual in effect; more suitably, it can be understood as a liberation from the old world into the new.

The most distinctive feature, however, is the heavenward progress of the high priest who passed through the greater and more perfect tent and entered once for all into the Holy Place. He is therefore in the heavenly sanctuary of which the Mosaic sanctuary, belonging to this created world, was but a copy and a shadow (Heb 8:5). Hence, in this progress Christ was moving from the world of shadow to the world of reality, though the shadow world (by reason of its origin) provides analogies for the real world. Just as the sacred place of Jewish worship comprised two tents (Heb 9:2–3), so also the heavenly world was perceived in two parts; first 'the greater and more perfect tent' (the heavens through which Christ passed, Heb 4:14), and then 'the throne of the Majesty in heaven' (Heb 8:1, 9:24).[38] In a short while it will appear that this ascension imagery is not simply a rhetorical device but makes a further contribution to the author's exposition of the death of Christ.

The Better Covenant

The author now uses his *new covenant* theme to present all else he must say about the heavenly priesthood. In Heb 9:15 Christ is described as the mediator of a new covenant, and in Heb 10:16–18 a second reference to Jeremiah's new covenant (see 8:6–13) concludes the exposition. The author's ingenious choice of the word *covenant* (used more frequently here than in the rest of the New Testament)[39] permits him both to retain features of early Christian tradition (such as the blood of Christ) and to move in a new direction (so that Christ opens up the eternal world). Properly speaking, a covenant *(diathēkē)* is a structure of social relationships that corresponds with the intention of God and with the constitution of the created world. It is not an agreement, compact or bargain. It is not the result of negotiation but is more like a declaration of intent, and it may be similar (if *diathēkē* leans towards 'testament') to the disposal of property. The terms of a covenant are revealed, not debated. They are the disclosure of what is the case, not a compromise of interests. Yet a *diathēkē* needs a mediator. The closely related Greek word *synthēkē* (which, indeed, means a 'compact') requires a mediator to arbitrate; but a *diathēkē* has a mediator who conveys the testator's intentions and assists in working them out (rather like an executor). For the old covenant the mediator was Moses (as Paul presumes in Gal 3:19–20), but he was no more than a faithful servant in God's household. Christ was faithful over God's household as a Son (Heb 3:5–6). Therefore, he is mediator of a better covenant for the benefit of those who are called by God. As Abraham was called by God to go out knowing not where in order to receive an inheritance (Heb 11:8), so people are now called to receive the promised eternal (i.e., heavenly) inheritance. Earlier, the author had addressed them as 'holy brethren, who share in a heavenly call' (Heb 3:1). This qualified offer of the heavenly inheritance to those who are *called* may be a delicate rebuke to members of the community who treated the tradition of Christ's death lightly; for the author must indeed include the death in the work of the mediator; his death took place to set them free *(eis apolytrōsin)* from contraventions of the first covenant. That is another way of saying, '[to] purify the conscience from dead works' and recalls the 'eternal redemption' *(aiōnion lytrōsin)* of verse 12. If the word *apolytrōsis* meant 'ransom' (see again Appendix A), the author could have devised a justification of Christ's death by picturing a condemned person being rescued by the death of another. But he does not. He tries to justify it from the situation of a testator (to which in a moment we must turn); so it is clear that *apolytrōsis* means 'liberation'. The death of Christ sets people free from the errors of their old existence and allows them to enter a new existence.

Why, then, is it necessary to enter the sanctuary *by the blood* of Jesus? The author has four arguments. First, the covenant is like a testamentary disposition that comes into operation only on the death of the testator (Heb 9:16–17). The analogy is fragile, since a testator could dispose of his possessions during his lifetime if he wished (think of the Prodigal Son); and we are supposing possessions being made available by the death not of the testator but of his Son.

Second, Moses inaugurated the first covenant by a blood sacrifice (Heb 9:18–21), though there are notable variations from the text of Ex 24:3–8.

Third, under Torah almost everything is purified with blood, and without bloodshed there is no remission of sins. That generalization is not excessive for the Pentateuch; but elsewhere penitence (Psalms and prophets), prayer (Daniel), fasting (Joel), and almsgiving (Ecclesiasticus) are sufficient. The whole argument is somewhat forced; but the author was determined to defend the presence of blood in the tradition—even though his own references to blood never go beyond having it sprinkled and regarding it as a means of cleansing (Heb 9:22).[40]

Fourth, since the cultic institutions on earth, which are only a copy of the heavenly, were necessarily purified with blood sacrifices, the heavenly things themselves require better sacrifices than those (verse 23). Most commentators are disturbed by such an extension of the parallel between earth and heaven. Some take it for granted that heaven cannot require purification and suggest a consecratory sacrifice; others suppose that heaven may need cleansing from the worshippers' defiled consciences or from the presence of satanic beings. But the author's meaning is rather different and begins from the Jewish perception that sacrifices were not intended to avert God's moral disapproval or to appease his moral indignation. Their purpose was to ward off the dangers associated with his power. Thus, when Christ begins the journey through the heavens, he removes obstacles on the approach to God—whether God's own jealous care of his possessions or hostile agencies or the unworthy, careless and sinful condition of the worshippers. It cannot be presumed that the spiritual or mystical approach to God is itself pure and sinless just because it is not physical or earthly. It must always be remembered that 'it is a fearful thing to fall into the hands of the living God' (Heb 10:31) and that 'our God is a consuming fire' (Heb 12:29; see Dt 4:24, 9:3). At the beginning of the Day of Atonement ritual, Aaron is told 'not to enter the Holy Place within the veil . . . whenever he chooses. *He may die,* for I appear in a cloud over the *hilastērion*', though he may enter with safety by bearing the sacrificial offering. Jesus did die: he took his own blood, and his death made the way safe for others. The ancient world pictured the ascent to God as full of evils and perils (see Eph 2:2, Revelation 12), and Heb 12:18-24 is no exception. But now at last Christ 'has appeared once for all at the end of the age to put away [the dangers from] sin by the sacrifice of himself'. He has entered into heaven itself and appears in the presence of God on our behalf (his continuing high priestly ministry of intercession, as in Heb 7:25). And just as the Jewish high priest emerged from the Most Holy Place to announce to the people their atonement, 'so Christ, having been offered once to bear the sins of many (see above, p. 263 and n. 25), will appear a second time, not to deal with sin but to save those who are eagerly waiting for him' (Heb 9:23-28).

What follows in Heb 10:1-18 is a coda. Torah has but a shadow of the good things to come, for it is repetitive, not definitive. The Day of Atonement itself is an annual reminder that animal sacrifices can never be finally effective. The intention of Christ's entry into the world is indicated in Ps 40:7-8 (LXX), the replacement of cultic sacrifices by a representative of mankind who performs the will of God in his own body. Therefore Christians 'have been consecrated through the offering of the body of Jesus Christ once for all'. Christ, now seated at the right hand of God, has offered for all time a single sacrifice; and by that single offering those who are consecrated are perfected, that is, fully exposed to the Divine Person. As the Holy Spirit says (in Jeremiah 31), there is no longer any offering for sin.

That last, dismissive statement has two consequences, which are taken up in the Final Appeal, 10:19-12:29. One I have already mentioned and need discuss no further, namely, the insistence, against those who profane the blood of the covenant, that there is no further sacrifice for sins. The other, in Heb 10:19-26, is an encouragement to profit from the author's exposition; and it conveniently sums up the benefits of Christ's death. The underlying sequence of thought must be somewhat as follows. Christians have a great priest over the household of God.

Where he is, Christians may also go; for he has inaugurated a new and living way from the earthly world to the heavenly world. As the Jewish high priest once a year, carrying sacrificial blood, went from the Holy Place to the Most Holy Place and therefore into the presence of God; so Christ once for all, presenting his own blood, went from the earthly world to the heavenly world, passing through his life of flesh to his existence as eternal spirit (see Heb 9:14), and is therefore seated at the right hand of the throne of the Majesty in heaven (see Heb 8:1). Christians may boldly follow the same route and confidently enter the sanctuary by the blood of Jesus, that is, by relying on his self-offering as their invitation and protection. Finally, this formal, theological conviction is matched by an experience of fulfilment and of both outward and inward purity; and this in turn becomes the basis of an appeal to the community as they see the Day drawing near.

This extensive appeal, as it draws to a close, changes the visual imagery of approach to God. Instead of a passage through the heavens represented as an entry into the inner sanctuary, it uses the ascent of the sacred mountain to the city of the living God, the heavenly Jerusalem. Amid the angels formed up in a sacred assembly and the first-born sons of heaven, there is the judge who is God of all and the spirits of just men already perfected (i.e., enjoying unrestricted access to God).[41] When Christians appear before him for judgement, the blood of Abel does not accuse them, but the blood of Jesus pleads for them. This highly rhetorical passage adds nothing to what has already been said about Jesus as mediator of a new covenant but exploits a new image for the death of Jesus in an appropriate manner for the eschatological mood of the *paraenesis*.

Community Instructions

The final section of community instructions, 13:1–25, is in the form of slogans, reminders and afterthoughts—most of which present no difficulty. But verses 9–14 look like a small unit, framed out of obscure allusions, one of them to the death of Christ. The ending is not too difficult. 'Let us go to him outside the camp and bear the abuse he endured' recalls 'Let us then draw near to the throne of grace' (Heb 4:16) and 'Let us draw near with a true heart' (Heb 10:22). This is the movement from earth to heaven that has been the main theme of the epistle; and it is combined with the shame of crucifixion, which was strongly felt in the community (Heb 12:2; and see p. 257). Why, indeed, if Christ had to die sacrificially, was a shameful death necessary, a death that put him outside the conventions of society, outside the gate?[42] This necessity is excused rather than justified by comparing Christ's death with Jewish sacrifices on the Day of Atonement. When the high priest had offered the sacrificial blood in the Most Holy Place, the slaughtered animals were not eaten but were carried outside the camp and completely burnt. Since the Jewish cult took place in a copy and a shadow of the heavenly sanctuary (Heb 8:5), the reality was displayed when 'Jesus also suffered outside the gate in order to sanctify the people through his own blood' (Heb 13:12). The parallel, of course, is imperfect; and the argument will persuade only those already persuaded. The author shows a touch of desperation in countering opposition to the death of Christ and throws in a reference to the tradition mentioned in Jn 19:20 that Jesus was crucified outside Jerusalem. It is very difficult to construct a profile of the opposition, and every suggestion is conjectural. But the author says, 'We have an altar from which those who serve the tent have no ability[43] to eat' (verse 10). 'Those who serve the tent' can be interpreted as Jewish leaders, who, after the destruction of the Jerusalem temple, carry out Mosaic legislation as best they can. The gifts and sacrifices they offer 'deal only with food and drink and various ablutions' (Heb 9:10), but they have no altar. We Christians, however, have access to an altar in the heavenly world where Christ is our high priest. From that altar we are sustained; but Jewish leaders have no access to it. Christians, therefore, should 'not be led away by diverse and strange teachings; for it is well that the heart be strengthened by grace, not by foods, which have not benefited their adherents' (verse 9).

Lastly there is the final blessing, written in collect form (Bruce):

> 20 Now may the God of peace,
> who brought again from the dead
> the great shepherd of the sheep,
> with [or by] the blood of the eternal covenant,
> our Lord Jesus

> 21 equip you with everything good that you may do his will,
> working in us that which is pleasing in his sight,
> through Jesus Christ,
> to whom be glory for ever and ever.
> Amen. (Heb 13:20–21)

The prayer must be traditional: in the first line, 'the God of peace', not elsewhere in Hebrews; in the second, the only mention of the resurrection of Christ; in the third, Christ is not elsewhere called 'shepherd'—this phrase comes from Is 63:11 (LXX); in the fourth, 'the blood of the eternal covenant' recalls Heb 9:20 and the epistle's use of 'eternal'; in the fifth, 'our Lord Jesus' is otherwise absent from Hebrews; so also are, in the seventh line, 'pleasing' *(euarestos),* and in the eighth, 'through Jesus Christ'. Thus, a traditional prayer, which used resurrection language instead of exaltation language, has been adapted by the addition of a phrase typical of Hebrews in the fourth line. The addition is loosely attached *(en haimati diathēkēs aiōniou),* and some ingenuity is required to give it meaning— perhaps, 'God brought the great shepherd from the dead in consideration of the eternal covenant established with mankind by Christ's blood' (following Michel's paraphrase). But it is artificial. The author wanted this resurrection blessing to make mention of his own insistence on the death of Christ. Since this is the only mention of Christ's resurrection in Hebrews, it is instructive to notice other occurrences of resurrection words. The verb *anistēmi* is used at 7:11, 15 of another priest arising in the likeness of Melchizedek, meaning no more than 'to take the place of' (the Levitical priesthood). The verb *egeirō* is used at 11:19 for Abraham's belief 'that God was able to raise men even from the dead', which, figuratively speaking, took place when Isaac was spared. In the common style of older Judaism, being raised means escaping the risk of death. In the same chapter, 'Women received their dead by resurrection [*anastasis*]', presumably the widow of Zarephath, whose son was revived by Elijah, and the woman of Shunem, whose son was revived by Elisha (Heb 11:35). In other cases, some Jews 'were tortured, refusing to accept release, that they might obtain a better resurrection [*anastasis*]'—the Maccabean martyrs. Thus, for the writer of Hebrews, resurrection has the conventional range of meanings: escape from the risk of death, revival after premature death and participation in Jewish national revival (in the imagery of Ezekiel 37). In his concern with the death of Christ, none of that was satisfactory or even useful—only exaltation would do—even though 'resurrection of the dead and eternal judgement' were included among the foundation teachings of the faith (see also n. 22).

Summary

When the reader of Hebrews asks the standard questions about authorship, destination, place of origin, dating, the character of the document and its religious background, the evidence almost always points in seemingly opposite directions. This suggests that the author has produced the most carefully balanced writing in the New Testament. He is not averse to pleading with his readers or warning or even threatening them, but he carefully accommodates both traditional and novel views. He was determined that they should hold fast their confession, that is, the *homologia*, or 'agreed tradition', which the Christian community publicly admitted. From references to the *homologia* it seems clear that the tradition spoke of Jesus as God's emissary, as Son of God, as a high priest who suffered, and included some indication of the Christian hope.[44] That hope gave rise to a characteristic confidence *(parrhēsia)* both in approaching the Divine Person and in preserving their social existence despite public abuse and hardship.[45] But they must not throw away their confidence; they must hold it fast. They must not shrink back but follow the numerous examples of heroic loyalty described in Hebrews 11 and ending with the crucifixion of Jesus in Heb 12:2. The author does not write these things because he must find something to say; he writes thus because their confidence is ebbing and their grasp of the confession is weakening. It is weakening at precisely the point where the author will not let go, namely, the death of Christ. Yet it may well have become increasingly irksome to continue acknowledging the shameful crucifixion as a central feature of the *homologia*, especially as entry into the Christian community was called enlightenment (Heb 6:4, 10:32), which might be expected (in the manner of competing spiritual religions) to turn the mind from repulsive features of earthly, to nobler features of heavenly, existence.

To meet this dissatisfaction, indeed to turn it to advantage, the author took hold of one element of the *homologia*, namely, the designation 'high priest', and developed it in such a way that the earthly death of Christ was in no way diminished (though the chief benefits were transferred to heaven). At one point he devises a very odd image: we have 'a sure and steadfast anchor of the soul, a hope that enters into the inner shrine behind the curtain [separating earth from heaven], where Jesus has gone as a fore-runner on our behalf' (Heb 6:19-20). But then ours is a very odd situation: not only do we endure, like Moses, 'as seeing him who is invisible' (Heb 11:27), but, unlike Moses, we 'see the Day drawing near' when God will finally 'shake not only the earth but also the heaven' (Heb 10:25, 12:26).[46] The primitive Christian eschatology is certainly maintained, both as regards definitive events of the recent past and expected events of the near future—but at a certain cost. For one thing, the traditional language of resurrection is dismissed in favour of words appropriate to exaltation.[47] For another thing, the work of Christ becomes a single, exclusive action; and its benefits are available once and once only to those who are enlightened and have experienced 'the powers of the age to come'. The benefits offered are very great, and the risk of losing them is very serious.

The chief benefit is a new and living way from earth to the heavenly world. It was opened by Christ, and we enter the inmost sanctuary 'by the blood of Jesus'. This is, of course, not a physical 'sanctuary made with hands, a copy of the true one' but heaven itself. We 'share a heavenly call' (Heb 3:1) and 'with confidence draw near to the throne of grace'. Christ appears 'in the presence of God on our behalf' to claim for us what is variously called salvation (Heb 2:3–4), membership in God's household (3:6), the promised rest of the people of God (4:1–11), the full assurance of hope (6:11), an eternal redemption (9:12), a purified awareness of the being and the service of God (9:14). It is of course true that we are expecting Christ to appear a second time 'to save those who are eagerly waiting for him' (9:28). It is equally true (with a different metaphor) that 'we must run with perseverence the race that is set before us' (12:1) and endure the discipline of hardship. But only our own dullness or carelessness can deprive us of the enlightenment that is affirmed in the words, 'By a single offering he has perfected [given full access to God to] those who are sanctified' (10:14).

Existence in the heavenly world means awareness of the living God; existence in the earthly world means awareness of death—awareness of the Devil, who has the power of death and hence awareness of separation from God. Thus, the two would seem divided by an impenetrable curtain. Yet God is not only concerned with the angelic inhabitants of the heavenly world; he has prepared the world to come for the descendants of Abraham, and he intends to bring many sons to glory. How, then, could it be possible to move from death to life, through the curtain from earth to heaven? In scripture by analogy there is a ritual pointing to that transition. On the Day of Atonement, the high priest, carrying sacrificial blood for himself and the people, passes through the curtain into the Most Holy Place, which is symbolically the meetingplace with God. Thus, for the whole people the destructive forces are neutralised and the new year inaugurated for the people in relation to God. This ritual is limited in value. It permits only the high priest to enter the divine presence, is valid only for one year, and cannot remove awareness of sin and death. In fact its very repetition makes it certain that the problem of human death (animal death is no problem) is not solved. Yet as a God-given ritual for this earthly world it is a shadowy copy of the true ritual, which must (1) provide access to God for all the sanctified, (2) be permanently valid, and (3) make human death a path to God. These conditions are fulfilled in the author's presentation of Christ as the great high priest.

First, Jesus becomes the representative for mankind. He was not an immortal angelic being but a mortal human being capable of learning from what he suffered, made like his brothers in every respect. As the consecrator he belonged to the same category of being as the consecrated. He is able to sympathise with our weaknesses; having been put to the test of death as we all are, he can help those who now fail the test. Therefore, in passing from earth to heaven he goes as our forerunner, as the pioneer and perfecter of our faith; that is, where he has gone we also go. What he has pioneered we also experience, what he has perfected we also inherit.

Second, the action of Christ as high priest has taken place once for all. The shadowy ritual of the old covenant was replaced by the true ritual of the promised new covenant, so that 'he has appeared once for all at the end of the age to put away sin by the sacrifice of himself'. What he presented was 'the blood of the eternal covenant', thus becoming 'the source of eternal salvation to all who obey him'. By his self-offering he secured 'an eternal redemption', 'so that those who are called may receive the promised eternal inheritance'. The way is open and safe to travel—with the warning that an offence against Christ's death turns its effects against the offender. The blood of Christ then becomes accusing blood, like the blood of Abel: 'If we sin deliberately after receiving the knowledge of the truth, there no longer remains a sacrifice for sins'. That judgement follows inexorably on the author's conviction that Christ's once-for-all, definitive act guaranteed complete access to God. It does not imply that in such a case, the community lacked techniques for dealing with faults, mistakes and passions—they would be no great problem—but it meant that the way to God for that person was closed and could not be reopened.

Third, Jesus died as our representative: he tasted death for everyone by the most shameful of deaths, but in dying he passed not into the power of the Devil but into the presence of God. Thus, he destroys him who has the power of death, the Devil, and delivers all those who through fear of death (as separation from God) were subject to lifelong bondage. If that were all, the death of Jesus might be understood as a heroic martyrdom, the triumphant suffering of a godly man who preserved his loyalty under extreme pressure. Chapter 11 has many such. But Hebrews requires us to understand the death of Jesus as the self-offering of our high priest. His death is a consecration ritual: we are consecrated by his blood and by the offering of his body (Heb 9:13–14, 10:10, 14, 29, 13:12). To be consecrated is to be brought near to God, to share his holiness (see n. 15). On the ancient principle that like recognizes like, we cannot see him and he cannot receive us unless we are consecrated. It is not an arbitrary condition for communion with God but a necessary consequence of the holiness of the Divine Being. Holiness destroys impurity unless impurity ruins holiness. Therefore consecration is allied to purification, and that leads to the question of sins. In Hebrews sins can be purified (1:3, 10:2), expiated (2:17), or dealt with by sacrificial offerings (5:1, 3; 7:27; 9:26; 10:4, 6, 8, 11, 12, 18, 26; 13:11; see also

9:15). But the author argues vigorously that the sacrificial blood of animals cannot remove the results and consciousness of sin. Yet if Christ is regarded as a high priest offering not animal blood but his own self and bearing the sins of many, his ritual action performed effectively and in reality what the tent ritual had ineffectively prefigured. That particular human death became a path to God because, suffering the results of human sin, the great high priest pioneered and made safe the passage through the curtain; and when God then crowned him with glory and honour, he was 'perfected' as our forerunner.

8

The Johannine Writings

The Gospel and epistles of John have their own distinctive style of language and theology. It is now widely held that these distinctive features developed within a particular early Christian community and that some part at least of their significance can best be understood if they are related to influences within the community. It is commonly thought that the Gospel was composed first and that the epistles followed later, but I have argued for a different view:[1] that the problems considered in 1 John arose while the Gospel was being composed and greatly influenced the final form of the Gospel. The arguments for that view, in my opinion, are numerous and strong; but even if they are not accepted, the epistle provides a necessary entry to the exegesis of the Gospel. So that is where I begin.

1 JOHN

1 John is concerned with the small-scale social relations of a community of Johannine Christians and their neighbours. The members of the community are 'those born of God', and their neighbours are the *kosmos*.[2] The differences between community and neighbours are expressed in a set of apparently sharp contrasts.

One is the universally familiar contrast between light and dark. God in his proper being is associated exclusively with light: he is light or is in the light (1 Jn 1:5, 7). Darkness—or walking in darkness—is the negation of God; but the darkness is passing away and the true light is already shining (1 Jn 2:8). Members of the community should therefore walk in the light, as Jesus is in the light; but if they walk in darkness, they behave falsely (1 Jn 1:6). Hence, the sharp distinction is already qualified when the writer admits that community members can choose between light and darkness. The community is not—or not yet—a centre of unshadowed light, and its members cannot give way to the perfectionist temptation of insisting that everything they do is right. The reason is simple: being in the light implies loving the brethren, being in the dark implies hating the brethren

(1 Jn 2:9–10). The contrast of love and hatred explains the sense in which light and darkness are to be understood; and the initial sharp contrast is replaced by a realistic gradation of light–shadow–darkness. Some members will stray into the shadow: what can be done to stop them going into the darkness is a prominent question in the epistle.

A second contrast is knowing and not knowing God. The *kosmos* of course does not know him (1 Jn 3:1), at least not in the sense that the community knows him (1 Jn 2:3, 5, 13–14); for knowing God is keeping his commandments, being in him, abiding in him, and walking as he (Jesus) walked (1 Jn 2:3–6). The Johannine Christian is under obligation to conduct his life as the Johannine Christ conducted his. It is clear that not all did so ('He who says he abides in him *ought* to walk in the same way in which he walked'). Yet they know the truth, and the truth is in them (1 Jn 1:8; 2:4, 21). They have an anointing *(chrisma)* that gives them knowledge and teaches them everything needful, but they can be deceived (1 Jn 2:20, 26–27). Indeed, they have discovered liars within the community who deny that Jesus is the Anointed One, who are therefore antichrists and do not possess the *chrisma* they claim (1 Jn 2:22).

Knowing God implies loving God (1 Jn 2:5), which requires us to keep his commandments and do what pleases him (1 Jn 3:22). There are only two commandments: 'that we should believe in the name of his son Jesus Christ and love one another as he has commanded us. All who keep his commandments abide in him, and he in them. And by this we know that he abides in us, by the spirit which he has given us' (1 Jn 3:23–24). But that, too, is ambiguous, for there is a spirit of truth and a spirit of error (1 Jn 4:6). In fact, the community had been in dispute—and to some extent still was—about possession of the spirit. A dissident group, which had left the community, claimed the *chrisma* for themselves while denying it to Jesus. They relied on the spirit and not on the flesh, and the *kosmos* had listened to their message (1 Jn 4:5). But, says the writer, God had sent his only Son into the *kosmos* so that we might live through him; and love for God is our response to his love in sending his Son to be the expiation *(hilasmos)* for our sins. If God so loved us, we ought to love one another (1 Jn 4:9–11). Hence, the death of Christ is present in the writer's mind as the obligatory pattern of our loving, for we should 'not love in word or speech but in deed and in truth' (1 Jn 3:18). But since this must be argued or pleaded, it is clear that not everybody agrees. Hence, the writer insists that Jesus Christ came 'not with the water only but with the water and the blood' and that to that fact 'the spirit is witness, because the spirit is the truth. There are three witnesses, the spirit, the water, and the blood; and these three agree' (1 Jn 5:6–8).

To us, that sounds like cryptic, riddling language; but the intended readers must have understood well enough. Presumably *water* refers to baptism, by which members entered the community. In Johannine tradition water baptism is associated with Jesus (though rather uneasily) and with his followers (Jn 3:5, 22, 26; 4:1–2). *Spirit* then refers to a complementary or second stage of community membership, when the divine gift is received or received in its fulness. *Blood* must refer back to its only other occurrence in the epistle, as the blood of Jesus, God's Son, which cleanses us from all sin (1 Jn 1:7). That earlier statement is put for-

ward to encourage members of the community to confess their sins. If they say
that they have not sinned—if, that is, they claim to be pure, unsullied spirit—
they deceive themselves and make God a liar. If they claim to know God and,
knowing him, to keep his commandments without fault—even to dispense with
his commandments because they are begotten of God (1 Jn 3:9)—the truth is not
in them. But if they confess their sins, he is faithful and just and will forgive their
sins (i.e., remit the punishment) and cleanse them from all unrighteousness (i.e.,
remove what contaminates the community's life). Hence, the death of Christ is
not only the pattern of Christian loving but the means of restoring the community's integrity when love has failed.

A third contrast, or a third way of presenting the contrast, may be added—
the contrast between the community and the *kosmos*. The whole *kosmos* is in
the power of the Evil One (1 Jn 5:19), that is to say, the Devil, to whom sinners
belong, 'for the devil has sinned from the beginning' (1 Jn 3:8). The *kosmos* can
be represented by Cain, for it is the scene of murder and death (1 Jn 3:12–15).
It is marked by lust and pride, so the community must not love the *kosmos*
(1 Jn 2:15–17), which in turn does not love the community (1 Jn 3:1, 13) and
harbours those who are inspired to deny that Jesus is the Christ (1 Jn 4:3). Thus
the *kosmos* is everything that the community is not—or rather that the community should not be. For the antichrists came from within the community,
where they had induced brotherly hatred, which is a kind of murder. In total
contrast to the *kosmos,* members of the community recognize love in that 'he
laid down his life for us; and we ought to lay down our lives for the brethren'
(1 Jn 3:15–16).

So far, the author's view of the *kosmos* is strongly disapproving. But it is not
consistently so. After all, Jesus Christ is the expiation *(hilasmos)* not only for the
sins of community members but also for the sins of the whole *kosmos* (1 Jn 2:2).
The Son of God appeared to destroy the works of the Devil (1 Jn 3:8). God sent
his only Son into the *kosmos* so that we might live through him (1 Jn 4:9); but
it is equally said that the Father has sent his Son as the saviour of the *kosmos*
(1 Jn 4:14); that is to say, the fate of the world does not depend on the false
prophets who have gone out into the *kosmos* and been well received by it but on
the Son of God and therefore on those who abide in him. Hence, 'as he is so are
we in this *kosmos*' (1 Jn 4:17), which is called *this kosmos* to express the conviction that it is passing away (1 Jn 2:17). The community exists as the nucleus
of a new *kosmos,* where the old commandment has new force because 'the darkness is passing away and the true light is already shining' (1 Jn 2:8). Although
the *kosmos* is still a threat, the community can be assured that they have overcome the *kosmos* by their faith in Jesus as the Son of God and are themselves
born of God (1 Jn 5:4–5). Their young men have overcome the Evil One by the
indwelling word of God (1 Jn 2:13–14). Since 'he who is in you is greater than
he who is in the *kosmos* (1 Jn 4:4), they are protected by him who was born of
God, and the Evil One does not touch them (1 Jn 5:18). They are the children
of God, unacknowledged by the world. But they know that when he appears,
they will be like him, for they will see him as he is. With that hope they purify
themselves, as he is pure (1 Jn 3:1–3). Since the divine seed abides in them, they

cannot sin, because they are born of God. They prove they are children of God by loving one another. They know they have passed out of death and into life because they love the brethren (1 Jn 3:9, 14). But they have to be commanded, urged and encouraged to practise love: it does not happen spontaneously. Just as the *kosmos* is not entirely the province of death, so the community is not entirely the province of life. Nevertheless, eternal life is the promise and indeed the gift (1 Jn 2:25, 5:11–12). The primary theme of the epistle is 'the word of life'—eternal life that has been disclosed, experienced and proclaimed (1 Jn 1:1–2).

So the writer of the epistle uses contrasts of life and death, light and dark, love and hate, truth and falsehood, knowing God and not knowing him, doing is commands and not doing them. None of these contrasts is in fact absolute. Even though there is total opposition between God and the Devil, that opposition cannot be exactly reproduced in the experience of the community. That, of course, is not a surprising discovery about the life of a religious sect; and there are several ways of trying to solve the problem. The writer of 1 John is pragmatic: members of the community are expected and required to show faith in Christ and love of the brethren within recognizable limits. The limits are generously drawn because the two commandments (that they believe on the name of God's Son Jesus Christ and that they love one another, 1 Jn 3:23) invite a range of responses to the demands of religious loyalty and social acceptance. For that reason not all responses may be satisfactory. They may prove sinful: they may have unfortunate consequences for community members (the 'penalties' of sin), and they may embitter community life (the 'uncleanness' of sinning)—even when those responses were not intended to harm the brethren or subvert the community. There is a world of difference (or so it seemed to the writer) between failure, misjudgement and moral immaturity on the one hand and explicit repudiation of Christ and of brotherly love on the other. That view he expresses by distinguishing sin that leads to death (i.e., that takes the sinner out of life in the community and into death in the *kosmos*) from sin that does not lead to death: 'Every kind of wickedness is sin, but not all sin leads to death' (1 Jn 5:16–17 NJB). the boundary between the community and the *kosmos* has, as it were, a certain width. Members may unfortunately stray into that no-man's-land and yet be recovered for the community; but if they pass right through the boundary, there can be no return. Then what prevents members who have strayed into the boundary area from becoming lost in the *kosmos,* and what recovers them for the community? The answer, in fact, is the death of Christ—which may be demonstrated in the following way.

In the epistle, the use of the verb *phaneroō* is very instructive. It means 'to make known, reveal, disclose'. At the beginning of the epistle, it announces the controlling theme: 'Life was disclosed, and we saw it, and we bear testimony and proclaim to you the eternal life which was with the Father and was disclosed to us' (1 Jn 1:2). As the winding, repetitive exposition develops, it becomes evident that the disclosure of life takes place in two ways. First, it was disclosed and became visible in Jesus Christ and is present in his word and commandments. Second, it is disclosed as a possibility for the community; indeed, the disclosure in the Son creates life in believers: 'God gave us eternal life, and this life is in his Son. He who has the Son has life; he who has not the Son has not life' (1 Jn 5:11–12). To be more explicit,

> He was disclosed [*ephanerōthē*] to take away sins,
> and in him there is no sin.
> No one who abides in him sins;
> no one who sins has either seen him or known
>   him. (1 Jn 3:5–6)

That is to say, 'by the disclosure of him in whom there is no sin, by abiding in him, by seeing and knowing him the habit of sinning is ended' (Grayston on 5:11–12). In this context, at least, the rare expression 'to take away sins' means removing not the consequences of sins but the choosing of sinful actions: 'For this reason the Son of God was disclosed, to destroy the activities of the devil' (verse 8). Thus, the coming of Jesus Christ removes the habit and disposition of sinning, it transfers believers from the sinful *kosmos* to the sinless community.

That use of the disclosure model produces an 'enthusiastic' view of the Christian community and, if unmodified, can only lead to trouble. But in fact it *is* modified:

> 9   In this the love of God was disclosed [*ephanerōthē*] to us,
>   that God sent his only Son into the *kosmos,*
>   so that we might live through him.
>
> 10   In this is love,
>   not that we loved God but that he loved us
>   and sent his Son as a *hilasmos* for our sins. (1 Jn 4:9–10)

In verse 9 the disclosure model is retained, and it carries with it the availability of life in which sin is inconceivable. But alongside it is placed the sending model, which (in verse 10) allows the writer to reintroduce traditional language about atonement for sins. As in 1 Jn 2:2, Jesus Christ is 'the *hilasmos* for our sins and not of ours only but also for the sins of the whole *kosmos*'. The word *hilasmos* occurs nowhere else in the New Testament and doubtless carries the general sense of a means of obtaining forgiveness.[3] In Septuagint passages giving instruction about the cult, it can refer to an atoning sacrifice; but in passages dealing with the give and take of common life, it can refer to the readiness of God to show mercy and forgiveness. The Good News Bible translation, 'means by which our sins are forgiven' suitably allows the exegete to devise his own interpretation. In my judgement, the old rendering 'propitiation' (which in itself can properly be applied to God) misreads the image in the writer's mind; and 'expiation' hides it behind a not-very-familiar technical word. Nor do I think it proper to introduce the word *sacrifice,* for the fact that *hilasmos* sometimes indicates a sacrifice does not demand that it should always imply one. There is no suggestion of cultic ideas in either passage. The second occurrence (in 4:10) presupposes God's love in sending his Son as a means of forgiveness and therefore is close to Septuagint passages that use *hilasmos* for the divine generosity. The first occurrence (in 2:2) follows on from the penitent sinner's confidence in having a Paraclete with the Father. It can be shown that the uncommon word *paraklētos* has the general meaning of sponsor or supporter.[4] It may appear in legal contexts, though not as a technical term, and is just as much at home in the governor's residence or the civil servant's bureau (as in, e.g., Philo). In the epistle, sinners who have admitted their wrongdoing approach the Father with Jesus Christ as their *paraklētos*. It is not sufficient that they are penitent (as it would normally be in Judaism), for those Christians are born of God and should not sin. But since Jesus, the One born of God, is their sponsor, he induces God to be favourable to the penitents and therefore is their means of forgiveness. Jesus stands by those who are his own and, since he is *dikaios* (i.e., acceptable to God), God grants his request. For God himself is faithful and *dikaios* (i.e., well disposed), and will forgive our sins and cleanse us from all wickedness (*adikia,* 1 Jn 1:9). But why so? If we sin, how can we claim Jesus as our sponsor? for we abide in him only if we do no sin (1 Jn 3:6). Or (with an essential shift of perception), 'If we walk in the light as he is in the light, we have fellowship with one another, and the blood of Jesus his Son cleanses us from all sin' (1 Jn 1:7). Presumably, the train of thought is something like this: If it is still your intention to walk in the light (and you have not abandoned it, as the dissidents have), the fellowship of our community remains intact; and when, as we all must, we confess our sins, the blood of Jesus removes all the defilement. By that means we may rely on God to forgive our sins and cleanse us from all wickedness.

Thus, the blood of Jesus has an important but limited function: it protects the outer boundary of the community against danger from the *kosmos,* and cleanses the community from the defilement of sins, committed accidentally or wilfully by members of the community—provided, of course, that the sinners have not rejected walking as he walked and have not denied that he came in the flesh. That is why the writer insists that he came by water and blood. A more ancient formula associated the death of Christ with conversion and entry into the community. When the Johannine writer adopted the disclosure model, he could say that the sins of the little children were forgiven for the sake of his name, that they had eternal life by believing in his name (i.e., the powers inherent in his status as Son of God, 1 Jn 2:12, 5:13). The references to water and name suggest baptism. Thus, in baptism community members received the regenerating power of his name. As fallible members of the community, they were protected and cleansed by his blood; and they had a modest expectation of benefitting from his Spirit.

The protective and cleansing effect of Christ's blood clearly draws upon very old beliefs about the power of blood. But it would be wrong to suppose that its efficacy was formal and impersonal. When the writer gives instruction in dealing with sinful members of the community, he makes it clear that one brother must intercede for another. The community must rely on its ability to ask from God and gain its request (1 Jn 5:14–16). It is not a matter of allowing the sinner to make what use he can of the means available but of involving the community in the process of restoration. Moreover, the disclosure of the love of God in the twofold sending of his Son (to be the means of life and the means of forgiveness) is a model of the community's behaviour:

> Beloved, if God so loved us,
> we also ought to love one another. (1 Jn 4:11)

The community must be the means of life and forgiveness for erring members:

> By this we know love,
> that he laid down his life for us;
> and we ought to lay down our lives for the brethren.
>   (1 Jn 3:16)

So much for the teaching of 1 John. When finally we turn to 2 and 3 John we find that neither epistle contributes anything to the theme of this study. That is not surprising with 3 John, which is simply a testimonial letter; but it is perhaps a little odd with 2 John, which attempts to sum up the main thrust of 1 John.

JOHN'S GOSPEL

There is very little agreement among scholars about the many disputed matters concerning the Gospel of John. Despite intense discussion, new initiatives, the exploitation of recent discoveries and periodic movement of opinion, it has not proved possible to find a generally persuasive answer to the familiar questions. That being so, it is sensible to make use of critical theories as a means of interpreting the Gospel as it now stands, not as it may have existed in a reconstructed form earlier in its development.

On one matter, fortunately, there appears to be substantial agreement, namely on the main structure of the Gospel:

1:1–18	Prologue
1:19–12:50	Jesus and the *Kosmos*
13–17	Final Gathering of Jesus and Disciples
18–20	Passion and Resurrection
21	Resurrection supplement.

At this point it is useful to make a comparison with the Synoptic Gospels. It cannot, of course, be confidently assumed that the Fourth Gospel was dependent on them or independent of them,[5] and it must be taken for granted that the Fourth Gospel is to be interpreted on its own terms. Yet comparison with the other Gospels shows up its distinctive features. Thus, all the evangelists give about the same proportion of the total length to the Passion and resurrection (15%–19%), but within the Passion and resurrection the proportion given to the resurrection varies greatly: Mark, 7%; Matthew, 13%; Luke, 30%, John, 41% (27% if John 21 is excluded).[6] The Fourth Gospel, as it stands, gives far more attention to the resurrection than Mark does and so confirms the Gospel's general emphasis on *life*.

By contrast, the Passion narrative in John 18–19 appears to be the shortest of the four. But the appearance is misleading, for earlier episodes of the Markan passion are placed even earlier in John. Corresponding sections in the two Gospels may be set out thus:

Mark		*John*
11:1–10	triumphal entry	12:12–19
11:15–17	temple protest	2:13–17
11:27–12:44	disputes with authorities	cf. 5–10
13	apocalyptic discourse, the coming of the Son of man	cf. 14:1–31
14:1–2	plot to kill Jesus	11:47–53, 55
14:3–9	anointing at Bethany	12:1–8
14:10–11	treachery of Judas	cf. 13:2, 27
14:17–21	prophecy of betrayal	13:18–19, 21–30
14:22–25	eucharistic words	cf. 6:48–59
14:27	shepherd and sheep	16:32; cf. 10:11–18
14:29–31	prediction of Peter's denial	cf. 13:36–38
14:32–42	Gethsemane	cf. 12:27; 14:31; 18:11

Thereafter, the Markan and Johannine Passions, with their many differences, keep more or less in step. Note that the Markan sequence of passages appears in a different order in John. Thus there is one Gethsemane reference where it would be expected in the garden at the beginning of the Passion and Resurrection section; another at the end of one farewell discourse in the Final Gathering section; and another almost at the end of the Jesus and the *Kosmos* section when the soul of Jesus is troubled. Other episodes appear in the Final Gathering section, mostly towards its beginning, others in the part of the Jesus and the *Kosmos* section that is clearly leading towards the Passion. But three episodes are transposed to an even earlier position. The temple protest becomes the very first action Jesus performs in Jerusalem. The disputes with authorities, as well as any suggestion of trial before the Sanhedrin (which might be expected after Jn 18:24), disappear from the final Jerusalem narrative but in their own distinctive form provide the main substance of chapters 5–10, where Jesus is almost continually on trial.[7] And the eucharistic words have no place in John's description of the final meal but appear rather surprisingly in the discourse on the Bread of Life in John 6. It is a matter of dispute whether these differences arise from John's independent tradition or from his variation of Mark's narrative but on John's own terms some of these placings are theologically significant. The aggressive temple protest at Passover initiates the whole series of festival conflicts that ends with the high priest's advice, 'It is expedient for you that one man should die for the people, and that the whole nation should not perish' (Jn 11:50). Consequently, the narrative heart of the Gospel is not a varied collection of miracles, parables, and community instructions (as in Mark) but an almost-continuous dispute about Jesus' authority to speak and act on behalf of God. And within that main section—in a seemingly intrusive episode—comes the insistence that eternal life is impossible without eating the flesh and drinking the blood of the Son of man. Thus, themes appropriate to the Passion have a dominating influence in the Jesus and the *Kosmos* section of the Gospel; and by contrast, the farewell discourses of the Final Gathering section—after chapter 13—are devoid of explicit references to the death of Christ.

Prologue

The Prologue may be read in two main ways: either as a coherent statement running from verses 1 to 13 with a supplement in verses 14–18 (e.g., Barrett) or as an originally coherent statement, now broken up and supplemented by intrusions and comments (e.g., Haenchen). If the second view is taken, the references to the Baptist (verses 6–8, 15) are obvious intrusions, perhaps intended to deny claims that had been made for him and possibly directed against a rival movement supported by his followers. On grounds of style and content other verses may also be excluded, though not everyone agrees what they are. Haenchen proposes an original form comprising verses 1–5, 9–11, 14 and 16–17. They were a rhythmical prose adaptation of the Jewish legend (based on Sirach 24 and—precariously—on 1 Enoch 42) of the divine Wisdom's entry into the world and failure to find acceptance there. Instead, however, of returning disconsolately to heaven, in the Prologue the divine Wisdom (called Logos) takes a new initiative by becoming flesh; and the Christian community ('we') testifies to its effectiveness.

It is a well-argued proposal, and has two consequences. First, the Prologue as it stands is marred by the intruded verses, which derive from polemics and from a theological misunderstanding of the original composition. Second, neither the original nor the Prologue as it stands says anything about the death and resurrection of Jesus. Precisely for that reason it is suitable as a foreword to the story of the earthly life of Jesus, at whose end it is then possible to speak of the death and resurrection (so Haenchen, vol. 1, pp. 129–30).

If, however, the other view is taken (Barrett)[8], verses 1–5 describe the primacy of the Logos in relation to created existence; verses 6–8 the prophetic manner in which the Logos was communicated to mankind, particularly by the latest and greatest prophet; verses 9–11 how the true light himself entered the world only to be rejected, except (in verses 12–13) by some who were born of God. In Johannine idiom that is a sufficient key to understanding the Gospel. In verse 14 the previous objective statements in the third person are replaced by *us* and *we*. The opening *And* might well be translated 'And indeed' to indicate that the becoming flesh of the Logos is not a new, additional statement but a concise summary of verses 1–11. Then follows the community's verification of verses 12–13, including the Baptist's personal evidence, not simply about the eternal Logos, but about Jesus and the consequences that follow for knowing God.

If that interpretation commends itself, there must be a hint in verses 10–11 that Jesus was rejected. In Jn 1:9 the Logos, through whom all things (*ta panta*) came into being, is entering the *kosmos*,[9] that is, the ordered and yet disordered world of social existence. The Logos, which provides illumination for all humanity, provoked no answering response from the institutions that the Logos had brought into being, which disregard, on standard Logos theory, must be irrational and incomprehensible. If we now suppose that mankind includes a chosen people who are familiar with the Word of the Lord and are taught to obey it (as Sir 24 implies), surely they should welcome the entry of the true light. But 'he came to his own home, and his own people received him not'. Thus, neither the Gentile world enlightened by the Logos nor the Jewish people instructed by Torah were able to receive and acknowledge the true light. Yet there were exceptions to this general failure: some accepted the true light and were given the ability to become children of God. In a world that fails to recognize the true light, the only remedy is to be born of God. If, therefore, these verses refer, however obliquely, to the rejection of Jesus—and consequently to his death and resurrection—they are setting these events within a context of self-destruction and rebirth.

Jesus and the Kosmos

Recognitions (1:19–4:54). Without hostility or debate Jesus is recognized as Messiah, the prophet promised by Moses, Son of God and King of Israel. He mysteriously associates himself with the Son of man. John Baptist calls him 'the Lamb of God who takes away the sin of the world' (Jn 1:29), and the Samaritans know that he is the Saviour of the world (Jn 4:42). After the initial recognitions, Jesus performs his first sign at Cana in Galilee (Jn 2:1–11); and after the Samaritan recognition, he performs his second Galilean sign (Jn 4:46–54—only these two are numbered). At the beginning of the section John Baptist bears testimony that he had seen the Spirit descend and remain upon Jesus and therefore identifies him as the Son of God (i.e., God's agent and plenipotentiary, Harvey, *Jesus on Trial*, pp. 88–92, Jn 1:32–34). Later, Jesus himself expounds birth from water and the spirit (Jn 3:5–8) and finally, in discussion with the Samaritan woman, insists that God is spirit and must be worshipped in spirit and in truth (Jn 4:23–24). Thus, the section, fairly well marked out, introduces leading themes of the Gospel and (as I will now show) introduces imagery that can be interpreted of the death and resurrection of Jesus.

In Jn 1:19–28 John Baptist denies that he himself is a figure of Jewish eschatological expectation: neither the Messiah (God's anointed agent who will perhaps introduce the new age or at least be the ruler when it is established) nor Elijah (the prophet who will return before the terrible day of judgement to restore harmony in Israel) nor the prophet promised by Moses (who would be the true teacher of the new age). The Baptist is no more than Isaiah's voice in the wilderness preparing the way for the Lord. John's questioners interpret his baptizing as a qualifying rite for sharing in the new age. But for John it is a ritual of illumination (Jn 1:31), a means of disclosing his unknown successor, a person so high in dignity that John is not worthy even to act as his slave. In Mark 1:4 it is said that John preached a baptism of repentance for the remission of sins—probably implying that forgiveness would follow on repentance of which baptism was a sign[10]—but in the Fouth Gospel John's function is simply to testify to Jesus. The word *repentance* does not occur; and forgiveness of sins is transferred to Jesus, though with a difference:

The next day he saw Jesus coming toward him, and said, 'Behold, the
Lamb of God, who takes away the sin of the world [*ho amnos tou
theou ho airōn tēn hamartian tou kosmou*]!' (Jn 1:29)

The next day again John was standing with two of his disciples; and
he looked at Jesus as he walked, and said, 'Behold, the Lamb of God!'
(Jn 1:35)

Since in scripture lambs are sometimes sacrificial animals and since sacrifice is
sometimes used to remove the consequences of sin, it can be argued that John
Baptist's earlier statement is an inspired though tacit prevision of the death of
Christ. If that seems implausible, it can be suggested that the Baptist has been
given the words of a traditional Christian formulation to express, perhaps, some
convictions of part of the Johannine community derived from the Baptist's disciples. But none of this stands the test of exegesis. The Gospel insists that John
came to bear witness to the light, to help the Jews identify Jesus as the Son of
God, and to understand what he offered. Presumably, his witness was true (see
Jn 5:33, 'You sent to John, and he has borne witness to the truth'), that is , not
misleading—central, not peripheral. Yet nowhere else is the Gospel concerned
with expiation, scarcely indeed with the forgiveness of sins. In Jn 20:22–23 the
Holy Spirit gives discernment of sins and the authority to remit or retain them.
That is no more than the internal problems of the Johannine community dealt
with in 1 John. But elsewhere in the Gospel sin is enslavement to the Devil among
those who belong to this *kosmos*. Their need is to be set free by the Son, who
cannot be convicted of sin; and their means to that end is to believe in him (Jn
8:23–46). It is not a matter of having sins expiated or even forgiven but of transfer to a new *kosmos*. If they stay as they are, they will die in their sins. They
must move from death into life. Such is the characteristic way by which the Gospel relates the death of Christ to human need and so gives the necessary clue for
interpreting the Baptist's proclamation.

'Lamb of God' is a unique phrase. The genitive could mean 'belonging to or
suitable for God' but most likely 'provided by God'. If so, Abraham's words to
Isaac come to mind. 'God will provide himself the lamb [*probaton* LXX] for a
burnt offering', though when God indeed provides a substitute for Isaac, it turns
out to be a ram (*krios*, Gen 22:8,13 LXX). It seems unlikely that the simple
phrase 'Lamb of God' (using not *probaton* or *krios* but *amnos*) would signify to
Greek readers the extensive Jewish speculation about the binding of Isaac.[11] But
some kind of lamb provided by God is necessary. When the phrase is used in
verse 35 it stands alone—not to avoid repetition (when did John every worry
about that?) but to allow it a much more obvious meaning: the lamb that becomes
leader of the flock: 'The two disciples heard him say this, and they followed Jesus'
(Jn 1:37). That is a common apocalyptic symbol and is prominent in Revelation:
the lamb standing as though it had been slain, who alone could open the sealed
scroll, is the shepherd who protects, guides and fights for the flock (Rv 5:6, 6:1,
7:17). He is called *arnion*, not *amnos*, and his apocalyptic aggression scarcely
suits the mood of the Gospel, though, Revelation certainly has links with the

Johannine community. The aggressive ram, leader of the flock, may lie behind the phrase in verse 35[12] but not so helpfully behind the longer phrase in verse 29.

The word *amnos* appears twice more in the New Testament: once when the Ethiopian eunuch quotes Is 53:7–8, which compares the unprotesting servant of the Lord to a lamb dumb before its shearers, and more significantly in 1 Pt 1:19, where Gentile Christians are ransomed from the futile ways inherited from their fathers not with perishable things such as silver or gold but with the precious blood of Christ, that of a lamb without blemish or spot. If the passage contains Exodus imagery, that may be a reference to the Passover, though Ex 12:5 says not an *amnos amōmos* ('lamb without blemish') but a *probaton teleion* ('animal in good condition'), whether sheep or goat. But some interpreters are drawn to 1 Pt 2:21–25, where Christ's sufferings are described in the language of Is 53. The servant of the Lord, who is compared to a dumb *amnos,* himself bore our sins in his body on the tree. With his wound we have been healed. Perhaps, therefore, the Lamb of God was meant to draw attention to the suffering servant even though he is not said to remove sins but to bear their consequences and even though Peter does not call him a lamb but the Shepherd and Guardian of our lives (see p. 246). It is not likely. It is much more likely that the evangelist, having thrown out this striking phrase and have repeated it, would provide his own indications of its significance. He has probably done so in his frequent and systematic reference to Passover: 2:13, 23; (4:45); 6:4; 11:55–57; 12:1; 12:20; 13:1–2, 29; 18:28, 39; 19:14.[13] In addition, there is the thought that John may have arranged his chronology of the final days to have Jesus crucified at the time when the Passover animals were killed (though he does nothing to draw attention to that). Also, in one of his few explicit Old Testament quotations, the comment that the soldiers' decision not to break the legs of Jesus fulfilled the scripture 'Not a bone of him shall be broken' (Jn 19:36), that is, the rule for the Passover sacrifice in Ex 12:46, Nu 9:12—though applied in Ps 34:20 to the hard-pressed godly man. Even if it cannot be said with total confidence that John's Lamb of God has the same meaning as 'Christ, our paschal lamb, has been sacrificed for us' (1 Cor 5:7), it seems more than probable.[14]

It is indeed true that Passover was an apotropaic, not an expiatory, sacrifice. That fact is not to be dismissed by pretending that all sacrifices could be thought of as expiatory; for apotropaic sacrifice alone makes possible a consistent understanding of the Baptist's words. Passover blood protected the people of Israel from the destroyer and made it possible for them to leave their enslavement in Egypt and become an autonomous company of God's people. M. Pesachim 10.5 says, 'In every generation a man must so regard himself as if he came forth himself out of Egypt.... Therefore are we bound to give thanks, to praise, to glorify, to honour, to exalt, to extol, and to bless him who wrought all these wonders for our fathers and for us. He bought us out from bondage to freedom, from sorrow to gladness, and from mourning to a Festival-day, and from darkness to great light, and from servitude to redemption'. Everything in that Passover liturgy is to be found in the Gospel.[15] What the Lamb of God does is change the condition of mankind. He makes his appearance to remove the practice of sinning, which is endemic in the *kosmos* (as is similarly stated in 1 Jn 3:5–6; see p. 280). Because the spirit comes down and *remains* upon him, it is he alone who can baptize with Holy Spirit (Jn 1:32–33) so that we are reborn not of flesh but of spirit (Jn 3:5–8) and can learn the possibility of worshipping God in spirit and truth (Jn 4:23–24). All of which is offered to the *kosmos;* so that the Baptist, who at first identified Jesus by an image from Jewish religion, now identifies him by a current image of Hellenistic religion, Son of God, if that is the correct reading (Jn 1:34). It is later disclosed that the Son's task is to save the *kosmos,* and the Samaritans indeed recognize him as 'the Saviour of the World' (Jn3:16–17, 4:42). Much more will be said on this theme later: at present it may simply be noted that by calling Jesus 'the Lamb of God who takes away the sin of the world', the first step has been taken for establishing the subtle relation between the death of Jesus and the fate of those who live in the world.

The First Sign. The subtlety appears again in Jn 2:1–11 when Jesus turns water into wine at Cana of Galilee. The first sign that Jesus performed at Cana is matched by a second in Jn 4:46–54 in which he brought back to life a boy who was in danger of dying. Moreover, there is a third Galilean sign in chapter 6 (so named in 6:14), namely, the feeding of the five thousand. The multiplication of the loaves is in some ways analogous to the vast provision of wine in chapter 2; and the discourse on the true bread surprisingly includes a passage on eating the flesh and drinking the blood of the Son of man.

In chapter 2 the mother of Jesus plays a significant part, is addressed by Jesus with surprising formality as *gynai* (woman)[16] and is apparently brushed aside with harsh or regretful words (depending on how *ti emoi kai soi* is translated.[17] The mother of Jesus does not appear again until, with two other women, she stands by the cross. Again, she is addressed by Jesus as *gynai,* this time to be given into the care of the beloved disciple and made part of his family. Thus, she saw the first sign of his glory and its completion when he died.

Jesus said to his mother, 'My hour has not yet come'. It contains a familiar Johannine thought and (as Barrett says) it is unthinkable that at this place *hē hōra* should have a different meaning from its meaning elsewhere, namely the hour of Jesus' death on the cross and exaltation in glory.[18] But the identity of meaning is not directly indicated and is best understood from a parable in Jn 16:21:

> When a woman is in travail she has sorrow, because *her hour* has come;
> but when she is delivered of the child, she no longer remembers the
> anguish, for joy that a man is born into the world.

Her hour is the time for bearing pain in order that a man *(anthrōpos)* may joyfully be born into the world. Thus when Jesus says 'My hour has not yet come', he indicates the time of death, which is transformed into the joy of resurrection. That indeed is not yet, but he points to it with a sign. The lack of wine is transformed into an overwhelming supply—the confusion and bitterness of that failure is transformed into vast rejoicing. Just as the conversion of water into wine is impossible, so is life from the dead; but (according to the Evangelist) they happen.

By the sign at Cana the Evangelist says that Jesus 'manifested his glory'. Since *doxa* (glory) in biblical parlance is often used for the high tension of God's power and presence, these are words of a divine epiphany (TWNT 2). But in John the *doxa* of Jesus, which is not his own but the glory that God gives him, is supremely disclosed in his death and resurrection (Jn 12:23–24).

If this disclosure breaks the mould of Jewish worship (and of all other worship, Jn 4:21–24), so it must be. There may be a hint of that thought when the water provided for the Jewish rites of purification is transformed into wine, but it is certainly significant that the next episode is Jesus' attack at Passover on temple worship.

If the Evangelist knew and used Mark (whether he did or not is still an open question), he doubtless perceived the significant function of the episode in Mark's Gospel; by this disruptive protest Jesus set in operation the forces that would destroy both the temple and himself. Although John, in his passion narrative, does not include the hearing of evidence against Jesus at a Jewish trial, he records in this episode something like the testimony (Mark calls it false testimony) that Jesus had said, 'I will destroy this temple that is made with hands, and in three days I will build another, not made with hands' (Mark 14:58). In John, Jesus says 'Destroy this temple, and in three days I will raise it up' (Jn 2:19). Thus, the temple protest, in John as well as in Mark, has the same traditional function of pointing to the death of Jesus; but in John it is placed near the beginning of the story, to give some indication of how the story will end and how it is to be read.

The protest takes place when 'the passover of the Jews was at hand', and is followed by signs (not described) that persuade many to believe in Jesus. But he finds their belief unsatisfactory (Jn 2:23–25) and shows why in a conversation with Nicodemus, who has been impressed by his signs (Jn 3:2). This is the first of three Johannine Passovers (unless of course it really forestalls the third in the series, the Jerusalem Passover of Jn 11:55–12.1). The second is associated with events in Galilee (Jn 6:4) and therefore has nothing to do with Jerusalem and the temple, but it ends (as the first Passover did) with questions about satisfactory and unsatisfactory faith. Now Passover is the festal commemoration of God's protective power: it once saved, and is still expected to save, God's people from the destroyer. In John that protective power is transferred to Jesus. When the Galilean Passover was at hand, Jesus saved the people from hunger as Moses had given them bread from heaven to eat. And Jesus declared, 'This is the will of him who sent me, that I should lose nothing of all that he has given me, but raise it up at the last day (Jn 6:31, 39). And again, just before the final Jerusalem Passover, Jesus reports to his Father, 'I have guarded them and none of them is lost but the son of perdition' (Jn 17:12). In accordance with Johannine Christology, the Son is the Father's agent in this world. Hence, he is charged with the safety of those whom the Father is willing to protect, so that the importance of Passover is diminished and the centrality of the temple (in any case somewhat artificial for the domestic festival of Passover) is called into question. When Jesus calls the temple not 'the house of God' but 'my Father's house' he clearly presents himself as God's agent, competent to decide how the temple should and should not be used. It is not to be used as 'a house of trade' (*oikos emporiou,* Jn 2:16). That is different from the corresponding synoptic saying (see pp. 198–99), and the difference should be respected. It is not proper to supplement the Johannine saying from the Synoptics, or to determine synoptic interpretation by the Johannine wording. In John it is often taken for granted that Jesus was protesting against commercial practices in a religious building, perhaps being of Sirach's mind that 'a merchant [*emporos*] can hardly keep from wrongdoing, and a tradesman will not be declared innocent of sin' (Sir 26:29). Did the Evangelist really think that Jesus was engaged in such a futile protest or that Jesus had any interest in reforming the cultus? When, a short while later, he has Jesus say, 'The hour is coming when neither in this mountain nor in Jerusalem will you worship the Father' (Jn 4:21), he was envisaging removal, not reformation. When the Evangelist brings Jesus back to the temple, as he does frequently, it is as a place of instruction where Jesus says nothing about trade or sacrifice.[19] Therefore a better interpretation must be proposed for *emporion,* and two suggestions may be offered. First, the sharp words of Jesus may be intended to apply the savage words of Ezekiel against Tyre to the Jerusalem temple (remembering that temple dues had to be paid in Tyrian coinage)[20]: 'By the multitude of your iniquities, in the unrighteousness of your trade [*emporia*] you profaned your sanctuaries' (Ez 28:18). If so, Jerusalem's temple was as corrupt as a heathen sanctuary (and therefore should be swept away?). Second and rather differently, the wisdom tradition of Prv 3:13 says, 'Happy is a man who has found wisdom and a mortal who has perceived understanding. For trading [*emporeuesthai*] in wisdom is better than in treasures of gold and silver'. That would be plausible as a Johannine conviction and would suggest a possible application of Ps 69:9, which the disciples called to mind: 'Zeal for thy house will consume me'. The Masoretic text reads 'has consumed me' as also does the probably original reading of the Septuagint. Hence, Jesus' zeal to make his Father's house a dwelling of the divine wisdom will consume him, that is, will bring about his death.

That inserted scriptural comment from the disciples leads into the final phase of this episode. The Jews ask Jesus what sign he will show them to justify his actions. But what are they asking for? Certainly not a supernatural prodigy that would compel their credulous assent. That is not the Johannine meaning of *sēmeion* ('sign'). Indeed on the only two occasions when the word is given to Jesus, he regrets the demand for *sēmeia kai terata* (i.e., signs regarded as mere marvels) and rebukes those who benefit from a *sēmeion* without seeing what it is pointing to (Jn 4:48, 6:26). It is true that the Evangelist has selected a number of *sēmeia* from a larger store, and has written them down so that his readers (1) may believe (or go on believing) that Jesus is the Christ, the Son of God, and so believing (2) may have life in his name (Jn 20:31). Therefore, they are open to consideration—to examination, dispute, acceptance or rejection. Many who see his *sēmeia* have faith in his name, conclude that God is with him, and follow him. They believe because 'when the Christ appears, will he do more signs than this man has done?'—though some will not believe however many signs he does.

Specific signs display his glory and cause disciples to believe in him, so that he is identified as the expected prophet. The giving of sight on Sabbath to a man born blind convinces some that he was not from God, others that he could not be a wicked man. When he brought Lazarus back from death, many Jews believe in him; but the hierarchy perceive him as a threat to the nation.[21] There are no bare signs, only signs with an expectation—often a scriptural expectation. So, for example, when those who have seen the sign of the feeding are reproached by Jesus for being satisfied with the food but unconcerned with the sign (i.e., what the food points to), they ask, 'What sign do you do, that we may see, and believe you?' and promptly offer the example of Moses when 'He gave them bread from heaven' (Jn 6:14, 26, 30–31). When Jesus related his own sign to the Mosaic sign, they are content, at least for the moment. A Johannine *sēmeion* is an indicator that points to what is expected of God, namely, his saving power and goodness; and it is effective for those who perceive the sign beyond the marvel.

The Jews of Jerusalem very properly ask Jesus for a sign that would indicate his reason and authority for disturbing the temple. In what sense was he objecting to 'a house of trade' and by what authority did he claim it as his Father's house? His reply is cast in the form of a divine pronouncement: 'Destroy this temple [*naos*], and in three days I will raise it up'. The imperative *destroy* means 'even if this *naos* is destroyed'. The *naos* is the sanctuary, the inner temple that signifies the presence of the Deity. It is not therefore a question of sacrificial animals and temple dues but of the presence of the Deity himself. That cannot be guaranteed by a temple that has been forty-six years building nor stirred to action by animal sacrifice. If the sanctuary is actually destroyed, it can be rebuilt in the shortest time—perhaps with a reference to Hos 6:2 'On the third day he will raise us up that we may live before him' (Lindars). The saying thus expresses God's total indifference to sanctuary and temple. The Son of God, as God's agent in the world, speaks in the person of God, which to the Jews is unthinkable and ludicrous. Nor was it readily comprehensible to the Johannine community, which found its way to another understanding: 'He was speaking', they said 'of the santuary of his body'. When he was raised from the dead, they remembered what he had said 'and they believed the scripture and the word which Jesus had spoken'. The scripture is quoted in verse 17 (Psalm 69 explaining the death of Jesus) and hinted at in verse 19 (Hosea 6 promising revival in three days), and the word of Jesus is fulfilled in his death and resurrection. As once the Jerusalem sanctuary signified the presence of God in the midst of his people (which presence would not be denied even if the temple were destroyed), so the body of Jesus now signified the presence of God (which was not contradicted by his death and was confirmed by his resurrection). Thus, the sign of death and resurrection is important both for Christology (in maintaining Christ's sonship) and for salvation (in assuring the presence of God for disciples).

Ascent and Descent. In the next episode Nicodemus, the recognized representative teacher in Israel (Jn 3:10), acknowledges Jesus as 'a teacher come from God'. His signs prove that God is with him. But Nicodemus has completely failed to grasp the dimension of what he is saying. It is not a matter of one teacher recognizing another but of the confrontation of below and above, flesh and spirit, earth and heaven. There is no path from one to the other except by birth from above. Consequently,

> No one has ascended into heaven
> but he who descended from heaven,
> the Son of man [who is in heaven].

> And as Moses lifted up the serpent in the wilderness,
> so must the Son of man be lifted up,

> that whosoever believes on him
> may have eternal life. (Jn 3:13–15)

The signs Nicodemus has welcomed are not indications that the saving powers of God are available on earth by the agency of Jesus; but an indication that they are available in heaven, to which Jesus alone commands access. Later in the Gospel it will be explained how believers can accompany Jesus to the heavenly world; but the present statement is written polemically against those associated with the Johannine community who claim that gift of the Spirit by detachment from Jesus.[22]

The Evangelist has already promised a vision of 'heaven opened and the angels of God ascending and descending upon the Son of man' (Jn 1:51). The image of a ladder reaching from earth to heaven, from Jacob's dream in Gn 28:12, is boldly used to represent the Son of man as the essential means of communication between the world below and the world above. That could be taken to mean that the Son of man conveys the needs of mankind to God and the knowledge of God to mankind—and so he does if, in the different image of Jn 3:13, he himself descends from heaven and returns there. This is a kind of explanatory myth, which says, in effect, of the work of Jesus Christ, 'It is as if the great king, dwelling on high, sent his son to meet the needs of his subjects dwelling far below and bade him, returning, do for them what they could not do for themselves'.

It is convenient at this point to survey the Johannine Son of man sayings[23] and their connexion with the characteristic use of verbs for ascending and descending (*anabainō* and *katabainō*) and lifting up (*hypsoō*). They fall naturally into two groups.

First, in Jn 5:24 Jesus says that those who hear his word and believe him that sent him have eternal life; they do not come into judgement but have passed from death to life. Thus, the word of Jesus provides knowledge of God and makes faith in him possible. Therefore, the judgement that separates believers from unbelievers is no longer necessary, an out-of-court settlement has already taken place, and believers already pass from death to life. Jesus then justifies these statements by saying, among other things, that the Father had given the Son authority to execute judgement because he is Son of man.[24] Thus Jesus as Son of man makes the judgement of life and death and so is the giver of life—simply by uttering his word. The Christological implication—that Jesus and his word are not separable—is made clear in two more sayings. In Jn 6:27 the Son of man gives food that does not perish but endures to eternal life. It is, of course, the bread of God that comes down from heaven, and Jesus himself is the bread. Thus the Son of man not only *gives* the bread but, in the person of Jesus, *is* the bread. Similarly in Jn 9:35–37 the Son of man gives sight to a man born blind. He not only *gives* illumination but, in the person of Jesus, is the light (Jn 12:34–35). Hence, the Son of man comes down from heaven to bring illumination, sustenance and life.

Second, the remaining sayings speak of the ascent of the Son of man. Jn 6:62 is polemical and should be linked with 'the glory which I had with thee before the world was made' (Jn 17:5). In Jn 12:23 the hour has come for the Son of man to be glorified (explained by the seed dying and then producing growth); and in Jn 13:31, when Judas has gone out into the night, Jesus says, 'Now is the Son of man glorified, and in him God is glorified.' In Jn 8:28 Jesus syas, 'When you have lifted up the Son of man, then you will know that I AM'; and finally, 'I, when I am lifted up from the earth, will draw all men to myself'—thus indicating, says the Evangelist, by what death he was to die (Jn 12:32–34). Thus, the ascent of the Son of man is not merely the termination of his work as revealer, nor is it merely the resumption of his former estate after a period of voluntary incognito. It is a necessary ascent by crucifixion for the benefit of those he came to save and to bring glory to the name of God.

The fulness of this scheme is not yet made known in chapter 3, but it is obliquely indicated by a remarkable image from Nm 21:6–9. According to the story, the Israelites were discontented with their nomadic condition and many were dying from snakebite. To quell their panic Moses made a bronze snake and put it on a military flagpole (a not uncommon meaning of Hebrew *nēṣ* and Greek *sēmeion*), thus imitating the familiar practice of providing a rallying point in battle. From that time the Israelites could rally to the God who might kill them but was equally able to heal. It is an effective example of folk psychology. Even if the Evangelist could not expect that all his readers would know the story (and its moral, *similia similibus curantur*), they would be sufficiently aware of the curative snake of Asclepius. It would not be difficult for them to read the symbolic words thus, 'As the destructive snake was publicly exposed so that injured Israelites could gaze at it and be saved, so the Son of man (apparently set on destroying Israel) must be publicly exposed so that people may believe on him and gain life.' Thus, the statement that 'the Son of man must be lifted up' (*hypsōthēnai dei ton huion tou anthrōpou*), with its resemblance to Mk 8:31 ('the Son of man must suffer many things') is a lightly veiled reference to crucifixion and its necessity. The verb *hypsoō* has a double meaning: to be lifted up ignominiously on the cross and to be exalted in the eyes of God and mankind.[25] The result of that lifting up is that believers may have eternal life (*zōē aiōnios*).

That important theme of the Gospel deserves at least a brief definition. In Greek, *aiōn* (from which the adjective *aiōnios* derives) is a long period of time in past or future, one of the epochs into which history may be split when history is regarded as stretching far back and far ahead. In apocalyptic writings it indicates the new epoch that will succeed this final phase of the corrupt and evil world. Hence *zōē aiōnios* is the life of the wonderful new epoch, as in Dn 12:2. But *aiōnios* can also mean the opposite of what is temporary and likely to pass away.

It can be used of the life of God: since the life of the deity must be perfection, it must be unchanging and therefore timeless.[26] So *zōē aiōnios* may mean the life of the other world, the world above—the life of spirit, since 'God is spirit' (Jn 4:24). So in the Evangelist's polemic against dissidents, it is asserted that the gift of eternal life does not come from arbitrary rapture by the Spirit but by entrusting oneself to the Son of man elevated on the cross. Believers have eternal life *in him* and, since he alone commands access to the world above, that life depends on his death. It should be noticed that in this Gospel, it is not said or implied that the Father *handed over* his Son to enemies and executioners; but it is firmly said that God *gave (edōken)* his only Son so that believers should not perish but have eternal life, and *sent* the Son not to condemn but to save the world (Jn 3:16–17). The giving and the sending express God's love and at no time do they escape from his purposive control. God gave his Son that he should descend with illumination, sustenance and life and ascend by means of the cross to prepare a way for believers who are in him.

The Recognitions section ends with 'the second sign that Jesus did when he had come from Judaea to Galilee' (Jn 4:54). It is the sign of the dying child who began to recover at the hour when Jesus had said, 'Your son will live'.

Anticipations of the Passion (5–12).   In this section, the Evangelist says more than once that the Jews in Judaea wanted to kill Jesus (Jn 5:18, 7:1) and tried to stone him (8:59, 10:31); that the hierarchy made unsuccessful attempts to arrest him—unsuccessful because his hour had not yet come (Jn 7:30, 32, 44–46; 8:20; 10:39)—but finally decided to put him to death and gave orders for his arrest (Jn 11:53, 57). Jesus himself says that the Jews want to kill him (Jn 7:19; 8:37, 40), though some of his hearers think he is deluded, and others think that the authorities would like to kill him but dare not (Jn 7:20, 25). The impression is given that Jesus usually forced the issue, that his hearers were shocked but divided in their response, and that the authorities became decisive only when persuaded that Jesus was endangering the Jewish community. Jesus, however, was acting under direct instructions from God, and nothing could injure him until he recognized that his hour had arrived. This is not, therefore, a teacher or even a prophet who may or may not be acceptable to the Jewish people but the agent of God, the Son of the Father, to whom judgement has been committed because he is Son of man.

The cure of the cripple in Jn 5:2–9 —a cure for which he had been waiting thirty-eight years—in itself provokes no discussion. That it took place on Sabbath causes objection, and Jesus' defence of his action leads to violent disapproval. Jesus not only asserts that God is continuously at work, without resting on Sabbath (a defensible view when referred exclusively to God); he also says that he himself, a man, imitates God. Therefore, he is both dispensing with Sabbath and making himself equal with God, that is, claiming the right to do what only God can do. The question of Sabbath is not at the centre of discussion, though it crops up again in Jn 7:21–24 with a rabbinic argument much less dismissive than the one offered here. But even if Sabbath keeping is not central, the matter should not be thought unimportant. If social rituals are attacked, that is easily seen as a threat to the community's self-definition. An attack on Sabbath observance (not merely a passing lapse [Schnackenburg, pp. 101–2]) would begin to destroy the community from within. Likewise, a claim to be equal with God, that is, to reject God's authority and do what only God may do, is blasphemy and therefore brings destruction on the community from outside. It is no surprise that the Jews are extremely alarmed and think of killing Jesus. Whereupon he, ironically, begins to speak about raising the dead and giving them life.

We are concerned only with the first section of the speech, verses 19–30. It makes two responses to the objectors. First, in verses 19–20, 30 Jesus insists that he is wholly subject to the Father, in no sense claiming independence. The Son can do only what he sees the Father doing; of himself he can do nothing. Not that Jesus is a mere servant, carrying out God's instructions without understanding, approval or initiative: he is not a servant but a Son: 'The slave has no permanent standing in the household, but the son belongs to it for ever. If then the Son sets you free, you will indeed be free' (Jn 8:35–36 NEB). And the Son, acting as the Father's agent[27] and knowing the Father's love, makes decisions that are accepted by the Father as his own decisions. Second, verses 21–29 are of particular importance for further teaching in the Gospel[28] and may conveniently be displayed as follows:

21 As the Father raises the dead and gives them life,
so also the Son gives life to whom he will.

22 The Father judges no one,
but has given all judgement to the Son,
that all may honour the Son,

23 even as they honour the Father.
He who does not honour the Son
does not honour the Father who sent him.

24 Amen, amen, I say to you,
he who hears my word
and believes him who sent me,
has eternal life;
he does not come into judgement,
but has passed from death to life.

25 Amen, amen, I say to you,
the hour is coming, and now is,
when the dead will hear the voice of the Son of God,
and those who hear will live.

26 For as the Father has life in himself,
so he has granted the Son also to have life in himself,

27 and has given him authority to execute judgement,
because he is Son of man.

28 Do not marvel at this;
for the hour is coming
when all who are in the tombs will hear his voice

29   and they will come forth,
      those who have done good, to the resurrection of life,
      and those who have done evil, to the resurrection of judgement.

Verse 21 states the main proposition; verses 22–23 justify the words 'to whom
he will'. Verse 24 makes the consequential claim that the word of Jesus gives
eternal life to the believer, in the sense that he has ceased living in the realm of
death and now lives in the realm of life. Verse 25 restates that claim in terms of
the old language of death and resurrection, which is continued even more vividly
in verses 28–29. Verses 26–27 state the presumptions that make all this possible.
One has already been mentioned above: that the Son of man who is the means of
communication between earth and heaven is the proper person to execute judgement (verse 27). The other is more far-reaching (verse 26): it may be taken for
granted that God has life in himself, he is self-existent Being. He is not dependent
on any thing or any being outside himself. All other existences are dependent on
him as their creator and life-giver—unless he chooses otherwise. And he has so
chosen in the case of the Son: 'He has granted the Son also to have life in himself'.
For 'granted', the Greek word is *edōken* ('gave'). In some parts of the Gospel what
the Father gave to the Son dominates the thought: God gave him words to speak,
deeds to perform and bring to completion and (as we have just seen) authority to
execute judgement. Indeed he gave all things into his hands and authority over
all flesh, particularly over those who would come to him by God's gift (strongly
emphasised). Moreover, God have him his name and his glory—and the cup that
he must drink.[29] None of these gifts is as comprehensive as the gift of having life
in himself; and we are reminded of the Logos through whom (or which) all things
were made, in whom (or which) was life (Jn 1:3–4). The statement is matched by
Jn 6:57, 'The living Father sent me and I live because of the Father' *(apesteilen
me ho zōn patēr kagō zō dia ton patera)*, and gives rise to the discourse on the
living bread in John 6 and to the raising of Lazarus in John 11. It stands behind
the saying 'I am the life' in Jn 11:25 and 14:6.[30] It should be remembered that
Jesus also says, 'Unless you eat the flesh of the Son of man and drink his blood,
you have no life in yourselves' (Jn 6:53).[31] That will need discussion later, for it
implies that believers can share in self-existent being; but at this point it raises
the question, What sort of death does Jesus die if he has life in himself? It cannot
be the extinction of his being but must be the cessation of his life on earth and
the resumption of his life in heaven.

Bread of Life.   The dispute between Jesus and the Jews of Jerusalem is interrupted by chapter 6, set in Galilee,[32] where Jesus has performed two signs and
demonstrated his glory. In Jerusalem he has dealt aggressively with temple worshippers, offended against Sabbath, and claimed equality with God. Should it be
inferred that Jesus gains acceptance in Galilee, where he behaves generously, but
meets with rejection in Jerusalem, where he acts abrasively? Not entirely, as an
example from Galilee will show. It is, moreover, an example (as Lindars remarks)
that takes up the reference in the final sentences of chapter 5 to Moses' testimony

to Jesus. On the approach to Passover, Jesus feeds about five thousand people with five barley loaves and two fish and satisfies their hunger. The people who see this sign acclaim him as the expected prophet; but (says the Evangelist) Jesus knew that they intended to take him by force and make him king.[33] So he withdrew— because, as we learn later (Jn 18:36), his kingship is not of this world. So here is the first disapproving response to a Galilean welcome. Then the disciples are caught in a storm while crossing the lake; but Jesus discloses himself to them, and at once they are safe. The moral of that is plain and it suggests that the disciples are not included—at least not yet—in the disapproval. When, however, Jesus again meets the people who had been fed, he reproaches them for seeking him not because they saw signs but because they ate their fill of the loaves. He bids them to labour not for the food that perishes but for the food that endures to eternal life, which the Son of man will give them, for on him God the Father has set his seal (Jn 6:26–27). If that is so—if Jesus is indeed 'the prophet like me' promised by Moses (Dt 18:15)—surely he will give authoritative and binding Torah, as Moses had done. So they ask, 'What must we do to perform what God requires of us? and Jesus answers, 'God requires you to believe on him whom he has sent' (Jn 6:28–29). That is definite and scarcely admits of qualification.

The dialogue discussion that follows is prompted by a question from the congregation (just as their subsequent responses help on each phase of the teaching). They ask what sign Jesus will perform that they may see and believe, what he will do (to persuade them). They are not asking for an even more impressive miracle of feeding but for an indication that the feeding already performed corresponds to the saving power and goodness of God (as I have argued above).

The discussion can be set out briefly. They propose a test case: according to scripture our fathers ate the manna in the wilderness (verse 31). Jesus says, We will talk about God who gave the life-giving manna, not Moses—and they agree (verses 32–34). Just as manna was life-giving food (*artos,* 'bread', now appears in its wider meaning 'food'), so is Jesus: 'He who comes to me shall not hunger, and he who believes in me shall never thirst'. Coming to him and believing in him depend on the will of God. But those who come will not be lost and those who believe will have eternal life—with the statement, emphatically repeated, that Jesus will raise them up at the last day (verses 35–40; also verse 44). The congregation are upset: how can someone whose mother and father they know claim to have come down from heaven? (verses 41–43). Jesus replies obliquely: coming to him is possible if they have been 'taught by God' (quoting Is 54:13), and believing in him is possible if they know he has seen the Father (verses 44–47). Jesus then repeats that he is the life-giving food (as in verse 35), returns to the fathers who ate manna in the wilderness (see verse 31) and yet died, and now offers such food from heaven that those who eat it will *not* die: 'I am life-giving food which has come down from heaven: anyone eating this food will live for ever, and—? What comes next? Surely we expect, '—I will raise him up at the last day'; but instead we have, '—the food I shall give for the life of the world is my flesh'. It is the most startling surprise, and we sympathize with the Jews who say, 'How can this man give us his flesh to eat?' (verses 48–52).

The last section of Jesus' teaching (verses 53–58) is even more surprising for it introduces the requirement of eating the flesh and drinking the blood of Jesus. It is much disputed whether these verses are properly a part of the Capernaum teaching and whether they contradict the main thrust of Johannine theology.[34] Whatever the answer may be, even if the Evangelist himself or an editor of the Gospel added these verses to an already-existing draft, the addition must have been intended to produce a shock. It is the business of the exegete to discover why. To do so, we may first consider two points that arise from the foregoing verses. First, the Jewish interjections are not mere literary devices; some at least serve a theological purpose. The words, 'Is not this Jesus, the son of Joseph, whose father and mother we know?' have the effect of defining Jesus by his birth, family and social group. If the congregation had comprised hellenistic scholars, doubtless they would have defined him as a *theios anēr* (a 'divine man'), but as it was, he was simply a country lad who had the gift but talked too bombastically. Thus, the one who claims to be the life-giving food is accepted by his peers as a genuine human being. Second, four times in this discussion (but nowhere else in the Gospel) Jesus says, 'I will raise him [or it] at the last day'. The presence of such traditional eschatological language is not satisfactorily explained as the residue of a view otherwise abandoned or as the deliberate intrusion of an older view to counteract the newer view. It is true that 'he who believes *has* eternal life' (Jn 6:47), that is, he already possesses the life of the other world even though the present world has not passed away. Nevertheless, this present world still exists. Believers still live within it (see 'I do not pray that thou shouldst take them out them out of the world' Jn 17:15), and they possess eternal life by faith. To hinder what might become an unqualified Gnostic understanding of eternal life, the older eschatological language is used: he who has eternal life will be raised up at the last day. Faith produces its immediate reward in nourishment, but finally the life of faith needs to be rescued from the death that awaits all human activity. Since Jesus himself is genuinely human, he, too—the giver of eternal life—is subject to death. Hence, the believer who possesses eternal life must come to terms with the death of Jesus if the 'eschatological reservation' is to be effective in his case.

Jesus says, 'The bread which I shall give for the life of the world is my flesh' (verse 51) and thus provokes a dispute among the Jews. He continues thus:

> 53   Amen, amen I say to you
>   unless you eat the flesh of the Son of man and drink his blood
>   you have no life in yourselves;
>
> 54   he who eats my flesh
>   and drinks my blood
>   has eternal life,
>   and I will raise him up at the last day [see verses 39–40].
>
> 55   For my flesh is food indeed,
>   and my blood is drink indeed.

56 He who eats my flesh
and drinks my blood
abides in me,
and I in him.

57 As the living Father sent me,
and I live because of the Father,
so he who eats me will live because of me.

58 This is the bread which came down from heaven [see verse 50],
not such as the fathers ate and died [see verse 49];
he who eats this bread will live for ever.

The teaching begins with the fathers eating manna in the wilderness in verse 31, returns to them in verse 49 after the first exposition, then shifts the level of the thought and finally disposes of them in verse 58. The shift of level occurs because the speaker in verses 35–51^b might well be the divine Wisdom, whereas in verses 51^c–58 he is a person of flesh and blood (the familiar Hebraism for genuine human existence). So, for example, in Proverbs the Lady Wisdom entices her hearers to find life and favour from the Lord: 'Come, eat of my bread and drink of the wine I have mixed' (Prv 8:35, 9:5); and in Sirach 24:21, more boldly, she says, 'Those who eat me will hunger for more and those who drink me will thirst for more'. Wisdom is presented by a woman because the noun is feminine in Hebrew and Greek; the Logos is suitably presented by a man, Jesus. Indeed, in many parts of the Johannine discourses Jesus could be heard as a representative voice of hellenistic Jewish thought. It might be that the thought carried its own inspiration, whoever the speaker happened to be. But the Evangelist does not think so. In the Prologue he says that the Logos became flesh; now he makes a stand on prepared positions: if you want eternal life, not only must you listen to his words but you must stomach his flesh and blood. The Son of man, who is the means of communication between heaven and earth, who gives you the food that endures to eternal life (verse 27) and who will be seen ascending where he was before (verse 62), is a being of flesh and blood.

That being said, it must next be noted that the words *flesh* and *blood* are not common in the Gospel outside this chapter. The Evangelist does not use *sarx* to develop his thought that the Logos became flesh, except perhaps in contrasting 'the will of the flesh' with the life-giving activity of God (Jn 1:13), and in making a sharp separation between flesh and spirit (Jn 3:6). *Blood* appears oddly in the plural in Jn 1:13 and associated with water in Jn 19:34, but there is nothing else. Since it is a question of *eating* flesh and *drinking* blood, it is impossible to ignore the synoptic words of Jesus at the Last Supper about his *body* and blood. It must be supposed that the Evangelist took the words (not part of his chosen theological vocabulary) from old cultic tradition, perhaps using a variant form or substituting *sarx* for *sōma* (which he used for a dead body). The words in verse 51^c, 'the bread which I shall *give* for [*hyper*] the life of the world' are something like the longer text of Lk 22:19, 'This is my body which is *given* for [*hyper*] you'; and the refrain 'I will raise him up at the last day' in verse 54 is perhaps reminiscent of Jesus' intention of drinking wine new with his disciples in the Kingdom of God (Mt 26:29). It is therefore likely that the Johannine community was familiar with a form of the cultic tradition and would realize that the Evangelist was using it in this context with a special intention. What is regarded in the Synoptic Gospels as an instruction for the closed group of disciples becomes here a public exposition; and when John's Gospel arrives at the final meal of Jesus with his disciples, it is replaced by the foot washing.

Why does the Evangelist transfer the cultic words of the community meal to this public occasion? Certainly not to add a second, ritual requirement to the single requirement of faith laid down in verse 29—though, of course, the ritual action may give expression (and, indeed, substance) to the faith. Nor, in my opinion, is it likely that the Evangelist used an odd contrivance to give teaching about the eucharist or to guard against a danger that the eucharist would become 'a mechanical repetition of the Last Supper'.[35] It seems to me that he makes striking use of his community's supper tradition to insist that faith in the life-giving Son of man means entrusting oneself to him who gave his body to suffering and death that others might have life. To the very end of the instruction the theme is *life:* you cannot have life in yourselves, eternal life, life in the typically Johannine sense of mutual indwelling (I in you and you in me), life dependent on the being of God unless you eat my flesh and drink my blood—for that is what food and drink should be. It is rightly pointed out that to eat human flesh and to drink human blood is socially repugnant, proper for vultures (Ez 39:17–20) but not for human beings. It should not need saying, however, that the words of Jesus do not require the believer to consume his flesh and blood either literally or symbolically (even when the verb *trōgō* is used).[36]

Since eucharistic practice builds up a powerful attachment to the symbolism of eating and drinking, it is worth pausing to explain that statement. All of us live at the expense of others. Sometimes our survival or even our advancement is secured by their death (e.g., of soldiers in war or of coal miners and asbestos workers). We may be said to consume them and their future, but we do not even symbolically eat them. Yet the metaphor of eating is not merely high-flown language. In ancient society, to eat someone's food is to share not only his hospitality but also his substance, to become part of the family group. We feed on him, share his life blood. We may even expect his protection, sharing in the fortunes of his house, whether good or bad. In the story of the manna it is said 'He fed you with manna . . . that he might make you know that man does not live by bread alone, but that man lives by everything that proceeds out of the mouth of the Lord' (Dt 8:3). In this Gospel, Jesus *is* the mouth of the Lord. He says explicitly that his flesh is the true life-giving food, his death is the true life-giving drink. In other words, there is no way of preserving eternal life for ourselves except by sharing his death in whatever corresponding way is presented to us. If that is your meat and drink, you will live for ever.

Not surprisingly, such teaching is offensive—to disciples as much as to others; and many withdrew from the company of Jesus (Jn 6:60, 66). If, indeed, there were Christians who regarded the eucharistic food as (in the phrase of Ignatius) 'the medicine of immortality' (Bultmann, p. 219), it would indeed distress them to be told that they must share the death of the Son of man. Presumably, they expected to receive eternal life, the substance of immortality, by consuming bread and wine—so they must be told that 'it is the spirit that gives life, the flesh has nothing to offer [except of course death]. The words I have spoken to you [about eating my flesh and drinking my blood] are spirit and they are life' (New Jerusalem Bible, my additions). In other words, there is no way from the life of earth to the life of heaven except by death and resurrection. The section ends with Peter's proper confession that Jesus has words of eternal life and with Jesus' response that one of the Twelve is a *diabolos* (who will betray him). Thus, it seems that Jesus knew from the first that his disclosures would provoke disbelief and betrayal, that his Father would give him some untrustworthy disciples, and that he himself would consciously choose his own betrayer. These extreme statements of foreknowledge (which may be defended as appropriate to the divine Logos), seem to dissociate Jesus from common human experience; but they indicate the conviction that the death of Jesus is not fortuitous or peripheral to his revealing work. It is foreseen by God and essential to what he purposes.

Dangerous Conflict. The dispute between Jesus and the Jews of Jerusalem is now resumed. Its course is confusing, and we cannot always be sure whether we are reading about questions for Jesus or questions for the Evangelist. However that may be, the provocative issues may perhaps be identified. First, in response to the charge that Jesus destroys Sabbath, he retorts that the Jews disobey Moses' law because they are intent on murder (Jn 7:19). Indeed, they belong not to God but to the Devil, a murderer from the beginning and the author of all untruth (Jn 8:44, 47). They are enslaved to sin, needing to be liberated by the truth (Jn 8:32–34). The Jews deny that they are slaves; they are sons of Abraham, and Jesus is a Samaritan and mad (Jn 8:48). They know he is mad when he says, 'If anyone keeps my word, he will never see death'. They say derisively, 'Are you greater than our father Abraham, who died?' To which Jesus replies that he is far greater in the unfolding purpose of God ('Abraham rejoiced that he was to see my day'), far greater in the honour due to God's primary agent ('Before Abraham was, I AM', Jn 8:56, 58). Some of these statements sound abusive, some overassertive: since they are absent from the synoptic tradition, they probably reflect conflict between Church and synagogue. The Johannine community accuses the Jews of murdering the incarnate Logos. Second, the denunciation of the Jews may be put differently: they do not know God. Jesus knows him, for he comes from him and was sent by him (Jn 7:28). But the Jews know him not. Therefore, they know neither Jesus nor his Father (Jn 8:19, 55). He has told them the truth from God, and they seek to kill him because his word finds no place in them (Jn 8:37, 40). He offers them not only his word but the promise that for anyone who believed on him, rivers of living water would flow from his heart (Jn 7:37–38), that is, the believer would possess both the external testimony of the words of Jesus and the internal experience of the Spirit. But the Jews were perplexed and divided by that offer, all the more because it followed the puzzling statement, 'I shall be with you a little longer, and then I go to him who sent me; you will seek me and you will not find me; where I am you cannot come' (Jn 7:33–34). This is the first of several ominous warnings about the limited time available.[37] It implies that Jesus, as God's agent, alone has access to God; that his hearers, needing his mediation, must take it while it is now on offer or discover, too late, that it is denied them. Not unnaturally, the Jews are disinclined to such a once-for-all demand and make a shrewd (or sarcastic) guess that Jesus intends his proclamation for the Greeks. And in a sense they are right, though they do not imagine that Jesus will reach the Greeks via crucifixion and ascension. Therefore, he will go away, they will seek him and die in their sins; where he is going, they cannot come. 'Why not?' ask the Jews. 'Does he intend to kill himself?' And once more they are almost right. Yet they will die in their sins unless they 'believe that I AM' *(egō eimi).*[38] When they have lifted up the Son of man, they will know that I AM—that he is acting not on his own authority but acting for God and speaking as God (Jn 8:21–28). Either they will exa't him by believing on him, or they will reject him and raise him up on the cross. Whichever they do, they will encounter the God who as yet is unknown to them as the Father of Jesus.

Throughout the dispute, questions of death and life are central. The Jewish response is divided: some are puzzled, many believe in him, others are openly hostile. As John tells the story, it seems as if Jesus is forcing the issue and driving them all to hostility: he warns the Jews that they are contemplating murder and becoming incapable of life.

The public dispute on these questions is interrupted in chapter 9 by the giving of sight to a man born blind. This sign sets out to exemplify the claim of Jesus, 'I am the light of the world' (Jn 8:12, 9:5) and signifies the enlightenment he brings to those who have been completely in the dark. Even though nothing is said about life and death, it provokes more conflict. The reason is simple: for John life *is* enlightenment ('In him was life, and the life was the light of men', Jn 1:4). The restoration of life to Lazarus exemplifies the possibility of recovering life while the light is still shining; and this main section of the Gospel includes in its elaborate ending an appeal to its readers to become sons of light before the darkness overtakes them (Jn 12:35–36). Hence, light means living in awareness of God, and darkness means the extinction of that awareness in death. If you have the light, you cannot be in darkness—unless the light is extinguished. The Prologue says that 'the light shines in the darkness and the darkness did not overcome it' (Jn 1:5). Is that true of Jesus, and of those who believe in him, or are enlightened by him?

The Good Shepherd.   The question seems to be answered, though obliquely, in chapter 10: first the exploitation of sheep-shepherd imagery in verses 1–18, then a reference back in verses 19–21 to the sign of chapter 9, and then a renewal of the dispute in verses 22–39 (with a reference back to sheep imagery in verses 26–29). The cross-linkings suggest that the material was consciously placed where it is, though most commentators find it rather refractory. If verses 7 and 9 (where Jesus says that he is the door) are omitted,[39] the image is lucidly presented: the sheep are penned within a fold, which has a door and its guardian.

The shepherd is admitted by the door, he calls his sheep by name and leads them out when they recognize his voice. They are threatened, it is true, by thieves and robbers who climb over the wall; but the sheep do not acknowledge the voice of strangers and so flee from them (verses 1–5). This is a riddle *(paroimia)* that needs interpretation (verse 6). All previous entrants to the fold, to whom the sheep paid no heed, were thieves and robbers intent on theft, slaughter and destruction. But not so Jesus: he came to give them life in abundance (verses 8, 10; see 10:28). Clearly the *paroimia* is polemical: Jesus defends himself as sole benefactor of the community and attacks his predecessors.[40] Who they are is a matter of conjecture, and for my purpose they need not be identified. It is simply necessary to notice that the community can be represented as liable to attack by destructive persons or forces. That is the necessary context for what is an unparalleled development of the shepherd image—that he gives his life for the sheep:

11 I am the good shepherd.
The good shepherd lays down his life for the sheep.

12 He who is a hireling and not a shepherd,
whose own the sheep are not,
sees the wolf coming and leaves the sheep and flees;
and the wolf snatches them and scatters them.

13 He flees because he is a hireling
and cares nothing for the sheep.

14 I am the good shepherd;
I know my own and my own know me,

15 as the Father knows me and I know the Father;
and I lay down my life for the sheep.

16 And I have other sheep,
that are not of this fold;
I must bring them also,
and they will heed my voice.
So there shall be one flock, one shepherd.

17 For this reason the Father loves me,
because I lay down my life,
that I may take it again.

18 No one takes it from me,
but I lay it down of my own accord.
I have power to lay it down,
and I have power to take it again;
this charge I have received from my Father.

There are two variations of the *paroimia,* and a theological comment. In the first variation (verses 11–13) the sheep are no longer inside the fold in danger from sheepstealers, but are in open pastures where a wolf may attack. The shepherd is their owner, who will risk his life to defend his property.⁴¹ Unlike the hireling, who will take care of himself, he is the good shepherd in contrast to the worthless shepherds of Zec 11:15, 17 (the Septuagint calls them inexperienced and useless). So the *paroimia* can be interpreted: the people of God are in safe hands. It is stated more explicitly than the Q saying, 'I send you out as sheep in the midst of wolves' (Mt 10:16//Lk10:3). If applied to the Christian community, the wolves may be a code name for false teachers (Mt 7:15, Acts 20:29). However that may be, the good shepherd risks his life to protect the sheep—so adding to the New Testament passages where the death of Christ is regarded as a protective action.

The second variation (verses 14–16) turns to the theme of enlightenment: the good shepherd knows his own, and they know him. To be aware of him to whom we belong and to be recognized by him (since awareness and recognition are both intended by *ginōskō*, 'to know') is enlightenment, walking or remaining in the light (as 1 John has it). It is based on the Son's awareness of the Father and the Father's acknowledgement of the Son (see Jn 10:38). Hence, our awareness of the Son is a mediated awareness of God, and God's recognition of the Son is a mediated recognition of us. The shepherd lays down his life for the sheep, not now to protect them but to make that mediation possible and to bring a special content to the enlightenment he provides. Compare Jn 6:56 'He who eats my flesh and drinks my blood abides in me, and I in him'. To be aware of God who acknowledges us is to be aware that death was necessary in completing the work his Son had to do. Moreover, if verse 16 is not to be read as an intrusion into the exposition, the shepherd's death is necessary to bring other sheep into the one flock. Thus, the death of Jesus is necessary for the enlightenment of his Jewish followers and for the spread of enlightenment among the Gentiles. The expectation that there will be 'one flock, one shepherd' of course presumes the resurrection.

The two variations are completed by a third section (versus 17–18), which is partly defensive, partly explanatory. Verse 17 is a defence against the suggestion that the death of Jesus was a mark of God's rejection, verse 18 against the view that Jesus died unwillingly. More positively, it gives some definition to this laying down of life, which later reappears in the elaboration of Jesus' commandment in Jn 15:13: 'Greater love has no man than this, that a man lays down his life for his friends'. First, Jesus lays down his life in order to demonstrate and initiate the power of resurrection. For those who believe in him he brings the gift of eternal life *(zōē aiōnios)*, but the conferment of the gift requires Jesus to surrender his psyche and take it again. Second, in surrendering his life, Jesus acts with freedom within the possibilities that are genuinely open to him. The Johannine understanding of his death is not a way of making the best of a bad thing but a conviction that what is done in freedom creates freedom. Third, what Jesus does, acting freely as the agent of the Father, evokes and therefore discloses the Father's love and is validated as the Father's commandment.

When Jesus again speaks publicly in the temple at the feast of the Dedication, the running fight between Jesus and the Jews breaks out again. Questions and answers from previous chapters are repeated, and some features of subsequent teaching are anticipated. The section Jn 10:22–39 sums up the basic conflict between Jesus and his disbelieving critics, records their verdict (in place of the Jewish trial in the Synoptic Gospels), and drops hints about his real defence, which is disclosed to believers: 'My sheep hear my voice, and I know them, and they follow me; and I give them eternal life, and they shall never perish, and no one shall snatch them out of my hand. My Father who has given them to me is greater than all, and no one is able to snatch them out of the Father's hand. I and the Father are one' (Jn 10:27–30)—in the sense that the Father is wholly committed by the actions of the Son and that the Son is wholly dependent on the decisions of the Father. And that applies particularly to the death of Jesus.[42]

The Sign of Lazarus. The death of Jesus having been indicated by the shepherd *paroimia*, his death and resurrection are provided with a paradigm in the *sēmeion* of Lazarus. In the movement of the Gospel, Lazarus is entirely a symbolic figure. Unlike the cripple in chapter 5 or the blind man in chapter 9, he speaks to no one and no one speaks to him. If he were allowed to speak, no doubt he would say, 'One thing I know, that though I was dead, now I live' (see Jn 9:25). As it is, we must discover the implications of this symbol of death and resurrection from what his friends and family say in response to him. In the first place, the report of Lazarus' death is not a mistake: it is not the invalid's healthy sleep when his sickness has abated (verses 12–14). He was four days dead, buried in a rock tomb (verses 17, 38–39). His death caused distress to relatives and friends, calling for consolation, mourning, weeping (verses 19, 31, 33). Jesus himself was upset and gave way to tears and protesting anger (verses 33, 35, 38);[43] and it was at once realized that answering the call for help would be at the risk of death for Jesus and his disciples (verses 7–8, 16).[44] Lazarus' death could have been prevented (so the sisters think, verses 21, 32) if Jesus had been there in time; yet, knowing their need of consolation, Jesus stayed away two more days (verse 6). Why the delay? Because the absence and the presence of Jesus is to be a special kind of Christological demonstration (see verses 41[b]–42). The illness of Lazarus is not 'unto death' (verse 4, in the sense that he will not be deprived of the divine power) but is 'for the glory of God, so that the Son of God may be glorified by means of it'. That is a very important statement, and it must be interpreted with care. Since nothing can be *added* to the glory of God, we must understand 'for the glory of God to be displayed', particularly in the actions of his Son.[45] The display of God's glory is not for his self-satisfaction but for his saving intention. It is not designed to terrify or coerce human observers, for it is presented as a display of his love (verses 5, 36). The glory of God is effectively demonstrated in human death, the situation in which it often seems to be absent or useless. As Haenchen remarks, 'God's glory does not consist in sparing the faithful life's difficulties; but, precisely in refusing to do so, he shows that he is able to make the impossible possible through his Son' (vol. 2, p. 57). Jesus does not simply ask God to reverse the death sentence (verse 22) or tell the faithful to wait patiently until the resurrection at the last day (verses 23–24). He makes an immediate offer to faith:

25   I am the resurrection and the life;
     he who believes in me,
     even if he dies
     shall live;

26   and every one living
     and believing in me
     shall not die
     finally.[46]

The carefully arranged repetitive structure holds together two senses of living
and dying. The believer may indeed die as human beings die, with all the painful
and frustrating distress implied in the Lazarus story. But if his way of living is
determined by faith in the Son of God, in another sense he will not die: he will
not undergo the final extinction of God's saving power. The opening words, 'I
am the resurrection and the life', are structurally balanced by the closing 'finally'
(in Greek, *eis ton aiōna*). What Jesus is *now* has *final* consequences, for he is both
the resurrection (i.e., the means by which death is overcome) and also the life that
thereby becomes possible.[47] And so Lazarus, bound hand and foot, unable to see
face to face is called forth from the tomb—to be set free (verse 44). What Jesus
does for Lazarus is provide illumination: he goes to awaken his friend who has
fallen asleep (verses 9–14). He performs in symbolic action what is indicated in
the words (already traditional when Eph 5:14 was written), 'Awake, O sleeper,
and arise from the dead, and Christ shall give you light'.

The response to this sign is divided (versus 45–46), and the theme of Jewish
hostility (Jn 10:39) is resumed. The Sanhedrin is at last persuaded to agree that
Jesus (who is the resurrection and the life) must die. If the signs Jesus performs
give rise to a new religious movement, the delicate balance that maintained
Roman toleration of Jewish life might be destroyed. In this predicament, Caiaphas gives prudent or perhaps cynical advice 'that one man should die for the
people, and that the whole nation should not perish' (verse 50). Not without
irony, John says that 'being high priest that (presumably fateful) year, he prophesied that Jesus should die for the nation and not for the nation only but to gather
into one the children of God who are scattered abroad' (versus 51–52).[48] The
implications are subtle and far-reaching. First, God is speaking through the high
priest, not chiefly as the political representative of Jewry but as the guardian of
the sacrifices and the presiding judge of the Sanhedrin. In order to preserve temple
and nation it is expedient that Jesus should die. This is strangely like the expiatory hope of the Maccabean martyrs: that God would accept the deaths of a few
victims as an appeasement for the sins of all the people and would then turn aside
the fury of the pagan oppressors (2 Mc 7:37–38; see p. 192). But in due course
(as the Evangelist no doubt knew) the disaster that Caiaphas hoped to avert in
fact took place: the temple worship was ruined and the life of that nation was
irrevocably changed. Thus, Caiaphas prophesied without understanding the real
import of his prophecy. The life of God's people became dependent no longer on
the hierarchy, the sacrifices, and the festivals but on him who had died for the

nation. Second, the death of Jesus therefore both overthrows Jewish life as it was and reestablishes it in a new fashion. The newness consists of extending the designation 'children of God' to those 'who are scattered abroad'. 'Jesus collects those who belong to him within and without Judaism, and lays down his life for them' (Barrett on 11.52). In other words, the death of Jesus breaks the familiar Jewish mould and makes possible a new unified gathering of God's people. This is the Johannine preparation for the moment when Jesus identifies the hour for the Son of man to be glorified, namely, when the Greeks who have come to the feast ask to see him (Jn 12:20–26).

The Anointing. In bringing Jesus to his final Passover in Jerusalem the Evangelist makes of traditional material (the anointing and the entry appear in the Synoptic Gospels, though in the reverse order and with different narrative features), provides his own introduction (Jn 11:55–57), and writes in three references to Lazarus (Jn 12:1, 9–10, 17). In agreement with Mark, a woman at Bethany uses 'costly ointment of pure nard' for anointing; the objection is made that it could have been sold for three hundred denarii and given to the poor; and Jesus replies by referring to his forthcoming burial and departure in contrast to the continuing presence of the poor. In agreement, however, with the superficially similar but different story of Lk 7:36–50, the woman anoints the feet (not the head) of Jesus and eccentrically wipes them with her hair; and it is consonant with Lk 10:38–42 that Martha should be serving supper and that Mary should be anointing the Lord's feet. The pervasive fragrance in verse 3 is a Johannine touch, and only John regards Judas as having raised objections to the waste of money and gives the reason why. It seems clear that an anointing story was part of the primitive passion tradition, and each evangelist made what he could of it. John used it partly to condemn Judas, partly to indicate a second unconscious prophecy (parallel to that of Caiaphas) of the death of Jesus (see Lindars). But the wording of verse 7 is obscure: 'Let her keep it for the day of my burial' would imply that the greater part of the ointment remained. Yet Judas's protest implies that it had already been wasted, and in any case Jesus is properly anointed by Joseph and Nicodemus in Jn 19:39–40. Perhaps an anticipatory anointing was intended, a kind of omen; but it is difficult to find that in the Greek text. No doubt the Evangelist was doing what he could to bind together traditional material by the death and resurrection theme announced in the Lazarus story.

The entry into Jerusalem is very briefly related. The festival crowd go out, carrying palm branches (only John so describes the greenery, perhaps hinting at a political demonstration like that given to the victorious Simon, 1 Mc 13:51) and shouting words of greeting to pilgrims approaching Jerusalem (from Ps 118:25–26) with an added acknowledgement of Jesus as Israel's king. Jesus takes no initiative (as he does in Mark's account) but simply responds to the crowd by mounting an ass, perhaps to indicate a peaceful intention. The Evangelist says explicitly that the entry was not recognized at the time as a royal acclamation, nor was the prophetic clue (mainly from Zec 9:9) picked up until Jesus had been glorified. In the light of what later happened, incidents in the life of Jesus demanded a new interpretation—even the disapproving remark of the Pharisees that the *kosmos* had gone after him, which introduces the Greeks in verse 20. But there is no direct bearing on the death of Jesus.[49]

The Arrival of the Greeks.   The symbolic appearance of the Greeks prompts Jesus to say,

> The hour has come for the Son of man to be glorified. (verse 23)
> *elēluthen hē hōra hina doxasthē ho huios tou anthrōpou*

If the Greek sequence is followed, it will be seen that *the hour (hē hōra)* is taken up in verse 27, *glorified (doxasthē)* in verse 28–33, and *Son of man* in verses 34–36. Hence, verses 24–26 precede the elaboration of verse 23 and indicate that what is true of the glory of the Son of man will also be true of those who serve him. This is done in three stages. First, just as the life of a plant cannot continue except by the death of the plant and its renewal in a buried seed, so the life of a Christian community cannot continue except by death and resurrection—first of Jesus and then of his servants.[50] The simile has some merit in suggesting that the death of Jesus can be fruitful rather than destructive, but it has the disadvantage of implying that death and resurrection belong to the normal course of the cycle of nature. Presumably the Evangelist wishes to say much more than that. Yet, second, he persists with a general rule of human life in asserting that 'he who loves his life loses it, and he who hates his life in this world will keep it for eternal life' (verse 25). It has the same proverbial generality as the saying of similar import in Lk 17:33. The parallel in Mt 10:39 has 'loses his life for my sake', and Mk 8:35 (with parallels) makes the reference equally specific. It is clear that John inherited the general form of the saying—implying 'This is the way the world goes'—and makes it specific by what follows. Third, verse 26 is a Johannine equivalent for Mk 8:34 and can be read in a similar fashion: any follower of Jesus must follow him in risking his life and relying on God's power to give life. But the Johannine saying has notable differences:

> If anyone serves me,
> he must *follow me;*
> and where I am,
> there shall my servant be also;
> if any one serves me,
> the Father will honour him. (Jn 12:26)

> If any man would come after me,
> let him deny himself
> and take up his cross
> and *follow me.* (Mk 8:34)

Both sayings are about following the Son of man. In John the consequences are being where he is (presumably with the glorified Son of man in his suffering and entry to the eternal world) and being honoured by God. In Mark the follower must repudiate himself, take up his cross, and see what happens when the Son comes in his glory. The Johannine saying offers the inducement of divine approval, declines to be explicit about crucifixion, and pictures the Son of man (who in Mk 10:45 'came not to be served but to serve, and to give his life as a ransom for many') as expecting to be served, by no means a Johannine theme.[51] Some part of the community tradition spoke confidently about the exaltation of Jesus and more obliquely about his death.

But that is a kind of aside: the hour announced in verse 23 is agonizingly present in verse 27 (a Johannine equivalent of the prayer in Gethsemane):

> Now is my soul troubled. And what shall I say? "Father, save me from this hour?" No, for this purpose I have come to this hour.

The words call to mind Psalm 6, especially 'troubled' (*tarassō,* already used in the response to Lazarus' death; used again in Jn 13:21; 14:1,27):

> My soul also is sorely troubled.
> But thou, O Lord—how long?
> Turn, O Lord, save my life;
> deliver me for the sake of thy steadfast love.
> For in death there is no remembrance of thee;
> in Sheol who can give thee praise? (Ps 6:3–5)

And the psalmist concludes in the confidence that God has indeed heard his plea and will transfer the trouble from him to his enemies. Jesus, too, is troubled by the nearness of death and by human anxiety (whatever the Pharisees may teach about resurrection) that God will be absent, beyond human reach. But unlike the psalmist, he asks God to glorify his name, that is, to exert his power and show himself the Lord of life and death. He does not ask to be saved from this hour, that is, this decisive confrontation with the destructive and alienating power of death, because the whole purpose of his life hitherto had been to force that confrontation. When the voice from heaven says, 'I have glorified it [the Name], and I will glorify it again', the meaning is that God validates what Jesus has already done and confirms what he must shortly do. Despite the misunderstanding and inattention of those who heard the sound from heaven, Jesus makes two announcements and two predictions. He announces (1) that the glorifying of the Name is for the benefit of the hearers, not for the benefit of Jesus himself[52] and (2) that the judgement of God on this *kosmos* is now decided, namely, that the dead shall hear the voice of the Son of God and, hearing, shall live (as was fully indicated in Jn 5:25). The edict has gone forth that death shall no longer be the destructive and alienating power dominating this *kosmos.* Hence Jesus predicts that (1) 'now shall the ruler of this world be cast out', and (2) that 'when Jesus is lifted up from the earth he will draw all men to himself.' These are programmatic predictions: they indicate what now becomes possible, that is, the exclusion of the tyrant and the allegiance of mankind, not what is actually expected to happen. The Evangelist did not suppose that the ruler of this *kosmos* had lost his power to inflict damage and destroy faith or that all hearers of the gospel would promptly believe. But he certainly thought that God's edict now made it possible to circumvent the ruler of this *kosmos* and that the exaltation of Jesus in crucifixion and ascension would claim adherents from Jews and Greeks. It is not of consequence to decide whether 'the ruler of this *kosmos*' refers to the power of death personified or to the mythical figure of the Devil. In the symbolic world of the Evangelist they are one and the same. In Jn 14:30 the ruler comes to lead Jesus to his death, though in fact the ruler no longer has power over him; and in Jn 8:44 the Jews are children of the Devil, a murderer from the beginning, because they are intent on killing Jesus. Somewhat as at Qumran (see Lindars), the Evangelist works with contrasting powers—here, life and death—and, when God glorifies the Name, death no longer has power to hinder life. That indicates the kind of death Jesus would die: a lifting up from the earth in crucifixion, which would in fact be the beginning of life.

At this point the final term of verse 23, namely *Son of man,* is brought into focus. Jesus says that the Son of man must be glorified, which means raised up in death. Since, as his hearers insist, Torah speaks of the perpetuity of Messiah, not his death, who can this Son of man be?—the Son of man who is to be lifted up, that whoever believes in him may have eternal life (Jn 3:14–15), the Son of man, in fact, whom *they* would lift up (Jn 8:28)? In reply, Jesus moves from the conflict between life and death to the conflict between light and darkness. The Son of man is the necessary link between heaven and earth and brings enlightenment. If, while you can, you walk in the light and believe in the light, you become sons of the light, that is, enlightened beings. If you walk in the dark, the darkness overcomes you. If you are sons of light, the darkness will not overcome you—as Jesus will show by his death (verses 34–36).

The Evangelist finds the darkness of unbelief portrayed in Isaiah, blames the Pharisees for discouraging belief, and finally allows Jesus to sum up his basic claim: that he came to bring light and by his words to save the *kosmos* from darkness. In demanding allegiance, he was claiming it not for himself but for the Father who sent him, the ultimate judge and giver of life (Jn 12:44–50). In this summary the death of Jesus is not explicitly present.

Final Gathering of Jesus and Disciples

Nor is it quite as prominent as we should expect when we come to the *third main section of the Gospel,* which presents the final gathering of Jesus with his disciples and his lengthy instructions to them. It is commonly said that whereas in the first half of the Gospel significant actions are followed by explanatory discourses, in the second half explanatory discourses precede and prepare for the Passion narrative. That may indeed be so, but it is remarkable that John 13–17 is never explicit about the suffering and death of Jesus. Even the saying, 'Greater love has no man than this, that a man lay down his life for his friends' (Jn 15:13) is part of a general commendation of love. No one can doubt that during these chapters, the death of Jesus is constantly in the writer's mind; but he refers to it with oblique rather than direct expressions.

That can be demonstrated at once by the supper narrative in John 13, which is placed *before* Passover and (unlike the Last Supper of the Synoptics) lacks all reference to the body and blood of Jesus (which has been treated in John 6). The supper takes place 'when Jesus knew that his hour had come to depart out of this world to the Father . . . , knowing that he had come from God and was going to God' (Jn 13:1, 3). That certainly refers to his death—specifically, to death regarded as a transfer from this world to that world, as a return after an absence. The significance derives not from the fact of dying but from the expectation of rising.

Since (as I remarked on p. 300) the Evangelist transferred the words about eating the flesh and drinking the blood of the Son of man from the community meal to a public occasion, it can be argued that on this occasion, he wished to write about something else. References to a common meal begin with verses 2 and 4, are interrupted by the foot washing, and then are continued in the disclosure of Judas's treachery in verses 18, 23, 25 (see 21:20), 26, 27 and 30. Since the latter part of the narrative deals with a member who excludes himself from the community, the former part may deal with members whose actions need not exclude them if properly dealt with. It is very widely held that the foot washing is a symbolic representation of the saving death of Jesus. It is difficult to support that view with direct evidence; perhaps the strongest argument is that foot washing was normally performed by domestic slaves, and crucifixion was a punishment for offending slaves. But why should Jesus choose foot washing—which was after all a commonplace social courtesy—as a symbol of the far-reaching consequences of his death? If 'the cleansing of the disciples' feet represents their cleansing from sin in the sacrificial blood of Christ' (Barrett, p. 436, though that is an uncommon Johannine thought),[53] would not a total cleansing have been more appropriate (as Peter suggests)? The common interpretation attributes a monumental silliness to Peter's outburst in verse 9 ('as though salvation lay in the quantity of the water' [Haenchen]); whereas the familiar Johannine 'misunderstandings' are not silly but are usually shrewd attempts to provoke further thought or explanation. In this case, exegetes usually misjudge the interchange between Jesus and Peter and misinterpret the symbolism. First, foot washing is a social courtesy. It is offered by the host to the guests to ease the discomfort of their feet; and it is offered by the guests to their host to avoid dirtying or polluting the furnishings. It is normally done when guests arrive. Since Jesus does it in the course of a meal, he suggests that his companions are fully accepted in his table-fellowship, apart from some less serious matters that can be put right by common civility. Second, Jesus, their Master and Lord, teaches his companions a lesson by personally washing their feet. He sets the example of putting right what is wrong. But Peter objects, *'Kyrie, su mou nipteis tous podas?'* The pronouns are emphatic by position: 'Lord, it's *my* feet—and *you* are washing them?'—not impulsively humble (Bernard), but very affronted. Peter cannot believe that anything is wrong with him or that Jesus of all persons should think so. Jesus replies that he will understand later—in effect, this is a dialogue constructed for the later Johannine community. Peter replies truculently, 'You will never wash *my* feet', and Jesus gives him a warning, 'If I do not wash you, you have no [more] part in me [and my company]'. To which Peter replies angrily, 'Why stop at my feet? Why not wash me entirely [since you think I am unclean]?' Peter, in fact, represents those in the Johannine community who say, 'We have no sin' (1 Jn 1:8); and the dialogue ends when Jesus, speaking with careful moderation, gives a ruling:

> He who has bathed
> has need only to wash his feet,[54]
> and is [otherwise] entirely clean;
> and you [plural] are clean,
> though not all of you. (Jn 13:10)

The members of the community are fundamentally clean, unlike Judas (and those he represents) who betrays Jesus, but they need to be cleansed from impurities acquired by accident and error. According to Jn 15:3 the disciples are already clean by the *logos* that Jesus has spoken to them, that is, his revelatory discourse has transformed their condition and fundamentally determined their new nature. But they still must live in the world, and their feet will (as it were) bring the world's dirt into the community. Hence, following 1 Jn 1:7 (see p. 281), they must rely on the blood of Christ to cleanse them from all sin. If that is so, the blood of Christ has an important subsidiary function, however great the extent of its main function. Third, in verses 12–17, Jesus himself interprets the symbolic action of foot washing, uses his office as God's agent to commend it to them, and promises God's blessing on those who follow his example. But example of what? Not surely of humble service, as is often said. Foot washing could be an example of respect for others (as when a disciple washed his teacher's feet); but one can scarely suppose that an act that in our society might be represented by a host's cleaning his guest's shoes is an appropriate model for the loving service Christians owe one another. If, however, Jesus has given a striking example of how his followers can help one another to remove the defilements of the *kosmos*, his explanation is apposite. In the Johannine community, as we know from 1 John, there were some who accepted only sinless members, others who found a way back for sinners who had not committed a sin unto death (1 Jn 5:14–17). On the instruction of Jesus, they are to wash one another's feet; and that is the lesson of this dramatic demonstration.

It is a remedy for features in the community that are damaging but not completely destructive. There are examples in John 14–16; but first in 13:21–30 Jesus must deal with Judas, who was clearly an embarrassing member of Jesus' company (Jn 6:71, 12:4–6). It is said (1) that Judas was the agent of Satan (Jn 13:2, 27) and did what he must do without the knowledge of the other disciples, and (2) that Jesus knew his betrayer and consented to the betrayal (because his hour had now come?), even pointing him out without allowing the Beloved Disciple and Peter to interfere. Jesus is 'troubled in spirit' at the nearness of death but resolutely takes the necessary action. Indeed, when Judas next appears (Jn 18:2, 3, 5) he is merely a marginal character in the garden, a disciple who has gone over to the enemy, and Jesus is in command of the situation, efficiently arranging his own arrest. Nevertheless, it may still have been possible to regard Judas's treachery as a humiliation for Jesus. Therefore that impression is overcome by what Jesus next says, with enormous emphasis:

> Now is the Son of man glorified,
> and in him God is glorified;
> if God is glorified in him,
> God will also glorify him in himself,
> and glorify him at once. (Jn 13:31–32)

In that saying, *glorify* means 'promote the honour, confirm the reputation', from one of the standard meanings of *doxa* in Greek usage. As the Agent of God, Jesus seeks not his own *doxa* but the *doxa* of him who sent him (Jn 7:18).[55] Thus, the embarrassing activity of Judas and the connivance of Jesus are marked with God's approval.

13:33–35. But two themes still need explanation: (1) the imminent departure of Jesus and the inability of his disciples to follow him (Jn 13:33, explained in 13:36–14:14) and (2) the new commandment of mutual love as the mark of discipleship (Jn 13:34–35, explained in 14:15–24).

When Jesus says, 'Yet a little while I am with you. You will seek me; and as I said to the Jews so now I say to you, "Where I am going you cannot come"', he is using what at first was a warning to the Jews (Jn 7:33–34, 8:21, 12:35) as an encouragement to disciples. Why the disciples need encouragement is made plain in a variant form of the present discourse, namely Jn 16:1–33: they will be excluded from the synagogues and even put to death as a supposed offering to God. In that case, what are they to make of the words 'A little while and you will see me no more; and again a little while, and you will see me'? As a woman in travail has sorrow because her hour has come but when she is delivered of the child no longer remembers her anguish, for joy that a child is born into the world; so the disciples weep and lament while the world rejoices, but their sorrow will be turned into joy. And that promise is made in the full awareness that 'the hour is coming, indeed it has come, when you will be scattered, every man to his home, and will leave me alone' (Jn 16:32).

If we turn back to the end of chapter 13, we find that Peter's desertion prompts the consideration of Jesus' departure and leads in a most surprising direction. Jesus says that Peter cannot follow him now but will follow afterward. Peter replies that he will lay down his life for him—presumably to save Jesus' life, not after Jesus' death. But Jesus does not accept that offer (even though a man laying down his life for his friends is the greatest kind of love, Jn 15:13); instead, he predicts that Peter will deny him—though the disciples are not to be troubled by that misfortune (Jn 14:1). It is possible to recover from denial, especially if it is seen to result from God's intentions rather than Peter's defects: 'Believe in God, believe also in me'.[56] God can be trusted to do what is necessary and to do it through his Agent, who goes to prepare for them a firm basis in God's household,[57] in other words, to give them membership (while still in this world) of the other world. The reason they cannot follow him *now* is that only the return of *Jesus* to his Father can secure that place for them: 'No one has ascended into heaven but he who has descended from heaven, the Son of man' (Jn 3:13). That done, he can return and take them to himself, so that where he is they may also be (verse 3). No one, he says, comes to the Father but by him (verse 6). He himself *is* the way. Not that he leaves them in order to find the way or mark out the way to God and their secure basis in God's household—for he *is* the path of communication with God (like the ladder between earth and heaven in Jn 1:51). He *is* the truth, that is, the knowledge of God (verse 7); he *is* the life, that is, existence with God ('I am in the Father, and you in me, and I in you', verse 20). Since he is the ascending and descending Son of man, he must go that he may come. Consequently, it appears that the *death* of Jesus in itself is not of central importance. It is the means by which he departs, the manner of his going. It is indeed important because when he withdraws from this world, he appoints the disciples as his agents (Jn 17:18) and they perform the greater deeds that must be done (verse 12); but withdrawal is in mind rather than death.

In Jn 14:15–24 Jesus reflects on the commandment of mutual love, already stated in 13:34–35. To remedy their apparently orphaned condition (verse 18 NJB) he promises to return when *they* keep his commandments (verses 15, 21, 23–24) and *he* secures the Paraclete for them (verses 16–17). If they love him (by keeping his word), he will disclose himself to them: he and his Father will come and make their *monē* (which I have rendered above by 'firm basis') with them. It is a matter of indifference whether Jesus prepares *monai* for them in God's household or joins the Father in making a *monē* with them. In either way, existence in the other world while living in this world is promised. In verse 19 there is presumably a reference to the (death and) resurrection of Jesus, though the logical connexion between 'because I live' and 'you will live also' is not immediately obvious. In this context it should doubtless be read thus: 'Because I live [in the power of my word], you will live also [if you keep my word]'. That interpretation is confirmed by the vine image of Jn 15:1–11, which moves from image to explanation with the words, 'If you abide in me, and my words abide in you' (15:7). It is further confirmed by 15:12–17, where keeping the commandment of love (which at its highest requires dying for your friends) is evidence that you are appointed as the effective agents of Jesus.

There is, however, another way of consoling the disciples for the departure of Jesus, namely, the promise of the Paraclete.[58] He is the Spirit of truth (i.e., the spirit that gives knowledge of God), unknown to the world but indwelling the community (Jn 14:16–17). Thus, the Paraclete has a positive function within the community though a negative function for the world. Positively, the Paraclete brings to mind all that Jesus has said to them and so confers peace—the removal of disturbance and fear (Jn 14:26–27). The Spirit of truth guides into all truth, declares the things that are to come, and glorifies Jesus by taking what is his and declaring it to the community (Jn 16:13–15). Even though Jesus' mouth has been stopped by the ruler of this world—to whom he submits only out of love for the Father (Jn 14:30–31)—the Paraclete brings the words of Jesus effectively before them and keeps alive within the community 'the Jesus affair'. As regards the world, Jesus is God's Agent, who showed the world its sin and therefore provoked its hatred, against both himself and his Father. That hatred extends to his disciples when they bear witness in his favour. It is thus to their advantage that Jesus should depart and allow the Paraclete to convict the world of sin, righteousness and judgement (Jn 16:6–11).

Throughout chapters 14–16 it is possible to detect disquiet in the Johannine community. They were taught that Jesus came as Agent of the Father, to reclaim for the Father those who belonged to him. He and the Father are one, though naturally the Father is greater than he; and they themselves are the people reclaimed for God. But for several reasons it is not easy to credit that teaching: some members of the community were slow-witted, like Thomas, Philip, the other Judas, and the unnamed ones in Jn 16:17; Peter denied him, and all abandoned him; Judas betrayed him. The community experienced hostility and suffering. While they wept and lamented, the world laughed; and Jesus had abandoned them to the world. If he had returned to the glory of the Father, they had not. What can be said to restore confidence?—That the humiliations inflicted on Jesus and on the community are within God's knowledge and control, that suffering will be turned into joy, and that Jesus will return and is indeed already present in the power of his words and the activity of the Paraclete. Nowhere is there direct reference to the death of Jesus: when a reference seems unavoidable in Jn 14:30–31 (the approach of the ruler of this world) the conversation ends abruptly: 'Rise, let us go hence'.

There is, however, a covert reference to the death in John 17, which presents another way of consoling the community: neither the present power of Jesus word nor the present activity of the Paraclete but their splendid appointment as agents of the Agent of God. In verses 1–5 the Agent reports that he has completed his commission: he has displayed God's glory, that is, his reputation and majesty, on earth (taking up Jn 13:31–32) and now asks to be restored to his precosmic glory. In verses 6–8 he reports that he has done for those God gave him what he was required to do and so (in verses 9–19) prays for those left behind in the world. Like Jesus himself, they are not this-world persons; so he asks God to consecrate them (i.e., to appoint them as divine agents) by the effective power of the Truth, which is God's word or command (verse 17). They are to be agents of God's Agent: 'As thou didst send me into the world, so I have sent them into the world' (verse 18).

> And for their sake I consecrate myself,
> that they also may be consecrated in truth
>
> *Kai hyper autōn egō hagiazō emauton,*
> *hina ōsin kai autoi hēgiasmenoi en alētheia.* (verse 19)

It is not easy to find a parallel to self-consecration, even to the expression.[59] In the Hebrew Bible those about to approach God—sometimes priests, sometimes the people—are told to consecrate themselves, that is, to undergo the proper ritual cleansings. The passages in Lv 11:44 and 20:7–8, 'Consecrate yourselves therefore, and be holy, for I am holy' were much used in rabbinic teaching to urge separation from uncleanness and obedience to God's commandments. In general, therefore, 'I consecrate myself' could mean that Jesus, in order to remove himself from this world and leave behind the disciples as his agents, separates himself from the contamination of the world and prepares himself to encounter God—perhaps by obeying God's commandment to surrender his life. Nothing *demands* a reference to the death, but the use of *hyper* suggests it (e.g., the shepherd dies *for* the sheep, one man *for* the nation, anyone *for* his friends, Jn 10:11, 11:51, 15:13—all *hyper*). In the Last Supper in Mk 14:24 the cup represents Jesus' blood poured out *for* many; and in 1 Cor 11:24 the bread represents 'my body *for* you'. No doubt, John also intends to recall the death of Jesus for the benefit of others; but he uses old tradition in a cautiously allusive manner.

To conclude, in John 13–16 Jesus has exposed the deficiencies of his disciples, consoled them for his absence, and given them ample instruction about their future conduct; in John 17 he has spoken confidently to the Father about the successful completion of his mission. So in this Gospel there is now no place for the Markan Gethsemane: Jesus is not dismayed at disciples who cannot watch one hour, nor does he (however submissively) plead with the Father to spare him the cup. The Johannine Passion beings—and ends—otherwise.

Passion and Resurrection

The narrative can conveniently be divided into seven sections:

18:1–14	Arrest
18:15–27	Denials
18:28–40	Pilate's Interrogation
19:1–16ª	Verdict
19:16ᵇ–30	Crucifixion
19:31–42	Death and Burial
20:1–31	Resurrection

In the arrest (18:1–14) the disciples are marginal figures. Even Judas, who has led the soldiers and police to the garden, is now a mere bystander. The encounter is between Jesus and the military. It is, of course, surprising that Roman soldiers have any hand in this business, even more surprising that there should be so many (at least two hundred, possibly six hundred), absurdly carrying lanterns and torches.[60] Even more absurdly, they are immediately in grovelling disarray as soon as Jesus discloses himself. In the encounter, it is the Roman power that is made to look foolish; Jesus is in control. He identifies himself (perhaps with a hint of the divine I AM, see n. 38), tells (not requests) them to let the disciples go, curtly refuses Peter's attempt to resist, and says he will drink the cup the Father has given him. Jesus is in command of the situation; the ruler of this world can do nothing against him except by his consent. No one takes his life from him; he lays it down of his own accord (Jn 10:17–18).

The next episode (18:15–27) takes place in the court of the high priest where Annas, Caiaphas' father-in-law, is in charge. Nothing is said about taking evidence from witnesses or about an attempt to press a charge of blasphemy associated with the temple and a claim to be Messiah (as in Mark). But it must be remembered that according to Jn 11:51 Caiaphas 'being high priest that year prophesied that Jesus should die for the nation'. So Annas formally questions

Jesus about his disciples and his teaching. Jesus makes no reply about his disciples; but we already know, as Annas does not, the teaching given to disciples in chapters 13–17. Moreover, the storyteller surrounds the Annas episode with Peter's three denials—which effectively means that Jesus has lost his strongest supporters. Yet he vigorously answers about his teaching, that it was given in public and can be cited by witnesses (and we can read it in chapters 1–12). Although Jesus is bound and ill treated, he maintains an ascendency over his questioners.

In 18:28–40 Pilate's negotiations with the Jews surround his central conversation with Jesus. The Jews are set on having Jesus put to death: though Pilate finds no charge against him, they refuse the offer to release him and demand Barabbas instead. If the Romans are persuaded to crucify him as a rebel, he will die in humiliation and shame; but—as the Jews do not perceive—crucifixion will raise him up, so fulfilling what Jesus had signified concerning his death (Jn 3:14, 8:28, 12:32). Thus, whatever dreadful thing befalls him, it is according to Jesus' will, not theirs. That, however, is only the setting for the definitive conversation, in which Jesus speaks to Pilate at least on equal terms. Indeed, he questions him sharply, lectures him, and takes him out of his depth. When Pilate asks whether Jesus is King of the Jews, Jesus easily demolishes the question (by showing that Pilate has picked it up from the Jewish accusers, without any enquiry of his own), and even declines *King* in favour of *witness*. If they must talk about Kingdom, his ruling authority derives not from this world but from the other world. The whole purpose of his existence is to bear witness to the truth, that is, to the knowledge of God. On that matter, which is fundamental to the Roman imperium as much as to the existence of Israel, Pilate can only ask a question. He therefore announces to the Jews that he finds no case against Jesus (the first of three such declarations, Jn 19:4, 6; see also verse 12), and offers a judicial compromise, that is, the release of Jesus according to the Passover custom.[61]

When the Jews demand Barabbas, Pilate has Jesus scourged, allows him to be dressed up and derided as a mock king, and dismisses the charge (whatever it was) but parades him before the Jews with the laconic remark, *idou ho anthrōpos* (perhaps, 'There's your man'). Tempting though it is to find Johannine subtleties in *man,* the dramatic significance of this episode is surely clear: the prisoner, accused of claiming kingship, is not convicted by evidence but can be punished for being a nuisance, and can be treated with contempt. He is not a king but a man.

Thus the beginning of 19:1–16ᵃ, but the Jews will not give up so easily. Now they demand crucifixion for a shocking religious offence: that Jesus 'made himself the Son of God'. The charge, of course, is true; and Jesus, in his ludicrously painful condition, can now alarm Pilate—even by refusing to speak. So Pilate wraps his authority round himself: 'Do you not know that I have power to release you, and power to crucify you?' And the Son of God replies,

> You would have no power over me
> unless it had been given you from above;
> therefore he who delivered me to you
> has the greater sin. (19:11)

That is the careful verdict of the Son to whom the Father has given all judgement (Jn 5:22, 8:16, 12:31), namely, that Pilate is guilty, though bearing less guilt than the Jewish accuser—and not Jesus but Pilate is on trial. Pilate tries to purge his guilt by releasing Jesus, makes one last dramatic appeal, 'Behold your King!', but collapses before a Jewish threat of a report to Caesar. So, in exchange for the astounding Jewish submission, 'We have no king but Caesar', he hands Jesus over to be crucified at the sixth hour of the day of Preparation of the Passover—on which (as Jews knew, though John is not explicit) the Passover lambs were sacrificed.

John's account of the crucifixion and death of Jesus, in 19:16^b–30, differs greatly from the synoptic portrayal. In comparison with Mark, there is no mention of Simon of Cyrene, the offer of wine and myrrh, the reviling and jeering by the crowd and by the two men crucified alongside, the darkness, the explicit loud cry to God, and the confusion about Elijah. Not only has the Evangelist chosen some episodes rather than others; he has changed the whole impression. Whereas Mark presents a scene of shouting, malice, confusion, distress and mysterious dread, John has a quiet, well-ordered scene. All involved do what has to be done with unhurried competence, including Jesus' carrying of his own cross. Pilate refuses to change what he has written, so Jesus is proclaimed in three languages as King of the Jews. That is the first of perhaps six instances of Roman actions that are not unfavourable to Jesus. The second comes immediately: the soldiers in the execution party, following custom, appropriate the prisoners' clothing but forbear to tear up the seamless tunic worn by Jesus. What else could they sensibly do? (one might say); but the Evangelist detects a reassuring significance in their decision: something that intimately belonged to Jesus has not been destroyed, as scripture (Ps 22:18) had foreseen.[62]

Then Jesus gives his mother as a new son the Beloved Disciple who thereupon becomes heir to the family tradition (and so can outrun Peter to the tomb in Jn 20:4). Then, to fulfill the scripture, he said 'I thirst', and 'they [presumably the soldiers, as Bultmann and Brown say]put a sponge full of the vinegar on hyssop and held it to his mouth. It is usually supposed that the scripture must be Ps 22:15 or Ps 69:21, though neither fits well; and it may be doubted whether the Evangelist wished to portray Jesus occupying the last moments of his life by fulfilling scraps from the Psalms. Another interpretation is possible. The Talmud explains that wine was given to numb the senses of a man led to execution, 'for it is written, "Give strong drink to him that is ready to perish, and wine to the bitter in soul"' (from Prv 31:6; Sanh. 43^a). This was a way of obtaining merit, according to Prv 25:21–22: 'If your enemy is thirsty, give him drink . . . and the Lord will reward you.' Hence, Jesus gives his executioners an opportunity of gaining merit, that is, he forgives them by accepting the wine they offer. His work is complete—by his own free choice he bows his head and gives up his spirit.

Since Jesus is already dead, his legs are not broken; but a soldier 'pierced his side with a spear, and at once there came out blood and water' (verses 32–34). The effusion of blood and water should be discussed with verse 35, for it is clear that the scripture quoted in verse 36, 'Not a bone of him shall be broken' (either from Ex 12:46 and Nm 9:12 dealing with the Passover lamb or from Ps 34:20 referring to the godly sufferer) takes up the soldiers' restraint; and the quotation in verse 37, 'They shall look on him whom they have pierced' (from Zec 12:10), takes up the soldier's spear thrust. These quotations are part of the apologetic for the crucifixion. Although it cannot be denied that Jesus endured pain and mockery, it can be said that he talked on equal terms with Caesar's representative, who proclaimed him king. By dying quickly he was spared protracted suffering and also the indignity of the *crurifragium* by God's direct intention, as shown in scripture.[63] And when he was indeed speared by the soldier, that too was foreseen by Zechariah; and it implies that the Romans who pierced him will look to him, that is, for salvation.[64] It is all of a piece with the Evangelist's desire, already mentioned, to speak quite well of the Romans.

What, then, is to be made of the effusion of blood and water so curiously confirmed and emphasised in verse 35? First, that verse resembles Jn 21:24. Both verses are authentication signals for the Johannine community, which presumably grasped their meaning. Hence, the statement about blood and water is not intended as the basis for some general Christian belief(e.g., the sacraments of eucharist and baptism) but as support for a disputed belief within the community. Second, we know that the community could talk of this matter in the form, 'This is he who came by water and blood, not with the water only but with the water and the blood. And the Spirit is the witness, because the Spirit is the truth' (1 Jn 5:6–8; see p. 277). I have argued that the community was plagued by dissidents who accepted baptismal entry into the community and then required a second stage of membership when the divine gift of the Spirit was received in its fulness. Once that had been received, Christians could not sin: therefore the blood of Jesus was not needed (as 1 Jn 1:7 says it is) to cleanse from all sin. The writer of the epistle insists on water (of baptism), the blood of Christ, and the Spirit as forming one indivisible truth. The writer of the Gospel speaks much more positively about the presence of the Paraclete who is to re-present the teaching of Jesus. In the symbolic meaning of blood and water he may intend to go further and suggest that the water of baptism would be of no avail had not the blood flowed first. As in chapter 6 the Christian cannot be fed by the divine words unless he accepts the death of Christ, so here he cannot be divinely washed unless Christ's blood has flowed.[65]

In chapter 20 the Evangelist develops and modifies traditional resurrection narratives: the early morning visit to the tomb, the discovery by Peter and the Beloved Disciple that the tomb is empty, an appearance by Jesus to Mary Magdalene followed by an appearance to the body of disciples, with a supplement in the response of Thomas. These stories are related more to help the community understand itself than to provide evidence of the resurrection. Three matters call for comment. First, when Peter and the other disciple visited the tomb, the other disciple 'saw and believed; for as yet they did not know the scripture, that he must rise from the dead' (Jn 20:8–9), that is to say, he believed because he saw the emptiness of the tomb, but for us there is the better way of believing by understanding the scripture. Since John assimilates rising from the dead to ascension, the scripture is probably Gn 28:12, which he used (in Jn 1:51) to establish the pattern of ascent and descent. There is, of course, no passage of scripture that says in so many words, that the Christ must rise from the dead (not even Ps 16:10, which is sometimes suggested); but if Jesus is the Son of man who descended from heaven, then of course he must also ascend into heaven, where he was before (Jn 3:13, 6:62).

In Jn 20:17 Mary is told, 'Do not hold me, for I have not yet ascended to the Father; but go to my brethren and say to them, I am ascending to my Father and your Father, to my God and your God'. By clinging to Jesus, Mary is (as it were) hindering his return to God. When, therefore, the disciples see the Lord in the two encounters of verses 19–29, he is the ascended Lord who confers the gift of the Spirit as he had promised ('If I go, I will send him to you', Jn 16:7),[66] and yet still shows them his hands and his side.

Thomas voices the need not only to be told by others but to see and touch. Wittingly or not, he speaks for future generations of believers. So he is invited to touch and handle what he now sees. Apparently he is content with seeing, and so makes his Christological confession, whereupon Jesus pronounces a blessing on 'those who have not seen and yet believe' (verse 29). What then has happened to the touching and why did he wish both to see and to touch? To see the wounds, perhaps at a distance, can convey information to an observer, who may indeed be moved though still remaining an observer. But if the observer touches the wounds, he himself responds to their meaning and effect. In subsequent generations, people could not see Christ's hands and side, but they could hear the apostolic testimony. If thereby they became (not indeed spectators) but mere auditors, where was the possibility of touching? In the demonstration of Christians washing one another's feet!

Summary

The teaching of 1 John is in some respects tentative, for it is being worked out in response to developments in the community and in fidelity to old traditions. But despite the involved presentation, it is relatively simple.

The Son was sent into the *kosmos* that people might live through him. His purpose was to disclose eternal life in place of death, to replace darkness by light, falsehood by truth, ignorance by knowledge of God, hate by love, and disobedience by obedience. He came to increase the realm of life and to diminish the realm of darkness—and to do so irreversibly, for the darkness and the *kosmos* were passing away. Those who responded to this disclosure—their response signified in the water of baptism—were children of God, born of him, indwelt by the divine seed, possessed of a special *chrisma,* and transferred from death to life. Every one born of God overcomes the *kosmos* by his faith, just as the young men of the community have overcome the Evil One by the word of God that resides in them. That is understandable because the Son was disclosed to destroy the activities of the Devil, that is, to remove the habit of sinning. Consequently, everyone born of God cannot sin; the community member therefore expects to remain in the Son and not to stray beyond the boundary of the light into the realm of darkness.

So far there is no mention of the death of the Son. Nor has it been necessary. The whole emphasis is on life: 'He who has the Son has life; he who has not the Son of God has not life' (Jn 5:12). The gift of life has been made possible by the word of God, the anointing, the Name of the Son of God—in other words, by disclosure or revelation. The only necessary reference to Christ's death—and it is indeed important—is in the delineation of love (which is one way of describing eternal life): 'By this we know love, that he laid down his life for us; and we ought to lay down our lives for the brethren' (1 Jn 3:16). Clearly, that is a piece of old tradition that appears a little awkwardly in this context. If the community needed no teaching because they possessed a permanent *chrism* (1 Jn 2:27), why should they be reminded of Christ's death on their behalf and urged to follow his example? The answer is plain. Members of the community were not sinless: they strayed beyond the boundary of the light into the shadowy area between light and darkness. Such wayward actions inflicted damage on the community (the penalties of sin) and embittered relations (the uncleanness of sin). Both could be remedied by confession and the blood of Christ. If they refused to admit that they had strayed, they made God a liar; for he had specifically provided the remedy. Christ would act as their sponsor, because he had given his life for them, and induce God to receive them back into the full life of the community (1 Jn 2:2, 4:10). Their sins would be forgiven for his sake (1 Jn 2:12)—not their sins only, but the sins of the whole *kosmos,* which he had been sent to save (1 Jn 2:2, 4:14). Therefore, it was necessary within the community to practise forgiveness by asking God to give life to those (but only those) whose sinning was not 'unto death', that is, whose words and actions had not taken them deliberately into the realm of death or darkness.

Those are excluded who deny the necessity of the death of Christ or its effectiveness in dealing with sin. The epistle insists that there are three witnesses to the truth, agreeing and inseparable: the spirit, the water and the blood. The writer adds the gift of the spirit to the community's possession of the word of God, the *chrism,* and the divine seed. But to distinguish the spirit of truth from the spirit of error he needs to invoke Jesus Christ, who came in the flesh; and to establish the content of love he must say that God sent his Son to be the expiation for our sins. At every point, therefore, the death of Christ both guards the outer boundary of the community and restores those who have carelessly strayed.

In the Gospel, the death of Jesus is far more prominent. The passion narrative presents tradition that cannot be ignored. Indeed, it plays an important part in Jesus' relation to the Jewish public, and appears with striking dramatic force in the final section of the Gospel. But it is almost absent from Jesus' relations with his disciples and is no more than a hint in the Prologue. Thus, the prominence of Christ's death is somewhat qualified.

The Johannine Passion narrative is very different from the Markan. The will of Jesus determines what happens. He is in control of his own arrest, exercises ascendency over his questioners, and dominates Pilate. Indeed Pilate finds him guiltless; and, throughout, the Romans are not wholly unfavourable towards him. It is Pilate and the Jews who are really on trial, and Pilate is less culpable than the Jews. When exalted on the cross, Jesus calmly settles affairs for his mother and the Beloved Disciple, makes room for the soldiers to gain merit, and gives up his spirit. He has done what had to be done, namely, prepare his return to the Father. Why it had to be done by crucifixion is suggested only by hints: two or three apologetic quotations from the Septuagint (explicit quotations are not common in this Gospel); the death took place tacitly when the Passover lambs were sacrificed; blood and water flowed from his side; even when he bestowed the Spirit, he showed his hands and side. Without being explicitly stated, it is implied that death by crucifixion was God's will and that Christians cannot enjoy the benefits of the Spirit unless they remain aware of the Lord's suffering.

In the farewell discourses it is taken for granted that what Judas does (with the connivance of Jesus) is within God's intention. Consequently, Jesus remains with the disciples only a short while longer before he dies; but suffering will be turned into joy—in a little while he will be with them again: in his words, in the Paraclete, and in the *monē* that they have in God's household or that he and his Father have with them. Because Jesus is both the way and the truth, he secures life in the other world for them only by returning to the Father, using death as a means of withdrawing from this world and so consecrating himself to encounter God.

Thus, in the Fourth Gospel the death of Jesus is not a private event belonging to the inner circle of disciples but a public event for all to see. Not everybody who sees understands, of course; and when Jesus sums up the significance of his public encounters, he makes no reference to his death. But in the long record of those encounters, many indications are provided of how to read the expectation of death.

The public encounters are dominated by conflict. Jesus appears to provoke the Jews, forcing them into resentful opposition and attempts to kill him. His prophetic actions in the temple at Passover seem to displace God's protection of his people. He breaks the Sabbath, thus weakening the structure of religious life from within; he purports to do what only God may do, thus risking the vengeance of God from without. In the Johannine perspective, however, the Jews wholly misread the situation: they are murderers because they know not God, seek to kill Jesus for disclosing the truth, and refuse to take advantage of a limited time for repentance and reformation. The death of Jesus is therefore inevitable, foreseen but not fortuitous or peripheral. It is not (as might be supposed) a mark of God's rejection but a component of his purpose and a disclosure of his love. Jesus dies willingly, and thus shows and initiates the power of the resurrection. Jesus and his Father are one—in the sense that the Father is wholly committed by the actions of the Son and the Son wholly dependent on the decision of the Father, especially in regard to death. Since God is the life giver, Jesus by his word gives eternal life because God has given him to have life in himself. The death of Jesus is therefore his cessation of life on earth and its resumption in heaven, thus making eternal life possible for those who now believe in him.

Hence, God, speaking through the unwitting Caiaphas, affirmed Maccabean martyrdom theology, that is, that one man should die and that the whole nation should not perish—and not the nation only but also the children of God scattered abroad. And Jesus thereby overthrew the Jewish manner of life, established it in a new fashion, and extended it to Gentiles. When Jesus was indeed troubled by the nearness of death, he begged God to glorify the Name, that is, to exert his power and show himself the Lord of life and death (for the benefit of his hearers, not his own), to pronounce his judgement on the *kosmos* (that the dead should hear the voice of the Son of God and live), and to circumvent the power of death and the ruler of this *kosmos*.

Even more striking, however, than these reasonably systematic statements is the symbolic exploration of Christ's death. It takes place by means of images and actions. The lamb of God is the Passover lamb whose blood protects God's people from destruction and marks the beginning of their changed condition when the habit of sinning is removed. The image of ascent and descent (prompted by Jacob's ladder) implies the descent of the Son of man to bring illumination, sustenance and life and his ascent (by crucifixion) for the benefit of those he came to save and to bring glory to the Name of God. The lifting up of the serpent suggests the manner in which the Son of man will give eternal life, that is, life of the eternal world, to believers. The death of the good shepherd both protects the sheep and mediates between mankind and God, enlightening the Jews and spreading enlightenment to the Gentiles. The buried seed suggests that new life comes after death, and proverbial wisdom—love life and lose it, hate life and gain it—suggests that the death of Jesus is the paradigm of human existence.

The lack of wine is replaced by an overwhelming supply, failure is replaced by rejoicing. Hence, the turning of water into wine is a sign of God's *doxa* displayed in Jesus, to be confirmed (as is suggested in the second sign when the official's son recovers from terminal illness) in the rising of Jesus from the dead. In the provision of bread for the five thousand the sign in fact is interpreted by the image of manna. The Son of man, much like the divine Wisdom, speaks as the bread of life (i.e., nourishment that ensures eternal life) but shows himself to be a person of flesh and blood who gives his body to suffering and death in order that others may have life. Those who desire life must not only listen to the divine Wisdom but must eat his flesh and drink his blood, thus feeding on the spirit-given words of Jesus. There is no way from life on earth to life in heaven except by death and resurrection. Hence, the story of Lazarus, which displays God's *doxa* when he seems to be absent or useless. The believer may indeed die with pain and distress; but if he lives in faith, he will not finally be deprived of God's power. Jesus declares himself to be the resurrection (the means by which death is overcome) and also the life that thereby becomes possible. Jesus goes to awaken Lazarus, to bid him rise from the dead and be enlightened. So, too, the death of Jesus is a transfer from the great darkness of this world to the light of the other world. The anointing at Bethany suggests that the death of Jesus is a life-giving fragrance that fills the world as the scent of Mary's ointment filled the house where Lazarus was present at table.

9

Revelation

No formal scheme can easily contain the exuberant imaginaton that produced the Book of Revelation. Nevertheless, discernment of its main structure is not too difficult.¹ It begins with a Prologue, or descriptive title, and a blessing on those who read and hear the words of the prophecy and heed what it says (1:1–3). Then it adopts the letter form to introduce the author to his readers, recounts the author's vision that has prompted the writing, and sets out Letters to Seven Churches in the province of Asia (1:4–3:22). The long section 4:1–22:5 contains the Apocalyptic Visions. After a change of scene from earth to heaven in chapter 4, six seals are broken with unpleasant consequences on earth and a seventh seal introduces seven angelic trumpeters. The trumpeters sound four times with rather similar effect, and then three times more—with increasingly disastrous and prolonged results. Even so, the seventh trumpet call leads up to seven plague angels who (this time briskly) pour out the nasty contents of seven bowls—after which the 'machinery' of the visions is discarded.² The Avenger of Blood arrives, and thereafter the narrative moves onward swiftly to the thousand-year reign of the martyrs and the creation of a new heaven and a new earth. The final movement is indeed swift in the telling, whereas the previous movement (during the triple seven of disasters) has been long drawn out and much interrupted. The interruptions are of two kinds: acclamatory hymns and close-up views (as it were) of particular episodes. They are placed at strategic points of the formal structure (e.g., 7:1–17 before the seventh seal is broken, 10:1–11:14 before the seventh trumpet, 12:1–14:20 after it and 17:1–19:10 after the seventh bowl). It is not difficult to conclude and the 'machinery' of seals, trumpets and bowls, though not insignificant, exists to make occasion for the 'interruptions'. (To that thought I shall return.) The Conclusion, 22:6–21, is also presented as a vision that serves to give authority to 'these words' and to make them public as the distinguishing test between the evil and the righteous, those inside and those outside. Those who hear and heed receive a blessing; and the whole work is protected by a curse on anyone who adds to it or subtracts from it. The book concerns 'what must soon take place'; and the speaker in the vision says 'I am coming soon' (22:7, 20), to which the answer is 'Amen. Come, Lord Jesus'.

There are numerous references to the death of Christ in Revelation. What they mean cannot be decided without some attempt at exegesis, although the exegetical problems of the book are notoriously perplexing. Fortunately, the problem is somewhat simplified when it is observed that almost all references to Christ appear either in the less-perplexing sections (the Prologue, Letters, and Conclusion) or in the passages that interrupt the Apocalyptic Visions.

Prologue

The Prologue says that God gave Jesus Christ a disclosure *(apokalypsis)*, to be shown to his servants, of what must soon take place. This is matched by the emphatic repetitions of 'soon' in the conclusion. Whatever impression is made by the recurrent machinery of the book, the author could not have said more plainly that he was writing about immediate concerns. On these matters Jesus Christ had made a disclosure through his angel to his servant John. In this writing John has borne witness to 'the Word of God and to the witness of Jesus Christ', namely the things which he saw (in these visions). The phrase in quotation marks is found also at 1:9 and 20:4, and (abbreviated) at 6:9—referring to the suffering of those who had spoken the word of God (in the standard missionary sense) and borne testimony to Jesus. So indeed they had, but in reliance on him who is the avenging Word of God (Rv 19:13) after having borne testimony as the slain Lamb. Then finally the 'disclosure' is renamed 'this prophecy' (as in 22:7, 10, 18–19) where *prophēteia* is understood not so much in the Greek sense (the gift of interpreting the will of the gods) but in the Hebraic sense of the prophetic oracle which discloses what your present situation really is, and what its consequences will be.[3]

Letters to Seven Churches

The Letters section (1:4–3:22) opens with a blessing, conveyed by John to the seven churches in Asia, from the Almighty and the seven (perhaps angelic)[4] spirits that are before his throne and 'from Jesus Christ the faithful witness, the first-born of the dead, and the ruler of Kings on earth' (Rv 1:4–5). At once the familiar death and resurrection pattern asserts itself by referring to Christ's martyrdom (*martys* has this uniform meaning in Revelation)[5], resurrection and consequent lordship. Moreover, it is given its selected biblical setting from the Book of Zechariah, which is much concerned with the prophetic intercourse with heaven and is developed in terms of Psalm 89,[6] which praises God for his steadfast love to Israel and David but reproaches him for concealing it so long. Then comes the first of the numerous doxologies of Revelation:

> To him who loves us
> and has freed us from our sins by his blood
> and made us a kingdom,

> priests to his God and Father
> to him be glory. (1:5^d–6)

If this were expressing Pauline teaching, whereby Christ's death was his
supreme act of love, we should expect *loved* instead of *loves*. That, indeed, is the
reading of two important miniscules and the numerous manuscripts that follow
the text of Andreas of Caesarea's commentary.[7] But divine love is not a prominent theme of Revelation, and its meaning here (as in Rv 3:9, *agapaō*, and 3:19,
phileō) is God's support and approval. Another variant is more difficult. Should
we read *freed us from (lysanti ek)* or *washed us from (lousanti apo)* our sins? The
textual and exegetical arguments favor *freed*,[8] but I doubt that it greatly matters—for two reasons. First, neither phrase is common, though there are parallels
of thought: to be freed from the restricting consequences of sin is expressed by
apolytrōsis (redemption) passages (e.g., Eph 1:7; Ti 2:14; 1 Pt 1:18–19; and to
be washed from the stain of sin by cleansing passages (e.g., Ps 50:4; Heb 1:3;
9:14, 22—23; 1 Jn 1:7). The freedom metaphor is perhaps more suitable to their
becoming a kingdom, and the cleansing metaphor to their becoming a priesthood
(see also Rv 7:14). Second, Revelation speaks much of repentance and restoration
to God's favour but without using the word *sins* (elsewhere only Rv 18:4–5, of
Babylon). Thus, it must be concluded that 'freed or washed from our sins by the
blood of Christ' was not devised by the author of Revelation but was used by him
to give a traditional cast to his dominant theme of the martyr death of Christ in
saying that it cancelled our sins.[9] In verse 7 the doxology has a scriptural
response: the coming of the king is announced (Dn 7:13) and his vengeance is
implied by the wailing of the tribes of the earth (from Zec 12). The solemn self-
declaration of the Almighty in verse 8, 'who is and who was and who is to come'
confirms the coming of the king as the coming of the Alpha and Omega.

Having thus established Christ's divine status, John can now use traditional
angel imagery to picture his risen majesty in Rv 1:12–20. The vision takes place
on the Lord's day, the day when Christians met to celebrate the presence and
power of the risen Lord.[10] The visionary experiences a mystical death and resurrection (Rv 1:17, Farrer), and the vision reassures him:

> Fear not, I am the first and the last [see 1:8],
> and the living one;
> I died, and behold I am alive for evermore,
> and I have the Keys of Death and Hades. (1:17–18)

Thus, Revelation begins with a vision of the Risen One and ends with the elaboration of two resurrections that are effected by the keys of Death and Hades.
One produces the Great Sabbath of a thousand years,[11] the other produces the
First Day of a new heaven and a new earth (Rv 20:1–22:5). Between the opening
and closing visions of the risen Lord, other visions explore the consequences of
his death.

But before we come to them, we must briefly notice the seven letters in 2:1–
3:22. Each letter is dictated by Christ who seems to have been acting like one of
the angelic inspectors of Zechariah. He presents his qualifications, somewhat ran-

domly selected from the foregoing verses, and then gives a visitation report on each community. In writing to Smyrna he is 'the first and the last, who died and came to life' (Rv 2:8); but that designation seems to have no priority over the self-designations of the other letters. All one can say is that death and resurrection is relevant to all the community judgements, though the fact is not formally indicated. Thus, the significant point to be noted is that the letters are not pastoral communications so much as warnings—before the scene changes to 'the hour of trial [*peirasmos*] which is coming upon the whole world' (Rv 3:10). It is the time for judgement to begin with the household of God, before it moves to those who do not obey the gospel (1 Pt 4:17).

Apocalyptic Visions

The case against the disbelieving world begins in Revelation 4 with a throne vision, more elaborate than the vision in Dn 7:9–28 but basically similar: 'This throne room is essentially a royal court where God is Judge, King, Magistrate and Executive' (Ford, p. 77). It is a trial court (Farrer, p. 89), or, better still, it is a religious court constructed on the principles of ancient Near Eastern theocracy. Hence, it is scarcely surprising that it fails to conform to modern Western judicial proceedings or that it experiences major enthusiastic and even riotous interruptions. Throughout Revelation, the throne means not simply the ceremonial seat of majesty but Majesty itself.[12] The twenty-four elders act as a kind of Privy Council whose chief duty is to articulate praise (and therefore approval) of the divine acts of creation and redemption (Rv 4:10–11, 7:11, 11:16, 19:4) and, by singing a new song, to confirm the divine approval of the deeds of Christ (Rv 5:8–9, 11–14; 7:13–17; 14:3–5). The four living creatures—lion, ox, man and eagle—symbolically represent the forms of living energy available in antiquity, to be placed alongside the nonliving forces of earthquake, wind, fire and water.

In the trial scene in Dn 7:10 'the court sat in judgement, and the books were opened'. In this case, however, the matter is not so simple. John saw, in the right hand of him who was seated on the throne, a scroll written within and on the back, sealed with seven seals'; and no one was qualified to break the seals and read its contents. It ought at once to be made plain that John *saw* no visual phenomena, not even in vision or imagination. It is true that the language of Revelation has immense visual quality; but that is a common Hebraic way of expressing intellectual perceptions without using abstract language.[13] If the student of Revelation tries to *draw* what John describes, the results are grotesque and incompatible. Western readers must learn the skill of translating the pictorial language into its propositional equivalent without losing too much of its imaginative vigour. Thus, it is beside the point to discuss methods of writing ancient documents or to debate the clumsiness of seven seals, or to ask why God himself could not have broken them. The writer means that the business of the court was absolutely inhibited for want of a competent and irreproachable witness to initiate proceedings (which was the ancient custom). Then comes the famous coup de theatre. A witness is indeed expected, namely, the victorious Lion of the tribe of Judah; but instead there appears the victimised Lamb (Rv 5:5–6).

If it is asked 'Why a Lamb instead of a Lion?' the simple and probably sufficient answer is that the lion terrifies the shepherds, seizes the lambs, and carries them off (see Jer 27(50)44–46 LXX).[14] The apparently defenceless lamb is the antithesis of the aggressive lion. This Lamb, however, was 'standing, as though it had been slain'—a statement difficult to make into a visual image but intelligible if understood as indicating an unblemished male person, wholly devoted in sacrifice to God (following the standard rules for choosing and offering the *ᶜolah*, the whole burnt offering).[15] Compare Jer 51:40, 'I will bring them down like lambs for the slaughter', and Is 53:7 of the servant, 'like a lamb that is led to the slaughter', though there is little enough of the servant's submissiveness in this apocalyptic Lamb. Indeed, for that reason exegetes have been inclined to discover the Lamb's origin in the aggressive horned rams who symbolically represent the forceful leaders of God's people in times of danger.[16] If that suggestion simply means that a member of the sheep family can aptly represent a leader, it is true enough; but little help can be expected from the farrago of Enoch and the versional problems of Test. Jos. (which say nothing about death or sacrifice) in identifying the Lamb who was slain. The Passover lamb, inappropriate at Rv 1:5 (see p. 326 and n. 9), is somewhat more acceptable here as the protector of God's people. But even so, the sacrificial blood was an apotropaic rite against the destructive forces unleashed by God[17] and a guarantee that God's people would not be harmed. In Revelation however, those who hold to the word of God and the testimony of Jesus must expect to suffer and die. Neither the Passover nor the other ritual animals of Jewish tradition fit the Lamb of Revelation, and perhaps John even chose the word *arnion* to keep him related but distinct.

The person called the Lamb is found worthy to open the sealed book. He is qualified because he has seven horns (i.e., he is fully empowered) and seven eyes (i.e., he is fully informed by the seven investigative spirits of God). But more than that: he had been slain and, by his blood, had purchased for God persons from every tribe and tongue and people and nation, and had made them a royal house to serve God as priests, (destined) to reign on earth (Rv 5:9–10). Thus the Lamb appears as the champion of God's people, uniquely qualified to represent them because he himself has suffered martyrdom. He presents the indictment against mankind's wickedness in the eastern part of the Roman Empire. Each seal opened is an item of the accusation. It is commonplace to interpret the seals simply as divine punishment, but there are strong arguments against that view. First, the fifth seal is not a punishment but a demand to be avenged from the souls of 'those who had been slain for the word of God and for the witness they had borne' (Rv 6:9–11). Presumably, that is why the twenty-four elders had been carrying bowls representing the prayers of the saints when they acknowledged the worthiness of the Lamb. Second, why indeed was the Lamb worthy to open the seals unless his martyrdom made him the proper witness and accuser? Surely he was not, as it were, allowed to press seven destructive buttons simply as a vengeful reward for martyrdom! Third, in the prophetic tradition, the prophet first indicts the sinful people before pronouncing God's verdict on them. But in Revelation the indictment is lacking unless Jesus presents it in himself and in the seals.[18] Fourth, it is commonly recognized that the contents of the seals are closely related to the synoptic Apocalypse (Court, pp. 49–54), which portrays disasters but not punishments. For these reasons it is necessary to interpret the first four seals as an indictment of (1) aggression and domination, (2) war and slaughter, (3) scarcity of food and the manipulation of markets, and (4) the policies that absorb a quarter of the world into the domain of death and Hades. And, as people of late antiquity believed, such disorder cannot be confined to human society: it spreads to other living creatures (Rv 6:8) and to the elemental forces of the world. Hence, the sixth seal shows the fabric of the world disintegrating and the consequent terror of all classes when exposed to 'the face of him who is seated on the throne' and to 'the wrath of the Lamb'.[19] They are terrified not by an arbitrary punishment from God or by the peculiar severity of the Lamb but by the terrible disclosure of what they are and what they have done. Of course, they have to be *allowed* by God to do these things—but that is the meaning of 'wrath' in Revelation just as much as it is in Romans (see Rm 1:18–32). They have not kept the word of God, they have defiantly gone their own way; and now they must take the consequences of their obstinate choice, which is indeed the divine punishment.

The Great Multitude

That view of God's nonintervening wrath has, of course, disadvantages for victims of the perversely wicked; but victims are not forgotten. The seventh seal is left unbroken until God's people have been given the protection of the court. The avenging angels are restrained until the servants of God have been sealed upon their foreheads and promised protection (Rv 7:1–3, 16–17). After 144 thousand out of every tribe of the sons of Israel have thus been sealed, John sees a vast throng whom no one could count, standing before the throne and the Lamb. Robed in white and shouting a victory song, they have passed through the great ordeal, and have 'washed their robes and made them white in the blood of the Lamb' (Rv 7:14). That apparently grotesque image is simple enough if we resist the attempt to visualise it. Washing robes is a ritual preparation for encounter with the divine (Ex 19:10; see rules in Lv 11:24–25, 28, 32, 40 and Leviticus 15 passim; also Rv 22:14). Is 1:18 had promised that 'though your sins are like scarlet, they shall be as white as snow'. Hence, John means that the blood of Jesus cleanses from all sin (prosaically stated in 1 Jn 1:7), implying that these Christians have regained their purity either by martyrdom or by reliance on the death of Jesus. It is important to notice that this conviction is not phrased in terms of the forgiveness of sins (which does not occur in Revelation) and envisages a different kind of relation between God and his people. Nor is the death of Jesus (to return to Rv 5:6, 9, 12) represented as an atoning death. Instead, by his blood he purchased (*ēgorasas*) people for God so that they became his precious possession (Ford, p. 94). The verb *agorazō* comes again in Rv 14:3–4, where the Revised Standard Version translates as 'redeemed' (here 'ransomed'). Neither this verb, nor its rare intensive form, carries explicit atoning significance:[20] the 'buying back' metaphor implies that Christ, at great cost to himself, recovered God's people for God and allowed them to become what God intended them to be. If we want to know how Christ's death effected such a rescue, the answer lies in the plan of the book. By making Christ a martyr and by inflicting the same cruelty on the saints, the hostile powers overreached themselves, they are trapped by their own wickedness, and the verdict in God's court goes against them.

After the reassuring hymn of the vast throng, the seventh seal is broken, and the court takes a recession: 'There was silence in heaven for about half an hour', perhaps for the presentation of petitions from the saints (Rv 8:1–5). Then the first four trumpets are sounded, announcing the disastrous consequences of human wickedness on the ordering of the universe as regards fire, earth (a volcanic eruption and tidal wave), water pollution, and air pollution.[21] A flying eagle screams a warning against three forthcoming woes, of which the first two are mass invasions of insects and cavalry, marked by the fifth and sixth trumpets. But before the seventh trumpet sounds, there is another significant interlude, a replanning of the business of the court. In Rv 10:1–11 seven thunders sound, presumably a further series of disasters; but John is forbidden to record them. Instead he is made to eat an open scroll, sweet to the taste but bitter to the stomach (like Ezekiel's scroll containing lamentation, mourning and woe, Ez 2:8–3:3). In other words, the time is over for viewing vast general disasters. There must be no more delay (verse 6): John must himself become a witness and undertake the disagreeable task of prophesying against peoples, nations, tongues and kings. This must therefore be a political denunciation even if (for discretion's sake) it is done in coded language.

The Imperial Threat

The political enemy, of course, is the Roman imperial power and its great urban centres, particularly Rome itself.[22] In Rv 11:1–2, John is required to measure the temple of God but not the outer court, which is given over to the nations who will trample on the holy city for 3½ years. Thus, the language of Dn 7 provides a covert reference to the Roman investment of Jerusalem from 66 to 70 C.E. Then the language of Zechariah 4 introduces the two witnesses, who, despite their prophetic power, are killed by 'the beast that ascends from the bottomless pit': 'And their dead bodies will lie in the street of the great city which is spiritually called Sodom and Egypt, where their Lord was crucified' (Rv 11:8). Since the crucifixion took place in (or near) Jerusalem, the Roman reader knows that John is writing about some far-off city of the Jews. But those who read the code (especially the words 'the great city') know that he is writing about the martyrdom of Peter and Paul in Rome and so can identify the beast from the pit. Perhaps he is writing about all Christian martyrs. Wherever Christians are martyred, there Jesus is crucified; but as they share his cruel death, so they share his spirit-given resurrection and the earthquake that marked it (verses 11–13; Mt 27:54). Not only does human wickedness disturb the fabric of the world; the resurrection of the saints shakes the structure of the empire. Hence, 'the rest were terrified and gave glory to the God of heaven' (Rv 11:13).

That apparent capitulation prompts the seventh trumpet and a triumphant response in the heavenly court: 'The kingdom of the world has become the kingdom of our Lord and of his Christ' (verse 15). It is time for the dead to be judged and for the saints to be rewarded and for destroying the destroyers of the earth (verse 18). But such rejoicing is premature and is brought to an end by a new outbreak of imperial violence. This is portrayed, sometimes with excessive obscurity, in Rv 12:1–14:20; but for this enquiry only a few references need be studied.

In Rv 12:1–6 a woman clothed with the sun is harassed by a red dragon who intends to devour the child she will bear. She brings forth a male child, who is to rule all the nations with a rod of iron. The child is caught up to heaven (verse 5), and the woman escapes to a refuge in the wilderness. In a Christian writing this looks like a reference to the birth and ascension of Christ. There is no mention of his death, which is not surprising if the woman represents Israel before and after the fall of Jerusalem (see Court).

The omission is made good in Rv 12:7–12, where Michael and his angels defeat the dragon and his angels in the war in heaven. The victory song celebrates the triumph of God's salvation, power, and kingdom and the authority of his Christ; 'for the accuser of our brethren has been thrown down; and they have conquered him by the blood of the Lamb and by the word of their testimony, for they loved not their lives even unto death'. This must be understood to mean that the Roman *imperium* has lost any divine sanction it may claim, having been exposed and humiliated by the devotion of the martyrs. Nevertheless, the war on earth has still to be won. In the fashion of Daniel 7, the beast from the sea in Rv 13:1–10 represents the supreme earthly Roman power, which people will still worship unless their names are written 'in the book of life of the Lamb who was slain from the foundation of the world' (Rv 13:8). That translation reproduces the Greek word order (as in AV, RV, NIV) but not perhaps its sense, for which we must consider the verses immediately following: 'If anyone is to be taken captive, to captivity he goes; if anyone slays with the sword, with the sword must he be slain' (God's refusal of indulgence to sinners, from Jer 15:2). Thus, John's intention is to say that people are or are not written in the Lamb's book of life by God's unvarying decision. That is a constitutive principle of his creation. Such an understanding is reinforced by 'dwellers on earth whose names have not been written in the book of life from the foundation of the world' in Rv 17:8. Hence in 13:8 RSV translates 'everyone whose name has not been written before the foundation of the world in the book of life of the Lamb that was slain' (so also NEB, NJB, GNB, TNT and Bousset, Swete, Allo, and Farrer). That translation appears to include an extreme form of predestination (not surprising in the polemical rhetoric of the writer) and to exclude the pretemporal significance of Christ's death, such as may be found in 1 Pt 1:18–20. To avoid those implications Charles, Caird, Beasley-Murray, Ford, and Sweet return to the Authorized Version.) But such a divisive judgement does not fully represent what is being said. If those destined to enter life are written in a book, it is the Lamb's book (see Rv 20:12, 15; 21:27). Entries into that book are decided by adherence to the Word of God and the testimony of Jesus. If the book is pretemporal, so also is the death of Jesus that Christians share (as Lohmeyer implies). John is asserting that the death of Christ, which calls for martyrdom and promises life, is not simply a historical accident that happens to be effective at that stage of the Roman *imperium* but a constitutive act for the creative order and human history. Unless he believed that, he would not be writing with confident hostility against the empire.

The attack continues in Rv 13:11–18 (the false lamb and the beast with its number 666, implying that the court has taken its true measure) and is balanced in Rv 14:1–5 with the true Lamb and the 144 thousand, who have been redeemed from mankind as first fruits for God and the Lamb. Amid much angelic activity in Rv 14:6–20 the time for judgement to be executed is announced, namely, the destruction of Babylon and the torment (in the presence of the Lamb) of those who worship the beast. Yet still the saints are called to endurance, and those who are still to die are assured of God's blessing, even as the order is given for the harvest to be reaped. It is a grape harvest: 'The great wine press of the wrath of God was trodden outside the city, and blood flowed from the wine press, as high as a horse's bridle, for one thousand six hundred stadia'. But whose blood? The imagery, clearly drawn from Is 63:3–6 and Joel 3:13, suggests the blood of God's enemies—with confirmation from the rider on the white horse in Rv 19:15, who treads 'the wine press of the fury of the wrath of God the Almighty' (Beasley-Murray). But some argue that John has recast the vengeful prophetic imagery (as often he adapts imagery to his own convictions) so that here it indicates the blood of Christ (suggested by *outside the city,* as in Heb 13:12, though some of the prophets have the Gentiles judged near Jerusalem [Ford]) and the blood of the saints and martyrs of Jesus as in Rv 16:6, 17:6 (Caird, Sweet). Perhaps John meant his symbol to be ambiguous, not more than a dramatic contrast to 'the river of the water of life', which flows from the new city in Rv 22:1. Whatever its meaning, it gives no clear answer to the question (which must be asked again later): how is the slain Lamb related to this picture of intolerable slaughter?

The theme is continued in what follows. Revelation 15 introduces the seven bowls containing the final wrath of God with a triumph song for the revealing of his judgements (verse 4). In Revelation 16 the bowls are poured out; though—compared with what is still to come—the writing is now rather perfunctory. An angel justifies the fourth judgement because 'men have shed the blood of saints and prophets' (verse 6). The seventh bowl contains a standard collection of disasters, made more specific in Revelation 17, which demonstrates the judgement of the great harlot (who is drunk with the blood of the saints and the blood of the martyrs of Jesus, verse 6), namely, 'the great city which has dominion over the kings of the earth' (verse 18). The taunt song in Revelation 18 against the imperial power and the mercantile, maritime economy of Rome ends with the accusation, 'In her was found the blood of prophets and saints, and of all who have been slain on earth' (Rv 18:24). The taunt song, of course, is intended to induce the destruction, not to celebrate it. The celebration begins in Rv 19:1–10 and includes the first intimation of the forthcoming marriage of the Lamb to his Bride.

The Rider on the White Horse

Before that, however, there is savage work to be done. The heavens open to disclose a rider on a white horse, no doubt intended to recall the first rider of the
first seal in Rv 6:2. But this rider explicitly revives memories of the risen Christ
who broke the seals. He is faithful and true, like 'the Amen, the faithful and true
witness, the beginning of God's creation' (Rv 3:14), who judges and fights in
righteousness (i.e., the right cause). His eyes are like a flame of fire (as they were
in 1:14 and 2:18) and on his head he wears many diadems (at least rivalling the
dragon and the beast from the sea in Rv 12:3 and 13:1). He bears a name (an
office or authority)[23] that no one knows but himself (see Rv 2:13); that is, he has
a task to perform that cannot be delegated to another and that no enemy or rival
can usurp by using his 'name'. His robe dipped in blood (or sprinkled with it,[24]
as Is 63:3) is perhaps a compressed reference to the martyrdom that had made it
possible for the saints to wash their robes and make them white in the blood of
the Lamb (Rv 7:14; see the white-robed armies of 19:14).[25] His name is the Word
of God, in the sense of Wis 18:15–16: 'Thy all-powerful word leaped from
heaven, from the royal throne, into the midst of the land that was doomed, a
stern warrior carrying the sharp sword of thy authentic command, and stood
and filled all things with death, and touched heaven while standing on earth' (a
hellenistic version of the killing of the first-born). Even if his destructive power
is only verbal (a sharp sword issuing from this mouth; see Is 11:4), his rule will
be ruthless (he will rule them with a rod of iron; see Ps 2:9 and the description
of Messiah in Ps. Sol. 17:26–27). The ruthless verbal command of an absolute
ruler can destroy thousands. The rider on the white horse will tread the wine
press of the fury of the wrath of God the Almighty, thus exercising his power as
King of kings and Lord of lords (as in 17:14). It is clear that this rider is the risen
Christ, and that he comes as the avenger of blood.[26] From now on, there is no
more mention of blood, not even in the gruesome feast of Rv 19:17–21, where
carrion birds, with customary ecological efficiency, dispose of the slain.[27] Until
this point blood has been a recurrent, almost obsessive theme. There was the
blood of Christ, which set us free and purchased us for God, in which the saints
made white their robes, by which they conquered (Rv 1:5, 5:9, 7:14, 12:11,
19:13). There was the blood of the saints shed by the Roman *imperium* (Rv 16:6,
17:6, 18:24), concerning which the martyrs pleaded, 'How long before thou wilt
judge and avenge our blood on those who dwell on earth?' (Rv 6:10), until vengeance was planned and carried out by the avenger of blood (Rv 14:20, 19:2).
All this bloodshed came from the prejudice, fear and misjudgement of human
beings and from the ruthless arrogance of political power. But there was more.
Blood entered into John's imaginative portrayal of natural disasters (with help
from Jl 2:30–31 and the first Egyptian plague, Ex 7:17–21) so that the four
constituent elements of the world (see above p. 327) are contaminated by blood:
earth by human blood as already mentioned, the air after an earthquake that
obliterates the sun and reddens the moon (Rv 6:12), fire mixed with blood from
a volcanic eruption (Rv 8:7) and water turned to blood (Rv 8:9, 11:6, 16:3–4).
It would be wrong to dismiss these statements as fanciful or to offer naturalistic

explanations for them. John's intention is surely to mark these violent disturbances of the elements as abnormally terrifying and destructive of human life.
An avenger of blood is needed not only for blood shed in slaughter but also for
blood shed in catastrophe. That is why, at the end, earth and sky fled away from
him who sat on the great white throne and a new heaven and a new earth, with
no more sea, was required. John's preoccupation with blood presents the fact of
destruction in its horrifying extent. His concentration on martyrdom shows how
the saints may hope to survive, but his repeated songs in praise of the Lord God
the Almighty are really a persistent plea that destruction should not be the last
word about human existence. And it is not. Although Death and Hades are repulsive and terrifying figures in Revelation, they are made to give up the dead (Rv
20:13—Christ having the keys, Rv 1:18); and in the end Death is no more (Rv
21:4). Indeed, the whole book is concerned with the God who lives for ever and
therefore with life: the tree of life (Rv 2:7; 22:2, 14, 19), the crown of life (Rv
2:10), the book of life (Rv 3:5; 13:8; 17:8; 20:12, 15; 21:27), the springs of living
water (Rv 7:17; 21:6; 22:1, 17), and the breath of life (Rv 11:11).

The New City

The myth of life, in the form of a city to replace the ruined 'Babylon', comes in
Rv 21:1–22:5. The new city is no more and no less than the structure of life
where God provides the only illumination by means of the Lamb. Therefore, it
excludes everything unclean, abominable practices and falsehood and includes
only those who are written in the Lamb's book of life (Rv 21:27, see 21:7–8).
But its gates are never shut. Its population is not confined to a predetermined
number. It is for the nations. The kings of the earth bring the glory and honour
of the nations into the city, and the tree of life produces leaves for the healing of
the nations. Its governing authority is the throne of God and the Lamb. The
saints and martyrs have had their share of reigning with Christ, but even their
symbolic thousand years is not sufficient to finish the problem of sin and Satan.
So Christ hands over the kingdom to God (as in 1 Cor 15:24–28), and a new
heaven and a new earth can be seen.

Summary

Revelation uses old traditional formulas and in particular works out the death
and resurrection pattern in apocalyptic imagery. Christ is the protomartyr, and
his death cancels sins. For those who die with him, purity is restored. By his
death, God's suffering people are recovered for God and assured of his protection.
Those who still must suffer and die will conquer by the blood of the Lamb, so
named because he is the prototype of an unblemished life wholly offered to God
in sacrifice. The martyrdom of Christ and his saints discredits the Roman *imperium* (which is charged with complete responsibility for their blood) and is established as the activity that constitutes a new creation and a new existence for
mankind. The Lamb therefore indicts Roman society, exposes the disorder of the
earthly and heavenly worlds, relentlessly pictures the blood that is shed, and

comes to avenge it. The power of the risen Christ does not simply promise rescue for suffering saints and new life after disaster; it promises compensation and the destruction of the destroyers. When John wrote down what would soon come to pass, he saw his violently disturbed world rapidly moving towards, first, the Great Sabbath rest expected by Jews and then (more comprehensively) a new First Day for the kingdoms of the world.

10

Survey and Summary

THE PROTO-PAULINE EPISTLES

Two features are common to all six epistles: on one hand the use of theological formulae about the death and resurrection of Christ and on the other hand the presentation of his death and resurrection as a model of Christian experience in the present age.

Theological Formulae

These formulae thrust themselves on the attention in several ways: some are explicitly identified as what Paul received and passed on, some as statements of what Christians believe or are expected to understand. Others (often in a formal setting) look like standard formulations, repeated with greater or lesser variations and regarded as familiar. Many stand out because of their stylistic features (e.g., compactness, parallelism, inversion), marking them off from the context and making them memorable. Some were taken over by Paul from those who were in Christ before him; sometimes he says they were, in other cases they clearly depend on expression not otherwise characteristic of his teaching. But others may well have been Paul's own formulations. He was certainly capable of composing strikingly pregnant expressions of the gospel; and it is significant that he adopted this method of teaching. One formula with variations, used in full or in part, is very common and can be set out as follows.

'He Died and Rose'. This appears in an overtly confessional form in 'We believe that Jesus died and rose again', but it is given to provide reassurances about those who have died: 'God will bring with him those who have fallen asleep' (1 Thes 4:14). In the same argument it is said that 'God has not destined us for wrath but to obtain salvation through out Lord Jesus Christ, who died for us so that . . . we might live with him' (1 Thes 5:9–10). This time the death is

336

not barely stated but is accompanied by *for us;* yet emphasis falls on his being alive and our living with him. In a rather more developed form there is Rm 5:10, 'We were reconciled to God by the death of his Son: much more, now that we are reconciled, shall we be saved by his life'. The phrase is end-heavy. Hence also, 'Christ died and lived again, that he might be Lord both of the dead and of the living' (Rm 14:9). Death may be a necessary condition of resurrection, or the death may achieve something of its own; but the quotations press in the direction of the living Lord and those who live with him.

'He Died. . . . He Was Raised'. This variation makes explicit what was understood in 'He died and rose', that Jesus was raised by the power of God. The piece of early tradition presented in 1 Cor 15:3–5—for the exclusive purpose of answering questions about the resurrection of Christ and believers—elaborates the simple formula. 'He died' is verified by 'He was buried'; 'He was raised' is verified by 'He appeared'. Both death and resurrection are referred to scripture: death being interpreted by scripture dealing with sins, resurrection by scripture foreseeing the third day. Here, too, the death of Christ carries its own significance; but the reason for this citation lies in the puzzle of resurrection. Much the same is true for other references of this type. Thus, 'He died for all, that those who live might live no longer for themselves but for him who for their sake died and was raised' (2 Cor 5:14–15). Death is the necessary stage in his life-giving operation. Indeed, baptism teaches the same: 'We were buried therefore with him by baptism into death, so that as Christ was raised from the dead by the glory of the Father, we too might walk in the newness of life' (Rm 6:4). It is true that the formula in Rm 4:25 attaches equal weight to both clauses: 'He was put to death for our trespasses and raised for our justification'; but it is quoted only to clinch the reference to resurrection in the previous verse. Rm 8:34 quotes, 'Christ Jesus who died, yes, who was raised from the dead,' only to add, 'who is at the right hand of God, who indeed intercedes for us'. The hymn in Philippians impressively makes Christ Jesus become 'obedient unto death, even death on a cross' but reserves the climax for his exaltation by God and his bestowal of the supreme Name (Phil 2:8–9). Only one reference leaves the two clauses equally poised: 'He was crucified in weakness, but lives by the power of God' (2 Cor 13:4)—that simply because it mirrors the apostolic weakness and power.

'God Raised Him from the Dead'. The resurrection part of the formula can even appear as a summary statement of Christian conviction: 'If you confess with your lips that Jesus is Lord and believe in your heart that God raised him from the dead, you will be saved' (Rm 10:9). The sentence 'We who believe on him that raised from the dead Jesus our Lord' is a description of Pauline Christians (Rm 4:24), not surprising since they knew that 'God raised the Lord and will also raise us by his power' (1 Cor 6:14). Paul's apostleship is 'through Jesus Christ and God the Father, who raised him from the dead' (Gal 1:1). And the sketch of Paul's missionary proclamation—at least its eschatological expectation—in 1 Thes 1:9–10 has the converts turning to serve a living and true God, and waiting 'for his Son from heaven, whom he raised from the dead, Jesus who delivers us from the wrath to come'. The saving power of Jesus seems to derive from his resurrection as God's Son and his expected arrival from heaven. No doubt Paul's words were pushed in that direction by the eschatological anxiety of the Thessalonian community; no doubt this emphasis on the resurrection is to be balanced by reference to the death of Christ, though that community too (see p. 337) is directed to the risen life. In all these references Paul presumes the death of Christ, but specifically theological interest in the death is not activated.

'He Died for Us'.   But here it is—to some extent. Christians can be described (perhaps by a familiar turn of speech) as brothers for whom Christ died (1 Cor 8:11, Rm 14:15). He 'gave himself for our sins to deliver us from the present evil age according to the will of our God and Father' (Gal 1:4). His self-giving arises out of love—either the love of Christ himself ('He loved me and gave himself for me', Gal 2:20; see 3:13, 'having become a curse for us') or God's love for us. While we were weak, ungodly and sinful Christ died for us (Rm 5:6, 8). Even these very positive statements about Christ's death are not really complete without reference to the resurrection. Christ gave himself for Paul, and Paul has been crucified with him; so that Paul himself no longer lives, but Christ *lives* in him. Christ, indeed, has died for sinners, who are thereby reconciled to God: much more, now that they are reconciled, will they be saved by his life. Thus, to make the divine love effective Jesus died for the benefit of sinners to deal with their sins and their plight in the present evil age, to reconcile them to God, and to make possible a risen life by means of which they will be saved.

'God Required Him To Die'.   In Rm 4:25 (see p. 337) the literal translation would be 'He was handed over', which may imply that *God* handed him over. Precisely that is said in Rm 8:32: 'He did not spare his own Son but handed him over for us all'. And in Rm 3:25 it is said that God put him forward (or intended him) as a means of atonement by his blood. Thus, not only is the death of Jesus a demonstration of God's love, it is also an effective means of restoring sinners to enjoyment of that love, in other words the business of atonement is not to rekindle God's love but to rekindle sinners' desire for it.

Thus, it becomes evident that Paul uses the dying/rising formula and its variations to maintain without reservation the theological importance of Christ's resurrection for believers. He is being faithful to a pre-Pauline emphasis, as can be seen from his description of the gospel of God at the beginning of Romans, where he justifies his apostleship. The gospel concerns God's Son, 'who was designated Son of God in power according to the Spirit of holiness by his resurrection from the dead, Jesus Christ our Lord' (Rm 1:4). The resurrection is explicit, the death is presumed. And that corresponds well with what we read in 1 Thessalonians, where the death of Christ is no more than the precondition of resurrection and the significance of the resurrection is interpreted by the Parousia. There is a strong sense of this present evil age (as it is called in Gal 1:4), of a world moving towards catastrophe. But God does not intend us to endure the dreadful consequences (his 'wrath'). We are to be saved from that by Jesus Christ, who died for us, so that whether we wake or sleep, we might live with him. The risen Lord is our protector. His dying for us is to secure benefits for the many by the sacrifice of the one, to overcome the alienation of our sins and make us acceptable to God, to reconcile us to God, to show the way that all must go, to instal us as members of the community that is to be protected from the wrath. But when the resurrection is central and is interpreted by the Parousia, everything has yet to be done. We wait for the catastrophe to happen; when it does, we shall see what Christ's lordship can do to protect us—and Paul does not abandon that conviction. We await a saviour from heaven who will transform our humiliation into glory (Phil 3:20–21). But Paul also begins to interpret the resurrection by the crucifixion and to discover what the Lord has already done.

The results can be seen in two notable passages in Romans. In chapter 5, Paul's difficulty in formulating what he has to say throws extraordinary emphasis on the point at which God's love is displayed: 'While we were weak and ungodly, not upright and sinful, Christ died for us. By his blood we have *now* been made acceptable to God; all the more shall we be saved by him from the wrath. When we were God's enemies, we were reconciled to him by Christ's death; all the more by his life shall we be saved. Not only that: we are boldly confident in God through our Lord Jesus Christ, for through him we have *now* accepted reconciliation' (to paraphrase the essentials of Rm 5:8–11). Moreover, that bold confidence is displayed in Rm 8:31–39, when God is reminded of what he has already done. Since God is on our side, he can give no hearing to our accusers. Since he spared not his only Son but handed him over for our sake, he will certainly give us all things. Since we are God's chosen people, he will support our cause. Since Christ, who died and, better still, rose for us, is at God's right hand, he will not condemn but intercede for us. Not all the terrors of a collapsing world (or, if you like, the wrath of God) can separate us from God's love in Christ Jesus our Lord.

The references to the blood of Christ recall the use of blood in protective rituals, and that is significant. Christ's protection is needed not only at the end of the age, when the wrath is unleashed: it is needed now. So when Christ said, 'This is my body which is for you', he meant 'for the benefit and protection of your community' (1 Cor 11:24). The Lord's Supper was the means of recalling the Lord Jesus, that is, making his power present—present in the power of his risen body and present in the power of his death. Hence, the Corinthian community may have felt unassailable. But power misused can forfeit protection and turn to destruction; and that was their danger. At the beginning of the letter Paul said that he decided to know nothing among them except Jesus Christ and him crucified (1 Cor 2:2), but perhaps he is belatedly saying what he *ought* to have decided. Clearly, he said much more than *Christ crucified;* by his own testimony in chapter 15 he talked at large about Christ risen and believers risen with him. No doubt his hearers became overconfident and were (as they said) reigning as kings. But to Paul they were exercising an ill-considered freedom. Christ died for all, he said, 'that those who live might live no longer for themselves but for him who died and was raised' (2 Cor 5:15). As often, therefore, as they ate the bread and drank the cup, they were proclaiming the Lord's *death* until he came (1 Cor 11:26). The words 'This cup is the new covenant in my blood' are probably reflected in the complex formula of Rm 3:24–25. Qualifying the freedom made possible in Christ Jesus is the statement that God intended him, by means of his blood, to be an atonement, that is, an inducement for God to do what he may properly do, namely, intervene and rescue sinners from the destructive consequences of their obstinate disobedience.

Finally, there are three or four examples of interchange formulae, in which two parties undertake complementary and opposite actions. So in Christ God was reconciling the world to himself (thus defusing the world's enmity towards God) and not counting their trespasses against *them* (thus abandoning his enmity towards the world). That would seem to be an important indication of how Paul came to perceive the atoning consequences of Christ's death, as an inherently dual movement: of sinners towards God and of God towards sinners. In different imagery but in the same context (2 Cor 5:19–21) the interchange is restated thus: 'Him who gave no recognition to Sin, for our sake God made the victim of Sin that we might become beneficiaries of God's saving goodness in him'. Such words presume a story: of human beings in the grip of a malicious power. His malice is diverted to God's representative, who shows him neither respect nor fear but allows himself to be entrapped so that the human prisoners can go free. It is possible, though not certain, that another Pauline remark belongs to the same story, namely, 'You were bought with a price' (1 Cor 6:20, 7:23). Christians who belong exclusively to God were bought at great cost, perhaps at the cost of sacrificing God's Son to the ruler of this age.

These interchange formulae are not competent to explain how atonement is possible but may provide ways of regarding the crucifixion so that it becomes significant rather then offensive or perverse. The view that the crucifixion is perverse was countered in 2 Cor 8:9: 'Though he was rich, yet for your sake he became poor, so that by his poverty you might become rich'. And the view that the crucifixion was offensive is not so much repudiated as defended in Gal 3:13–14: 'Christ redeemed us from the curse of the law, having become a curse for us . . . that in Christ Jesus the blessing of Abraham might come upon the Gentiles'. Human beings are pictured as unprivileged and offensive to God: to secure for them privilege and God's approval, Christ surrendered his own riches and blessing.

Protection Formulae

I have already drawn attention to the theme of divine protection, which already appears in 1 Thessalonians as the work of the risen Lord and elsewhere (most notably in 1 Corinthians) is attributed to his blood. In 1 Cor 1:18 the proclamation of the cross is God's power, perverse to those on the way to destruction, supportive to those on the way to salvation. As baptism gives ritual protection, so proclamation of the cross gives verbal protection. It is not a verbal formula, to be uttered and controlled by the believer, and its power can be destroyed if it is used to create social *sophia;* but, if it colours our understanding of life, in time of weakness, fear and trembling we may receive a demonstration of the Spirit and of power (1 Cor 2:1–5). As Passover blood protected against the destroyer, so Christ's blood protects against destructive forces outside the boundary of the community (1 Cor 5:7). The brother for whom Christ died (1 Cor 8:11) has been released from allegiance to the gods and brought within the protective lordship of Christ. As the Israelites, when passing through the sea, were protected from the enemy by a cloud and, when passing through the desert, were protected from death by water and manna (provided that protection was not abused, 1 Cor 10:1–13), so believers were protected from the gods by the cup of blessing and the one loaf (1 Cor 10:16–22). To seek the Lord's protection by eating and drinking at his table activates the most ancient and profound of human relations with the Deity. In Hebrew idiom, to drink someone's blood means to benefit by the sacrifice of another, to be protected from the danger to which the other succumbs. When Christians are told to drink the cup in remembrance of Christ they should indeed remember that they receive benefit from his death; but, more than that, they have his protection not only when they escape his sufferings but also when they share them.

Midrashim

The protection formula in 1 Corinthians 10 is attached to an exploration of the Exodus story. In 2 Cor 3:6–11 Paul relies on Moses' experience when receiving Torah on Mount Sinai to sustain his argument that the old covenant that kills (the written code) had its own veiled glory but nothing like the greater and unconcealed glory of the new covenant, which raises from the dead (the Spirit). So he manages to move from the veiled glory of the crucifixion to the open glory of the resurrection. A little later he uses the creation story of Genesis 1. When God had separated the darkness from the light, he made Adam in his own image. So when God brought Jesus from the darkness of the cross to the light of the resurrection, he displayed his glory in the person of Christ, who is the image of God (2 Cor 4:6). Then, prompted no doubt by Isaiah's reference 'to the new heavens and the new earth', Paul again interprets Genesis 1 somewhat in the manner of Wis 19:6: 'The whole creation, with all its elements, was refashioned in subservience to thy commands' (to ensure Israel's safety at the Exodus). By the death and resurrection of Christ, God rearranges all things, and anyone in Christ is a new creation (2 Cor 5:17). In Romans 4 there is a determined exposition of Abraham's faith as, in effect, faith in God who raises the dead and in Romans 10 passages from Leviticus and Deuteronomy on the giving of Torah are used to establish the principle of faith in the caring activity of God, thus leading to the formula, 'If you believe in your heart that God raised him from the dead, you will be saved' (verse 9).

It should be clear that using these midrashic treatments, Paul is not trying to discover scriptural proof for his theological convictions; nor does he suggest that Christ's death and resurrection in some way fulfils Old Testament promises. Instead, he is providing contexts for his theological convictions, namely, the creation, the foundation story of Abraham, the Exodus and the giving of Torah.

Thus, the theological and protection formulae are not simply the passwords and talismens of a newly successful religion but the key words of a fundamentally new initiative by God.

Explanatory Myths

Sooner or later it becomes necessary to ask, What kind of procedure is in mind when we say, for example, that the death of Christ effects atonement or has saving power? The best we can do is to say, 'It is rather like the following situation', and then tell a story about human beings and God. The story, preferably containing traditional images with emotive force, is intended to provide not objective information but a viewpoint from which the death of Christ can be comprehended. I call these stories explanatory myths because the discovery of truth by means of imaginative stories was of great importance in antiquity. Paul's writings contain some examples.

In 1 Cor 2:6–8 he produces a piece of *sophia* contained in a *mysterion* disclosed by the Spirit. In effect, it is the story of a king who desires to benefit his people but, being surrounded by the corrupt and hostile members of his court, cannot act directly. He must appear to yield to the opposition; but he sets a trap which will use their lack of loyalty and perception to destroy them. As a result they crucify the Lord of glory (i.e., the Agent who carries the royal glory incognito) and so bring about their own downfall. Then, benefits from the king can reach his people. This is a story not about how the king is persuaded to change his mind but about his plan for defeating enemies and achieving his good intentions.

In 2 Cor 2:14–16 the king has won a notable victory, which has brought the prospect of relief to his loyal people and destruction to his enemies. We see him leading the victory procession (in the style of a Roman emperor's triumph), we smell the ceremonial incense—pleasing or displeasing according to our position as loyal or disloyal people. Sharing the victory is Christ, by means of whom the victory was won; and among the captives is Paul himself—a captured enemy, used by the king in the campaign—through whose recent sufferings the war was continued. The king needs no persuasion—in fact, he is almost ruthless in pursuing his benevolent purposes.

In Gal 3:23–4:7 the imagery of warfare is replaced by the imagery of social constraint: children repressed by child minders, minors restricted by guardians and trustees—no better off than slaves even though the family property is rightfully theirs. When the time had fully come for the children to enter into their inheritance, it was necessary to relieve the guardians of their authority and cancel their power. To do so God sent his Son (that is, his natural representative) born of woman, born under Torah (hence unchallengeably legitimate) to redeem those under the authority of persons appointed by Torah. The guardians are obliged to free the children and let them receive recognition as God's children. Indeed, they receive the spirit of his Son (which raised him from the dead) and greet God as 'Abba, Father'. The story implies that God was being resisted by the guardians of his own Torah, so he adopted the device of sending his Son to buy back the captives and set them free. What he did to buy them back is not an element of the storytelling; presumably, he paid the extreme penalty required by Torah and yet was brought back to life. In fact, the whole story illustrates the movement from imprisonment to freedom, from death to life, and, like the previous stories, suggests how God makes his will effective, despite opposition.

In Romans 5 the hostility comes not from hostile subjects or disloyal people or misguided guardians of Torah but from Sin. In verses 12–21 there is a new reading of the traditional story of Adam—which might well count as a midrash or even as an extended interchange formula. It goes like this: By one man's disastrous misjudgement the evil, unseen power of Sin was let into the world of human affairs, accompanied by its destructive agent Death. When, as a result of Sin's presence, all men sinned (i.e., yielded to the promptings of Sin) Death reached out to take hold of all men. So from Adam to Moses Death became dominant in human life, and Sin thus exercised its rule even over those who had not made Adam's disastrous error. Thus, human beings were *victims* of sin, not held responsible for their wicked actions. They were under an irremovable inner complusion, like slaves sold to an evil master. They were thereby condemned to death and alienated from God. When Moses received Torah from God on Mount Sinai as a counter to the law of Sin and Death, there was inserted into the deplorable human situation the possibility of a change in their condition. At first, indeed, it seemed to make that condition worse, for wickedness became exposed as defiance of God's will, and more acts of defiance took place: Sin seized upon God's commandment intended to be life-giving and perverted it into desire that brought death. Nevertheless, Torah did not fail, for it taught God's people to look forward to him who should come. When, indeed, he arrived, Adam's disastrous misjudgement was reversed and its consequences overwhelmed. By one man's generous self-giving the divine Grace was made abundantly available for innumerable sinners. Consequently those who receive the abundance of Grace and the gift of acceptance by God will reign in Life with Jesus Christ as their Lord. Whereas Adam's fault constituted mankind as victims of Sin, Christ's self-giving constituted mankind as accepted beneficiaries of God's Grace.

That, of course, is not an assessment of religious history but a story about the human situation—almost a morality play. Two nonhuman powers, Sin and Grace, are in contention, accompanied by their executive arms, Death and Life. The significance of each human wrongdoing lies not so much in itself as in the opening it makes for destructiveness and enslavement. Once they have taken hold, subsequent repentance, pleas for forgiveness and promises of amendment are of little use. The significance, therefore, of Christ's particular act of 'rightdoing' (whereby he is captured by Sin and suffers Death) lies in the opening it makes for the far more powerful influences of Grace and Life. They are inherently more powerful because Sin has no resources of its own: it can only feed on, corrupt, and destroy what is good. When, therefore, the power of Grace is effectively deployed, Sin is overcome and Death gives way to Life. Hence, Christ's death is the effective deployment of Grace in God's contest with Sin.

That characteristically Pauline way of viewing Sin as the relentless spoiler of goodness is clearly not the same as the conventional meanings of sins, trespasses, transgressions and the rest. When Paul cites the traditional phrase, 'Christ died for our sins according to the scriptures', he doubtless means that Christ died to deal with the consequences of our sins. What are those consequences? That God is offended? or is obliged to punish us? or needs persuasion to restore us? or must be diverted from his intention by a cruel death? None of those replies is suitable. The consequence of our sins (which are accidental, careless, or deliberate contraventions of the rules by which God governs mankind and the universe) is the introduction of instability and danger into normal life, even leading to disbelief in the power, goodness and presence of God. A powerful remedy is needed to avert the danger and to restore confidence, namely, the death of God's chosen agent, Jesus Christ. That aversion sacrifice for individual sinning moves in the same direction as God's victory in the conflict between Sin and Grace, but the conflict story transfers the saving power of Christ's death from individual sinning to the social context of the human situation.

Two more fragments of the myth appear in later chapters. In Rm 7:7–13 God utters a prohibition that Sin fastens on, argues with Adam (or whichever person happens to be Adam), deceives him, and destroys his life. (That feature I tacitly included above.) In Rm 8:3–4 God sends his Son to act in the social situation that is the operational centre of Sin, equipped with full authority to deal with Sin, to condemn it to be dispossessed from its apparently impregnable position in human society. When that happens, the Spirit raises Christ from the dead and equally brings to life those put to death by Sin's power. Finally, in Philippians 2 there is another development of the myth or—perhaps more probably—the adaptation of another myth. In my view it is used to establish the lordship of Christ, but it is 'compatible' with the myth I have been considering. According to the Philippian hymn, Christ exchanged the powerful condition of a god for the powerless condition of a slave in order that God might exercise his power from within the human situation at its lowest point. Indeed, by enduring the slave's death—the repulsive death of crucifixion—he became obedient (i.e., subject) to the destructive forces that dominate mankind. Thus, God's special Agent becomes mankind (Adam) at its most humiliated and then receives the supreme name and universal lordship by which humiliated mankind may be saved.

So it may be said that cross, resurrection and Parousia belong together; and together they have the same magnitude as the giving of Torah, the Exodus, the call of Abraham, and the creation itself. When the resurrection is related chiefly to the Parousia, Christ is the risen Lord who protects his own and sustains them in their ordeal. When the resurrection is related more fully to the death of Christ, it is possible to draw on the protective and atoning power of blood, which is effective both with people and with God. The Son of God, his chief representative and agent, becomes poor, offensive and unprivileged so that sinners may be set free and enriched. Sinners are reconciled to God, and God is induced to do what he may properly do for them. On the human side, the death of Christ restores confidence in God and the safety and stability of life, which is endangered by sins. On the divine side God, determined to help his people, finds a way of defeating Sin by sending his Son to be a victim of Sin's power and by releasing divine Grace within the most humiliating situation. So sinners are made acceptable to God, brought within the Christian community, and protected by the cup of blessing and the one loaf even when they share the sufferings of Christ.

That last remark brings me to the second main strand in Paul's teaching, namely, his presentation of the death and resurrection of Christ as a model of Christian experience in the present age. Perhaps it would be better to describe it as a critical and interpretative principle.

A Critical Principle

Since the cross was offensive to Jews and socially destructive to Greeks, a gospel that proclaimed the crucifixion of Jesus would inevitably disturb some valued features of Jewish and Greek life. It should not be supposed that Paul was hostile to Jews or Greeks, but it should be understood that he was correcting certain impulses in hellenistic Jewish Christian communities.

To some extent the death of Jesus belonged to the story of Israel's ill treatment of the prophets (1 Thes 2:15) and therefore could be used in Jewish self-criticism. Much more sharply, it could be used against those who made circumcision a central demand on Gentile converts (Gal 6:12–13). It is not difficult to guess some reasons for that demand: if Gentiles accepted the operation (demeaning, as they would think) of circumcision, they would have made an abject surrender to Torah—to which, therefore, their presence in the community would be no challenge. But Paul was determined, in some respects, to challenge Torah; and he demanded submission to the cross of Christ, which in different ways was equally objectionable to Jews and Greeks.

Now, in the same passage in Galatians Paul says that he glories only in the cross of our Lord Jesus Christ, through whom the *kosmos* has been crucified as far as he is concerned and he has been crucified as far as the *kosmos* is concerned. That rhetorical phrase-making conceals a serious decision: Paul is content to accept the scornful judgement of the world on him for promoting a religion that honours a crucified person; and he in turn judges the world by its response to God's self-disclosure in the crucifixion. So against the *sophia* of the world that is passing away Paul sets the logos of the cross, which is the power and wisdom of God. It is no valid objection that to Greeks the proclamation of the cross is *mōria* (stupid social destructiveness), for it is the function of the cross to begin the dissolution of this age. Hence, the death of Christ brings a critical assessment to bear on the Christian community's relation to the society around it (1 Cor 1:18, 23–24).

So it does to relations within the community. Even when conscience places no restraint, the cross helps a Christian to avoid sinning against a fellow Christian by carelessly using Christian freedom if he remembers that the fellow Christian is a brother for whom Christ died (1 Cor 8:11, Rm 14:15). The cultic disputes within the Roman community are met with the reflection that 'none of us lives to himself, and none of us dies to himself. If we live, we live to the Lord, and if we die, we die to the Lord; so then whether we live or whether we die, we are the Lord's. For to this end Christ died and lived again, that he might be Lord both of the dead and of the living' (Rm 14:7–9). And finally, the Philippian community was influenced by a Gentile group that used circumcision as initiation into a kind of mystery rite, claiming perfection in this life. Being already raised from the dead (according to their gnosis) they were naturally hostile to the cross. Against this, Paul presented his own gnosis of the Lord, renounced his privileged Jewish identity in order to gain Christ as Lord, to experience the power of his resurrection, and to share in his sufferings and be conformed to his death (Phil 3:2–21). He has no doubt of the resurrection hope but shows reserve in claiming it. For we await a Saviour who will transform our humiliating condition into the condition of Christ's glory.

An Interpretative Principle

That brings me to participation in Christ's sufferings or (which may be another way of putting it) interpreting one's own sufferings as a share in his.

The theme of Christian suffering runs throughout the epistles, but it varies in significance. In 1 Thessalonians the sufferings of the end-time are required by God and provided by Satan. The Lord experienced suffering in carrying out God's instructions; so must we (1:6, 2:18, 3:3). That simply accepts the well-known fact that established structures resist change and inflict suffering on those who try to promote change. But Paul found himself with a leading role in promoting change. In the interim between Christ's resurrection and Parousia Paul was an apostle of Christ whose activities in preaching and performing the signs of an apostle were consciously disruptive of the existing *kosmos*. Apostles were fools for Christ because they spread the *mōria* of the cross. God, it seemed, had exhibited them as last of all, as men sentenced to death. So they suffered deprivation, social rejection, and self-sacrifice (1 Cor 4:9–13).

That was suffering *for* Christ; next comes suffering *with* Christ. Paul's experience of distress was his abundant share in Christ's sufferings, as he relied only on God who raises the dead (2 Cor 1:8–9). His humiliation was the means by which God was victorious (2 Cor 2:14–16). Paul was constantly handed over to death—namely, the attrition of his energies—so that life might be at work in his churches. The relentless description of his ministry as a catalogue of dangers and ambiguities produces the striking words, AS DYING, AND BEHOLD WE LIVE (2 Cor 4:7–15, 6:3–10). For Christ's sake he was content with weaknesses, insults, hardships, persecutions and calamities; for when he was weak, he was strong (2 Cor 12:10). In Gal 6:17 the stigmata of Jesus in his body probably imply just such experiences in his apostolic ministry; but earlier in that letter he had explicitly formulated his experience by means of participation words. In a highly personal variant of the standard formula he says, 'Christ loved me and gave himself [was crucified] for me' and 'Christ lives in me'. And he also says, 'I have been crucified with Christ', which must be linked with, 'I died to the law'. Put differently, Paul's ego, formed by Torah, was destroyed. His new ego fully appropriate to the new age had not yet replaced it, but God was pleased to reveal his Son to him; so

Christ lives in him. Paul, therefore, is confident no longer in Torah but in the cross of Christ (Gal 2:19–20).

In Philippians Paul is in prison for Christ's sake. He is encouraged to bear suffering by the help of the Spirit of Jesus Christ (i.e., the Spirit that raised him from the dead). His apostolic task in this situation is to magnify Christ (i.e., proclaim his lordship), whether by living or by dying—more perhaps by dying than by living (Phil 1:12–14, 19–21). According to 3:7 he had already died one death—the loss of his Jewish identity—and was again prepared to 'share his sufferings, becoming like him in his death' that he might know the power of his resurrection. For Paul, sharing his sufferings was the apostolic certainty, sharing his resurrection the apostolic hope.

In some measure that was true not only for the apostle but also for other Christians: 'It has been granted to you that for the sake of Christ you should not only believe in him, but also suffer for his sake, engaged in the same conflict which you saw and now hear to be mine' (Phil 1:29–30). What is more, Paul seeks to persuade the Christian communities to evaluate present hardship in terms of future glory (2 Cor 5:1–10). Human existence in the body pulls in two ways, towards the flesh (the enticing but weak and fallible conditions of social life) and towards the Spirit (the exciting but elusive conditions of new life). The painful predicament is not the temptation to make *unspiritual* decisions when spiritual decisions are required but the discovery that spiritual decisions can be flawed and even destructive. In the second half of Romans 7, Paul presents the devout Jew (or indeed the upright Greek) who delights in God's rule and strenuously does good but discovers that his success is not what he intends and promotes what he despises. He is caught between two antithetical powers: one he respects and consciously esteems, the other he despises and rejects but finds himself carrying out its demands. So (in Paul's language) he is the captive of Sin. The only remedy is for the lordship of Christ to overcome Sin, for the life-giving Spirit to set him free from the rule of Sin and Death (Rm. 7:14–8:2). The perception of Death as Sin's enforcement officer, as a part of the evil tyranny that oppresses mankind, deters Paul in this context from giving positive valuation even to Christ's death. Death is something that must be destroyed (hence the persistent theme of resurrection) and it is the lordship of Christ established by his resurrection and the Spirit who brought him to life that finally dominate the human predicament.

Nevertheless, Christians were baptized and they ate the Lord's Supper. When they were baptized, they were brought into allegiance with Christ as participators in his dying and rising again. By this ritual the pattern of dying and rising was transferred from Christ to the baptized person. He appropriated the benefits of Christ's self-giving and had a share in the new constitution of mankind. His old identity was concrucified with Christ in order to end his servitude to Sin (Rm 6:3–10). Thereafter, he participated with fellow Christians in the body and blood of Christ. As by sharing bread the community participates in the risen life of Christ, so in sharing wine it participates in his death. In some measure it experiences his risen power, provided that the eucharistic gathering shows forth his death till he comes (1 Cor 10:16, 11:23–26).

THE DEUTERO-PAULINE EPISTLES

Like the main epistles, Colossians and Ephesians develop their arguments by means of theological formulae. Some are brief borrowings from the earlier writings, such as the allusion to being buried and raised with Christ in baptism—though borrowed with a difference. In Romans 6 the baptized know that they will be raised with Christ: in Colossians 2 they have already been raised. They have died, and their life is concealed in the heavenly world with Christ in God. Two elaborate formulae, structured rhythmically rather like hymns, set forth that understanding. The Colossian wisdom hymn makes the Son of God the prototype of all created being and portrays him as ruling and sustaining the universe in his resurrection power, since by his death he had pacified the dissident power on earth and in heaven. The Ephesian hymn celebrates the blessings available in the heavenly world, our redemption by the blood of Christ (namely, the forgiveness of sins), knowledge of the mystery (namely, the union of all things in Christ), and a more comprehensive redemption of what belongs to God, for which the Spirit is the guarantee.

In these formulae, tradition has secured a modest place for the death of Christ. By and large, however, they are resurrection hymns. In the impressive credal formula of Ephesians 4, death and resurrection are not included, though the formula is supported by an ascension midrash on Psalm 68. But discussion of the unity and ministry of the Church does not draw upon Christ's death (as it certainly does in 1 Corinthians). That observation is not entirely offset by the presence of explanatory myths: the deception of God's enemies and his triumphal procession (already familiar from previous epistles); the bond that God blotted out and the indebtedness that he transferred to Christ; the magical revival of the dead Lord and his installation with plenary power in the two worlds; and the prince who chooses as his bride a shamefully discredited maiden and presents her to himself radiant, without spot or blemish. In each of these, explicitly or tacitly, the prime mover is God; and each story turns on the humiliation or suffering of Christ and his astonishing recovery and triumph. His resulting lordship is always clear; it is sometimes, but not always, clear that his death is more than the necessary precondition of his lordship. And that corresponds to the baptismal teaching, namely, that in baptism we are buried and raised with Christ, being brought to life by the saving initiative of God in Christ's resurrection. The range of resurrection consequences is variously expressed: the new creation, the new being, the reconciliation, the obliteration of enmity, the announcement of peace, access to the Presence and the new temple. The death of Christ is preliminary to a major initiative by God in grace, love and wisdom. We are reconciled to God in the resurrection body that death made possible.

In that body, so far as it exists in the lower world, we learn the truth of the gospel, the hope laid up in heaven. We are party to the mystery of the divine will, namely, to unite all things in Christ, in heaven and on earth. Therefore, 'sleeper awake, and arise from the dead; receive the gift of illumination and knowledge from Christ'. God rescues us from the power of darkness and transports us into the realm of his beloved Son, so that we share the inheritance of the saints in light. From darkness to light is probably baptismal language. Darkness implies death—alienated from God, hostile to the truth, in the grip of dark powers, helplessly culpable of trespasses and sins, unable to play a part in maintaining the structure and fabric of the world. Yet Christ is able to reconcile such people by his death in a physical human body and present them holy, blameless and irreproachable, freed from the dark forces and given a share in Christ's risen lordship in the heavens. *How* his physical death could effect that extraordinary transformation is not explained, though echoes of the deception story and the triumphal procession probably indicate the imaginative level on which the question rested. In any case, nothing is said or implied about the cross as a means of influencing God, only about the cross as God's way of achieving his purposes. In consequence, chief stress falls on resurrection.

The death of Christ has some independent significance. It stands behind the portrayal of Paul as the supreme mystagogue who suffers disproportionately at the hands of earthly enemies, who have still to be pacified (i.e., reduced to impotent obedience). In that way Paul makes God's intention fully known. The suffering Paul of the earlier epistles, suffering alongside, and on behalf of, fellow Christians, becomes the isolated apostle who makes up for what is lacking in Christ's afflictions for the sake of his body, which is the church. Members of the church may dispute and quarrel, but they do not suffer. The death of Christ is not for them a model of the vulnerable sufferer, for they are equipped with the whole armour of God for a ritual battle in the heavenly places.

It is, indeed, implied that confidence in the blood of Christ replaces the blood of circumcision as a condition of admission to the community and that his death as a fragrant offering to God is in some ways an example to Christians. But it seems to play little part when ethical instruction is under discussion. Not even the resurrection, which secures for Christians their life hidden with Christ in God, is directly influential in shaping their continuing life in the world.

1 PETER

It is instructive to consider 1 Peter alongside the deutero-Pauline epistles. Like them, it contains several elaborate theological formulae, as well as reminiscences of explanatory myths. Like them, it shows the importance of resurrection, both in direct statements about rising and regeneration and in equivalent images. But whereas Colossians and Ephesians are occupied with knowledge of the divine mystery in the heavenly world, 1 Peter is concerned with the precarious situation of Christians in this present world. Resurrection is therefore a living hope, available only when present suffering is replaced by expected glory. Those who adhere to Christ share his fortunes; first the suffering, then the glory.

The sufferings of Christians are social disabilities. It is therefore necessary to work out a strategy for the suffering community. It falls into two parts: firstly, to explain their present situation; secondly, to tell them how to handle it.

Their present necessity arose because God had liberated them from their ancestral pattern and from the unchanging ways of ancient society. As in a covenant ritual, they had been sprinkled with the sacrificial blood of Christ, so making them a new people. That had aroused the natural hostility of their fellow Gentiles, so that Christians experienced humiliation or practised humility—according to the response they made to Christ's death on their behalf. In fact, to accept the benefits of Christ's suffering made it difficult—even dangerous—for them to sin, as they waited for their exile to cease and the age of glory to begin.

That stressful situation can be contained and dealt with if it is realized that God's blessing is called forth by innocent suffering. If you do right and suffer patiently, God is moved by your plight: he bestows his blessing, which is more powerful than the malice of tormenters. That is proved by Christ; and his suffering is the master copy, the line of footprints to be followed. He is the representative sufferer and shows that innocent suffering can compensate for wickedness. Moved by Christ's suffering, God conferred his blessing and raised him from the dead. Moreover, God extends his blessing to those who belong to Christ, those for whom he suffers, those who are in the same case as he is. Further, because Christ suffered on account of sins, he has established the pattern for dealing with sins—with the remarkable consequence that he who by innocent suffering attracts God's favour brings benefits for those who know only God's disfavour, gives them access to his presence, and transfers them from the world of flesh to the world of spirit. These are not chiefly individual benefits but are a powerful means of healing social disorder, so that Christians can become good stewards of God's varied grace (1 Pt 4:10).

MARK

As much as the Pauline epistles, Mark's Gospel was devised to deal with problems within a Christian community that needed theological instruction and moral guidance. The method, of course, was different. The evangelist seldom used theological formulae and, when he did, presented them as part of a narrative episode (e.g., the ransom saying in Mk 10:45). Protection formulae may be recognized in some apocalyptic sayings (e.g., these who will not taste death, in Mk 9:1) and the apocalyptic chapter Mark 13 (with its numerous links to the Passion narrative) is an extensive reassurance of final safety; but Mark makes no use of the traditional association of the Last Supper with Passover to establish a protective meaning. Mark seldom used standard midrashic methods but had his own way of enlisting scripture in his service (e.g., the use of Psalm 22 to structure part of the Passion narrative). And instead of intepretative myths he used actuality—as it was, or as it was seen by his community; that is to say, his theological instruction and moral guidance is carried by a large collection, carefully arranged, of meaningful stories: about the Baptist, about exorcism and the restoring of sight, about conflicts with opponents and disputes among disciples, about fasting at weddings or rebellion in vineyards, about a premature anointing and a final supper. There are no stories of the risen Lord, for everything turns on the death of Christ, the suffering of his disciples, the sufficiency of the Galilee storytelling and the promise of his coming. Thus, the same tension between death and resurrection appears in Mark as in the *corpus Paulinum.*

Mark begins with the fate of John Baptist as spokesman for God, not simply declaring the divine intention but so speaking God's word that its intention is initiated. When, therefore, John is put to death the danger of uttering the prophetic word and the cost of initiating the divine intention is made plain. Since Jesus was at least a prophet, as many agreed, his effective announcement of the Kingdom of God—his declaration of the immediacy of the divine power by means of healing and exorcisms—was straightway at risk, and it becomes clear that this Agent of God may well be rejected. The activities of the kingdom are a matter of dispute to the actual contemporaries of Jesus, as no doubt they also were to some members of Mark's community. Therefore, Mark tells his story as if it were a drama taking place on a front stage and an inner stage. On the front stage we see conflict with religious and political authorities; on the inner stage we see elucidation of the conflict for disciples.

On the front stage Jesus responds to hostility by forcing a confrontation with the hierarchy. Making use in prophetic manner of symbolic action, cursing and parable, he sets in train forces that will destroy the temple and the power of the hierarchy. Before the Sanhedrin he claims position at the right hand of the divine Power and prophesies that he will come against them with the overwhelming authority of that Power. To the hierarchy this may be blasphemy or the malign word of a false prophet; so it must be proved false and neutralised. The Sanhedrin act effectively. Jesus is helpless to save himself from their condemnation. They induce the Romans to crucify him, and he is mocked by the hierarchy, the people of Jerusalem, the Roman soldiers, and his fellow sufferers. His disciples abandon him. When Jesus appeals to God (in the opening words of Psalm 22) to reverse his desperate situation by conferring his blessing, no answering voice is heard. Yet ironically, the truth about Jesus is spoken by those who deride him: the soldiers rightly salute him as king of the Jews, the passers by correctly observe that he saved others but cannot save himself. When he dies, the tearing of the temple curtain is an omen of God's abandonment of the sacred place, and the centurion's confession that Jesus was Son of God an omen of how he will be acknowledged in the Roman world. The opponents of Jesus are enticed from their defences by his vulnerability: they use their power against the chief agent of God, the divine Son; and by destroying him they destroy their own power.

On the inner stage the disciples are made to understand that Jesus is not only to be seen as a prophet and is not usefully to be named *Christos:* he is the Son of man, of whom God requires humiliation, suffering and death for the benefit of others, to whom God will restore power and glory. They are present for all three announcements of the death and resurrection of the Son of man (and each announcement helps to inaugurate the events); and they also learn that their personal fate and status and activity within the community are to be moulded by the death of the Son of man. When the sons of Zebedee request positions of honour alongside Jesus in his glory, they are offered a share in his suffering and are taught a lesson in serving others; the Son of man himself is a servant not by helping others from a position of security but by giving his life that others may live. This is Mark's nearest approach to an interchange formula and derives from Jewish martyrdom teaching—on the lines that God may shorten the days and accept the freely offered life of his agent and not exact the full measure of lives that are now forfeit. It must be admitted that the ransom metaphor is derived from an explanatory myth; but neither the word nor the myth is to taken literally. The 'many' are not literally prisoners of God, and God does not set them free in literal exchange for a more eminent victim, whom he wishes dead. The emotional force of the saying, in this context, is that all are to be servants of others even at the cost of being powerless; but that only the Son of man is required by God to die in their services.

That striking interpretation is confirmed by the explanatory words of Jesus at the Last Supper. By designating the bread and the wine as his body and his blood, he indicates that the disciples will eat and drink at the cost of his life. With these prophetic words and cultic actions he sets certain events in train. He intends to sacrifice his life, according to the covenant that God has made with him, so that the Kingdom of God (so far postponed in its fulness) may arrive. Then the disciples and the many may eat and drink with him in that kingdom. But in order that Jesus may be joyfully united with friends at the great banquet of the kingdom he must now be isolated from them, despite the protests of disciples and Jesus' own hopes. And even when he is risen from the dead, he does not rejoin them. They must go to find him in the unfulfilled promise and hostility of Galilee.

THE PASTORAL EPISTLES

The writer makes striking use of theological formulae of the Pauline type, but with a difference. For one thing, they give dramatic power, polemical force, moral earnestness and solemnity to his instruction, but they seldom lend theological support to an argument. For another thing, they are brought within a scheme that offers the great benefits of religion by means of epiphanies. That might perhaps be an adaptation of the Pauline theology to a common hellenistic description of salvation if Christ's death retained its prominence in the formulae. But it does not. When the author confesses 'the mystery of our religion' (1 Tm 3:16) in concise formal phrases that move from his manifestation in the flesh to his being taken up in glory, nothing is said about Christ's death.

The author, indeed, says that Christ Jesus in his testimony before Pontius Pilate made the good confession and no doubt assumes knowledge of the Roman crucifixion; but the reference is intended to encourage heroic endurance. It is said that we have died with Christ (presumably in baptism) and that he is risen from the dead and we shall live and reign with him. We do not yet reign with him for our resurrection has not yet taken place (as some falsely teach). He is to be judge of the living and the dead by his epiphany and his kingship. To all of this a reader of Paul's letters cannot take exception. Nor can he lightly refuse a Pauline gloss to the statements that Christ gave himself to redeem us from all iniquity, gave himself as a ransom *(antilytron)* for all. *Gave himself* ought surely to be a reference to Christ's death. But the author never says so. We are saved by Christ from ignorance, called with a holy calling, made into a people of his own, zealous for good deeds by regeneration and washing in the Holy Spirit. At the beginning is the epiphany of our Saviour Christ Jesus, who abolished death and brought life and immortality to light through the gospel; at the end is the epiphany of our great God and Saviour Jesus Christ. In the Pastoral Epistles the death of Jesus is an example to us all; all other aspects of the wholesome doctrine can best be understood as benefits of his self-disclosure.

MATTHEW AND LUKE

If the Q tradition is examined (i.e., the sayings of Jesus that are recognizably present in Matthew and Luke, though usually absent from Mark), it is found to contain no direct reference to the death of Christ. The tradition is well supplied with sayings about extreme hardship for disciples, an experience that they share with the Son of man himself and more generally with the ill-treated prophets and messengers of God. So there is a single indirect reference to Christ's death in a saying (with a variant in Mark) that defines a disciple as someone who takes up his cross and follows Jesus. Hence, for this layer of tradition, the death of Christ is a marker that firmly identifies discipleship as belonging to the story of the suffering prophetic agents of God. This was discipleship in a time of crisis: the days of the Son of man were soon expected, when all would be changed. What disciples needed was an encouragement to stand firm, following the example of Jesus, rather than attempted explanations of why the suffering had been necessary. God's agents had suffered in the past and were still suffering—but a change was on the way.

Matthew keeps to the main structure of Mark, adds a great deal of Jesus' teaching, and rearranges miracle stories and teaching in convenient blocks of similar material. His collection of sayings lays great stress on the hardships of disciples: given the times in which they live, the suffering and death of Jesus is a model for Christians. To the Q saying about the sign of Jonah he adds his special comparison: that the Son of man's three days and three nights in the heart of the earth correspond to Jonah's imprisonment in the belly of the whale. Some devoutly ingenious mind, having learnt that Christ's sufferings were the climax of prophetic sufferings, had hit upon this comparison—on the principle that a shocking event having been once rehearsed is not so disturbing on its actual performance. To a more questioning mind, however, there must be the fairly disturbing implication that Jesus, like Jonah, was reluctant to obey God's command.

Matthew is a collector and arranger of material. When his liturgical tradition presents him with the statement that Christ's blood is poured out for the forgiveness of sins, he records it even if it does not match his previous teaching—reflected in the liturgical prayer of Jesus—that forgiveness is available to those who repent and forgive others. In the passion narrative Matthew tells anecdotes against the Jews; the centurion and soldiers are convinced by prodigies. Nothing really engages Matthew's attention until he takes the eleven disciples to Galilee, where Jesus claims full authority from God and authorises them to make Gentile disciples, who will obey his teaching. Matthew's interest lies not in the death and resurrection but in the *teaching* of Jesus.

Luke gives an eminent place to John Baptist at the beginning of his Gospel but without suggesting that his death is an omen of the prospect for Jesus. Indeed, no omen is needed when the scribes and Pharisees are filled with murderous fury against Jesus for his announcement of God's grace for Gentiles as well as Jews. In the part of the Gospel where Luke makes the most extensive and least systematic alterations to Mark's scheme, he retells the events that led to the passion so that they can be seen in a different light. The transfiguration becomes an impressive conference with the Jewish immortals about the 'exodus' of Jesus, which then becomes his 'assumption' (both words that dignify the thought of death). Then Jesus begins a stately progress towards Jerusalem, following a plan determined by God, undeterred by Herod and other opponents. In due course he will experience the hostility of the Jerusalem hierarchy and share the prophetic fate. But he will finish his course and be perfected, that is, he will attain the proper fulfilment of his predetermined progress. He is not the hapless victim but is himself the disturber of his enemies. It is all written and must be accomplished—a theme often repeated. Luke does not follow Mark in reporting that the Son of man gives his life as a ransom for many but has no doubt that Jesus died in obedience to God's predetermined plan.

If, at the Last Supper, the shorter text is original, Jesus renounced wine until God's kingdom came and gave them bread as his body—without explaining what was meant. If the longer text is original, his body was a memorial sacrifice and his blood (represented by the outpoured wine) achieved the new covenant implanting Torah. But the reader cannot be quite sure that Luke is thinking theologically about the body and the blood, for his additions to the supper narrative convert these statements about the death of Jesus into instruction for church leaders. Luke's Gethsemane removes the distress, warns against the temptation to give up when under pressure, and presents Jesus as a model of goodness and nobility. On trial Jesus is noble-spirited in his innocence and then goes like a great prince to his execution, not flogged or mocked by the soldiers. On the cross there is no reproach addressed to God: Jesus magnanimously forgives his executioners, promises paradise to a repentant criminal, and finally commends himself to the hands of God. His nobility sustains him to the end and is recognized by the Roman centurion.

Luke ends his Gospel with the risen Christ showing gracious condescension. He makes two epiphanies to depressed and uncomprehending disciples. In principle he takes them through the law of Moses, the prophets, and the Psalms and shows how the Christian proclamation is justified by the written intentions of God. It was God's will that the Messiah should suffer and rise on the third day. It is God's will, and if you look you will find the proof in scripture; but what proof is necessary when nobility is its own commendation?

In Acts the dominant interest is the resurrection and the continuing activity of Christ by means of his 'Name', which produces healing, salvation, forgiveness of sins and the gift of the Holy Spirit. No scriptural passage is cited to explain the death of Christ, but several passages are cited as relevant to his resurrection. It is said that God foretold by the prophets that his Messiah would suffer, as he did by hanging on a tree. On one hand, therefore, he was delivered up by God's plan and foreknowledge, but on the other his death was accomplished by the Jewish people. Behaving like their fathers who killed the prophets, they killed him by the hands of lawless men. For Luke, the death of Jesus is the final proof that the Jews need salvation, and the gospel begins with the resurrection. Apart from the speech at Miletus, Luke ascribes no positive benefit to the *death* of Christ. He died in order to be raised from the dead. If his death, predetermined by God, contradicted God's attestation of him by wonders and signs, the guilt falls squarely on the Jews (the Romans are only instruments). The Jews' ignorance may justify a plea for mercy but does not remove guilt and blame. So Jesus died in obedience to God's will, but Luke does not attribute to him an atoning death.

HEBREWS

When the religion of Christians is described as 'enlightenment', it is not surprising if the death of Christ becomes something of an embarrassment. If Christians are participating in the miraculous activities and scriptural illuminations of the Holy Spirit and are experiencing the energies of the coming age, what are they to make of the crucifixion? There are indications in the Epistle to the Hebrews that some Christians regarded the cross as shameful and gave no sacred meaning to Christ's death. So, with great boldness, the author of the epistle firmly grasped the unpleasant fact of death—our death and Christ's death—and sought a way of restoring significance to this central feature of the primitive Christian proclamation, at the cost of setting aside the standard teaching about resurrection.

He set about his task by looking for a situation in which the shedding of blood was a necessary and acceptable method of remedying human need. Not surprisingly, he found it in cultic sacrifice and, for special reasons, chose the Day of Atonement. He then constructed an elaborate explanatory myth saying, in effect, 'It is as if Christ were a unique high priest, acting in the extracurricular manner suggested by Melchizedek's encounter with Abraham, and introducing the new covenant foreseen by Jeremiah'. He implies that he found the designation *high priest* in the confession of his community; and it seems clear that he allowed the ancient cultic language not only to justify the sacrificial death of Christ but also to create a worldview in which that death became necessary.

Human beings live in the earthly world, which contains shadowy representations of the heavenly world but is marred by the fear of death and the Devil. It permits no full access to the Divine Being who exists beyond the heavens. Yet the heavenly world is intended as the possession of human beings not exclusively of the angels who serve the Throne. The Son of God, who is the eternal Wisdom, properly belongs to the inmost reaches of the heavenly world above the servant angels; but for a little while he was made lower than the angels that he might become mankind's representative and forerunner. Like all human beings, he had flesh and blood. By suffering, prayer, and death he underwent the *paideia* required for full access to God; that is, he passed through the realm of death, destroyed the Devil's power, and entered into full awareness of God. He experienced death for every one of us, being perfectly equipped as the pioneer of mankind's salvation.

It is an axiom that seeing the Lord requires a state of holiness or consecration that is achieved by the purification, expiation, or sacrificial removal of sins. Consciousness of wrongdoing is not compatible with consciousness of the glory of God. Hence, the passage of the Son from the earthly world to the heavenly world can be compared to (or is prefigured by) the passage of the high priest on the Day of Atonement from the outer temple into the inmost sanctuary. The great high priest makes a final unrepeatable sacrifice by offering his own blood in a consecration ritual that overcomes hostile powers, purifies the sinful condition of human worshippers, and equips them to enter the presence of God, who is a consuming fire. So Christians can confidently enter the heavenly sanctuary, relying on the self-offering of Jesus as their invitation and protection.

This epistle, written with great energy, has power to stir the imagination and to leave the impression that everything would be lucid if the reader were not so dull of hearing. It is intended not as an essay in speculative theology but as a word of encouragement for a community in need of help and rebuke. It rests upon the conviction that although no date has been set for the appearing of Christ a second time to save those awaiting him, believers can be assured of full access to the Throne in heaven. Holiness, the vision of God and perfection are within their grasp unless they spurn the Son of God and profane the blood of the covenant by which he was sanctified and outrage the Spirit of grace. If a community member commits that sin, he leaves the community and there can be no return. Thus, the doctrine of the epistle makes it possible for the community to endure indefinitely in the shadow world, confident in its access to the real world, exclusively by the blood of Christ.

JOHN: EPISTLE AND GOSPEL

It is a fundamental conviction of Johannine theology that all the benefits of God's salvation are conveyed by the disclosure of the Son as the divine Agent. In principle, the descent of the Son of man disclosed the possibility of eternal life, and his subsequent ascent guaranteed possession of it. Since life is the dominating theme, death need scarcely be mentioned except as that which is overcome by life—not that death is taken lightly. Death is ignorance of God, falsehood, hatred and wickedness—in other words, not simply physical death with all its misery (and it certainly is that) but also the comprehensive ruin of what human life should be. But that has been overcome, at least for those who believe in the Son. The Son of man had descended to bring illumination, sustenance and life. He had destroyed the activities of the Devil and the practice of sinning had ended. Those who believed in him had passed from death to life—at least in principle.

But the principle needed to be qualified. It was not true that members of the community had ceased from sinning or that the darkness had passed away. The present *kosmos*, especially the Jewish part of it, was in conflict with the community and causing distress—thus continuing what it had begun when it put the Son of man to death. Both from the hatred of enemies and from the community's own tradition the fact of Christ's death had to be admitted. The Son of man had not simply revealed eternal life; he had also died to make it possible.

What then could be said about his death? A proverb might ease the pain: If you love your life, you lose it; if you hate this world's life, you keep it for eternal life. A natural analogy might suggest a meaning: if a grain of wheat dies it bears much fruit. The burden of anxiety may be lifted by a determined attempt to put the responsibility on others. Thus, the death of Jesus is not directly present in the farewell discourse but is frequently and openly discussed with the Jews. No doubt, Jesus forced the Jews to a decision, by dispensing with Sabbath and doing what only God may do. By prophetically miming the end of temple sacrifice and its replacement by the divine Wisdom, he drove the Jews to hostility—not that he wished to destroy Jewish life but that he wished to replace it at a higher level incorporating the Gentiles. But the Jews cannot seize the opportunity he gives them. So if they will not exalt him in faith, they will lift him up on the cross. But from their perversity good may yet come; for, as the high priest rightly says, 'It is expedient for you that one man should die for the people, and that the whole nation should not perish'. Thus, an expression of martyrdom theology, taken from the emotional traditions of Jewish nationalism, is extended to the 'children of God who are scattered abroad' and something is gained from the disaster.

But something more positive may be said if we consider the cultic tradition of the community, which identifies bread and wine with the flesh and blood of the Son of man. In hellenistic Judaism it was a familiar thought that the divine Wisdom provided immortal food by which mankind lived and attained the life of heaven. When, therefore, the Son of man (who is the agent of the divine Wisdom) descends with the gift of life, it is by the obedient but freely chosen gift of his life, by his surrender of flesh and blood in death, that the *kosmos* is able to live. This self-giving discloses the divine love and indeed becomes the pattern of the community's loving. Just as five barley loaves were replaced by countless baskets full of bread and purifying water was replaced by immense quantities of wine, so the death of this one man was replaced by the glory of God.

One is tempted to say, 'this dreadful death', and so it was. But in John's Gospel it is not described as dreadful. The suffering and humiliation are there, but they are always within the control of God—indeed, even of Jesus himself. He is not so much the victim as the key person in a carefully defined plan. Each group of persons plays its unwitting part. Jesus calmly does what he has to do and then (as it were) goes through the door and shuts it.

Thus, the death of Jesus is the means by which he departs from earth to heaven. As it could be said that a man's being born blind or the illness of Lazarus was for the glory of God (namely, that they were occasions when the glory of God had to be displayed), so it could be said of the death of Jesus. He was lifted up on the cross to begin his ascent. It was the withdrawal from this world of one

who (by God's gift) had life in himself and his return to the heavenly world. By
this ascent Jesus gave access to that upper world for those he came to save. By
laying down his life he demonstrated and initiated the possibility of resurrection.

So much for an attempt to give meaning to the death of Christ in the movement of descent and ascent. Once it had achieved its essential purposes, what
further significance had it in the community's life? It certainly provided a model
for Christian living if members would accept it—but there was more. It could be
described by the ritual analogy of the Passover lamb, which protects God's people
from the perils of the hostile and sinful world. Since community members are not
sinless (as in principle they should be), some remedy is needed for accidental,
careless, and wilful sin—provided that the sinners have not deliberately rejected
Jesus and refused to walk as he walked. Sinners can appear before God with Jesus
as their paraclete: by repentance and the blood of Christ their sins are forgiven
and they are cleansed from their wrongdoing. So the blood of Christ protects the
community from the follies of its members and protects the members from the
condemnation of the community.

REVELATION

The author of Revelation is not embarrassed by the death of Christ and has no
intention of excusing it, defending it, or explaining it away. Instead, he uses it
to attack the social and political structures of his day. The slain Lamb stands
before the divine court and relentlessly accuses Roman society for its ill treatment
of God's people and its radical disturbance of mankind and the world. The
author of Revelation, in fact, adopts and revises an old perception of Christ's
death, namely, as a protective device that turns away evil. The powers of evil
cannot easily be restrained, and the saints must suffer as Christ had suffered; but
they are not abandoned, they will not go unavenged, and their time of dominance will come. Shed blood calls for vengeance, and vengeance will soon fall
upon those who are not written in the Lamb's book of life. In the great conflict
between the creative powers of the heavenly world and the destructive powers of
the abyss everything turns on the martyrdom of Christ—even when the author
allows his imagination to run wild with bitter images of catastrophe and chaos.
In the end, shed blood makes no further demands. The theme is life. There is a
new heaven and a new earth. The marriage of the Lamb to his bride is foreseen,
and trees are planted for feeding and healing the nations.

11

Conclusion

In the course of a long exegetical enquiry we have always been occupied with the apparently simple statement, 'He died and rose'. Those words, when appearing in early confessions of faith, indicate a double event, not separate events in sequence. When one part is stressed and the other neglected, the faith of the Christian community, with its prayers and its morals, becomes distorted. Neither part of the double event can be accepted without reservations. The resurrection is literally improbable but metaphorically attractive; the crucifixion is certainly probable and in all senses unattractive. Hence, it was sometimes necessary to restrain Christians who made extravagant application of resurrection imagery to themselves and ignored the significance of crucifixion imagery. Indeed, it was sometimes necessary to find good reasons for allowing crucifixion imagery a continuing role within Christian self-definition. Hence, the primary question to ask about any New Testament statement of Christ's death and resurrection is not what theory of atonement it supports but how it effects the community. In turn, that question can be divided: How does the statement affect their worship of God and how does it affect their moral behaviour?

That being so, it is worth recalling the only explicit remarks of Jesus (according to Mark's Gospel) on the significance of his death. In one he says that the Son of man came not to be served but to serve and to give his life a ransom for many. The saying is intended as a model of behaviour for his followers, that is to say, in order to serve others, he (the special Agent of God) throws himself on God's compassion and surrenders his life. Others should do likewise. It is taken for granted that God will respond, and there is no need of theological calculation about how and why. The only thing is that Christ's self-giving is a model of Christian service.

The other Markan saying, at the Last Supper, also indicates that Christ will give his life for many, now not for their instruction but for their benefit. This is his blood of the covenant, namely, the blood that activates the covenant made by God with Jesus: If he sheds his blood, he will drink wine new in the Kingdom of God. Since the disciples at supper eat bread knowing it to be his body and drink wine knowing it to be his blood, they are united with Jesus and so share the benefit of his covenanted death—which is life in the Kingdom of God. As the first Markan saying links the death of Christ with the moral behaviour of the community, so the second saying links his death with their worship of God.

Since the death of Christ is a preoccupation of Mark (not supported by additional sayings in Matthew and Luke), it is often considered precarious, by critical scholars, to attribute these two sayings directly to Jesus. But the saying of the Last Supper is certainly traditional (on the evidence of 1 Corinthians 11), and the ransom metaphor had become popular for assessing the heroic deaths of the martyrs (on the evidence of 2 and 4 Maccabees). Since Mark awkwardly attaches the ransom saying to the dominant thought of service, he may have plucked *ransom* out of the air; but he may equally well have used a traditional ransom saying to identify the extent of service that Jesus intended. It is significant that Mark's threefold prediction of suffering, death and resurrection for the Son of man offers variations on a traditional formula, which in turn elaborates the simplest confession: 'He died and rose'. The most striking feature of that elaboration is not the intrusion of eventful details but the assertion that the death of the Son of man is God's doing. *Son of man* is peculiar to the sayings of Jesus: it indicates the person required by God to experience both suffering and triumph. Thus, Mark allows Jesus to speak twice explicitly and three times implicitly about his covenant with God whereby the willing death of Jesus has promised consequences for him, and for his disciples in relation to their moral behaviour and worship of God. If that is Mark's invention, we have no better guide to the mind of Christ, but it might very well be Mark's chosen way of preserving the tradition of the words of Jesus.

We can gain some impression of what the early Christians thought by studying the pre-Pauline formulae. In the first phase, since resurrection was more striking and encouraging than crucifixion, emphasis fell almost exclusively on the second member of 'He died and rose'—so much so that confessional formulae referring only to the resurrection became established: 'If you confess with your lips that Jesus is Lord and believe in your heart that God raised him from the dead, you will be saved'. Of course, death is presumed or mentioned in passing as the necessary precondition of resurrection. 'He rose' implies that he is still alive, and the implicit 'He died' tacitly links the living Lord to an actual event and person. When the formula took the variant form 'He died and was raised', it added the implication that the resurrection was God's doing—duly made explicit in other formulae.

Quite early, however (though perhaps in a second phase), the death attracted particular attention: 'He died for us'. That produces the characteristic Pauline description of a Christian, 'the brother for whom Christ died'; and it appears in Paul's earliest letter (as I regard it) 1 Thessalonians: 'He died for us so that we might live with him'. But the thought is already present in the pre-Pauline formula by which the simple twofold event becomes a fourfold event: 'He died, he was buried, he was raised, he appeared'. In that formula both the death and the resurrection are found in the scriptures (they have divine approval) and have consequences for the community: the death deals with sins and the resurrection with commissioning. Paul quotes this outline of his gospel in order to answer questions about the resurrection. Although he does not develop 'died for our sins' (the plural noun is not characteristic of Paul), he must have accepted the phrase as meaning 'he died in order to deal with the consequences of our sinning'.

That thought may perhaps be advanced by the words about the cup of the Lord's Supper in the Pauline tradition: 'This cup is the new covenant in my blood'. That presumably refers to Jeremiah's promise that God would forgive his people's wrongdoing and remember their sin no more. Finally, if Paul's remark, 'Christ our paschal lamb has been sacrificed', can be taken as old tradition, it implies that the blood of Christ protects the community when God's destructive power is at work, namely, when his wrath operates on a world moving towards destruction.

Thus, I suppose the pre-Pauline understanding to have been something of this kind: The death and resurrection of Jesus foreshadow the end of the old age and the beginning of the new. We are waiting for Jesus to come from heaven when God will bring his wrath to bear on the evil of this present age. Since we acknowledge our sins and Jesus who died for us is our protector, we shall be safe from the wrath and shall live with the risen Lord in the new age.

Throughout the epistles Paul does not depart from that scheme, though he expands it and discusses its many implications. Constantly he bears in mind the glory of the risen Lord and its promise for believers. The transition from the veiled glory of the cross to the open glory of the resurrection is like the transition from darkness to light at the creation. Indeed the death and resurrection of Christ are a new initiative by God, and those who are in Christ are part of a new creation.

It is a powerful and exciting set of images. If taken in some manner as literally true it could lead to aggressive self-confidence among successful communities or to disheartenment among unsuccessful ones. Hence, Paul found it necessary to bring the cross more explicitly into the centre of his pastoral instruction. His announcement of a new creation was not mere rhetoric: if a new social life for humanity was becoming available, former social structures must give way. But that is what they cannot easily do. Social structures under attack seek either to seduce their attackers or to hurt them. Among successful Christians at Corinth, who were being seduced by the *kosmos* to the conformist *sophia* of urban life, and to excessive self-regard at the Lord's Supper, Paul opposed the disruptive quality of the cross. Among Christians in Galatia, who wished to complete Christian initiation by the addition of circumcision, he opposed a death that carried the curse of Torah. By contrast, among Christians at Thessalonica and Philippi, Paul found it possible to interpret their suffering *for* Christ as suffering *with* Christ. Present hardship is the necessary condition of future glory, as the cross was necessary for the resurrection. And of course, Paul himself was the test case: he had painfully sacrificed his former social identity and now was sharing Christ's sufferings in the hope of sharing his resurrection.

Thus, the death of Christ provides a critical principle for assessing the formation of Christian conduct in relation to fellow Christians and to the world;

and an interpretative principle for assessing the kinds of suffering that Christians and Christian communities must endure. But how is the death of Christ to be understood in itself, as an action that in some way changed the prospects of Jews and Gentiles, as an action marked by God as definitive and binding?

Following old tradition, Paul says that Christ gave himself for our sins to deliver us from this present evil age. The present age was evil because God himself had allowed the enemies of goodness and truth to prevail. This is Paul's doctrine of the wrath. God warns mankind by law and prophets—which law they break, which prophets they kill—and in the end hands them over to the unpleasant consequences of their obstinate disobedience. Hence, the evil of the age and the plight of mankind.

Only God himself can change the course of the evil age. Since Christ gave himself to deliver us from the evil age, he came with God's approval as his Son and plenipotentiary. His self-giving was wholly within the divine intention, the effective means of a new initiative in place of the wrath.

The self-giving was done for the benefit of sinners and in full knowledge of their sins. When we were weak, ungodly and sinful Christ died for us and thereby showed God's love. God himself intended Christ's blood as a means of atonement, that is, as an inducement for God to do for us what he may rightly do on our behalf. In his saving goodness he may properly intervene to rescue us from the disastrous consequences of our disobedience.

God is induced by Christ's blood to rescue sinners from their plight because he deals with them in Christ Jesus. Towards those who are joined to Christ what God does corresponds to what he did for Christ. Therefore, Christ at the right hand of God intercedes for us, and says, 'These are mine'. So Paul reads the early formula, 'He was put to death for our sins and raised for our justification'.

Thus, Christ's death shows a movement of God towards sinners. It is equally a movement of sinners towards God. There is a twofold movement: in Christ God was (1) reconciling sinners to himself (awakening in them a desire for his love) and (2) not counting their trespasses against them (so demonstrating his love). The theological problem is not how God can justify the impulse of his mercy against the demands of his justice but how he can persuade sinners to rely on his love when they are enduring the consequences of his wrath.

The dual movement may be regarded as an interchange: the Holy One comes under a curse, sinners receive a blessing; the rich Agent of God becomes poor, the poor enemies of God become rich. Christ, who made no concessions to Sin, becomes the victim of Sin, whereas we who are bound up with Sin benefit from the saving goodness of God. The idea of ransom is not far off, the exchange of an eminent hostage for a host of lesser prisoners. The death of one is substituted for the death of many. But the beneficiary of this exchange is not God, but Sin; and even Sin loses when Christ is raised from the dead.

Alternatively, the process of atonement may be regarded as if there were a great king who plans to relieve the plight of his people. But his court is filled with hostile and treacherous officers (rather like the shepherds in Ez 34). He cannot, therefore, proceed directly but must act deviously. He sends his son to outface the court establishment. Failing, through self-interest and malice, to recognize the son, they make the disastrous mistake of putting him to death. Now the king proceeds against them, wins his victory, brings the son to life, and rescues his people. The story—not of course to be read literally—dramatically indicates the problem. It is not a problem of God's offended dignity or of his obligation to inflict punishment. It is a problem of overcoming hostility. Pleading, warning, threats and punishment are ineffective. How can God plan to overcome hostility without increasing hostility? His chosen method is to overcome his opponents by the force of their opposition.

For all its usefulness, that explanatory myth leaves us too much as spectators, as beneficiaries from but nonparticipants in the drama of redemption. The truth of the matter is that the grace of God and the gift of life are not opposed merely by a powerful conspiracy but are attacked and ruined by a parasitic contagion called Sin. Sin, with its companion Death, seizes on whatever is good and ruins it. Human beings, almost before they know what they are doing, are caught up into the compulsive service of Sin. It is not simply that my limited wrongdoing spreads out and has untold consequences (though it has) or that my acts of sin diminish my awareness of God (though they do) but that my sincere, energetic attempts at goodness can turn out disastrously. In these circumstances, talk about the forgiveness of sins is rather unimportant (and Paul does not talk about it). Anyone who has found a social situation running out of control (a strike, a race riot, a politicoreligious procession) knows what Paul means: everything fans the flames—the use of power, the offer of concessions. When such is the case, only two things are possible: either something so shocking happens that the violence collapses, or the violence runs its course into mindless bitterness and then a totally new beginning is made. For Paul the shocking event was the humiliating death of the Son of God, which could be regarded as an aversion sacrifice to avert danger and restore confidence. But more an aversion sacrifice to avert danger and restore confidence. But more profoundly, the death and resurrection of Christ could be understood as the death of one who represents the first Adam and the creation of one who constitutes the new Adam. Moreover, we are united with him in a death like his and so shall certainly be united with him in a resurrection like his. Our self-identity belonging to Sin is destroyed, and our new identity is in process of formation by the Spirit that raised Jesus from the dead.

Obvious traces of Paul's teaching can be seen in Colossians and Ephesians. The same explanatory myths and some new ones carry forward that kind of storytelling theology. But two striking changes are evident. The death of Christ is no longer the focus of attention: it has been replaced by the sufferings of Paul, the chief mystagogue. The death of Christ, indeed, makes possible the resurrection benefits in the body of the Church, but interest is directed to the blessings of the heavenly realm, where Christ in risen power sustains and rules the world. Indeed, Christians are already raised with Christ, their life is hid with Christ in God. They experience forgiveness of their trespasses through his blood, they have knowledge of the mystery, and they look forward to the redemption of all that belongs to God.

In a theological sense, Colossians and Ephesians have learnt rather little from Paul. They present Christianity as a New Life religion, and do it very well. Inter-

est in the church is well advanced. Moral instruction is dignified and something
more than conventional. The hymnic language is high flown, suggesting exciting
possibilities in the supernal world. But in both epistles there are only enough
significant references to the death of Christ to give the enquiring mind a chance
to begin with Christ's suffering on earth before rising to the mysteries of heaven.

Something very different was produced when 1 Peter was written on the
periphery of the Pauline mission area. The death and resurrection of Christ are
basic for the content of faith and the existence of the community. But because
the communities were suffering or in danger of suffering, Christ's death becomes
the model for understanding their unhappiness, for invoking God's blessing in
their need and for preventing further social disadvantages. The writer's very positive engagement with Christ's death is theologically instructive, but it has to be
noticed that Peter's use of the cross to encourage social conformity (and even
justify social subjection) contradicts Paul's use of the cross to promote a kind of
social disturbance.

Like 1 Peter, Mark's Gospel both supports the basic affirmation, 'He died and
rose', and also puts most emphasis on Christ's death. There is no resurrection
appearance, though Christ has certainly risen and Christians are instructed about
life between his rising and his coming again. It is in the Galilean actions and
stories that disciples may expect to see Jesus. But those actions are provided with
a framework, namely, the imprisonment and death of John Baptist. The cruel
fate of the prophet, the ambiguity of prophetic signs, the conflict between
prophet and the shepherds of Israel are in prospect for Jesus and for those who
would follow him. In the end Jesus is rejected by the leaders and the people,
mocked by the Romans, reviled by his fellow sufferers and abandoned by his disciples and apparently by God. In this remorseless narrative only one possibility
of hope appears: the opponents of Jesus by their actions allow the truth to appear.
Thus, the truth of Mark's Gospel does not exist apart from the cruelty and suffering, does not triumphantly leave distress for others to bear; rather, truth must
be learnt and experienced within that situation. He indicates that this has higher
authority than the evangelist's opinion by using traditional sayings of Jesus to his
disciples, deliberately repeated or carefully located, to give a special character to
the latter half of his Gospel (see pp. 359–60). So the Gospel, which begins with
the prophet's courting danger to announce the Kingdom of God in words and
deeds, ends with the Son of man as ransom and victim, so that others should join
him when God displays the kingdom that comes not by power but by suffering.
In that way, also, Mark is not remote from 1 Peter—though Mark suggests sub-
version by submission.

If we return to the Pauline corpus and consider the Pastoral Epistles, we are
in another world. The Christian communities for which Timothy and Titus are
responsible are encouraged to behave sensibly and modestly, to live sober,
upright, and godly lives. They are provided with models of sound teaching incorporating several unexceptionable Pauline phrases, though the heart has gone out
of them. The writer prefers to speak of Christ's epiphany by which life and
immortality was brought to light. The very appearance among us of our great

God and Saviour sufficed to release us from sin, cleanse us by the Holy Spirit, and purify us as a people for himself. Anyone familiar with Pauline language, on hearing that Christ gave himself as a ransom for all would remember Christ's death; but the author is not explicit. When some Christians say that the resurrection had taken place already, they are told that if we die with him, we shall live with him; if we endure, we shall reign with him. That almost sounds genuinely Pauline until one realizes that the good confession that Christ Jesus made before Pontius Pilate is really an encouragement to heroic endurance.

Luke is also impressed by the heroic quality of Christ's death, and he describes the events that led up to it with dignified expressions. Jesus is indeed a great person, who, in moving towards his end, provides models of goodness and nobility and displays magnanimity and graciousness. In a culture different from our own this could have been an acceptable way of moderating the offence of the cross and of demonstrating the greatness of Jesus' person and work. But one may wonder whether Luke had much sympathy with what Mark was saying. If he grasped it, he took some pains to modify it.

Luke was chiefly interested in the resurrection, which he thought (perhaps rightly) to have been the chief burden of the primitive proclamation. When he had to justify the death of Jesus, he said two things: that he died in accordance with God's predetermined plan contained in scripture and that it was entirely the fault of the Jews. What could be a more Jewish verdict than that? And given the devout moralism of Luke's Jewish heritage, it is not surprising that Jesus died an obedient but not an atoning death.

Matthew is a collector, arranger, and preserver of traditional material. What Mark says about the death of Christ he accepts. His modest additions to the Passion narrative make it more distinct: the motives of Judas and the carelessness of the guard supply human interest; the earthquake concedes a liking for prodigies. But when Matthew adds great blocks of teaching to Mark's narrative, the character of Jesus is altered, and so is the characeter of his death. Whereas in Mark Jesus is a prophet who works wonders by his healings and his parables and in the end does the greatest marvel by his crucifixion, in Matthew Jesus is an inspired teacher and healer whose work cannot be destroyed by death but is continued by those to whom he gave authority. Consequently, even persecution and death cannot ruin their teaching mission.

Paul's explanatory myth of a great king scheming to outwit and ruin his enemies by the death and resurrection of his son is replaced by something superficially similar but very different in the Epistle to the Hebrews. The divine being is an all-consuming fire, existing beyond the heavens in the world of reality. Those who live in the world of shadows can approach him only through sacrificial purification and the overcoming of many dangers. The shadow world, constantly threatened by death, which moves even further from God, is not wholly lacking in indications of the real world. It has cultic rules and practices that will not allow the link between shadow and reality to be forgotten. But the daily performance and necessity of the cult cannot finally, decisively, overcome death and plot a route through the heavens to the fearsome presence of the deity.

Therefore, the divine Wisdom, the Son of God, becomes the pioneer of our salvation by acting as the mysterious Malchizedek high priest on an archetypal Day of Atonement. Sharing in flesh and blood, he sacrificially offers himself (thus neutralising the Devil's power of death), makes expiation for the sins of the people, and becomes our forerunner by passing through the heavens. There he is seated at the right hand of the Throne and lives to intercede for those who draw near to God.

The argument of the epistle is done with enormous skill. Old traditions are modified—even transformed—by new perceptions, but not abandoned. When it seems that the great high priest is beyond the reach of ordinary Christians, we remember that he shared, in fact, our own sufferings and temptation and that he is our forerunner. Yet despite the writer's spiritual ingenuity and persuasion, the reader may uneasily gain the impression that the death of Christ is over and done with, that his living presence is remote beyond the heavens, and that reality is more like a cultic sanctuary than anything else.

The Johannine community shared a view with Hebrews that the gospel offers enlightment and with the Pastoral Epistles that the benefits of the gospel were achieved by disclosure. Unlike Hebrews, it was not moved to develop its views by means of a Jewish cultic pattern (indeed, it was rather hostile to the Jewish cult); nor was it content, as the Pastoral Epistles were, to rely on the formal recital of traditional phrases (though 1 John shows that it was not indifferent to formal tradition). Instead, it adopted a pattern of descent and ascent (rather like Philippians) and worked out the consequences by reexploring the traditional material of the Jesus story. In consequence, it is made plain that the death of Jesus is entirely within his own control, the means by which he makes his return to the Father and makes possible for others the gift of eternal life. By faith in Jesus the transition from death to life takes place, provided that faith includes the constant awareness that the life of the many depends on the death of the Son of man. If Jesus is rejected and his death ignored, condemnation follows. That is true as much for the community as for the hostile world. Hence, the distinction between the community and the world is defined and guarded by the death of Jesus. As the Passover lamb once guarded the people of God, so the death of Christ protects sinners within the community and restores them so long as they confess their sins and remain faithful. Thus, the problem of the Epistle to the Hebrews is solved with more acceptable results: the death of Christ is constantly present when the community (in the eucharist) eats the flesh and drinks the blood of the Son of man; the living presence of the Father and the Son is guaranteed when they abide with the believer or welcome him to his place in the Father's house; and reality indeed is more like a country farm than anything else, with bread, and vines, and sheep, and a road, and life, and light.

Revelation speaks of the death of Christ in relation to the martyrdom of Christians and the sufferings of mankind caused by the frightening and tyrannical power of the state. Christ the protomartyr is both witness and accuser in the divine court, and the prophet himself is drawn into the process of denunciation. The twofold conviction of this writing is that by suffering martyrdom

Christians are rendered guiltless and that by causing martyrdom the state has become diabolic and must be destroyed. If the blood of Christ is still (in a sense) the protective blood of the Passover, equally significant is the belief that shed blood calls for vengeance—in the hope that beyond the vengeance there will be a new world.

During the exegetical work and throughout the survey and summary, I have been forced to the conclusion that the most instructive passages in the New Testament about the death of Christ are the six Pauline Epistles, the Gospel of Mark, and the First Epistle and Gospel of John. The variety of those writings, the different situations to which they are addressed and the theological intensity of their discursive and narrative skills give a profundity to their understanding of Christ's death such as is not observable elsewhere. The one addition that might be made to that list is Hebrews. It is a remarkable tour de force, almost to be regarded as a prize composition: entrancing to hear, difficult to live with. For those who do not naturally share the author's experience of sacrifice and his view of the shadow world and the real world, the epistle presents an intellectual puzzle that can perhaps be solved. Those more sympathetic to the author's views, who make the intellectual effort that the epistle deserves, may in the end conclude that the argument concedes more than it gains.

At a secondary level of intensity other New Testament writings take their place. It is entirely proper that voices should be heard (whether we agree with them or not) that stress the commissioning of Christian teachers and the dangers they run in being the disciples of Jesus (Matthew); the majestic dignity of Christ's suffering in the midst of Jewish (though not really Roman) hostility and the consequent encouragement for missionaries to the Gentiles (Luke-Acts); or the evidence that Gentile Christian communities could regard themselves as risen with Christ in a stable but exciting world (Colossians and Ephesians), that faithful sayings of Jesus could be preserved in the tradition of the community when the gospel had delicately been rephrased (Pastorals), or that the death of Christ could be an encouragement both to Christians who needed to avoid the attention of the authorities and to those who wished to denounce the state (1 Peter and Revelation).

I stress these social connexions of the death of Christ (not confined, of course, to the writings just mentioned) in order to counter the impression that his death belongs chiefly to a transaction with God carried out on our behalf, that it is primarily a feature of atonement theology concerned with the removal of guilt. To act on that impression is to deprive Christ's death of a great deal of its significance. In discussing Paul, I have described his use of the death and resurrection of Christ as a critical and interpretative principle, that is, a principle examining and assessing the life of the community and allowing the community to interpret its own sufferings as participation in the sufferings of Christ. The Gospel of Mark is so arranged as to give direct teaching about taking up one's cross and following Jesus; and (after the resurrection) it sends disciples to find the risen Christ in Galilee, where he provoked the hostility that led to his death. In the Johannine tradition the death of Christ protects the boundary between the community and the world, and the Evangelist deliberately placed his tradition about eating the flesh and drinking the blood of the Son of man in a discourse attached to the feeding of the multitude. It ought to be clear from baptism and eucharist that the death of Christ is the structural principle of Christian life, not merely the formal principle of forgiveness.

It is true, of course, that Christ's death has to do with forgiveness. There are two reasons why the standard Jewish practices of atonement and forgiveness were thought to be insufficient, why, indeed, the simple rule that God forgives our trespasses if we forgive those who trespass against us would not suffice. First, the Christian communities needed to demonstrate that they were indeed God's people. So admitting their sins and repenting of them, they asked God's forgiveness through Jesus Christ their Lord; that is to say, Jesus, who died for them now vouched for them before God, saying, 'All mine are thine, and thine are mine, and I am glorified in them'. Second, being forgiven was not simply a matter of hearing God say, 'Your sins are forgiven', but also the additional words, 'Rise, take up your bed, and walk'. Since forgiveness is requested through Jesus Christ, who died and rose again, the very act of forgiveness transforms the sinner's condition. The death of Jesus discloses the seriousness of sin. It is not a case of people fundamentally good who occasionally (or even habitually) go wrong, but of people whose social existence is in the grip of Sin. They need a remedy not only for their wickedness but for the inevitable corruption of their attempted goodness. They need to be released from the compulsion of Sin and brought to a position where they can fight it. If they are in Christ, if the Son abides in them, if they follow Jesus it is possible—because he died and rose.

In speaking of the seriousness of sin I refer not so much to the horror, shock, and anguish of particular wicked actions or to the appalling extent of some ruthless political judgements, some callous economic decisions, some misguided scientific plans. I refer to the power that ruins what is good in human life. There are two ways of dealing with such an evil power. One is to resist it, fight it, and perhaps destroy it; the other is to allow it to win and encourage it to overreach itself and produce its own destruction. It is significant that the dominant images of the explanatory myths depend on the second method—significant, too, that they are images of divine victory. In Western tradition it is constantly asserted that Paul's doctrine of atonement is juridical, that his metaphors are forensic (who are we to accuse Jews of legalism?). If we want to see a divine court in action, we have only to read Revelation. But at the back of Paul's mind—even, indeed especially, in the passages dealing with 'justification'—is the imagery of God, who is contriving a means of liberation for his captive people. Or (if we cannot resist the courtroom drama) when God justifies the ungodly, he gives the verdict, in the trial of strength between the sinner and his adversary, to the sinner—because the sinner does not fight on his own account but is represented by his champion, Jesus Christ, who died and rose again.

The New Testament writers have, at the back of their mind, 'a story that is told'. That story decides what kind of God they are talking about. Paul's story points to a God who cunningly uses his power to win liberation for his people,

John's to a God who delegates his authority so that by descent and ascent he can enlighten his people, Mark's to a God who commissions his Son to display the divine gifts of healing and suffering for the renewal of his people. All the stories essentially contain the death and resurrection of Jesus not only as the story of a unique person but as the story of us all: AS DYING, AND BEHOLD WE LIVE.

APPENDIX A

The *Lytron* Group of Words

This excursus is a summary of information available in AGB (s.v. *lytron* TWNT 4 (s.v. *lyō*), and Hill, *Greek Words and Hebrew Meanings* Chapter 3 and a brief survey of the New Testament occurrences as a whole.

	New Testament Occurrences	Septuagint Occurrences
lytron	2 (1 each in Mk and Mt)	19 (11 in Pent, 2 in Pr, 1 in Is)
lytroō(mai)	3 (1 each in Lk, Ti, and 1 Pt)	104 (37 in Pent, 25 in Ps, 11 in Is)
lytrōsis	3 (2 in Lk, 1 in Heb)	10
lytrōtēs	1 (in Acts)	2
apolytrōsis	10 (1 in Lk, 7 in Paul, 2 in Heb)	1
antilytron	1 (in 1 Tim)	1
apolytroō	—	2

In classical Greek, *lyō*, meaning 'to loose', gives the sense of releasing persons on receipt of a ransom, hence 'to ransom'. So *lytron* is a means of releasing, or ransom money, thus giving rise to the verb *lytroō, lytroomai,* 'to ransom'. The remaining words of this group are rare or very rare. With few exceptions, *lytron* words are associated with ransom, often with mention of a price.

In the legal codes of the Septuagint, *lytroō* has the classical, profane sense of ransom by payment. In other parts of the Septuagint, however, the verb means simply to release or deliver and lacks any association with ransom or price. The verb *lytroō* is used forty-five times to translate the Hebrew verb *g'l*, which means to act as a kinsman in recovering what is lost, to redeem. The verb *g'l* itself on forty occasions has God as subject, often for his redemption of Israel from Egypt or Babylon. Here there is no obvious ransom suggestion and the Hebrew verb means 'to redeem'. The verb *lytroō* (when representing *g'l*) can indeed be used for the payment of a ransom price in a legal context with a human subject (eighteen times). Or it can be used as God's deliverance of Israel and of individuals (twenty-seven times). The verb *lytroō* is used forty-three times to translate the

370

Hebrew verb *pdh,* which means to transfer a person or things to one's own own-ership from another's by giving an equivalent or a substitute. The verb *pdh* itself on thirty-three occasions has God as subject, meaning 'to recover' with no idea of an equivalent. The verb *lytroō* (when representing *pdh*) can be used in the Pentateuch of ransoming by individuals with the provision of a substitute; or of God's deliverance of people (thirty-three times). Thus, the Hebrew verbs, outside the legal codes, move away from the legal or cultic area into the wider area of 'releasing', and in the Septuagint the Greek verbs follow them. So in 1 Mc 4:11, 'There is one who redeems [*lytroumenos*] and saves Israel'. It may be added that *g'l* and *pdh* are sometimes represented in the Septuagint by other Greek words without any notion of ransom; and that sometimes *lytroō* translates other Hebrew verbs that lack the idea of ransom. Philo interprets *lytroō* even in ransom passages as 'release'. In the Dead Sea Scrolls *pdh* means deliverance: if the notion of price is required, it is specifically added. In rabbinic texts, *g'l* is frequently used of the deliverance from Egypt, without any ransom idea.

The New Testament occurrences can be considered in five groups.

Luke. In Lk 1:68, in Zacharias' liberation song at the birth of John Baptist, 'The Lord God has visited his people and made *lytrōsis* for them', namely rescue from enemies and salvation (see verses 71, 74, 77–79). In Lk 2:38 the prophetess Anna spoke of Jesus to all who were looking for the *lytrōsis* of Jerusalem, namely, its liberation. In Lk 21:28, perhaps Luke's rewriting of a section of apocalyptic signs, the Son of Man appears as controller of the heavenly powers and the bringer of *apolytrōsis,* that is, redemption. In Lk 24:21, on the Emmaus journey, they had hoped that Jesus would have been able to redeem *(lytrousthai)* Israel, and in Acts 7:35 God made Moses a ruler and *lytrōtēn* (redeemer) of Israel.

Thus, in Lukan writings this group of words has no hint of ransom or of any *means* of liberation or redemption, and it is not connected with the death of Jesus. The words are always *prospective,* that is, they point to something promised or expected, not to anything in the past.

Hebrews. In Heb 11:35, Jews endured torture refusing to accept *apolytrōsis,* that is, release. In Heb 9:12, 15, Christ appeared as high priest of the good things that have come and, taking his own blood, entered once for all into the Holy Place, thus securing an eternal redemption *(aiōnion lytrōsin).* Moreover, for this reason he is the mediator of a new covenant, since a death has occurred which redeems *(eis apolytrōsin)* them from the transgressions under the first covenant.

Here the words are explicitly associated with the death of Christ and its affect on transgressions. Whether the argument permits the general sense of 'deliverance' or demands the more-specific sense of 'ransom' is discussed in the examination of Hebrews (see Hill, *Greek Words and Hebrew Meanings* pp. 68–69f); but it appears that *apolytrōsis* is still in part prospective. Its meaning is shaped by the writer's awareness of the two structures of earth and heaven and the need to be released from one to the other. It is likely that the word indicates rather *what* Christ achieved than *how* he achieved it.

The Pauline Epistles.   In Rm 3:24 sinners are justified by his grace as a gift through the redemption *(apolytrōsis)* which is in Christ Jesus. Here the death of Christ is a *hilasterion;* what Christ makes available is liberation from law into grace. In Rm 8:23, possessing the Spirit as first fruits, we look forward to our sonship, namely, the *apolytrōsis* of our body, that is, the liberation of the body from its present constraints and possession of the liberty of God's sons. In 1 Cor 1:30 Christ is our wisdom, which is our righteousness and sanctification and *apolytrōsis,* that is, our deliverance from the folly, weakness and so on of verses 26–28. So for the main epistles, *apolytrōsis* available in Christ Jesus is once linked with his death and once offered at the end of the age.

In Col 1:14, we have *apolytrōsis* in Christ, which might seem to give a name to our rescue from the power of darkness and transfer into the kingdom of God's beloved son (in verse 13); but it is more conventionally described as the forgiveness of sins in verse 14ᵇ. In Eph 1:7, in Christ we have *apolytrōsis* through his blood, the forgiveness of our trespasses. In Eph 1:14, the Holy Spirit is the guarantee of our inheritance, for the sake of the *apolytrōsis* of God's property. And in Eph 4:30, we were sealed by the Holy Spirit for the day of *apolytrōsis,* that is, the day of final liberation. Ephesians seems to split the 'already but not yet fully' sense of the Pauline *apolytrōsis* into an initial redemption comprising the forgiveness of sins, associated with the blood of Christ (but only seldom elsewhere, namely, Mt 26:28; 1 Jn 1:7, 9) and a second redemption perhaps associated with the resurrection.

In Ti 2:14, Christ gave himself that he might redeem *(lytrōsētai)* us from every prohibited action and might purify a people for himself. The Septuagint parallel in Ps 130:8 suggests the meaning that Christ, by his death or his epiphany, released us from the consequences of prohibited actions. In 1 Tm 2:6 there is one mediator between God and men, the man Christ Jesus, who gave himself as an *antilytron* for all. *Antilytron* is no more than a hellenistic elegant variation from *lytron.* The saying is constructed in the tradition of Mk 10:45 and in the context of the Pastoral Epistles says little more than Ti 2:14.

1 Peter.   In 1 Pt 1:18, they were ransomed *(elytrōthēte)* from the futile ways inherited from their fathers, not with silver or gold, but with the precious blood of Christ, like that of an unblemished lamb. Thus, the Revised Standard Version speaks of ransom, presumably because precious metals are mentioned; but except metaphorically it is unsuitable. The sacrifice of lambs, to which Christ's sacrifice is compared, was not a ransom ritual; and no ransom had been paid to the patriarchal futility from which they had been liberated.

Mark and Matthew.   In Mk 10:45 (= Mt 20:28), 'the Son of Man also came not to be served but to serve, and to give his life as a ransom for many [*lytron anti pollōn*].' This saying is formally parallel to 1 Ti 2:6:

dounai	*tēn psychēn autou*	*lytron*	*anti pollōn*
[to give	his life	a ransom	for many] (Mk 10:45)

dous	*heauton*	*antilytron*	*hyperpantōn*
[He gave	himself	a ransom	for all] (1 Tm 2:6)

It is usually said that the Hebraic phrase *his life* (napšo) is replaced by the Greek word *himself* and that the Hebraic inclusive use of *many* (as the opposite of *few*) is made plain in Greek by the word *all*. If indeed *gave himself* is a sufficient interpretation of *gave his life*, the force of the Markan saying is much reduced, for *gave himself* in the Pastoral Epistles may refer to the first epiphany of Christ rather than to his death. If so, the Markan saying might mean that the Son of Man gave himself to be a liberator on behalf of mankind, so that he was offering them not servitude (see the first half of verse 45) but freedom. If, however, the traditional view is maintained that Jesus knowingly died in order to set mankind free and regarded his surrendered life as comparable to a ransom price, it is scarcely possible to picture God as a tyrant who will release captives for payment. The death of Jesus would have to be explained from Jewish martyrdom teaching, namely, that God in his generosity accepts the death of one dedicated leader as sufficient substitute for the death of many.

Thus, whenever the *lytron* words are used in the New Testament, they indicate release from some kind of constraint. Release is usually referred to Christ— secured by or available in him. Clear reference to the death of Christ is found in Heb 9:12, 15; Eph 1:7; and 1 Pt 1:18; and more or less probably in Rm 3:24; Ti 2:14; 1 Tm 2:6; and Mk 10:45 = Mt 20:28. There are three references to the Holy Spirit in Rm 8:23 and Eph 1:14, 4:30. That release is from wicked activities is indicated in Heb 9:15, Col 1:14, Eph 1:7, and Ti 2:14 and perhaps implied in Rm 3:24. If 'ransom' is ever a suitable translation, it is only in a metaphorical sense: 'It is as if a great price had been paid at great personal cost'.

APPENDIX B

Justification:
The *Dik*–Group of Words

Dikaios

The study of this term, and the group of words that express it, best begins with the Greek adjective *dikaios*. It means what corresponds to *dikē*, which, in turn, means accepted social behaviour, sanctioned by the gods or prescribed by law (Hill, *Greek Words and Hebrew Meanings,* pp. 98–100). Hence, it means observant of one's duties to gods and men, fair, lawful, just and right. In the Septuagint it translates *çaddiq* (in the vast majority of cases), an adjective from the noun *çedeq,* which means conformity to accepted standards. It is close in meaning to *dikaios.*

In the New Testament it is used in the following ways:

What is right. Mt 20:4, Lk 12:57, Jn 5:30 and 7:24, Rm 7:12 ('The commandment is sacred, right and good'), Eph 6:1, Phil 1:7 and 4:8, Col 4:1, Ti 1:8 and 2 Pt 1:13.

A person who is devout, observant, and practising. Mt 23:28, and 25:37, 46; Lk 1:6; 14:13–14; 18:9, 12; and 20:20; Acts 4:19; Rm 2:13 ('It is not those who listen to the law who are devout towards God; it is those who practise who will be treated as devout'); 1 Tm 1:8–11; and Heb 11:4.

A person in good standing with the community (and with God). Mt 1:19, 10:41, 13:17, 23:29; Heb 12:23; Jas 5:6, 16. It is joined with 'holy' (Mk 6:20), with 'devout' (Lk 2:25), with 'good' (Lk 23:50), and with 'god-fearing' (Acts 10:22). It is the opposite of 'unjust' (Mt 5:45), of 'evil' (Mt 13:49), of 'disobedient' (Lk 1:17) of 'sinner' (Mt 9:13 ‖ Lk 15:7), of 'those who do evil' (1 Pt 3:12, quoting Ps 34:15–16), of 'the impious and the sinner' (1 Pt 4:18, quoting Prv 11:31 LXX), of 'sons of the evil one' as compared to 'sons of the kingdom' (Mt 13:43), of weak, ungodly, sinful (Rm 5:7).

Paul seldom uses *dikaios,* but at Rm 3:10 he introduces it into the initial stage of his catena of quotations to demonstrate that the scriptures, which primarily apply to Jews, show that no Jew is *dikaios,* and a fortiori no Gentile. In Gal 3:11

and Rm 1:17 it appears in his very important quotation from Hab 2:4. The sense in which Paul reads that quotation must be decided by studying the two passages and by comparing them with other forms of the quotation:

ho de dikaios ek pisteōs zēsetai (Gal 3:11, Rm 1:17)

ho de dikaios mou ek pisteōs zēsetai (Heb 10:38)

çaddiq be emunato yihyeh (Hab 2:4 MT)

ho de dikaios ek pisteōs mou zēsetai (Hab 2:4 LXX)

See DSS Com. Hab 8.1–3: 'Interpreted this concerns all those who observe the Law in the House of Judah, whom God will deliver from the House of Judgement because of their suffering and because of their faith in the Teacher of Righteousness' (DSSE, p. 287).

The Hebrew can be translated, 'The godly man will live by his faith' (or faithfulness). The Septuagint says 'by my faithfulness', and Hebrews says, 'My godly man will live by fidelity' (or faith). Paul's quotation could be, 'The godly man will live by faith' or, according to a recent fashion, 'The man who is godly because of faith will live'. Stendahl (p. 26) proposes, 'There will come a time when the righteous will live by faith'. So far, I have avoided *righteous,* but we must come to it later.

Innocent. Mt 23:35 and 27:19 (in 27:4 a variant for *athōos* and in 27:24 a variant addition to 'blood'), Lk 23:47 and 1 Pt 3:18.

An epithet for Jesus and God. Jesus is given *ho dikaios* as a title of eminence in Acts 3:14, 7:52, and 22:14; and in 1 Jn 2:1 Jesus Christ *dikaios* is our sponsor *(paraklētos)* before God precisely because he is in good standing *(dikaios)* with God. He may again be so named in 1 Jn 3:7; but in 1 Jn 1:9 and 2:29 the word is probably applied to God. It is so applied in Jn 17:25; Rm 3:26 (God is both *dikaios* and the one who 'justifies' him who has faith in Jesus); 2 Tm 4:8; and Rv 16:5 (see 15:3), 7 and 19:2, where God's judgements are *dikaia,* as in 2 Thes 1:5–6.

It is important to realize that in Hebrew religion *dikaios* is a standard liturgical epithet for God, his actions, his decisions and his law. Thus, in Psalm 35 the speaker attempts to enlist God on his side against his opponents: 'Give me the decision according to thy "rightness", and let them not rejoice over me' (verse 24). If God does, the supporters who desire the speaker's 'rightness' will shout for joy and be glad and will say evermore, 'Great is the Lord, who delights in the welfare of his servant', and he himself will tell of God's 'rightness' and praise him all the day long (verses 27–28). To tell God that he is *dikaios* and to stress the rightness of his decisions (as is done constantly in Psalm 119) is a method of submission and persuasion. It is not a reminder that he is an impartial judge administering a law by which he too is bound but a submission to a God who will deal with his people as he sees fit whether they are good or bad, needy or wicked—even if it is not theologically satisfactory to hold such a view of God.

Dikaiosynē

From the adjective we can move to the noun *dikaiosynē,* which means the quality possessed by the *dikaios.* Hence it suggests acceptable social behaviour, according to which each person plays his or her proper part and gives to others their appropriate rights—the opposite, in Aristotle, of *pleonexia,* or ruthless self-assertion. In the Septuagint it translates the vast majority of occurrences of *çedeq* and *çedaqah* (the two forms being masculine and feminine respectively, without difference of meaning), the originating nouns for all words of the *çdq* root. The meaning is (1) the *condition* of a covenant community where each member plays his proper part, no one takes more than his due, and no one receives less; (2) the *activity* needed to maintain that balance or to restore it when it is disturbed. These two aspects correspond, more or less, to the distributive and corrective justice of Greek civic life, though the abstract concept of justice is not fully appropriate to Israelite communal relations. In Israelite communities it was the king's chief duty to maintain the covenant relations, and correspondingly Yahweh was regarded as chief supporter of the covenant. Hence, any community that demonstrated the *condition* of item 1 possessed *dikaiosynē;* and anyone who looked after needy and ill-treated members of the community and restrained aggressive members and protected the community from its enemies was showing the *activity* of item 2 and was thus practising *dikaiosynē.* By an idiom that diverges from Greek usage, Yahweh's successes in repairing the covenant are his *çidqot* ('triumphs', Jgs 5:11; see 1 Sm 12:7; Mi 6:5); and in the special situation from which Deutero-Isaiah was composed, Yahweh's *çedaqah* is the vindication and salvation of his people, Is 46:12–13.

In the New Testament *dikaiosynē* is used in the following familiar senses:

What the dikaios *does.* Heb 5:13 (as a mature person); 1 Jn 2:29, 3:7–8 (the opposite of 'doing sin'), 10; Rv 22:11.

What makes a judgement dikaios, *that is, expert and authoritative in its field.* Acts 17:31, Rv 19:11.

A name for religious duties, the practices of piety. Mt 3:15, 5:20, 6:1, 21:32 (all M); Lk 1:75 (L, joined with *hosiotēs,* 'holiness'); 2 Cor 9:9–10 (quoting Ps 112:9); Eph 4:24 (joined with *hosiotēs*); 2 Tm 4:8; Ti 3:5; Heb 11:7, 33; 2 Pt 2:5 (see 2:7), 21.

Rather more generally, in the sense of practical goodness. Mt 5:6, 10 (both M); Jn 16:8, 10 (opposed to sin, linked with judgement); Acts 10:35, 13:10, 24:25; Rm 14:17; 2 Cor 6:7; Eph 5:9 (linked with *agathōsynē,* 'goodness'), 6:14; Phil 1:11; 1 Tm 6:11; 2 Tim. 2:22, 3:16; Heb 1:9, 12:11; Jas. 3:18; 1 Pt 2:24 (opposed to *hamartia,* 'sin'), 3:14; 2 Pt 3:13.

Membership in the new age. The word does not appear in 1 or 2 Thessalonians, Colossians or Philemon; but elsewhere in Paul it has special theological importance in Romans (thirty-three times), 1 Corinthians (once), 2 Corinthians (five times), Galatians (four times). The most instructive way to examine these theological uses of *dikaiosynē* is to observe how they are regularly associated with one group of words and opposed to another group. *Dikaiosynē* is linked with consecration and liberation (1 Cor 1:30), with the Spirit and faith (Gal 5:5), with the believing heart and the confession of Christ that leads to salvation (Rm 10:10). It belongs with peace, joy and the Spirit in the Kingdom of God, and is contrary to food regulations (Rm 14:17). It appears when Christ is the end of the law for everyone who believes (Rm 10:4–6). Hence, joined with faith (Phil 3:6, 9; Rm 9:30–31) and with life (Gal 2:21, 3:21), it is opposed to sin and to Satan (2 Cor 5:21, 6:14, 11:15). So joined with life, spirit, freedom and consecration it is opposed to death, sin, uncleanness, lawlessness and the old Adam (Rm 5:17, 21; 6:13, 16, 18–20; 8:10). Even this formal list indicates the chief significance of *dikaiosynē* in Paul's teaching: it is the condition of the community of people who have emerged from the old age and have taken the first step or so into the new age. Whereas in Judaism *dikaiosynē* could suitably be the possession of those who were pleasing to God by fulfilling their religious duties, carrying out the practices of piety and giving themselves to practical goodness; in the Pauline communities *dikaiosynē* was possessed by those who put their faith in Christ Jesus and so experienced his risen life and his gift of the Spirit. We are concerned therefore with two divergent qualifications for God's approval and help. Paul had boldly transferred the divine approval from performance of the law to the practice of faith. If the believer or the believing community exercised faith—to avoid misunderstanding, 'faith working through love' (Gal 5:6)—it experienced *dikaiosynē*, namely, the active divine favour.

Abraham and dikaiosynē. Paul justified this transference by using the story of the forefather Abraham in which God promised him innumerable descendants: 'He believed the Lord; and he reckoned it to him as righteousness' (Gn 15:6, quoted in Gal 3:6 and elaborated in Rm 4:3, 5, 6, 9, 11, 13, 22; see Jas 2:33). Following the Septuagint almost exactly, Paul has *episteusen de Abraam tō theō kai elogisthē autō eis dikaiosynēn.* (See the commentaries of Cranfield and Käsemann on Rm 4:2 for Jewish interpretations.) The overliteral rendering in the Septuagint, still preserved in the wooden translation of the Revised Standard Version, presumably arose from failure to understand Hebrew cultic language (see Rad, *Old Testament Theology,* Gn 15:6) and from timidity with Hebrew idiom. The Massoretic text means that Abraham accepted and trusted God's promise, whereupon God regarded him as qualified for the divine favour. There is no suggestion that God observed Abraham's fidelity and accepted that in lieu of good deeds. Paul means that Abraham put his faith in God and was therefore treated as qualified to receive *dikaiosynē*, namely, God's blessing. The verb *logizomai* is rather a favourite of Paul's, and it can often mean to treat somebody or something as he or it really is (Rm 2:26, 6:11, 9:8; 1 Cor 4:1; 2 Cor 3:5, 10:2, 7, 11). The word *imputation,* which means attributing to a person a character he does not really possess (SOED), is quite unsuitable. Paul thinks that Abraham's faith makes him a proper object of the divine *dikaiosynē*.

God's dikaiosynē. In the preceding section I have taken the view that *dikaiosynē* means not a condition of the believing person, as it does when it means membership in the new age, but an activity of God. This therefore brings us to passages where Paul speaks of *dikaiosynē theou,* eight times in Romans (1:17; 3:5, 21, 22, 25, 26; 10:3, 3), in 2 Cor 5:21, and in Phil 3:9 where 'my own *dikaiosynē* which is from the law' is contrasted with 'the *dikaiosynē* from God

based on faith'. Elsewhere in the New Testament the thought is present in Mt
6:33 ('Seek his kingdom and his *dikaiosynē)*, and in the wording of Jas 1:20 ('the
anger of man does not work the *dikaiosynē* of God') and of 2 Pt 1:1 ('in the
dikaiosynē of our God and [the] saviour Jesus Christ'). It is much debated
whether *dikaiosynē theou* means (1) the *dikaiosynē* or condition of godliness that
God makes possible in the believer or (2) the *dikaiosynē* that God himself exercises on behalf of the believer. The words in Phil 3:6, 9 certainly refer to a *dikaiosynē* that comes from God and so suggests the first meaning; but the significance
of that conclusion depends on the meaning of *dikaiosynē*. If it means a condition
of moral acceptability, Paul distinguishes between achieving it by performance of
the law and receiving it as a special concession from God—though whether such
a concession is comprehensible and whether it introduces by a formula what Paul
has sought to exclude are questions worth discussing. If, however, *dikaiosynē*
means God's saving goodness, Paul is distinguishing between attempting to claim
it by devotion to the law or expecting to receive it by faith in Christ and identification with him. If the latter understanding is right, Phil 3:9 can be included
under the second meaning, as also can Rm 10:3, where the Jews, zealous indeed
for God but without proper understanding, are inadequately aware of his *dikaiosynē*. Hence, they seek to establish their own *dikaiosynē* (by devotion to the law)
instead of submitting themselves to God's *dikaiosynē*, which he demonstrated by
raising Christ from the dead. (This interpretation does not accept the Revised
Standard Version translation of verse 3 'the righteousness which comes from
God' but agrees with Käsemann on Rm 10:3; see also Käsemann's "The Righteousness of God" in Paul, pp. 168–182. It is opposed to Cranfield's translation,
'God's proffered gift of a status of righteousness in his eyes' vol. 2, on Rm 10:3
and to his argument in vol. 1, pp. 96–99 that *theou* in the phrase *dikaiosynē
theou* is a genitive of origin.) To these references must first be added Rm 1:17:
in the gospel the *dikaiosynē theou* is being revealed from God's faithfulness *(ek
pisteōs)* for our faithful response *(eis pistin)*. The apparently rhetorical figure
'from faith for faith' is a kind of rabbinic exegesis based on the variation in wording between the Masoretic text of Hab 2:4 and the Septuagint (see p. 375): the
dikaios lives by God's *pistis* (acc. to the Septuagint) and by his own (acc. to the
Masoretic text). Paul ingeniously includes both. Finally, the remaining references
in Rm 3:5, 21, 22, 25, 26 are to be understood in the second meaning. Although
the expression *dikaiosynē theou* does not appear in the Septuagint, it is wholly
appropriate to Old Testament piety, as frequently in the Psalms (e.g., Ps 143:1):
'Hear my prayer, O Lord; give ear to my supplications: In thy faithfulness
(*ᵉmunah, alētheia*) answer me, in thy *dikaiosynē*' and (verses 11–12) 'For thy
name's sake, O Lord, preserve my life: In thy *dikaiosynē* bring me out of trouble:
And in thy steadfast love [*ḥesed, eleos*] cut off my enemies, and destroy all my
adversaries, for I am thy servant'. It is more fitting to look for the meaning of
dikaiosynē in the prayers of Israel than in its legal procedure; or rather, it is
necessary to understand the forensic language, when applied to God and his people, as one of Israel's forms of devotion.

So far I have avoided using *righteousness* which R. A. Knox called 'a mean-
ingless token word' (p. 11). However useful it may once have been to hard-

pressed translators, it has now no currency in normal English. Together with *righteous,* it is an archaism in modern translations and should be abandoned. However difficult to find suitable modern expressions, there is no point in using words that either produce no impression at all or muffle the sound of a passage by the booming echoes of antique moralistic disputes. Since *dikaiosynē* is used to express both the activity of the divine being and the situation of human beings, it may be understood primarily as *God's saving goodness* and secondarily as what makes us objects of his saving goodness, namely, our *acceptability* to God.

Dikaioō

Leaving aside the nouns *dikaiōma* and *dikaiōsis,* which are best treated in the exegesis of Romans, we come finally to the verb *dikaioō.* It means to regard *something* as right; or to give *someone* his rights (the opposite of *adikeō,* to deny him his rights), or, more commonly in secular Greek, to give him his deserts. In the Septuagint it is much less common than the adjective and the noun, and it usually renders the Hebrew verb *çdq,* which is also less common than adjective and noun. It is frequently said that the verb belongs to the law courts. That, however, is likely to mislead, for it readily allows us to import modern ideas about law and judges and to forget that the ancient world did not have 'law courts' in our sense. It is safer to say that the verb is at home in dispute procedures. Hebrew usage, unlike Greek, leans towards giving someone his rights rather than his deserts and hence is in favour of supporting the disadvantaged. For example, 'Give judgement for the orphan and the poor; give the depressed and the destitute their right' *(dikaiōsate,* Ps. 82:3); and Is 5:23, which curses those who for a bribe declare that the wicked disputant is in the right, and deprive the good disputant of his rights. Ancient disputes are contests between claimants who try to persuade their hearers—the elders or a jury of fellow citizens—that they are in the right and their opponents are in the wrong. In the New Testament, *dikaioō* means to admit or show that somebody (something) is right in the judgement of men or God. Interpretation varies according as the verb is active or passive and according to the subject of the verb.

People as subject of the active verb. Lk 7:29, 'All the people and tax-collectors *edikaiōsen ton theon* ['justified God', i.e. admitted that God was in the right] but the Pharisees and lawyers rejected the purpose of God for themselves'. For the thought (though the passive verb is used) see Mt 11:19 and Lk 7:35, where the divine Wisdom is shown to be right by its results or adherents; and also Rm 3:4, quoting Ps 51.4. In Lk 10:29 and 16:15 one party tries to persuade the other party that it is in the right (it 'justifies itself').

People as the subject of the passive verb. This usually means that people are admitted or declared to be in the right. In Lk 18:14 the tax collector sent home *dedikaiōmenos par' ekeinon;* that is, he, not the Pharisee, was shown to be in the right. In 1 Cor 4:4 *dedikaiōmai* means 'cleared of charges', and in Rm 6:7 *dedikaiōtai apo tēs hamartias* means 'freed from the accusation of sin' (acc. to Hill, p. 142, a Jewish martyrdom cliché). (See Lk 18:3, *ekdikēson me apo tou antidikou mou* means, 'Get me my rights from my opponent'.) In a very few passages the passive verb, though perhaps used in no more than a conventional way, may mean that a person's words or actions will be properly judged (by God) at the final judgement, namely, Mt 12:36–37 and Rm 2:13 (see verse 16). The important formula in Ti 3:4–7 conveys characteristic Pauline teaching in the statement that God 'saved us, not because of deeds done in righteousness, but in virtue of his own mercy'; the means are 'the washing of regeneration' and the aim 'that we might be "justified" [i.e., freed from the consequences of the sins described in verse 3] by his grace' (see 1 Cor 6:11).

The majority of passages, however, refer to the actual condition of the Pauline missionary communities and almost exclusively to the debate about the qualification for the entry of Jews and Greeks into those communities: Is it by faith or conformity? The passages in question are Rm 3:20 (Jews and Greeks in Rm 3:9) and Gal 2:16, 17 (Jews and Gentiles in Gal 2:15), both referring to Ps. 143:2; Rm 3:24, 26, joined with four occurrences of the noun ('all' in Rm 3:23 means 'both Jews and Greeks'); Rm 3:28 (with Jews and Greeks in 3:29–30); Rm 4:2 referring to Abraham (Gn 15:6) and arguing in Rm 4:9–12 that Abraham is a type of both the uncircumcised and the circumcised (see Jas 2:21, 24, 25, which gives a variant—in fact a traditional Jewish—exegesis of the Genesis passage); Rm 5:1, 9, where Jews are enemies of God (like Gentiles) and are 'justified' by faith and the blood of Christ; Gal 3:11, referring to Hab 2:4 (Jews in 3:10 and Gentiles in 3:14); Gal 3:24 where 'we' are justified by faith (*we* means 'all' in 3:22, 26; Jews and Greeks are abolished in 3:28); Gal 5:4, which is specifically concerned with Gentile Christians who are under persuasion to become Jewish Christians.

In this special discussion it is misleading to imagine God conducting a trial or operating a law code. Paul is arguing about the quality that will persuade God to admit that somebody is *dikaios* in the sense 'devout' or 'acceptable'. He is not describing a legal subterfuge for pretending that guilty persons are innocent but is discussing how Jews and Gentiles belong to the Church. (The passive verb in 1 Tm 3:16 must be studied in relation to the formulaic statements of the Pastoral Epistles.)

God as the subject of the active verb. I can find no difference in meaning between this form and the passive use just described. Thus, God 'justifying' the Gentiles by faith in Gal 3:8 (arising from the quotation of Gn 15:6) is equivalent to 'no one is justified before God by the law' in Gal. 3:11; and what follows in verse 8 makes it clear that 'justification' is an aspect of the divine blessing. In Rm 3:26, 30 the verb belongs to the Pauline debate already mentioned. In Rm 4:5 God is startlingly described as *ton dikaiounta ton asebē* 'him who declares the ungodly claimant to be in the right' (see the quotation from Is 5:23 above), thus overthrowing any proper legal character to the dispute. The same effect is achieved in Rm 8:33: 'Who shall bring any change against God's elect? It is God who justifies; who is to condemn?' If, by the process set out in Rm 8:30 from foreknowing to preordaining, calling, 'justifying' and glorifying, God has set himself wholly on our side and made us his own, the dispute is over—on his terms, not ours.

The use of *justify* has crept in, carrying with it *just* for *dikaios,* and *justification* for *dikaiōsis* in Rm 4:25 (and perhaps Rm 5:18) and for *dikaiōma* in Rm 5:16. Unlike the *righteousness* words, the *just* words are commonly used in modern English with familiar meanings. *Just* and *justice* express the idea of treatment in accord with what is morally right or proper; *to justify* is to demonstrate that something said is right or that something done is appropriate. These meanings are tangential to the Pauline usage. According to context one must use a phrase: to secure a person's rights for him, to declare that he is in the right or to set him in the right.

As a guide to the extensive discussion of this subject, besides the commentaries on Galatians and Romans, see Hill, *Greek Words and Hebrew Meanings,* chapter 4 and Ziesler, *The Meaning of Righteousness in Paul.* A partisan guide to the crater-strewn battlefields of Lutheran scholarship may be found in Käsemann's *Commentary on Romans* and 'Justification and Salvation History in the Epistle to the Romans' in *Perspectives on Paul.* Also see Reumann, *Righteousness in the New Testament.*

APPENDIX C

The *en Christō* Formula

This formula, well known to be a distinctive feature of Pauline writing, has been frequently discussed; but the discussion has suffered from two defects. First, Deissmann's pioneering treatment of 1892, which insisted on the local sense of the preposition *in* and compared being in Christ to being in the atmosphere which we breathe, has been taken altogether too prosaically. For many Pauline uses, it would be possible to defend this analogy by remembering the experience of joining a group of people and discovering a friendly or a hostile atmosphere in which to work. Instead, many students have taken refuge in mysticism from the necessity of providing a full account of Pauline usage. Since they usually presume a subjective definition of mysticism in terms of religious experience (see Tannehill, p. 4, n. 7) instead of an objective definition in terms of the mystery of God's love (see Louth, 'Mysticism'), they make *en Christō* an anthropological, not a theological, conception. Second, to avoid that result, or for dislike of individual mysticism, some interpreters have followed a different path. They have interpreted *en Christō* as an expression of our corporate life in union with Christ. Thus, it becomes an ecclesial, not a theological, conception; and it relies largely on passages that speak of 'being in Christ'. But such passages are in the minority. The vast majority of passages imply physical or mental activities by God or by other persons. The proof of that observation will now be set out, and consequences will be drawn for a comprehensive interpretation of the formula.

The two most thorough investigations in recent years have been Neugebauer, 'Das paulinische "in Christo"'; idem, *In Christus;* and Bouttier, *En Christ.* To these must be added the wider survey of formulas, Kramer, *Christ, Lord, Son of God.* Shorter statements may be found in Best, *One Body in Christ* (pp. 1–33); Conzelmann, *An Outline of the Theology of the New Testament* (pp. 208–212); Ridderbos, *Paul* (pp. 57–64); Sanders, *Paul and Palestinian Judaism* (pp. 458–61). In my own judgement, the following excerpt from Lohse, *Grundriss der neutestamentlichen Theologie* (p. 96), as far as it goes, is sound:

The new life is no longer life *en tō Adam* but *en tō Christō* (1 Cor 15:22). The formulaic expression *in Christ* does not suggest a mystical relationship, as was widely supposed in earlier studies (by Deissmann and others). Paul is not a mystic, since for him Christ is the *Kyrios* who stands face to face with the believer and is always distinguishable from him. Hence, there can never be a mystical union of I and Thou. On the contrary, in using the preposition *en,* Paul indicates that humans are shaped at any given time by forces that have power over them. It is not in humans' nature freely to dispose of their affairs in complete independence. He who stands alongside Adam is dominated by sin, law and death. He lives *en sarki* (Gal 2:20, Rm 8:8–9, etc.) or *en nomō* (Rm 2:12, 23), that is, the law dominates humans as long as they live (Rm 7:1). But he who belongs to Christ perceives himself within the scope of Christ's authority. By baptism *eis Christon* (Rm 6:3) he is admitted into the body of Christ and, as a member of the eschatological people of God, is saved from the power of sin, the law and death. If anyone is in Christ, he is *kainē ktisis* (2 Cr 5:17). Hence, for Paul, being in Christ is simply the new life, so that *en Christō* can be used in a formulaic manner for various purposes, including the meaning of our own adjective *Christian.* But it is always to be understood from the fundamental position that *en Christō* indicates our shaping by the Christ event.

Dia Christou

It is well known that the *en Christō* formula is often similar in meaning to the corresponding use of *dia* with the genitive. It will be an advantage to study the use of *dia* before examining the wider and more frequent uses of *en.* (See Kramer, pp. 84–90, 146–47.)

The following *dia*-plus-genitive formulas are found in Paul:

1. the *kyrios* formula (*through our Lord Jesus Christ, through Jesus Christ our Lord, through the Lord Jesus Christ,* and *through the Lord Jesus— through the Lord* does not occur), thirteen times
2. *through Jesus Christ* (sometimes with Christ Jesus as a variant), seven times
3. *through Christ,* four times
4. *through Jesus,* twice, including one variant reading

The formula 1, from its elaboration, is likely to have been liturgical; and formula 2 possibly pre-Pauline since it is not confined to Paul (see Acts 10:36; 1 Pt 2:5; 4:11; Heb 13:21; Jn 1:17; see also Jn 3:17, Jude 1:25) and the word order *Jesus Christ* is not distinctively Pauline). The other formulas may be shortened forms of formula 1. There is no occurrence of *through the Lord,* which would be ambiguous.

All formulas are used to qualify actions performed by God or by other persons. It is not easy to see any distinction between the formulas in this respect. Those qualifying God's actions (formula 1 except as indicated) are

Romans

1:5	Paul received grace and apostleship through (see verse 4)
2:16	God judges the secrets of men through [formula 2]
5:1	We have peace with God through
5:2	We have access to grace through
5:9	We are saved by Christ's blood through [formula 3]
5:11	We have received reconciliation through
5:21	The divine grace will reign and give eternal life through
7:25	(Presumably) deliverance is available through

1 Corinthians

8:6ᵃ	All things (are caused by God to) exist through
8:6ᵇ	We are brought into existence (by God) through
15:57	God gives us victory through

2 Corinthians

1:5	Consolation is abundant through [formula 3]
1:20	There is the Amen to God through [formula 2] (see verse 20ᵃ, '*In* him is the Yes')
5:18	God reconciled us through [formula 3] (see verse 19, '*in* Christ')

Galatians

1:1	Paul was made an apostle through [formula 2]
6:14	The world was crucified to Paul and he to the world through

Philippians

1:11	We are filled with the fruits of righteousness through [formula 2]

1 Thessalonians

4:14	We have died through [formula 4] (see 1 Cor 15:18, '*in* Christ')
5:9	Salvation is available through

These expressions describe the divine activity—especially that which is for the benefit of God's people—as taking place through Jesus Christ our Lord, very often with a clear notion of the direct, personal action of Christ. Those qualifying actions by other persons (formula 1 except as indicated) are

Romans

1:8	Paul thanks God through [formula 2]
5:11	We boast in God through (see Phil 3:3, '*in* Christ Jesus')
5:17	Those who receive the abundance of God's grace will reign through [formula 2] (but see Rm 5:21 under God's actions)
8:37	We succeed exceptionally through
15:30	Paul encourages them through (see 1 Thes 4:1, '*in* the Lord Jesus')
[16:27	Glory is offered to God through] [formula 2]

2 Corinthians

3:4	We are confident through [formula 3]
4:5	We are slaves through Jesus [formula 4] (a variant reading; 'Jesus Christ as Lord' precedes it)

1 Thessalonians

4:2   Paul gives instructions through

Whereas it is easy enough to see how formula 1, 'through Jesus Christ our Lord', can describe the saving activities of God, it is not easy to see how they can describe the actions of Paul and other Christians. The clue is provided by a solitary example of 'through the name of our Lord Jesus Christ' in 1 Cor 1:10. Paul appeals by using the 'name' (i.e., the authority) of Christ. In 2 Thes 3:6 Paul commands '*in* the name of our Lord Jesus Christ' (see 1 Cor 5:4), and in this instance there is no difference between *dia* and *en*. That *name* signifies authority is shown by Phil 2:10, 'In the name of Jesus every knee should bow', and that it includes power is presumed by 1 Cor 6:11, 'You were justified in the name of the Lord Jesus Christ and in the Spirit of our God'. Thus, the name formula elucidates the meanings of both the *en*-formula and the *dia*-formula. As regards the latter, it may be said (with perhaps too-bold simplicity) that God acts through the power given to Christ, that Paul and others act through the authority bestowed on Christ, and that Christ himself acts by means of the authority and power he possesses from God.

This may be pressed a little further if attention is given to Rm 5:17–19, where Adam and Christ are contrasted. 'The one man' (Adam) is placed alongside 'the one man Jesus Christ':

> So therefore the result *through* the wrong act of one man
> was condemnation for all men;
> and the result *through* the right act of one Man
> was acquittal and life for all men.
> For as *through* the disobedience of one man
> the many were constituted as sinners,
> so also *through* the obedience of the one Man
> the many will be constituted as righteous.

Through represents *dia* with the genitive. To that passage must be added 1 Cor 15:21–22:

> For since *through* man comes death,
> also *through* Man comes resurrection of the dead.
> For as in Adam all die,
> so also in Christ shall all be made alive.

Two things are clear: that the actions of Christ produce their results because he is the counterpart to Adam and that *through* Christ is very similar in effect to *in* Christ. In 1 Cor 15:44–49 Paul says more about the parallel and contrast between Adam and Christ and lays it down that 'just as we have borne the image of the man of dust, we shall also bear the image of the man of heaven' (see 'conformed to the image of his Son', Rm 8:29). Jesus Christ is to be understood as the heavenly Man, God's Son created in the image of God (2 Cor 4:4), who represents both the divine being and the intended nature of humanity. When, therefore, God acts 'through our Lord Jesus Christ', he is acting through his representative thus acknowledged by the Christian community; and when Christians act 'through the Lord Jesus Christ', they are acting in conformity with their redeemed nature.

En Christo

Now we can turn to *en Christō* and *en kyriō*. The *Christos* formula, has two main forms: (1) *en Christō Iēsou*, thirty-three times of which three (possibly four depending on the text) are expanded to *in Christ Jesus our Lord* (though the expansion is stylistic to mark the end of a 'paragraph'), and three represent *en autō*; (2) *en Christō*, twenty-eight times of which one is textually uncertain, and three represent *en autō*. In general, *in Christ Jesus* is preferred, except that in 2 Corinthians only *in Christ* appears, plus two *in him* phrases referring back to *in Christ Jesus*. It can scarcely be shown that there is any difference in meaning or emphasis between the two forms (see 1 Cor 4:15[a–b], though see Kramer, pp. 143–44). Like the *dia* formula, they describe the actions of God and of other persons, but in addition they are used to indicate the condition or status of persons. The *kyrios* formula, though similar, follows a distinctive path. Of God's actions there are

Romans

3:24	God justified us by his grace through the redemption which he provides in Christ Jesus.
6:23	God's free gift of eternal life is offered in Christ Jesus our Lord.
8:2	The Law of the Spirit of life in Christ Jesus sets us free.
8:39	God's love, from which we cannot be separated, is (effected) in Christ Jesus.
16:7	God performed the creative act (the assumption behind the verb *gegonan*, see Gal 3:14) for Paul and others in Christ.
16:10	God approves someone (who is therefore *dokimos*) in Christ.

1 Corinthians

1:2	God consecrates us in Christ Jesus.
1:4–5	God gives us his grace in Christ Jesus and enriches us in him.
1:30	From God we have our existence in Christ Jesus.
15:22	As in Adam all die, so in Christ he will bring all to life.

2 Corinthians

1:19–20	God confirms all his promises in Christ Jesus.
2:14	God leads us captives in his triumph in Christ.
3:14	In Christ God removes the veil.
5:17	God makes a new creation for anyone to be in Christ.
5:19	In Christ God was reconciling the world to himself.
5:21	As Christ is made to be sin … we are made (by God) to be righteousness *in him*.

Galatians

2:17 God justifies us in Christ.

3:14 God extends his blessing of Abraham to gentiles in Christ Jesus.

5:6 In Christ Jesus neither circumcision nor uncircumcision has any force with God, but only faith acting in love.

Philippians

2:1 (Presumably, God provides) encouragement in Christ.

3:9 We are 'found' (by God) in Christ.

3:14 God offers a further prize in his upward call in Christ Jesus.

4:7 God's surpassing peace guards us in Christ Jesus.

4:19 God fulfills our needs by his wealth and splendour in Christ Jesus.

1 Thessalonians

5:18 God's will, that we should rejoice in all things, is expressed in Christ Jesus.

As for Paul's actions, Paul speaks the truth in Christ (Rm 9:1; 2 Cor 2:17, 12:19), boasts in Chirst Jesus (Rm 15:17; 1 Cor 15:31; Phil 3:3); his apostolic preaching fathers Christians in Christ Jesus (1 Cor 4:15[b]), and so he is their father in Christ (1 Cor 4:15[a]); he has his ways of behaving in Christ Jesus (1 Cor 4:17); he conveys his love for them all in Christ Jesus (1 Cor 16:24); he is weak *in him* (sharing Christ's crucifixion), though he will live with him from the power of God (2 Cor 13:4; compare the ironical confession, we are fools for Christ's sake, weak and dishonoured; you are wise in Chirst, strong and honoured, 1 Cor 4:10); he has freedom in Christ Jesus (Gal 2:4); his bonds are shown (by his actions) to be in Christ (Phil 1:13); and he boldly commands in Christ (Phlm 1:8).

Actions by other persons are

Romans

6:11 Christians are to reckon themselves dead to sin but living to God in Christ Jesus.

16:3, 9 Paul has fellow workers in Christ Jesus.

1 Corinthians

4:15[a] The Corinthian community have countless guides in Christ.

15:19 They hope in Christ in this world and the next.

Philippians

1:26 They also may boast in Christ.

2:5 They are to think among themselves as they think in Christ Jesus.

Philemon

1:20 Philemon, acting in Christ, can relieve Paul's anxiety.

1:23 Epaphras shares Paul's imprisonment in Christ Jesus.

In all there are sixty-three occurrences of *in Christ (Jesus)* in Paul: twenty-six referring to God's actions, fourteen to Paul's, and eight to others'—in total forty-nine. The remaining fourteen will be considered under *being in Christ (Jesus)*, but they are only a fifth of the total. It would need very striking evidence to prove that 'being in Christ' was primary and 'acting in Christ' secondary.

The divine actions fall into two broad groups. First, there are his saving acts, sometimes expressed in the familiar words *redemption, justification, reconciliation,* and *liberation;* sometimes by the more striking images of God's captives and his new creation. God gives us our new existence in Christ Jesus and we look forward to its completion in resurrection life. According to Paul's exposition of these themes, it must be clear that *en Christō Iēsou* means 'by the death and resurrection of Christ Jesus'. Second, there are the acts by which God supports and enhances the community life of those he has redeemed—through his approval and consecration; through the confirmation of his promises and the removal of the veil from the scriptures; through his grace, enrichment, wealth and splendour, and peace and protection; through his inalienable love and the ability to rejoice; and through his encouragement to reach forward to the prize ahead. This remarkable list of the divine activities within the community carries one, invariable condition—that the community's life is moulded by the death and resurrection of Christ Jesus. If it were not, there would be no need to make repeated use of the phrase *in Christ Jesus.*

Paul's actions arise from his apostolic commission and are shaped by his conviction that he himself lives no longer, but Christ lives in him. Insofar as he, and fellow Christians, are in Adam they all die; insofar as they are in Christ, all will be made alive. Christians can reckon themselves as dead to sin but alive to God in Christ Jesus. Their hope in this life is rooted in him and has its proper consequences in the life of the new age. In this present age they must therefore make their community decisions as people who are determined by the death and resurrection (or exaltation, in the Philippian context) of Christ.

Thus, already, the formula has developed a community dimension. That is further expressed in the final group of uses. Being in Christ Jesus is represented by

Romans

8:1   There is therefore no condemnation to those who are in Christ
            Jesus

12:5   We, though many, are one body in Christ

1 Corinthians

3:1   Men of the flesh, babes in Christ

15:18   Those who have fallen asleep in Christ

2 Corinthians

12:2   A man in Christ . . . caught up to the third heaven

Galatians

1:22   The churches of Judaea that are in Christ (to distinguish from
            Jewish assemblies)

3:26–28   In Christ Jesus you are all sons of God, through faith. For as
            many of you as were baptized into Christ have put on Christ.
            There is no (special privilege for) Jew or Greek, slave or free,
            male or female; for you are all one in Christ Jesus.

Philippians

4:21   The saints in Christ Jesus

1 Thessalonians

1:1   The church of the Thessalonians in God the Father and the Lord
              Jesus Christ

2:14   The churches of God that are in Judaea in Christ Jesus

4:16   The dead in Christ will rise first

2 Thessalonians

1:1   The church of the Thessalonians in God the Father and the Lord
              Jesus Christ

It is not by chance that only one of these references is in the singular, and that some are inescapably communal. The essential clue is provided in what is probably the earliest extant Pauline sentence: the address to the Thessalonians. To distinguish the *ekklesia* from civic assemblies it is described as 'in God', and to distinguish it from Jewish synagogues it is described as 'in the Lord Jesus Christ'. The address may be written out more meaningfully in this way: 'The community in the jurisdiction of God (see 'In God is the result, not in me', Dem 292.21; Acts 17:28; Philo *De Fug.* 102) and in the lordship of Jesus Christ'. Not, of course, that God's jurisdiction is confined to this and other Christian communities, but these are the people who know him in whose jurisdiction they lie. When they were baptized, they entered into Christ's lordship, they took upon themselves the authority of Christ and, since Christ is the heavenly Man, the last Adam (i.e., the 'new humanity'), there can be no special privilege and no exclusive authority for Jews or Greeks, slaves or freemen, males or females. Since they are within Christ's lordship, they are no longer condemned to remain in the power of sin or even in the power of death. They may be immature, but they are sons of God, his consecrated agents under instructions from Christ, one community of believers under his lordship.

This interpretation gives meaning to all the references (even to 2 Cor 12:2, where Paul went on a secret mission to prepare him for the torment of his weaknesses). But two observations are necessary. First, even though this interpretation keeps Christ distinct from believers, it does not destroy the intimate association that is customarily read into this use of the *en Christo* formula. By the Jewish doctrine of agency, what the agent does is accepted as the act of his lord; and the lord honours the acts of his agent. Second, since the interpretation relies on the lordship of Christ, which is linked chiefly with his resurrection, his sufferings and death are not readily present to the mind. Hence, the being formulas are more limited than the formulas of God's, Paul's and other persons' actions and in fact are closer to the *kyrios* formula to which we now come.

En Kyrio

The formula occurs thirty-five times (one representing *en autō*), frequently in Romans, 1 Corinthians, and Philippians, infrequently elsewhere. Of these uses,

four times it occurs as *in the Lord Jesus,* once as *in the Lord Jesus Christ.* Those
descriptive of God's actions are

God's Actions

Rm 16:13	Chosen *(eklekton,* by God) in the Lord (though the adjective may mean no more than 'eminent')
1 Cor 7:22	Called (by God) as a slave or freeman in the Lord
2 Cor 2:12	A door was opened (by God) in the Lord
2 Thes 1:12	So that the name of the Lord Jesus may be glorified in you, and you (may be glorified by God) *in him*

These are rather formal phrases. Only one of them hints at saving activity; the
others belong to encouragement.

Paul works as an apostle in (1 Cor 9:1), loves in (Rm 16:8; 1 Cor 4:17; Phlm
16), hopes in (Phil 2:19), requests in (1 Thes 4:1; 2 Thes 3:12), rejoices in (Phil
4:10) and is confident in (Rm 14:14; Gal 5:10; Phil 2:24; 2 Thes 3:4) the Lord,
namely, with the Lord's instruction and support.

Others stand fast in (1 Thes 3:8, Phil 4:1), are faithful in (1 Cor 4:17), rejoice
in (Phil 3:1, 4:4), agree in (Phil 4:2), greet one another in (Rm 16:22; 1 Cor
16:19), welcome others in (Rm 16:2, Phil 2:29), love in (Phlm 16), marry in (1
Cor 7:39), join in the apostolic work in (Rm 16:12[a,b]; 1 Cor 15:58; 1 Thes 5:12),
refresh the apostle in (Phlm 20), and constitute the seal of his apostleship in (1
Cor 9:1, 2) the Lord.

Being in the Lord is especially represented in Romans 16, which contains an
exceptionally long list of people to be commended and greeted and employs literary variation to deal with the problem: 'In the Lord' (five times) as well as
'beloved or chosen in the Lord', 'those who work hard in the Lord', those who
stand alongside 'in Christ (Jesus)', 'brethren', 'saints', 'kinsmen', and 'beloved' as
suitable descriptions of community members. This has little bearing, outside the
greeting context, on the meaning of the formula. Since 1 Thes 1:1 has been discussed above (p. 389), the only two additions are

1 Cor 11:11	Nor is a wife (to be considered) apart from her husband; nor is a husband (to be considered) apart from his wife in the Lord (where *the Lord* means the heavenly Man, the new humanity; see above, p. 389)
Phil 1:14	The majority of brethren in the Lord (i.e., who accept the Lord's authority)

In summary, God acts for the benefit of his people *through* Jesus Christ by
the power given to him as the heavenly Man, and his people act *through* Christ
in conformity with their redeemed nature. Even more comprehensively, God acts
in Christ when performing his saving work by Christ's death and resurrection.
Correspondingly, the community acts *in* Christ when their life is moulded by his
death and resurrection. Moreover, the community can be described as *being* in
Christ or in the Lord when they are knowingly living within his authority and
acting under his instructions.

When the same formulae are examined in Colossians and Ephesians, some differences appear: (1) the formulae are much more frequent in Colossians and Ephesians than in Paul *(en Christo* 4½ times as frequent, *en kyriō* 1½ times as frequent; (2) *en Christō* is used four times as often as *en kyriō*; in Paul the ratio is three to two; (3) the distribution of the formulas between the three main uses is notably different. In Colossians 69%, and in Ephesians 79%, of *en Christō* uses refer to God's actions (42% in Paul). In Colossians 12%, and in Ephesians 17%, of *en Christō* uses refer to the actions of other persons (39% in Paul). *Being in Christ* claims 19% in Colossians, 4% in Ephesians and 19% in Paul. Thus, generally, in Paul the activity of Christians claims almost as much attention as the activity of God; in Colossians and Ephesians the divine activity receives much more emphasis. If *en Christō* and *en kyriō* formulas are amalgamated, the figures are as follows:

	Percent	
	Paul	*Colossians and Ephesians*
God's actions	29	60
Actions of other persons	51	26
Being in Christ	20	14

In Colossians activities in Christ Jesus (1:4, 2:6) and in the Lord (3:18, 20; 4:17) are conventional; being in Christ (1:2, 28) and in the Lord (4:7) are formal. The actions of God, however, are displayed in long, involved paragraphs or summarising phrases.

1:14–20   Here the *en*-formula and two uses of the *dia*-formula relate not to Christ Jesus but to the Son of God's love, the image of the invisible God. *Through* him and *in* him all things were created, *in* him all things cohere; *in* him the whole *plērōma* dwells and *through* him reconciles everything to him(self), having made peace by the blood of his cross. *Through* him we have the redemption, namely the forgiveness of sins.

2:3   The mystery of God, namely Christ, *in* whom all the treasures of wisdom and knowledge are hidden (by God).

2:7–15   Christians have received Christ Jesus as Lord and have been rooted and built up *in* him and established in the faith (by God). *In* Christ the *plērōma* of deity dwells bodily; *in* him Christians are filled full (by God)— they experience a special circumcision. They were buried *with* him in baptism, *in* which (or *in* whom?) they were raised from the dead (by God). God has disarmed the powers, having triumphed over them *in* him (or *in* the cross).

3:3   Your life is hidden *with* Christ *in* God.

In Ephesians the activities of Christians are a little more prominent. They hope in Christ, hear in him the word of truth, have boldness and confident access in him, glory in the church and in Christ Jesus is given to God, and they are taught and instructed in him as the truth is *in Jesus* (Eph 1:12, 13; 3:12, 21; 4:21). They are faithful in the Lord Jesus, they grow into a holy temple in the Lord, they obey and produce faithful service in the Lord, and are strong in him (Eph 1:15; 2:21; 6:1, 10, 21). This is a livelier use than Colossians provides; and the uses of *en Christō* to indicate God's actions are not so formally disposed as in Colossians. The Ephesian passages are

1:3–14	*In* Christ Jesus God has blessed and chosen us, destined us to be sons *through (dia)* Jesus Christ; he has bestowed grace upon us *in* the Beloved, and given us redemption *through* his blood, namely, the foregiveness of trespasses. God set forth his purpose *in* Christ, to sum up all things in him, *in* whom we have received our inheritance and been sealed with the promised Holy Spirit.
1:20	God displayed the activity of his might *in* Christ by raising him from the dead.
2:5–10	God made us alive together *with* Christ (v. l., *in*), raised us up *with* him, and made us sit *with* him in the heavenly places *in* Christ Jesus, that in the coming ages he might show . . . his grace . . . towards us *in* Christ Jesus. We were created *in* Christ Jesus for good works.
2:13–18	*In* Christ Jesus God has brought us near in the blood of Christ, who has created *in* himself *(en autō)* one new man, bringing the hostility to an end *en autō* (*in* himself or *in* the cross?), and *through (dia)* him we have access to God.
2:21–22	*In* Christ Jesus the whole structure grows into a holy temple *in* the Lord, and you also are built for a dwelling place of God in the Spirit.
3:6	Gentiles are partakers of the promise *in* Christ Jesus.
3:9–12	The mystery was hidden *in* God, and his ageless purpose performed *in* Christ Jesus our Lord in whom we have access to God by faith.
4:32	God *in* Christ forgave you.

Further examination of these passages belongs to the exegesis of the two epistles that clearly make a rather distinctive use of the *en Christō* formula.

In the Pastoral Epistles the formula appears seven times in 2 Timothy, twice in 1 Timothy, and never in Titus. Seven times it is preceded by the article in referring to life, grace, faith(fulness), love and salvation, which are in Christ Jesus. (2 Tm 1:1, 13; 2:1, 10; 3:15; 1 Tm 1:14, 3:13). People who are living a godly life in Christ Jesus (2 Tm 3:12) expect to find these virtues conveyed to them by God (so explicitly in the credal formula 2 Tm 1:9). Clearly, the Pauline phraseology is remembered, but it is becoming less useful in rallying the post-Pauline communities.

1 Peter speaks of (1) God's actions in that he called them and gave them peace in Christ (5:10, 14) and (2) the actions of other persons who rejoice *in God,* and behave well in Christ (1:6, 3:16).

Acts shows some lingering memories of the formula. Paul's preaching at Pisidian Antioch includes the appropriate words:

> Through [*dia*] this man forgiveness of sins is proclaimed to you,
> namely from all [offences] from which you could not be justified by the
> law of Moses *(en nomō),*
> *in* this man everyone who believes is justified. (Acts 13:38–39)

Possibly we should include Acts 4:2, though the most likely interpretation is 'proclaiming in [the proof case of] Jesus the resurrection from the dead'. In Paul's Areopagus sermon, he quotes 'In him [God] we live and move and have our being' (Acts 17:28, already mentioned on p. 389 above).

Two of the early Fathers are more generous in their use of *en Christo* than these later canonical writers. 1 Clement (eleven references) refers to God's calling in Christ Jesus (32.4, 46.6) and uses *en Christō* to distinguish God's Christian, apostolic agents from the Jewish prophets (43.1). Otherwise he is concerned with the behaviour of the Christian community: its piety, faithfulness, conduct, righteousness, love, instruction, good repute and common interest are all marked as 'in Christ' (1.2, 22.1, 28.1, 38.1, 47.6, 48.4, 49.1, 54.3). Ignatius often uses *en Christo* in greetings (Letters to Ephesians, Magnesians, Romans) and farewells (Eph. 21.2, Trall. 13.2, Philad. 11.2, and Polyc. 8.3, which also has the solitary example of *in the Lord*). To ask what Ignatius meant by these and other variations of his epistolary beginnings and endings is pointless: anything would do as long as it sounded splendid. In the body of his letters he used the formula, occasionally with an imaginative touch, (1) of God's activity—being raised up in Christ by the Father (Trall. 9.2) who is faithful in Christ to fulfil prayers (Trall. 13.3), being found, perfected and blessed in Christ (Eph. 11.1, 3.1; Philad. 10.2); and (2) of the activity of Christians who do all things in Christ (Eph. 8.2), are one in faith and in Christ Jesus (Eph. 20.2) and love one another in him (Mag. 6.2), sing to the Father (Rom. 2.2), rejoice (Polyc. Phil. 1.1), display purity and sobriety (Eph. 10.3) and compassion in him (Philad. 10.1); while Paul, a prisoner in Christ Jesus (Trall. 1.1, Rom. 1.1) makes mention of the churches to God in him (Eph. 12.2). It is not unlike Paul's use, but in Ignatius it can scarcely be said that *en Christō* is a leading indicator of his theology or a verbal expression that allows theological views to develop. In Ignatius, *en Christō* is traditional and sound; in Paul (with *en kyriō*) it is expressive and innovative.

And that is all. There is nothing else (according to Goodspeed's *Index patristicus* and *Index apologeticus*) in the Apostolic Fathers or the Apologists. Two apparent examples in Justin's *Dialogue* prove to be no exception. In *Dial.* 73.3, Justin says that a scriptural prophecy came true for no Jew but 'we can show that it happened *en tō hēmeterō Christō*' ('in the case of our Christ'); and in Dial.

121.1 'blessed in him' is modelled on Ps 71:17 (LXX). The very occurrence of these two forms emphasises the total absence of the Pauline formula. It is plain that in Paul's hands, the formula is theologically constructive, that it undergoes a significant development in Colossians and Ephesians but thereafter becomes a fragment of tradition that, fifty years after his death, is forgotten.

APPENDIX D

Judas

Judas is confined to the narrative part of the New Testament—Mark, Matthew, Luke/Acts, and John, roughly in order of increasing frequency—and he never *appears* before the passion (earliest in John) though he is *mentioned* earlier. When the Twelve are named in Mark 3:16–19 par. there seem to be three quartets, one headed by Simon Peter, one by Philip and one by James ben Alphaeus. Judas is in the third (variable) quartet, but before the passion he is as inactive as most of the others. In the early Church he is never mentioned, but no one else is except Peter and John Zebedee. In the apostolic list he is 'Judas Iscariot who also betrayed him' (Mk 3:19) and in Jn 6:70 he is anonymously denounced as a *diabolos* (i.e., a slanderer or informer).

The verb *paradidōmi* (betray, hand over) is a stock feature of the passion. In Paul, Jesus is handed over by God (Rm 4:25, 8:32), which may also be the meaning in the two passion predictions in Mk 9:31 par.; 10:33 par. Matthew, as well as Mt 26:2 and Lk 24:7. Or Jesus hands himself over (Gal 2:20; Eph 5:2, 25, see Jn 10:17–18, 'I lay down my life and take it up again. Nobody takes it away from me but I lay it down of my own accord'). Similarly, 'When he suffered, he did not threaten; but handed himself over [*paredidou* 'trusted' RSV] to him who judges justly' (1 Pt 2:23). In the Gospels the chief priests hand Jesus over to the Roman authorities (Mk 15:10 par. Matthew, Jn 18:30, 35 and 19:11), as is admitted by the hierarchy, Pilate and Jesus and accusingly stated by Peter and Stephen (Acts 3:13, 7:52). And then there is Judas, who hands Jesus over not to death or to the Romans but to the Jewish Sanhedrin. It was thus thought appropriate to speak of Christ's death as a *paradosis*, a betrayal or handing over. Death did not win in fair fight or by superior power: Jesus was handed over by God's prior decision, with Jesus' own consent and initiative, by the agency of Judas, the Sanhedrin, and the Romans. If it is said that Judas was prompted not by God but by Satan, Satan was acting in his ancient role as agent of God.

In the Passion narrative Judas appears in three Markan episodes: the betrayal, the Last Supper, and the arrest.

The Betrayal (Mk 14:10–11)

Judas, one of the Twelve, goes to the chief priests to betray Jesus. They are pleased and promise him money, whereupon Judas seeks a convenient way of betraying him. Luke writes somewhat more formally and adds the pious comment that Satan entered Judas (Lk 22:3–6). Matthew reports a fragment of the negotiation between Judas and the priests and deduces the agreed payment from the price of a slave in Zech 11:12 (see Ex 21:32). John knows nothing of this conspiracy or the payment of money. Mark, followed by Luke and elaborated by Matthew, is providing behind-the-scenes information that would more suitably fit between the Last Supper and Gethsemane, that is, in the place occupied by Mark 14:27–28. In view of John's silence and Matthew's elaboration, we can only wonder how much was really known.

The Last Supper (Mk 14:18–21)

In the course of the meal, Jesus makes an *amen* statement, saying that one of those eating with him will betray him *(paradōsei mē)*. That upsets the company, and each says, 'Surely not me!'; but Jesus insists that one who dips with him into the dish is indicated. Then he utters a prophetic woe:

> For the Son of man goes as it is written of him,
> but woe to that man by whom the Son of man is betrayed!
> It would be better for that man if he had not been born. (verse 21)

Luke shortens and rewrites the episode and shifts its position but says nothing new; Matthew adds an interchange between Judas and Jesus: 'Judas who betrayed him said, surely not me, rabbi! He said to him, You have said so' (Matthew 26:25). Mark, as it stands, implies that *someone* will betray him, but Jesus does not (yet) know who it will be. The Son of Man (i.e., the person concerned), goes as it is written of him in Ps 41:9: 'My bosom friend in whom I trusted, who ate of my bread, has lifted up his heel against me'. A particular betrayer is not identified, but a doom is pronounced: the (unknown) instrument of God is destined to be used and then rejected, to become an execrated nonperson. Matthew adds Judas's question, phrased as a claim to special intimacy from a man who could *not* be given the unlikely task. Jesus replies *su eipas,* a formula declining comment or (if interrogative) throwing the question back. When we read this exchange after reading the account of the betrayal, we are convinced not only that Judas is dishonest but that Jesus knows he is. These two are fencing, and the others are baffled. But if the betrayal properly belongs *after* this conversation, a more natural meaning of the passage could be maintained. It would also help with John's account of the final supper (which is not a Passover). Before the Passover the devil had already made up his mind that Judas should betray Jesus (Jn 13:2; see Barrett, St. John, p. 439). In the course of the meal Jesus asserts, with reference to Ps 41:9, that one of the company will betray him and indicates to

Simon Peter and the beloved disciple whom he has chosen. He dips a piece of bread and gives it to Judas b. Simon Iscariot and says, 'What you do, do quickly'. The company thinks that Judas is going to buy necessities for the festival or to give Passover alms; but the devil has now entered Judas, and he goes out into the night (Jn 13:21–30). Jesus has sealed his own fate.

The Arrest (Mk 14:42–46)

In Gethsemane Jesus rouses his disciples because his betrayer *(ho paradidous mē)* is at hand. So Jesus knows what is afoot; and indeed 'Judas, one of the twelve' immediately arrives with an armed band from the hierarchy. He gives them the prearranged sign, that is, he kisses Jesus, who is promptly arrested. In Mt 26:50 Jesus says to Judas, 'Friend, do that for which you have come' (RSV mg), that is, he accepts Judas's intention and aim, perhaps with agreement, perhaps with a resigned shrug. Luke shortens the account but gives Jesus the dramatically reproachful words, 'Judas, would you betray the Son of Man with a kiss?' (Lk 22:48). John 18:2–9 knows nothing of the kiss. In that account, Judas knows where Jesus often meets his disciples. So he goes there with soldiers and some officers from the chief priests and Pharisees. Then he has carried out his necessary task and needs do nothing more. Jesus discloses himself to them with Judas standing there, forbids his disciples to resist, and is arrested. It looks like either fatalistic resignation or a predetermined plan. Judas perhaps is the agent of Satan, used by Jesus in his plan of self-giving. He is 'the son of perdition [*apōleia*]' (in the ancient phrase): either the agent of destruction in the last times (see 2 Thes 2:3) or—since John's thought seldom runs on those lines—the destructive agent who fulfills God's intention and is expendable. There seems to be a double theological theme: what happens to Jesus happens by his own choice and arrangement; but since it is a cruel fate, it is attributed to Satan by the agency of Judas, to rid God of *direct* responsibility. A good deal depends on the balance between Jesus' consent and his initiative (to return to the words I used above).

That curious conclusion is not contradicted by subsequent denunciations of Judas. It goes without saying that Judas loses his reputation—that is what being a 'son of perdition' means. In Mt 27:3–10 Judas repents, makes restitution in an odd kind of way that perversely fulfils scripture, and takes his own life. Could repetance, restitution and death perhaps meet the conditions for atonement, that is, for securing a place in the world to come for a person—whether sinner or not—who had performed dreadful deeds? Luke thought not: in Acts 1:18–19 he writes up the field of blood story to give Judas an appalling and shameful death. That tradition is followed and grotesquely exaggerated by Papias (Fr. 3 in *Ancient Christian Writiers* no. 6, (p. 119). According to Tertullian *Adv. om. haer.* 2 the heretical Cainites defended Judas either because he stopped Jesus from betraying the truth or because he made salvation possible by thwarting the powers who opposed Christ's suffering (Foerster, *Gnosis*, vol. 1 p. 42; see Hennecke,

Schneemelcher, and Wilson, *New Testament Apocrypha*, vol. 1, pp. 313–314, vol. 2, pp. 62–64).

In 'How the Story Grew: Judas in fact and fiction' M. S. Enslin came to the conclusion that the death story 'certainly reflects no fact in the life of Judas, even if, which seems to me at best uncertain, this unfortunate man ever existed before his mention in Mark's narrative'. It is difficult not to feel some sympathy with that judgment, but there is one more kind of evidence to be considered: the actual naming of Judas, which is shown in the following table (following Nestle-Aland[26]):

The Twelve

Mt. 10:4	Judas the Iscariot who in fact betrayed him
Mk 3:19	Judas Iscarioth who in fact betrayed him
Lk 6:16	Judas Iscarioth who became a traitor
Jn 6:71	Judas son of Simon Iscariot who was to betray him

The Anointing

Jn 12:4	Judas the Iscariot, one of his disciples, who was to betray him

The Conspiracy

Mt 26:14	one of the Twelve called Judas Iscariot
Mk 14:10	Judas Iscarioth, the one of the Twelve
Lk 22:3	Judas called Iscariot, being of the number of the Twelve

The Prediction

Mt 26:25	Judas who betrayed him
Jn 13:2	The Devil had made up his mind that Judas son of Simon Iscariot should betray him (but 'Judas Iscariot son of Simon' is strongly supported and is preferred by some editors)
Jn 13:26	He gave it to Judas son of Simon Iscariot
Jn 13:29	Judas had the money box

The Arrest

Mt 26:47	Judas one of the Twelve
Mk 14:43	Judas one of the Twelve
Lk 22:47	the one called Judas, one of the Twelve
Lk 22:48	Judas
Jn 18:2	Judas who betrayed him
Jn 18:3	Judas
Jn 18:5	Judas who betrayed him

The End

Acts 27:3	Judas who betrayed him
Acts 1:16	Judas who became guide to those who arrested Jesus
Acts 1:25	Judas

This information may be illustrated as follows:

	Mt	Mk	Lk	Lk/Acts	Jn
Judas Iscarioth	—	2	1	—	—
(the) Iscariot	2	—	1	—	1
son of Simon Iscariot (v.l. 'Iscariot son of Simon', 'from Caryoth' and 'Scarioth')	—	—	—	—	3
(the) one of the Twelve	2	2	2	—	—
one of the disciples	—	—	—	—	1
who betrayed him	3	1	—	2	4

These facts suggest:

1. that the various Judas episodes did not belong to the original Passion narrative, since Judas is reintroduced almost every time by a descriptive phrase
2. that the early Christian community knew more than one Judas and that the betrayer had to be distinguished from the others
3. that 'the one who betrayed him' is the latest identifier when the stories moved to communities that did not know the significance of the earlier names, that Iscarioth/Iscariotes belongs to an older stage, the odd name being remembered (though variably) when the father's name had dropped out and that 'Judas ben Simon Iscariotes' (preserving the full family name) is the oldest tradition

Hence, it must be concluded that Judas was indeed a member of the Twelve and was not Mark's invention. But his actions were not essential to the Passion narrative. The earliest Christians were entirely concerned with the death of Jesus at the hands of the Romans, and they offered a theological justification: God had so decreed and Jesus chose to be handed over. When the Christian group found itself in conflict with the Jews at an official level, the handing over was attributed to the temple hierarchy. Judas plays a part in the narrative only when the danger to Christian communities came from within. Mark merely recalls the fact of the betrayer; he suggest no reason, explores no motive. Luke seizes on journalistic touches and includes the horror story of Judas's death. Matthew uses Judas as a dreadful warning and adopts his favourite device of a pretext in scripture. John shows mixed motives; partly an attack on monetary wealth in the Church, partly an old tradition that Judas did what he did by prior arrangement with Jesus or at least on an impulse from Jesus.

Notes

Chapter 1

1. For a Jewish view, see Vermes, *Jesus the Jew.*

2. For a very clear outline of views discussed mainly in the seventies, see Ziesler, *The Jesus Question.* For a solid contribution to NT study, see Dunn, *Christology in the Making.* Contributions to the debate can be very diverse, e.g., Robinson, *The Human Face of God;* Hanson, *Grace and Truth;* Moule, *The Origins of Christology;* Lampe, *God as Spirit.*

3. Taylor's Christological interest began with *The Virgin Birth* in 1920, but his main work came later: *The Names of Jesus, The Life and Ministry of Jesus* and *The Person of Christ in New Testament Theology.* His soteriological books were *Jesus and His Sacrifice, The Atonement in New Testament Teaching* and *Forgiveness and Reconciliation*—expressing views summarised and revised in *The Cross of Christ.*

4. Cf. the treatment and bibliography in Hans-Ruedi Weber, *The Cross: Tradition and Interpretation.*

5. Cf. Evans, *Resurrection and the New Testament,* chap. 1: 'A doctrine of resurrection does not appear to be one of the tenets long established in Judaism by New Testament times, but rather a comparative newcomer to it' and 'Resurrection is both limited in scope, and is not in itself positive and saving' (pp. 11, 16). See also Schillebeeckx, *Jesus,* pp. 518–23. A search through OTP suggests no revision of my statement. 1 En 51:1–5 might be slightly interesting if chapters 37–71 could be regarded as relevant to NT study; and Abimelech as Rip Van Winkle in 4 Bar. 6:6–10 is pure farce.

6. See Gubler, *Die Frühesten Deutungen des Todes Jesu.*

7. James is mildly defended by Laws, *A Commentary on the Epistle of James* and David, *The Epistle of James.* In *The Cross in the New Testament,* pp. 309–16, Morris sets out what James might have written had he turned his mind to atonement: it does not improve my view of James. Cf. Morris's earlier technical study, *The Apostolic Preaching of the Cross,* and the less technical form of it in *The Atonement—Its Meaning and Significance.*

8. The *general* point I am making would be much the same if some other view were taken of synoptic relations.

9. In Gubler's book (see n. 6), which is a comparison of recent scholarly attempts to trace the prehistory of NT references to the death of Christ, the word *umstritten* appears

everywhere. When preferring to follow A and B against C, the author is trudging through quicksand.

10. Bultmann, 'Das Verhältnis des urchristlichen Botschaft zum historischen Jesus' p. 111; Marxsen, 'Erwägungen zum Problem des verkündigten Kreuzes', p. 163. Cf. also the cautious assessment in Kümmel, *Theology of the New Testament,* pp. 85–95 ('Jesus' Suffering and Death').

11. Jeremias, *New Testament Theology;* Schürmann, *Jesu ureigener Tod,* esp. chap. 1. Cf. also Léon-Dufour, 'Jésus face à la mort menaçante'.

12. I comment on the technical problems at the beginning of each writing or group of writings. In the Pauline corpus I need not give an opinion about the authenticity of 2 Thessalonians, which includes no reference to my subject. I am not persuaded by arguments that put Galatians first or by others that place Philippians in midcourse. In mentioning 'certain affinities', I refer to (1) indications which have led some scholars to propose that Luke wrote the Pastoral Epistles or that Luke and Hebrews have a family likeness, (2) the view that Luke and Matthew were modifying Mark at about the same time, (3) the fact that questions of church order and morality are prominent in the Pastorals and Matthew; and (4) the fact that both Matthew and Hebrews, though in very different ways, draw on strongly Jewish-Christian traditions. The placing of Revelation with John simply recognizes that it had some link with Johannine Christianity and appeared at about the same time.

Chapter 2

1. The formal structure of Pauline letters is now a thriving field of study: Doty, *Letters in Primitive Christianity;* J. L. White, *The Body of the Greek Letter.*

2. 1 Thes 4:1. See the commentaries and MM ad loc.

3. In 2 Thes items 1–3 are again quickly disposed of in 1:1–2. Item 4 begins with the thanksgiving of 1:3, is reinforced by 2:13, and ends at 2:15. The prayer (5) follows in 2:16–3:5, and other matters (6) in 3:6–15.

4. Here and in 2:14 *mimētai egenēthēte* has a passive sense. So perhaps the verb is a divine passive, as Best suggests: 'God caused you to have the same experience'. In other occurrences of *mimeomai* and *mimētēs,* the meaning is active: 'copying or modelling oneself on someone's example', 2 Thes 3:7, 9; 1 Cor 4:16, 11:1; Eph 5:1; Heb 6:12, 13:7. Even 3 Jn 11 means 'Do not copy the bad example but the good'. Cf. Best ad loc. The view of Rigaux and Marshall that the imitation consists in the manner of receiving the hardship, viz., with joy prompted by the Holy Spirit, is scarcely convincing: Paul implies that they were (or were expected to be) so exalted by experience of the Spirit that sufferings could be taken in their stride. That is not the meaning of Heb 12:2 or the testimony of the synoptic Passion narratives. The Johannine Passion has Christ die with philosophic dignity but not with 'enthusiasm'.

5. For the evidence that these verses contain a pre-Pauline formula, see Best, pp. 85–87; Marshall, pp. 56–57. Since the formula contains no reference to the cross, Munck argues that it cannot be regarded as a summary of the missionary preaching ('1 Thess. I.9–10 . . .' pp. 104–10).

6. There is a development of similar ideas in Jn 5:19–29.

7. *rhyomai* appears as (1) to be rid of someone, Rm 15:31; 2 Thes 3:2; (2) to be delivered by God, Rm 7:24; 2 Cor 1:10; (3) an OT designation of God the Redeemer, Rm 11:26. See also Is 47:4; 49: 7, 26; 54:5, 8.

8. Cf. Psalm 7, where God is asked to deliver his own and to let his anger fall on the suppliants' enemies, as in 2 Thes 1:5–12.

9. There are also numerous *kyrios* references in 2 Thessalonians, where neither death nor resurrection is mentioned.

10. In 2:15 Paul uses the verb *apokteinō*, possibly because LXX uses it at 1 Kgs 19:10 (which Paul quotes in Rm 11:3), where Elijah complains that they 'have killed the prophets with the sword'. So, in Acts 3:15, Peter speaking to the Jews, says, 'You killed the Author of life'. There are similar accusations, using different verbs for killing, in Acts 2:23, 36, 4:10; 5:30; 7:52; and 10:39—though Paul phrased the matter more justly in 13:28: 'They asked Pilate to have him killed'. Cf. Lk 24:20. The Passion predictions in Mark use *apokteinō*: the shortest form in 9:31 merely says that the Son of Man will be handed over to men who will kill him whereas the form in 8:31 says that he must suffer much and be rejected by the Jewish Sanhedrin and be killed. The most elaborate form in 10:33–34 has the Sanhedrin condemn him to death and hand him over to the Gentiles, who ill-treat and kill him. The form in 8:31 arises from a conflict between Christians and Jews; the Pauline version comes from a less-comprehending phase of that conflict.

11. On this passage, see Steck, *Israel und das gewaltsame Geschick der Propheten*, pp. 274–78. Also Schoeps, 'Die jüdischen Prophetenmorde'; Fischel, 'Martyr and Prophet'. Since the historical evidence for maltreatment of the prophets is slender (e.g., 1 Kgs 18:4–13, Jer 26:20–23, the harassment of Jeremiah himself, and the death of John Baptist), the formula is clearly rhetorical: (1) confession of sin, Neh 9:26, and later rabbinic tradition; (2) complaint addressed to God, 1 Kgs 19:10, 14 quoted in Rm 11:3, cf. the resigned saying of Jesus, Lk 13:33; (3) a reproach by or on behalf of God threatening punishment: Jer 2:30 (cf. Jos. *Ant.* 9.265–6 and 10.38–39), Mt 22:6 and 23:37 (= Lk 13:34), Acts 7:51–52, 1 Thes 2:15 (cf. Mk 12:1–9); and Rv 11:4–13, 16:6, 18:24, where the reproach is addressed to pagans; (4) polemic against a rival religious group: the martyrdom of Isaiah in Asc. Isa 5.1–14 (see Nickelsburg, *Jewish Literature between the Bible and the Mishnah*; (5) heroic loyalty, Heb 11:35–38, Mt 5:11 (cf. Js 5:10. See also Schippers 'The Pre-Synoptic Tradition in 1 Thessalonians II 13–16' and Chilton, *A Galilean Rabbi and His Bible*, pp. 133–136.

12. So also perhaps is Paul; see 'drove us out', verse 15.

13. Strong support has recently been given to the view that 1 Thes 2:13–16 is a later intrusion. Pearson ('1 Thes 2:13–16: A Deutero-Pauline Interpolation?') argues that verse 16ᶜ must refer to a past, eschatological public event, that verses 15–16 contain stock Greco-Roman anti-Semitism and an accusation (of killing the Lord Jesus) that was developed after 70 C.E. Paul could not have written these words, which are the only NT evidence for the belief that churches in Judaea were harassed by the Jews between 44 and 66 C.E. These verses contain traditional material that were given a new shape after 70 C.E., when the Church and the rabbis of Jamnia were engaged in polemics. Boers ('The Form-critical Study of Paul's Letters', p. 151) finds that the excision of this paragraph enables the epistle to conform more closely to his structural proposals. The excision is accepted by Koester, '1 Thessalonians—Experiment in Christian Writing', p. 38. Boers's structural demands could be seriously entertained if Pearson's arguments were sound; but (1) he *assumes* that the decisive incidence of the divine wrath must be a public event, but that is not a necessary or (on Paul's lips) a possible assumption; (2) he notices that the main charge against the Jews is not that they killed Jesus but that they prevented evangelism and has to explain it as the Paulinizing of anti-Jewish polemic; (3) his tentative suggestions for explaining the introduction of this polemic cannot account for the very low-key reference to Christ's death at a time when Paul's other epistles had put the crucifixion in a dominant position. It is therefore better to leave these verses where they are.

14. *ei* with indicative *pisteuomen* means 'if we believe [as we do]', equivalent to the English 'since we believe'. The verbs *apethanen* ('He died', rather than 'He was put to death'), and *anestē* ('He rose', rather than 'He was raised' or 'God raised him') are simply factual, without theological overtones. Apart from *anastēsontai* in verse 16, which depends on this verse, and two LXX quotations in Rm 15:12, 1 Cor 10:7, *anistēmi* is not Pauline (but cf. Eph 5:14). The simple name Jesus (as in 1:9–10) belongs to an early stage of Christology. There is no agreed view whether 'through Jesus' belongs to 'have fallen asleep' or is a second complement of 'will bring' (see Best, ad loc). In 4:2 Paul has said, 'You know what instructions we gave you *through the Lord Jesus*' (cf. Best, Marshall). As so often with Paul's prepositional phrases, words must be supplied to convey his meaning: here probably 'through [my fidelity to] the Lord Jesus'. Cf. Gal 1:1, 'Paul an apostle, not from men nor through a man but through Jesus Christ and God the Father', which can be written out thus: 'Paul an apostle, not from [the commission of a group of] men, nor through [my allegiance to] a man, but through my allegiance to the Lord, etc.' So here: 'those who have fallen asleep through [their fidelity to] Christ'. See Klijn, *1 Thessalonians 4:13–18 and Its Background.*

15. Paul clearly supposes that God has marked out a path and a goal for Christians. In the case of the Thessalonian community, it was marked by the initial effectiveness of the preaching of the gospel (1:4–5, demonstrating their *eklogē*, i.e., God's choice of them), by subsequent suffering *(eis touto keimetha)*, and by their deliverance from the wrath at the day of the Lord (1:10). To say that God has marked out the path does not imply that all will follow it who should. Equally, if there is another path leading to destruction, it does not imply that the rest of mankind must certainly perish on it; though God may make them drunk, or send 'upon them a strong delusion, to make them believe what is false' (2 Thes 2:11). The drunkenness of hostile unbelievers comes from the cup of staggering in OT, e.g., Is 51:17–23; and it is to be regarded in the same sense as the epigram, *Deus quos vult perdere, dementat prius.*

16. Comparison of the Thessalonian and Corinthian communities: (1) their great gifts and energy, 1 Thes 1:2–5, 3:6–7, 1 Cor 1:4–9; (2) waiting for the Lord, 1 Thes 1:10, 1 Cor 1:7; (3) the problem of those who had died, 1 Thes 4:13, 1 Cor 15:29; (4) conditions at the Parousia, 1 Thes 4:14–17, 1 Cor 15:23–28; 35–53. Both epistles refer to Paul's instructions (1 Thes 4:1–2) or traditions (1 Cor 11:2): embryonic in 1 Thessalonians, much developed and argued in 1 Corinthians.

17. The most recent phase of discussion about *gnosis* at Corinth begins with Wilckens, *Weisheit und Torheit* in 1959, condensed in his article on *sophia* in TWNT 7, esp. pp. 519–525. And see Grant, *A Historical Introduction to the New Testament,* pp. 204–6; Barrett, *Essays on Paul,* pp. 6–14; Scroggs, 'Paul: Sophos and Pneumatikos'; Wilson, *Gnosis and the New Testament,* pp. 51–55; idem, 'How Gnostic Were the Corinthians?'; Ellis, *Prophecy and Hermeneutic in Early Christianity,* pp. 45–62; Barbour, 'Wisdom and the Cross in 1 Corinthians 1 and 2', pp. 57–71.

18. See p. 18

19. Betz, *Der Apostel Paulus und die sokratische Tradition,* p. 66.

20. Wilckens, TWNT 7, p. 520, line 26 on *Heilsgut.*

21. Bab. M. 59b. *Rab Anth,* pp. 340–41; and see Loewe's discussion, pp. 690–93.

22. TWNT 7, *sēmeion,* pp. 220–21. Cf. Grayston, 'Not with a Rod', p. 15.

23. Mk 8:11–12 (= Mt 16:1–4); Mt 12:38–42; Lk 11:16, 29–30; Jn 2:18, 6:30.

24. Hengel, *Atonement,* chap. 1.

25. Theissen, *Social Setting of Pauline Christianity,* pp. 99–102.

26. Demosthenes, *Or.* 9.54 associates *mōria* with *paranoia* (derangement), and a 6C

papyrus couples it with *akatastasia* (anarchy). In Aesch. *Pers.* 719 the verb imples reckless action. See TWNT 4 s.v.

27. *Kosmos* in Paul means (1) the created world (Rm 1:20) with its stars and planets (Phil 2:15) and variety of sounds (1 Cor 14:10), composed of angels and men (1 Cor 4:9) heaven and earth (observing the parallel between 1 Cor 8:4 and 8:5). Hence *kosmos* can be used to mean (2) everybody (Rm 1:8; 1 Cor 4:13, where *kosmos* is parallel to *pantes*), human existence (Rm 5:12–13) and in effect everything (Rm 4:13; 1 Cor 3:22). When God judges the world he is judging mankind (Rm 3:6, 19; 1 Cor 11:32). These are stock meanings and not very interesting. The word begins to capture attention when it means (3) society at large: 'In such and such manner we behaved in the world, and especially towards you' (2 Cor 1:12). Christians are clearly a distinguishable part of the *world* of Corinthian society from which they should not wholly withdraw themselves (1 Cor 5:10). They are in fact to use the world as though they had no full use of it (1 Cor 7:31, Barrett's trans). In any case, the form of this world is passing away, and in due course the saints will judge the world (1 Cor 6:2). Sometimes the world and its affairs are in contradiction with God's concerns (2 Cor. 7:10) or in competition with the Lord's demands (1 Cor 7:34, prob. Gal 6:14 and poss. Gal 4:3). Hence, in 1 Cor 2:12 'the spirit of the world' is contrasted with 'the spirit which is from God' and the contrasts in 1:20–21, 27–28; 3:19. Finally, just as in Rm 11:12, 15 the Jews are singled out for attention and the rest of mankind are the world or the Gentiles, so in 2 Cor 5:19 we Christians, having been reconciled to God through Christ, are charged with the ministry of reconciling the world to God.

28. When liturgical formulae are set aside, the use of *aiōn* is mostly for 'this age' (Rm 12:2; 1 Cor 1:20, 2:6, 8, 3:18; 2 Cor 4:4; and Gal 1:4). The phrase *eis ton aiōna* in 1 Cor 8:13, 2 Cor 9:9 is conventional and means 'for ever'. 1 Cor 2:7 refers to God's decision 'before the ages', and 1 Cor 10:11 speaks of Paul's contemporaries as those upon whom *ta telē tōn aiōnōn katentēken*. The meaning is not certain (see Barrett *ad loc.*) but probably implies that the ends towards which previous ages were directed have reached their appointed limit with the first generation of Christians. Since Paul does not use *aiōn* for the situation that follows 'this age', he probably uses some such scheme as before the ages–the ages–the reign of God (the last phrase in 1 Cor 4:20; 6:9–10; 15:24, 50).

29. English translations now abandon the stolid rendering of RV, 'in wisdom of words', and offer 'in eloquent wisdom' (RSV), 'in the language of worldly wisdom' (NEB), 'in the language of human wisdom' (GNB), 'in words of human wisdom' (NIV), 'in the plainest possible language' (TNT), 'by means of wisdom of language' (NJB, with a note to say: 'human wisdom [here philosophical speculation and tricks of rhetoric]'). The BFBS version in modern Greek rather naughtily has *ochi me sophian rhetorikēn*. Some translators have not decided whether Paul is mentioning the manner of his speaking or the content of what he said.

30. Metzger, ad loc.

31. On the confidence expected of a sophist, see Munck, *Paul and the Salvation of Mankind*, p. 158, n.2.

32. An interpretation of this passage in terms of hellenistic religion such as is found in the mysteries and in Gnostic sects, has followed the initiative of Reitzenstein, *Hellenistic Mystery-Religions*, pp. 426–36. The attempt would have been more persuasive if Paul had used *teleios* language in other epistles and with a clearly religious sense and if Reitzenstein's view could be related less to general features of hellenistic religion and more to the particular features of this argument with the Corinthians.

33. For information about *mysterion,* see Conzelmann, p. 62, and TWNT 4 s.v. The word appears five times in 1 Corinthians (six if 2:1 is included; but the arguments of Barrett and Conzelmann in favour of *martyrion* are to be preferred to Metzger on the grounds that *martyrion* is appropriate to the gospel and in Paul is the less-common word, whereas *mysterion* is aptly introduced only when Paul discloses the special information of 2:7), twice in the singular, three times in the plural. It also appears once in Romans (excluding the dubious final paragraph of the epistle), once in 2 Thessalonians, and ten times in Colossians and Ephesians, always in the singular. It seems likely that there were special reasons for its rather frequent use in 1 Corinthians. It can scarcely have been a dominant feature of Corinthian Christianity, since it fails to appear in 2 Corinthians. It must therefore have been relevant to a debate taking place when 1 Corinthians was written. It is clear that *mysterion* has nothing to do with the general Greek meaning of sacred rites: in four out of five places it refers to written or spoken statements. Paul's use is close to that of LXX (only Daniel and the Apocrypha), where it means an ordinary secret that a friend ought not to betray, a royal intention that the king alone decides on and discloses to whom he wills (Jdt 2:2 *to mysterion tēs boulēs autou,* cf. Eph 1:9), the hidden divine plan for vindicating the tormented righteous man (Wis. 2:22), and apocalyptic information concerning God's wisdom and power, which is disclosed to the seer (Dan 2:20). In Rm 11:25, 1 Cor 15:51, and 2 Thes 2:7 the *mysterion* is put forward to solve a perplexity arising from Paul's eschatological expectation. In 1 Cor 2:7 the *mysterion* overcomes the conflict between wisdom and the proclamation of the cross. When the plural is used (as in 1 Cor 13:2 between prophecy and knowledge), items of knowledge are being indicated that solve problems in the community or perhaps individual problems (1 Cor 14:2), 1 Cor 4:1 implies that Paul and his associates, by faithfully serving Christ, have access to the secret intentions of God. By and large, *mysteria* are statements or perceptions that solve community problems. The case is different with *mysterion* in Colossians and Ephesians. The hidden mystery of 1 Cor 2:7 is elaborated in Col 1:26–27 and becomes a means of transition from sufferings to glory, a method of becoming *teleios* in Christ, in whom all the treasures of wisdom and knowledge are hidden (Col 2:2, 4:3). In Eph 1:9 a stock phrase 'the mystery of his will' (see Jdt 2:2 mentioned above) occurs in a stream of high-sounding phrases and is doubtless much the same as 'the mystery of the gospel' in 6:19. In Eph 5:32 the mystery is an interpretation of Gn 2:24 that applies it to Christ and the church. In 3:3, 4, 9 the hidden secret of 1 Cor 2:7 is reapplied to the inclusion of both Jews and Gentiles within the one Christian community. Ephesians in its use of *mysterion* (as in many other respects) is close to the language of the Qumran sectarians (see Murphy-O'Connor, *Paul and Qumran,* contributions by Kuhn, Coppens and Mussner).

34. For the so-called parties at Corinth, see Hurd *The Origin of 1 Corinthians,* pp. 96–107; Conzelmann, pp. 33–34.

35. Paul's use of *teleios* is scarcely technical. In 1 Cor 13:10 *to teleion* is contrasted with *to ek merous* (what is complete with what is incomplete) in the context of an illustration from childhood and manhood. 1 Cor 14:20 speaks for itself: 'Do not be children, ... but in thinking be mature [*teleioi*]'. So probably Phil 3:12, 15. Eph 4:13–14 'mature manhood ... no longer children'. Col 1:28, 4:12 may be different, but there is nothing else in Paul (except Rm 12:2, where *teleios* is ascribed to the divine intention).

36. It is widely held that the rulers of this age are hostile supernatural powers (Barrett, Conzelman) or, more cautiously, that they are political rulers and the malign spiritual powers prompting their policies. In view of Carr, *Angels and Principalities,* pp. 118–120, which is an abbreviated form of 'The Rulers of This Age', this conviction (going back to M. Dibelius *Die Geisterwelt im Glauben des Paulus* of 1909) must now be revised. It is no

argument that the mythical context requires the rulers to be demons: it depends on the myth, and many myths by their very function include human figures. If the wisdom in question is social wisdom, the rulers are likely to be political rulers. Conzelmann's question, 'What should earthly powers have to do with supernatural wisdom?' adds a welcome touch of comedy to a normally solemn debate.

37. 1 En 22:14; 25:3, 7; 27:3, 5; 36:4; 40:3; 63:2; 75:3; 83:8—most of which are certainly pre-Pauline.

38. As Allo (p. 43) appears to think.

39. On glory see Ramsey, *The Glory of God,* chap. 3; on Lord see Taylor, *The Names of Jesus* p. 43.

40. See Rm 6:3–4, Col 2:12. The reference to baptism in the formula of Eph 4:4–6 notably lacks any mention of the death of Christ; but the two may be linked in Eph 5:25–26. There are more or less probable references in 1 Jn 5:6, 8; Mk 10:38–39; Lk 12:50; and Acts 8:32–38.

41. See Barrett, pp. 362–64; Conzelmann, pp. 275–77.

42. Cf. 1 Pt 3:21.

43. Cf. Lampe, *The Seal of the Spirit,* pp. 3–18; TWNT pp. 949–50.

44. As Jeremias observes in *Eucharist Words* (pp. 59–60), suggesting an early Christian passover haggadah; cf. Allo, pp. 126–27. For his treatment of 1 Cor 5:7 see pp. 205–6.

45. For the apotropaic significance of the Passover cf. Heb 11:28; Jub 49:15 and see Gaster, *Passover,* chap. 2. On the destroyer, Noth, *Exodus,* pp. 91–92 and Caird, *Principalities and Powers,* p. 36. For theories and exegesis of Passover, Childs, *Exodus,* pp. 178–214.

46. The last words of the prayer that supplements the Dayyenu prayer; in Roth's edition p. 33. Dalman, *Jesus-Jeshua,* p. 123.

47 Justin, *Dial.* 72.1.

48. The old-fashioned renderings of RV are here a safer guide to the Greek.

49. See BDB s.v. *basar* 4; 'own flesh' Gn 37:27, Is 58:7; near of kin Lv 18:6.

50. In LXX *sōma,* renders *basar* in some of its uses, and various words meaning 'corpse' or 'body'. The Hebrew language did not possess a common word that had the functions of *sōma* in Greek. The Greek books of LXX account for a large part of the occurrences of *sōma.* See Ziesler 'SOMA in the Septuagint' for an examination of passages where *sōma* may mean something more than the physical body, and also for information about discussion of the Pauline usage.

51. For this distinction, see below, p. 32–33.

52. However, when Paul deals with the consequences of eating, especially in 10:16–17 and 11:24–29, he qualifies his ruling that eating is an indifferent matter.

53. Cf. the common Platonic *melē kai merē.*

54. Senft, ad loc.; TWNT 1 s.v., p. 126.

55. Even though the Delphic slave by the act of liberation became technically *hieros,* under the protection of the god, he was a freedman of the private owner and under no obligation to the god; see Bömer, *Untersuchungen über die Religion der Sklaven,* p. 136. The verb *agorazein* does not suit the Delphic texts (Conzelmann, p. 113).

56. Dion. Halicarn. *de Thuc. jud.* 8.3.

57. Fr. 97 (Schweighaüser). This and the previous reference come from Pierce (*Conscience in the New Testament*), whose examination of Greek usage, as well as his discussion of New Testament passages, is instructive but not satisfactory. See TWNT 7 s.v. The use of *conscience* in 1 Cor 10:25–30 is similar: see Barrett, ad loc. and idem, 'Things Sacrificed to Idols'. It is, of course, not always clear that 'conscience' is the appropriate translation of *syneidesis.*

58. Elsewhere Rm 2:22; 2 Cor 6:16 Gal 5:20; and 1 Thes 1:9; add Col 3:5 and Eph 5:5.

59. 1 Cor 12:2 refers to their *past* participation: see Barrett and Conzelmann, ad loc.

60. For the appearance of idolatry in lists of vices, see also 1 Cor 5:10–11, Gal 5:20, Col 3:5, Eph 5:5 and cf. 2 Cor 6:16. The idolater in 1 Cor 5:10–11 is associated with the fornicator, ruthless person, rapacious man, abusive person, and drunkard, i.e., with compulsive aggressors. Hence, idolatry is one of the destructive forms of inner rage.

61. For the Greek use of *daimonion,* see Acts 17:18.

62. Test. Reub. 2:1–2, 3:3–6, and see the note on the 'vast demonology' of the Testaments in Charles, APOT 2 p. 296. *Daimonion* is uncommon in LXX: Five passages echo Dt 32:17; Tobit and three other passages refer to demons of destruction and desolation, which are commonplace in the Synoptic Gospels (but not in John). Apart from this present reference to demons, Paul has nothing about wicked or evil spirits. His perhaps overblown reputation as a demonologist must mainly depend on seven references to Satan, in each of which, however, we may be confronted with a manner of speaking rather than a matter of belief.

63. The cup and table of the demons may come from Is 65:11 (LXX): 'You who forsake me, and forget my holy mountain, and prepare a table for the "demon" [i.e., the deity], and fill a cup of mixed wine for Fortuna'. Contrast Prv 9:2–6, where Wisdom has mixed her wine and set her table and invites to life, gnosis and understanding: 'Come eat of my bread and drink of the wine I have mixed'.

64. For varieties of sacrifice, see de Vaux, *Ancient Israel* pp. 415–21. On p. 453, like most commentators, he assumes that 1 Cor 10:18 is a reference to the peace offering; but apparently 'the peace offering was a rare event in public sacrifice, occuring regularly only on the Feast of Weeks' (Schürer and Vermes, Vol. II p. 296, n. 14). If Paul has a particular sacrifice in mind, it is likely to be Passover.

65. Despite the short text in Lk 22:19ᵃ and the Didache. See Conzelmann, ad loc.

66. 'Cup of blessing' might imply 'cup that conveys a blessing'. To avoid that Paul adds 'which we bless', following hellenistic, not Jewish, custom—though he does not replace the Jewish *eulogein* by the hellenistic *eucharistein.*

67. Jos. and Asen. 8.5, 15.5, 16.16.

68. 1QS 6.4–5 repeated in 1QSa 2.17–21; Vermes, DSSE, pp. 69, 102. For the supposed relation between the Qumran meal and the Lord's Supper, see Leaney, *The Rule of Qumran,* pp. 182–184.

69. The difficulty is shown by the hesitations in AGB s.v. Under meaning 1 can probably be placed Phil 3:10, sharing Christ's suffering, and Phlm 6, sharing faith; under meaning 2 sharing the gospel (Gal 2:9, Phil 1:5), sharing the relief of the Saints (2 Cor 8:4), sharing the Holy Spirit (2 Cor 13:13, Phil 2:1 cf. 2 Cor. 6:14 [verb Rm 15:27, Gal 6:6]); and under meaning 3 Rm 15:26, 2 Cor 9:13 (verb Rm 12:13, Phil 4:15).

70. Col 1:20 has the stange phrase 'blood of his cross'; and Eph 1:7 and 2:13 refer to redemption (defined as the forgiveness of trespasses) and removal of alienation to the blood of Christ. Apart from the Last Supper saying in the Synoptic Gospels (and consequence references to blood guilt in Matthew and Acts), statements about his blood are confined to 1 Peter, Hebrews, John, 1 John and Revelation. It may be suggested that two separate impulses account for this distribution of references, the first being Jesus' eucharistic words, the second the development of violence against Christians.

71. So Atkinson ('Blood of Christ'), who also mentions the improbable view that *blood* can signify 'life preserved and active beyond death', suggested as long ago as 1883 by Westcott, *Epistles of St John,* pp. 35–36. On the passages that say 'the blood is the life' (Gn 9:4; Lv 17:11, 14; Dt 12:23), see Grayston, 'HILASKESTHAI'.

72. The Hebrew of 2 Sm 23:17 lacks the verb *drink* (see NEB, n.) which is supplied from LXX and 1 Chr 11:19. Cf. Driver, *Notes on the Hebrew Text*, p. 367. Was the Hebrew scribe reluctant even to write a phrase for David about drinking blood? Did the writer of 4 Mc overcome a similar reluctance by saying, 'reckoned as equivalent to blood'?

73. Lietzmann (pp. 49–50) has an excursus on the cultic meals of antiquity, to be read with Kümmel's corrective note 5.48 Z.21 v.25 (as well as Angus, *Mystery Religions*, pp. 127–33). Scarce ancient parallels to the eating of the deity are neither convincing nor relevant. TWNT 3 s.v. *Klao* pp. 737.1.37–740.17.

74. Apart from the references in 1 Cor 10:16, 11:24 (my body), 11:27 (the body and blood of the Lord), 12:27, cf. 12:12, there are only Rm 7:4 (cf. Rm 12:5, one body in Christ) and Col 2:17, Eph 4:12, Heb 10:10 (the offering of the body of Jesus Christ). In the Gospels 'the body of Jesus' meant his dead body.

75. Cf. Senft, ad loc.

76. Cf. Moule, *Birth of the New Testament*, pp. 37–38.

77. Cf. Rordorf, *Sabbat et dimanche*, pp. 133–4, n. 3. Spicq, *Notes de Lexicographie*, pp. 446–48.

78. Bornkamm 'Lord's Supper and Church in Paul'. Jeremias, *Eucharistic Words;* Dalman, *Jesus-Jeshua;* Käsemann, 'The Pauline Doctrine of the Lord's Supper'; Bultmann Theology 1, pp. 144–152.

79. Jeremias, p. 249–50. Cf. Justin, *Dial.* 70.4: 'In this prophecy it appears that he speaks about the bread which our own Christ gave us to do as a memorial [*poiein eis anamnēsin*] of his having been made corporeal for the sake of those who believe in him, for whom also he became subject to suffering; and about the cup which he gave us to do [*poiein*] with thanksgiving as a memorial [*eis anamnēsin*] of his blood'.

80. Jeremias on Greek and Jewish memorials and God's remembering, *Eucharistic Words*, pp. 238–55; Wainwright, *Eucharist and Eschatology*, pp. 64–68, follows Jeremias but mentions some objections—which seem to have persuaded most scholars.

81. On *hyper* see AGB s.v. 1 a; Moule, *Idiom Book;* Goppelt, *Theology of the New Testament*, vol. 2, pp. 92–98. The relevant use of *hyper* in LXX appears almost always in the later writings:

1. *People*. A solitary Wisdom passage: a person standing surety for you has staked his very self *for you (edōken tēn psychēn autou hyper sou*, Sir 29:15); Is 43:3–4, quoted in my discussion of 1 Cor 6:20ᵃ, has *hyper sou*, and Ez 45:17, 22 records various sacrifices offered *for* the house of Israel, *for* all the people; there are similar phrases in 2 Mc 1:26, 3:32, 12:43–44.
2. *Sins*. Apart from 1 Kgs 16:19, where Zimri died *for* (= because of) his sins, this group is dominated by the animal sin offering or guilt offering. (Ez 40:39; 43:21, 22, 25; 44: 29; 45:17, 22, 23, 25; 46:20; this is *ta hyper hamartias*, a variant for the more familiar *peri hamartias* (a v.l. in Ez 43:21); indeed, *hyper hamartias* in 1 Esd 7:8 and 9:20 is *peri hamartias* in 2 Esd 6:17 and 10:19; the choice of preposition depends on fashion; cf. also Mi 6:7.
3. *Religious virtues*. The Maccabean martyrs die *for* the holy laws, their country and Judaism (2 Mc 6:28, 7:9, 8:21, 14:38; 3 Mc 1:23) or for virtue, beauty and goodness and godliness (4 Mc 1:8, 10; 6:22; 11:2; 15:32; 16:13. The same theme in Jos. *Ant.* 13.5, 6

82. See Nicholson, *Preaching to the Exiles*, pp. 82–84; von Rad, *Old Testament Theology*, vol. 2 pp. 212–217.

83. It is unnecessary to assign different meanings to *diakrinō* in these two verses. 'To assess properly', is also the meaning in 1 Cor 14:29 (as in Mt 16:3). Only in this epistle

does Paul use the active verb: in 1 Cor 4:7 'to give too high an assessment' and in 1 Cor 6:5 'to give a [legal] ruling'. For 1 Cor 11:29, the meaning in Pindar, O.10(11) 46 is tempting: 'to set a place apart for the purpose of a religious meal'.

84. For discussion of the structure and origins of the formula, see Conzelmann, pp. 251–54.

85. Conzelmann, p. 256 very properly discusses the origin of 'on the third day'. To his suggestions the following observation may be added:

1. There were two early forms displayed in the synoptic Passion predictions: *meta treis hēmeras anastēnai* in Mark and *tē tritē hēmera egerthēnai* in Matthew (Luke), cf. Mt 27:64 and Paul. In Lk 18:33, Mark's verb is preserved, but it is changed to Matthew's verb in Lk 9:22. Mark's *anastēnai* is changed to *egerthēnai* but always with strong dissent from ms. variants. Obviously *egerthēnai* was necessary in circles that said 'God raised' rather than 'Christ rose'. Matthew and Luke change Mark's 'after three days' to 'on the third day' (also Acts 10:40); but D and it usually dissent, and Matthew himself has 'after three days' in 27:63, which of course he understands to mean 'until the third day' in the next verse.

2. It is commonly said that the two expressions mean the same, but that 'after three days' is less correct and so is usually altered. Field (*Notes,* pp. 11:13) showed that 'on the third day' means 'the day after tomorrow', as in Lk 13:32–33. But he failed to show that *meta treis hēmeras* is only another form of 'on the third day' since his only example of the *meta* construction was Hos 6:2: 'He will heal us after two days, on the third day we shall be raised up'. The two clauses are parallel and equal, and 'after two days' means 'on the third day'. It must be held that Mark's odd form was original, and that Matthew's form has changed its phrasing to something more familiar.

3. It may be suggested that Mark's form comes from Daniel's half-week, and that the three or four days correspond to the time, two times and half a time during which the saints are given over to the oppressor (Dn 7:25).

4. That obscure apocalyptic thought did not occur to those who were accustomed to the fairly common narrative use of 'on the third day' in LXX. It is used to conceal a gap in the narrative or to allow time for some necessary things to happen. The deliberate pause before God's activity is shown in 2 Kgs 20:5, 8 and Hos 6:2; but the most impressive example is Ex 19:11, 16, when the Lord comes down upon Sinai on the third day and Moses ascends to meet God. Some imagery of that kind is necessary to account for early Christian emphasis on 'the third day'; a simple historical reminiscence is quite insufficient.

86. *Hamartia* appears in Romans forty-seven times, in 1 Corinthians three times, and in the remaining epistles seven times; *Hamartanō* in Romans and 1 Corinthians seven times each. The verb means acting wrongly, but the noun seldom indicates a wrong action: certainly in Rm 7:5, 1 Thes 2:16 and with the classical sense of 'mistake, error' in Rm 14:23 and 2 Cor 11:7; in OT quotations in Rm 4:7, 8 and 11:27; and in formulae in 1 Cor 15:3,17 and Gal 1:4. The great majority of uses in Romans are of Sin personified. The uses of the verb in 1 Cor 6:18; 7:28, 36; 8:12, 12; 15:34 fall comfortably within the meaning, 'to act in a religiously offensive way'.

87. See Derrett, *Law in the New Testament,* pp. 418–20.

88. See Gray, *I and II Kings,* ad loc.

89. Forgiveness by direct appeal to God, Ps 25, 32:1–5, 103:6–14; Is 43:25, 55:6–9; Dn 9:3–19. Atonement by repentance, Sir 5:4–7, Ez 33:10–16; by loyalty and faithfulness, Prv 16:6; by study of the Law, Sir 39:5; by honouring parents, Sir 3:3, 14–15; by

forgiving neighbours, Sir 28:2; by almsgiving, Sir 3:30, against sacrifice, Ps 40:6; 51; Mi 6:7–8; Sir 34:19.

90. Scrutiny of *Loci citati vel allegati* in Nestle-Aland[26] p. 761 is instructive. When overlapping information is discounted there are eight explicit quotations of Is 53, but only 1 Pt 2:22–25 refers to any statements that have a direct bearing on 'died for our sins'. (Mt 8:17 indeed quotes verse 4[a], but in the MT form, and refers it to Jesus' healing activity). There are twenty-two allusions, varying from probable to imaginative, only five of which refer to relevant verses (verse 4[a] in 1 Jn 3:5, verse 5[a] in Rm 4:25, verse 11[c] in Rm 5:19 and verse 12[c] in Mt 26:28 and Heb 9:28). As for 1 Cor 15:3, it is placed against verses 8[d] and 12[e,f]. Contrast this with Hengel, *Atonement,* p. 60, and compare Hooker, *Jesus and the Servant.*

91. Hengel, *Atonement,* p. 36. On pp. 57–60 he admits that we lack clear evidence from pre-Christian Judaism of vicarious suffering of the Messiah based on Isaiah 53 but then argues that Isaiah 53 had an influence on the origin and shaping of the earliest kerygma. He even says that we should know Isaiah 53 to be messianic if more prerabbinic texts had been preserved.

92. See Gärtner, *The Temple and the Community in Qumran and the New Testament.*

93. 1QH 9. See Price on pp. 29 and 35 of Charlesworth, *John and Qumran.*

94. Hengel, *Atonement,* pp. 1–32.

95. D = statement about Christ's death only; R = about his resurrection only; D + R = the two joined.

	D	R	D & R
1 Thessalonians	2:15	1:10	4:14; 5:10
1 Corinthians	1:(13)17–18, 23; 2:2, 8; 5:7; 8:11; 10:16; 11:23–27	6:14; 15:4–8, 12–26, 29, 32, 35–57	15:3–4
2 Corinthians	1:5 (2:15, 5:21)	1:9	(2:15–16), 4:10–14; 5:14–15; 13:4
Galatians	1:4; 3:1,(3:13, cf. 4:5), 5:11; 6:12–14	1:1	2:19–20; 5:24–25
Romans	3:25; 5:6–9; 8:(3), 32; 14:15	1:4; 8:11 (but cf. 8:10); 10:9	4:24–25; 5:10; 6:3–11; 7:4; 8:17, 34; 14:9
Philippians	3:18		1:20–21; 2:8–9; 3:10–11

96. See n. 95.

97. On the rendering of *parakaleo, paraklesis* in the New Testament, see Grayston, 'A Problem of Translation'.

98. Cf. Schweitzer, *The Mysticism of Paul the Apostle,* chap. 7 'Suffering As Dying with Christ'. For the interpretation of 'the sufferings of Christ' see TWNT 5 pp. 930–32, and Kümmel, p. 196 s.99 Z.27. Also Bultmann, pp. 28–29.

99. Russell, *Method and Message of Jewish Apocalyptic,* pp. 271–76. Cf. Allison, *The End of the Ages Has Come,* p. 66.

100. If this conjecture is right, it might give point to the otherwise puzzling section 2 Cor 6:14–7:1.

101. See Appendix C on the *en Christō* formula. The other three passages are 2 Cor 3:14 (God removes the veil), 5:17 (God performs a new creative act), and 5:19 (God was

reconciling the world). In 2:17 and 12:19 Paul speaks 'in Christ'; and in 12:2 he is 'a man in Christ'. The longer form 'in Christ Jesus' does not appear in 2 Corinthians, and 'in the Lord' only once (2:12)—perhaps because lordship language would too much encourage Corinthian pretentions.

102. See below, n. 206.

103. The phrase is taken from George, *Communion with God in the New Testament,* p. 184, where it partly elucidates the language of Hauck in TWNT 3 p. 807. For the care needed in defining mysticism, see Tinsley in *A New Dictionary of Christian Theology.* Mysticism, when defined by examination of mystical experience, is of limited application in the New Testament.

104. Cf. Bultmann, ad loc. For the discussion whether *nekrōsis* in verse 10 is merely a variant for *thanatos* or has a significantly different meaning (as Bultmann and others think it has), see Collange, pp. 154–55.

105. For references, see Collange, p. 174.

106. RSV translates *to parautika elaphron tēs thlipseōs* (in 2 Cor 4:17) by 'this slight momentary affliction', and other versions do much the same. If that is correct, an unconvincing cheerfulness has replaced Paul's excessive alarm at the beginning of the epistle. But *parautika* can mean 'for the present' (so Collange), and the figurative sense of *elaphros* can be 'bearable' (AGB, s.v.).

107. LS puts it, '*the body* as the tabernacle of the soul'. See TWNT 7 pp. 383–85. Wis 9:15 has 'perishable body', parallel to 'earthly *skēnos*'.

108. For the textual problems of verse 3, see Barrett, ad loc. Alternative translations are provided for the variants *ekdysamenoi* and *endysamenoi*; the meaning is much the same.

109. Collange, besides examining this section in detail and arguing himself into a difficult position, surveys the great variety of interpretations. Bruce, who thinks that 5:1–10 deals with the intermediate state between death and resurrection, lists important studies of the passage.

110. The meaning of *synechei* ('controls') is settled in TWNT 7, p. 881.

111. Cf. also 2 Mc 1:5, 8:29. Also Josephus, *Ant.* 6:143 as well as 7.153, 295 which use *diallassō*. *Katallassō* and the noun are not significant in Greek and Hellenistic religion (TWNT 1), and appear in LXX only once more beside the three passages of 2 Maccabees.

112. See Hering, p. 53.

113. Collange's suggestion (p. 269) that Paul is using a formula known to the Corinthians is scarcely persuasive.

114. See Hooker, 'Interchange in Christ'; idem, 'Interchange and Atonement'.

115. Pauline *charis* may be considered under the following headings:

1. as part of the epistolary formulae of greeting and farewell: Rm 1:7; 16:20, 24; 1 Cor 1:3, 16:23; 2 Cor 1:2, 13:13; Gal 1:3, 6:18, Phil 1:2, 4:23; 1 Thes 1:1, 5:28; 2 Thes 1:2, 12, 2:16, 3:18; Phlm 13:25

2. as a thanksgiving formula: Rm 6:17, 7:25; 1 Cor (10:30), 15:57; 2 Cor 2:14, 8:16, 9:15.

3. as generous activity, gift, or kindness: Rm 4:4 (the opposite of an obligation); 1 Cor 16:3; 2 Cor 1:15 (v.l., *charan,* 'pleasure'), 8:4, 6, 7 19.

4. as a generous gift bestowed by God and received by Christians—(a) the apostolic commission to Paul: Rm 1:5, 12:3, 15:15; 1 Cor 3:10, 15:10; 2 Cor 1:12, 12:9; Gal 1:15, 2:9; Phil 1:7; and (b) the actual existence of a Christian community: Rm 5:2, 6:14–15, 11:5–6, 12:6; 1 Cor 1:4; 2 Cor 6:1, 8:1, 9:8, 14; Gal 1:6, 5:4.

5. as God's gracious act in Christ: Rm 3:24, (4:16), 5:15, 17, 20, 21, 6:1; 2 Cor 4:15, 6:1, 8:9; Gal 2:21.

116. On 2 Cor 8:9 and 1 Cor 10:4, cf. Dunn, *Christology in the Making,* pp. 121–123, 183–84.

117. For attempts to discover what the rivals taught, see Barrett, *Essays on Paul,* pp. 68–72. In the same volume, the essays 'Christianity at Corinth', 'Paul's Opponents in 2 Corinthians' and 'PSEUDAPOSTOLOI (2 Cor. 11.13)' deal with the wider questions.

118. The prepositions *apo* and *dia* (the latter applied both to Christ and to God) cause no trouble if the necessary supplements are inserted. O'Neill's omission (p. 19) is unnecessary. Cf. the gospel in 1:11–12, which was not *kata anthrōpon* ('on human authority'); he did not receive it *para anthrōpou* ('from a human source'), nor was it dependent on human teachers.

119. O'Neill, pp. 19–20 gives the variants and proposes to omit *ponērou.* The variant quoted could, of course, be translated 'from the age of [= marked by] the present evil', and the other variants could have arisen by a perplexed misreading of its sense under the influence of *ho aiōn houtos.*

120. See Jn 9:24 and Bultmann, *John,* p. 336, n.1.

121. But see Betz, ad loc. and Lüdemann, *Paul, Apostle to the Gentiles,* pp. 50–53.

122. See n.2.

123. This corresponds to the *narratio* section in Betz's analysis. It is not necessary here to discuss whether his understanding of the *narratio* (p. 30) is persuasive. The next section corresponds to his *propositio.*

124. *Dikaiosynē* has theological significance in 1 Cor 1:30 and 2 Cor 3:9, 11:15; *dikaioō* in 1 Cor 6:11; *dikaios, dikaioma, dikaiōsis* none. According to Stendahl (*Paul among Jews and Gentiles,* p. 26), 'Paul's doctrine of justification by faith has its theological context in his reflection on the relations between Jews and Gentiles, and not within the problem of how *man* is to be saved, or how man's deeds are to be accounted, or how the free will of individuals is to be asserted or checked'. See Appendix B. Reumann (*Rightousness in the New Testament*) insists that *dikaiosynē* is the heart of Paul's gospel: see therefore the reservations of Fitzmyer in the same book.

125. See Appendix B.

126. Cf. Betz, 'Rechtfertigung in Qumran'.

127. Cf. Moule, 'Death "to Sin", "to Law", and "to the World": A Note on Certain Datives'.

128. E.g., in trying to represent a hypothetical dilemma: 'My doctor has given me three months to live. What shall I do?'

129. Cf. TWNT 2, s.v. *egō,* p. 354–55. Bonnard says that Paul describes not his personal experience but the general condition of Christians (without saying why); hence, this verse must not be read as an exceptional spiritual condition reserved for certain mystics (of course not!). Betz, p. 121 says that Paul uses himself as the prototypical example of what applies to all Pauline Christians but offers only an unconvincing string of passages and admits (p. 123) that the Pauline ego has not been satisfactorily investigated. Bruce says that the *I* in verse 18 is not primarily personal, that a note of personal experience may be found in verse 19, and that intense personal feeling is recognizable in verse 20.

130. Schweizer, 'Dying and Rising with Christ'; Tannehill, *Dying and Rising with Christ;* Dupont, SYN CHRISTŌI: *L'union avec le Christ suivant saint Paul,* pt. 1 (' "Avec le Christ" dans la vie future'). This enquiry takes account of Pauline phrases with *syn Christō,* etc., and Pauline uses of numerous syn-compounds: *syn apothnēskō, basileuō, dox-*

azō, egeirō, zaō, zōopoieō, thaptomai, paschehō, stauroō; and *synklēronomos, symmorpheō(os)*, and *symphytos,* (Bouttier, *En Christ,* pp. 38–53).

131. Cf. Betz and Bruce, ad loc. Other *in*-references (Gal. 4:19, 6:17) have already been noted in the discussion of ego. Add also 2 Cor 4:10–12 (where Paul carries about in his body both the dying and the life of Jesus) and 13:3–4. There are explicit references to God's indwelling in 1 Cor 14:25 (which an unbeliever or outsider can recognize by the effect on him of prophecy in the community) and Phil 2:13, 'God is at work in you [i.e., within the community], both to will and to work for his good pleasure'. There are no examples of Christ in you (or in me) in 1 and 2 Thessalonians, 1 Corinthians, or Philippians.

132. Kramer, *Christ, Lord, Son of God;* Dunn, *Christology in the Making,* pp. 33–46.

133. Schoeps, *Paul,* p. 146.

134. There is nothing in the deutero-Pauline writings nor in the general epistles to link *Son of God* to the death of Christ. The Johannine writings contain 1 Jn 1:7 'the blood of Jesus his Son cleanses us from all sin' and (indirectly) 4:10, '[God] sent his Son to be the expiation for our sins'. But the Gospel is more reticent about the Son (preferring the Son of Man): '[God] gave his only Son' (3:16—*edōken,* an echo or an avoidance of *paredōken?*), 'Glorify thy Son' (17:1), and 'He ought to die, because he made himself the Son of God' (19:7). The beloved son in Mark's parable of the vineyard (Mk 12:6) may suggest, but does not qualify as, a theological statement. We are perhaps brought closer by 'Abba, Father' in Gethsemane (Mk 14:36), by 'I am' when the high priest asks, 'Are you the Christ, the Son of the Blessed?' (though Jesus immediately reverts to Son of Man, Mk 14:61–62), and by the centurion's admission, 'Truly this man was [a] son of God' (Mk 15:39; cf. the additions in Mt 27:40, 43 and Dunn, p. 48). Even Hebrews is less explicit than might be expected: Jesus, 'although he was a Son, learned obedience through what he suffered' (5:8). Those who commit apostasy crucify the Son of God on their own account (6:6); and a Christian who deliberately sins spurns the Son of God and profanes the blood of the covenant (10:29). Yet when the writer brings his main argument to a head and speaks of the self-sacrifice of our great high priest, he refers to the blood of *Christ* (Heb 9:12, 14). It appears that the earliest Christian writers did not spontaneously associate explicit references to the death of Christ with the title *Son of God.*

135. Betz, p. 131. Contrast TWNT 1, 771–72.

136. Betz (p. 145) says of Paul's position that 'for him the Torah was not given to be faithfully obeyed as the covenant, but for the purpose of breaking it and generating sin'. If by *it* in the last clause he means 'the covenant', I think that both statements in that clause are false. Betz depends too uncritically on some of Bultmann's thoughts in *Theology,* vol. 1 sec. 27, esp. p. 267.

137. Cf. Koch, *Prophets,* vol. 2, p. 94: 'The apostle rightly sensed that commandments and laws played no part at the beginning of the history narrated among his people'.

138. Rad, *Deuteronomy,* p. 167.

139. See Bonnard, p. 152, referring to p. 75, and TWNT 7 s.v. *synkleiō.* The verb occurs also in Rm 11:32 and Lk 5:6. The image of inhabitants of a town confined (and protected) by a wall or confined (and threatened) by besiegers is common (e.g., Jos 6:1; 1 Kgs 11:27; 1 Mc 5:5: *synekleisthēsan hyp' autou,* 'They were shut up by him in their towns').

140. On the fence, see Feldman, *Parables and Similes of the Rabbis,* pp. 35–38; *Rab Anth,* p. 156; Foerster, *Palestinian Judaism,* pp. 36–37, 169. Feldman remarks that the fences 'aimed at preserving throughout the generations the social order of the world'.

141. Paul's extraordinary treatment of the story of Sarah and Hagar in Gal. 4:21–31 is another attempt to provide imagery from Torah itself for the contrast between law and

promise. Hagar's son is associated with slavery, flesh, and persecution; Sarah's son with freedom, spirit, and promise. They are two covenants; they correspond to Jerusalem as it now is and as God intends it to be. And Torah itself orders the rejection of Hagar's son.

142. In NT, *exagorazō* occurs in Gal. 3:13, 4:5 and also in Col 4:5, Eph 5:16 in 'redeeming the time', which RSV understands as 'making the most of'; cf. Dn 2:8, the only LXX occurrence. For its secular use, see Lyonnet, *L'emploi paulinien de exagorazein.* Betz's note, ad loc., is clear and exact. Paul's use of the simpler *agorazō* is not helpful (see on 1 Cor 6:20, 7:23).

143. For linguistic details, see Bruce, ad loc. BDB, s.v. *qelālāh* on Dt 21:23 says 'object of a curse' and adds 2 Kgs 22:19, 'that they should become a desolation and a curse'. But 'cause of a curse' is more likely. Cf. Jer 24:9 and seven more times and esp. Zec 8:13. On the curse, see Pedersen, *Israel,* I-II pp. 437–52, esp. 443.

144. See Becker, pp. 185–86; Harvey, *Jesus and the Constraints of History,* p. 22 and Hooker, 'Paul and Covenantal Nomism', against Sanders, *Paul, The Law and the Jewish People,* pp. 25–27.

145. Paul had 'died to the law' (Gal 2:19), but compare his behaviour to Jews and Gentiles, when he is *ennomos Christou* (1 Cor 9:20–21). To avert rash generalisation, see Sanders, *Paul, The Law and the Jewish People* and Räisänen, *Paul and the Law.*

146. For the aim and exegesis of sections Gal 5:1–12 and 13–24 Betz is particularly instructive.

147. Bultmann, *Theology,* vol. 1, secs. 22–23; TWNT 7; Käsemann, *Perspectives on Paul,* pp. 25–27; Jewett, *Paul's Anthropological Terms,* pp. 95–116; Kümmel, *Theology of the New Testament,* pp. 174–178; Ladd, *A Theology of the New Testament,* pp. 466–75; Drane, *Paul—Libertine or Legalist?*

148. In 1QM 4.3 the group elected by God is opposed by 'the assembly of [sinful] flesh' and the opposition to the sons of light is called 'all sinful flesh'. Cf. Kuhn, 'New Light on Temptation, Sin and Flesh in the New Testament', p. 103.

149. See TWNT 7, s.v. *stigma.* If the stigmata suggest a protective marking, to prevent molestation of someone who belongs to the Lord, it is apt to recall the 'talismanic' virtue of the cross already noticed in the discussion of 1 Corinthians on p. 45.

150. Cf. Cranfield, p. 695.

151. See Beker, *Paul the Apostle,* esp. chap. 5 on the structure and argument of Romans.

152. I take the view (with Cranfield and Barrett) that chapter 16, except for the doxology, belongs to Romans. In trying to influence an independent community that he had not founded, Paul sends numerous greetings to members who might be persuaded to his view.

153. Dodd, p. 23. Cf. Hanson, *The Wrath of the Lamb,* p. 110. Paul adopts a popular Jewish view: cf. Wis 11:16, 12:23; Prv 5:22; Philo *Her.* 109 Test Gad 5.

154. 'Awareness of [the power of] Sin' is intended to translate *epignōsin hamartias.* RSV's 'knowledge of sin' might suggest several interpretations, as in other versions: 'consciousness of sin' (NEB), 'to make a man know that he has sinned' (GNB), 'tell us what is sinful' (JB). Since in Rm 3:9 (the first of forty-seven appearances of *hamartia* in Romans) *pantas hyph' hamartian* can be translated, 'all men are under [the power of] sin', it is reasonable to suppose that the same intention is present here. When Paul uses *hamartiai* in the plural, the meaning is not in doubt, namely 'wrongful actions'; so Rm 4:7 (and the singular in 4:8), 11:27 in OT quotations, as well as 7:5. But his use of the singular is by no means so straightforward. In principle, *hamartia* may indicate a wrongful act (equivalent to *hamartēma,* as in 4:8); or it may suggest the practice of sinning, i.e., *hamartēsis,* as might be possible in 5:20; or it may be a loose replacement for *hamartōlos,* sinful, as in

14:23. In the great majority of uses, in Romans 5–8, however, *hamartia* seems to be personified as a malign power. See n. 8. 156. Räisänen (*Paul and the Law*, p. 100) casually dismisses the notion of Sin as a power.

155. *Apolytrōsis* ('redemption') of verse 24 comes again in Rm 8:23, 1 Cor 1:30. *Protithēmi* ('put forward') of verse 25 was used in Rm 1:13 (in the sense 'intend', which may be the meaning here); *hilastērion* ('expiation') here only (elsewhere in NT, only Heb 9:5; cf. the verb, Heb 2:17); *en tō autou haimati* ('by his blood') is repeated in Rm 5:9 and may be eucharistic language (cf. 1 Cor 11:25); *paresis* ('passed over' unless it is a variant for *aphesis*, 'forgiveness', which is not a Pauline word) only here in NT; *proginomai* ('former') only here in NT; *hamartēma* ('sin'), elsewhere 1 Cor 6:18. *Anochē* ('forbearance') of verse 26, elsewhere Rm 2:4. Other indications that the formulation was non-Pauline are the participle *dikaioumenoi* in verse 24 instead of the expected finite verb and the string of connectives in verses 24–26: *dia, en, hon, dia, en, eis, dia, en, eis, to.* This produces a diffuse and overqualified set of statements that, in the end, are difficult to interpret. The complexity of the formulation may perhaps be explained if Paul made use of an existing formula (his quotation beginning with *dikaioumenoi* into which he inserted another formulaic element *(en tō autou haimati)* and two characteristic themes: 'by his grace as a gift' and 'by faith'. Cf. Hunter, *Paul and His Predecessors*, pp. 120–22 and especially Reumann, *Righteousness in the New Testament*, pp. 35–38.

156. *Parabasis* is an infringement of the law (Rm 2:23) or Adam's disobedience of God's command (Rm 5:14). Where there is no law, there is no *parabasis* (Rm 4:15; Gal 3:19). A *parabatēs* is someone who infringes the law (Rm 2:25, 27; Gal 2:18). Adam's *parabasis* in Rm 5:14 is described as his *paraptōma* in 5:15–18, and in verse 20 it is said that 'law came in, to increase the *paraptōma*' (the word being perhaps chosen in contrast to the *dikaiōma* of Christ in verses 16 and 18). Apart from this context, *paraptōma* is a rather more general word with the classical sense of 'blunder' (i.e., a culpable misjudgement). In Gal 6:1 a church member may be detected in a *paraptōma* or may impulsively commit a *paraptōma*. In Rm 11:11–12 *paraptōma* means Israel's disastrous misjudgement in not responding to the gospel. In Rm 4:25 Jesus was handed over for our *paraptōmata*, and in 2 Cor 5:19 God, acting in Christ, was not taking count of their *paraptōmata*. In Rm 5:19 Adam's *paraptōma* becomes his *parakoē* ('disobedience'). The word is joined with *parabasis* in Heb 2:2; it has no further theological significance in its only other NT appearance (2 Cor 10:6). Although *apeitheia* and its verb ('disobedience') have some theological function, they are not used in atonement contexts (Rm 2:8, 10:21, 11:30–32, 15:31). The *anomia* ('lawlessness') group of words, very common in LXX, is uncommon in Paul (Rm 2:12, 4:7, 6:19; 1 Cor 9:21; 2 Cor 6:14; 2 Thes 2:3, 7–8) and never in atoning contexts. Thus, together with *hamartia* and *hamartēma* (see nn. 154–55), Paul used a range of words for acts of wrong, in addition to the naming of specific actions; but they play only a small part in his teaching about the death of Christ. On *parabasis*, see TWNT 5, p. 736; on *paraptōma* see TWNT 6 pp. 172–73.

157. Koester, (*Introduction to the New Testament*, vol. 2, p.93) regards Rm 4:25 as belonging to the Antioch community; but whether preserved in the Antioch tradition or created by it (which, from the emphasis on 'our' *paraptōmata*, is less probable) is not clear.

158. Since *paradidōmi* belongs to the Passion tradition, it is not necessary to search Isaiah 53 to find a reason for its use. The verb is common in LXX, but its use in Is 53:6, 12 is somewhat odd. In verses 6 and 12^d it represents *pg'*, as it does in one other passage, Is 47:3. That passage is obscure in MT, and LXX is a guess. In 53:6 it means 'caused to fall upon'; and in 53:12^d 'to make entreaty'. In 53:12^c *paradidōmi* renders ^{c}*rh*, 'to pour out', translated, 'He poured out his soul to death' (RSV), or 'he exposed himself to face

death' (NEB). *Dikaiōsis* is not an LXX word (only once); and the confessional formula uses *paraptōmata*, not *hamartiai*. The only strong case for linking Rm 4:25 with Isaiah 53 is that the thought of the former is as indefinite as the thought of the latter is obscure.

159. See p. 64 and n. 115.

160. Since RV, modern versions (except NIV) take *dia pisteōs* as parenthetical. Gifford argued the case in 1886 *(Epistle of St. Paul to the Romans)* and decided for the modern translation. In addition to the modern suggestion that Paul is adapting an already existing formula, the argument runs thus: That *pistis en* is a possible construction is demonstrated by Eph 1:15, 'your faith in the Lord Jesus' and Col 1:4, 'your faith in Christ Jesus' (though almost certainly not in 1 Tm 3:13; 2 Tm 1:13, 3:15, where the phrasing implies 'your faith [and love] which is available in [the community of] Christ Jesus'). But these passages (though not likely to be directly Pauline) tally with the Pauline habit of referring to faith in God or Christ rather than in some object (even if 'his blood' stands for 'him who died'). Moreover, there is no parallel to 'faith in his blood'. True, there is also no parallel to *dikaiōthentes en tō haimati autou* in Rm 5:9 ('justified by means of his blood'); but it is more satisfactory to reserve the noun *faith* for a relation between persons and therefore to dissociate it from *en tō autou haimati*, which can then be translated here as it is in 5:9 (and not 'faith in the atoning efficacy of his blood'). More generally, it may be urged that it is unsuitable to the importance of *blood* to introduce it off-handedly as a qualification of *faith* and that the emphatic position of *autou* supports this consideration ('by his own blood'). Finally, there is a good parallel to parenthetical *diapisteōs* (with a similar uncertainty about the presence of the article) in Gal 3:26, *pantes gar huioi theou este dia tēs pisteōs en Christō Iēsou,* 'For you are all sons of God through faith in Christ Jesus', where modern versions since RV (except NIV, JB) rearrange the order as, e.g., in RSV, 'For in Christ Jesus . . . '. If the argument is good, *through faith* is a parenthesis, and the original ran, 'as an atonement by means of his own blood'.

161. *Apolytrōsis* in NT appears at Lk 21:28; Rm 3:24, 8:23, 1 Cor 1:30; Eph 1:7, 14, 4:30; Col 1:14; Heb 9:15, 11:35. In the Letter of Aristeas, Diodorus Siculus, Josephus, Plutarch and an inscription from Kos it signifies the release of prisoners or slaves on the payment of compensation. In Dn 4:34 LXX (Rahlfs; verse 30^c Swete) no payment is possible, and in Philo none is implied (TWNT 4, p. 354). Perhaps in factual narratives 'ransom' or 'compensation' was usually intended; in religious discourse the word meant 'release'. RSV hides behind *redemption* and *redeem* (except Heb 11:35, where *release* is used, and Eph 1:14, where it seems not to appear in the translation). NEB uses variously *liberation, set free, release, deliverance* and, at Eph 1:14, *redeem*. Cranfield, vol. 1 p. 207 gives other 'price' passages. See Appendix A.

162. See Grayston; HILASKESTHAI.

163. The verb is conventionally translated 'boast, glory, exult', which is well enough in its way but not really happy in our culture. *Self-confidence* comes close if it means not brashness but the inner confidence (usually with a social origin) that enables people to avoid embarrassment and despondency. Its opposite, namely, *disappointment* is common in Psalms, e.g., *mē kataischunēs me apo tēs prosdokias mou,* 'Do not disappoint my expectation', Ps 119:116.

164. Romans 9 is troubled by the problem of God's love for Israel; so 9:13 quotes Mal 1:2–3, 'Jacob I loved, but Esau I hated', and 9:25 quotes Hos 2:23, 'her who was not beloved I will call "my beloved"'. 2 Cor 9:7 quotes Prv 22:8^a (LXX) 'God loves a cheerful giver'. In 2 Cor 13:11, 13 there are 'the God of love' and 'the love of God'. In 1 Thes 1:4 community members are 'brothers beloved by God'. In OT style, God's love is his spontaneous choice of those whom he accepts as his people. In 2 Cor 5:14 and Gal 2:20 Christ's death for others is ascribed to his love.

165. *Eph ho pantes hēmarton.* The construction *eph ho* has been endlessly discussed, esp. by theologians who suppose that profound statements can be established on prepositions. Paul uses *eph ho* again in 2 Cor 5:4; Phil 3:12, 4:10. The general connective phrase *in that* is sufficient everywhere.

166. *Aphiēmi* in the sense 'forgive' occurs only in Rm 4:7; *aphesis* (forgiveness) only in the deutero-Pauline Eph 1:7, Col 1:14. Rm 4:7–8 quotes Ps 32, where sins are forgiven, covered, and not reckoned. For the last, see also 2 Cor 5:19. In Rm 11:27, quoting Jer 31:34, sins are taken away. Cf. Taylor, *Forgiveness and Reconciliation,* pp. 2–10.

167. Rad, *Old Testament Theology,* vol. 1, p. 387; see also vol. 2, pp. 349–50; Johnson, *The Vitality of the Individual in the Thought of Ancient Israel,* pp. 94, 107; Köhler, *Old Testament Theology,* pp. 148–50; Eichrodt, *Theology of the Old Testament,* pp. 221–23.

168. So also Ps 6:5, 30:9, 88:10–12, 115:17; and see the authorities mentioned in n. 167. Although Hebrew people pleaded with God to preserve them from Sheol, they accepted death calmly from the hands of God when their time came in the natural order of things (e.g., Job 2:10). Only tentative steps were taken (perhaps as in Ps 73:23–28) to assert a relationship with God beyond death, and only apocalyptic writers began to regard death as the great enemy (e.g., Is 25:7–8; 4 Ezr 8:53; Test. Lev. 18:10). Paul, in the Pharisaic tradition, expresses strong hostility to death and is fully conscious of its power to break communion with God. The apparently otiose comment in Rm 5:13 that 'sin is not counted where there is no law' is understandable when it is remembered that *sin* (as distinct from error, fault, crime) must always be defined in relation to God. If knowledge of God has ceased and if God's law is unknown, *sin* goes unrecognized; but its disastrous consequences are still present.

169. The phrase *spontaneous and unmotivated* is suggested by Nygren, *Agape and Eros,* pt. 1, trans. A. G. Hebert, p. 87, though Hebert's *uncaused* has been replaced by *unmotivated* in pt. 2, trans. P. S. Watson.

170. See Betz, *Galatians,* pp. 186–89.

171. See on Gal 2:19 (p. 72) for the phrase *dying to sin.*

172. On the interpretation of Romans 6 see esp. Tannehill, *Dying and Rising with Christ,* pp. 21–43.

173. If we set aside Mk 1:9 ('baptized *in* the Jordan') and Mt 3:11 ('baptize in water *for* repentance'), almost all the remaining *batizeineis* references mean 'baptize into allegiance to—' Moses (1 Cor 10:2); whatever John's baptism demanded (Acts 19:3); Paul (1 Cor 1:13, 15); the Lord Jesus (Acts 8:16, 19:5); Christ (Gal 3:27); Father, Son and Holy Spirit (Mt 28:19); probably also 'the one body' (1 Cor 12:13). *Into Christ Jesus* cannot be interpreted in the same way, because *into his death* can scarcely be stretched to mean 'into allegiance to his death'. Hence, Paul has adapted the formula. In Paul, *eis Christon* is rare and has no fixed meaning: 'believe *on* Christ Jesus', Gal 2:16 (and v.l. in Phlm 1:5); but 'against' (1 Cor 8:12), 'directed towards' (2 Cor 11:3), and 'until' (Gal 3:24).

174. *United with him* represents *symphytos,* 'grown together', but the organic meaning is not essential unless required by the context. *A death like his* represents *tō homoiōmati tou thanatou autou.* The noun *homoiōma,* 'likeness, copy, figure' can mean anything from what is exactly like something else (e.g., Jd 8:18b; 2 Kgs 16:10; Rm 5:14; Hermas M.4.1.9) to what seems like or is suggestive of something else (as in Ezekiel's perplexed visions). When two similar objects are compared, features in one correspond to features in the other, and *homoiōma* can indicate the correspondences (cf. Sir 31(34):3 and Mk 4:30 v.l., *en tini homoiōmati parabalōmen autēn?* 'With what corresponding thing shall we compare it?' Hence, in this passage, Paul speaks of Christian experiences that correspond to the death and resurrection of Jesus.

175. 'Consider' (RSV) or *regard* yourselves' (NEB) is not satisfactory for *logizesthe* because it implies an *as if*, contrary to fact. *As if* is not required until the *hōsei* in verse 13.

176. Cf. Tannehill, pp. 43–47. I cannot accept Tannehill's view that the body of Christ refers to the collective body of the old epoch, 'the body of sin'.

177. M; also (with commentary) Oesterley, *Sayings of the Jewish Fathers*, where the paragraphs are differently numbered. For Midr. Ps., see *Rabb. Anth.*, p. 125 and the whole chapter.

178. The modern interpretation of Romans 7 begins with Kümmel's *Römer 7 und die Bekehrung des Paulus* of 1929; cf. idem, *Man in the New Testament*, pp. 49–61. Influential interpretations are Bultmann, 'Romans 7 and the Anthropology of Paul' and Bornkamm, 'Sin, Law and Death'. Neither is free of high-minded obscurity, in contrast to the perhaps overpersuasive lucidity of Nygren. Sanders (*Paul, the Law and the Jewish People*, pp. 73–75) contends that Paul connects law and sin in a new way in Rm 7:7–25.

179. For the meaning of the Hebrew word, see Childs, *Exodus*, pp. 425–28. On the Greek word see AGB, s.v. and TWNT esp. p. 170.

180. See Nickelsburg, *Jewish Literature between the Bible and the Mishnah*, p. 226. The quoted verses are from Charles APOT, vol. 2.

181. Cf. Bornkamm, p. 90: Paul allows room 'for the possibility that "desire" can express itself nomistically just as much as anti-nomistically, i.e., in the zeal for one's own righteousness'. This is rejected by Räisänen, *Paul and the Law*, pp. 111, 141–148, though without suggesting what Paul's *epithumia* was. Yet we have Paul's own testimony in Gal 1:13–14.

182. Bornkamm, 'Sin, Law and Death', pp. 95–96; Käsemann, p. 196; Wilckens, p. 79. Räisänen, *Paul and the Law*, p. 230, n. 9 disagrees.

183. For the relation among *good, life* and *blessing*, see Dt 30:15–20 (equally among *evil, death* and *curse*), where blessing depends on obedience to Torah. Paul examines the significance of this section of Deuteronomy in Romans 10. In Romans 7 he draws attention to a situation where the blessing operates apart from Torah.

184. See Barr, *Escaping from Fundamentalism*, pp. 33–34.

185. The immense debate, still not settled, about Rm 7:14–25—is this a description of Christian or pre-Christian experience?—is mostly prompted by the view that Paul is writing a treatise (see above, p. 86). If, however, he is arguing the case with Jewish Christians that salvation no longer requires the acceptance of Torah, the interpretation of Romans 7 is simplified. Rm 7:7–13 says that Torah, for all its sacred character, has provided the base of operations for Sin. Rm 7:14–25 insists that devotion to God through Torah is bound to fail and implies that Jewish Christians will press Gentile Christians into that pattern of failure if they demand of them full adherence to Torah. The present tenses in verses 14–25 describe not what Christian life is but what it will become if Jewish Christians press too hard. Following Kümmel (see n. 178), most commentators agree that Paul writes about pre-Christian experience with Christian hindsight. The view that Paul describes Christian experience is argued persuasively by Nygren and formally by Cranfield. To me it seems impossibly difficult to suppose that Paul could have described a Christian as 'sold [as a slave and put] under [the power of] Sin' (Rm 7:14) when he had already said 'you have been set free from Sin, and have become slaves of God' (Rm 6:22) and will go on to say, 'The law of the Spirit of life in Christ Jesus has set thee free from the law [i.e., authority and rule] of Sin and death' (Rm 8:2). Nor, I think, can it refer to Christian moral effort (cf. Rm 6:12, 8:12), which has quite a different tone and is performed in the Spirit. If this were the converted Paul speaking of himself, it would be a self-consciously introspective passage, totally at variance with 'I live no longer' (Gal 2:20). Best (pp. 86–

87) ingeniously combines the two views. My proposal is rather different: Paul is warning Jewish Christians, on the basis of his own Jewish experience, about the situation into which they are in danger of falling back.

186. See Community Rule 3.13–4.46 (DSSE, pp. 64–67) and the discussion by Leaney, *The Rule of Qumran and Its Meaning,* pp. 37–46 and commentary.

187. *Rab Anth,* chap. 11. The sentences quoted appear twice (with different attributions) on pp. 125 and 296. In rabbinic hands, teaching about the two impulses is little more than a dramatic way of talking about temptation to do what you know you ought not to do; and the bad impulse is mostly about sexual temptation. Davies (*Paul and Rabbinic Judaism,* pp. 21–27) argued unconvincingly that this rabbinic manner of speaking (scarcely a 'doctrine') lies behind Romans 7.

188. In verse 24 *ek tou sōmatos tou thanatou toutou* is both compressed and ambiguous. First, does *toutou* belong to *sōma* or *thanatos*? Since death is a main theme of this passage (verses 5, 10, 13, 13) and body is not, it goes better with its immediate neighbor *thanatos.* The complaint is not against mortality but in being put to death by the deceptions of Sin. Second, the genitive joining *death* to *body* is best translated by a phrase that recognizes the quasi personification of *death.*

189. Cf. Kramer 19c,d.

190. The Greek *nomos* (usually translated 'law') is again rendered by 'rule' (see p. 108), because Paul is talking about opposing powers capable of exercising control and authority. The phrase *to pneuma tēs zoēs* sounds deplorably vague if translated 'the Spirit of life' and so is correctly rendered 'life-giving Spirit'. 'Set *thee* free' follows the text of Nestle-Aland[26] and the reluctant decision of Metzger, ad loc. (see above p. 104). I suspect that Paul turns from I-formulations (partly generalizing and partly self-expression) to a sudden *thee* in order to bring home what is said to each reader or hearer of his epistle.

191. Cf. Rm 14:17, 'the kingdom of God . . . is righteousness, and peace and joy in the Holy Spirit'; Rm 5:5, 'God's love has been poured into our hearts through the Holy Spirit, which has been given to us'.

192. For the view that 'Paul changed the tradition of the sending formula, which stresses the incarnation of the pre-existent Son, in order to draw attention to the act of salvation' (Käsemann, p. 217) see Dunn, *Christology,* pp. 44–45.

193. See Käsemann p. 216.

194. Schneider (TWNT 5 s.v. *homoiōma,* pp. 195–96) is a good example.

195. So also Gal 1:16, 'to reveal his Son in me' and 4:19 'until Christ is formed in you'; 2 Cor 13:3, 5, 'Christ is speaking in me', 'Jesus Christ is in you'.

196. The formulation of this sentence is tiresome. Does *to pneuma* mean 'your spirit' (JB, TNT, NIV), even 'your spirits' (RSV), or 'the Spirit' (GNB)? NEB, with RV, evasively prints 'the spirit'. Recent commentators decide for 'Spirit'. It causes no difficulty that the adjective *nekron* ('dead') is balanced by the noun *zoē* ('life'); but if *dia hamartian* means 'because of sin', does *dia dikaiosynēn* mean 'because of God's saving goodness' or 'so that you may live rightly' (NEB mg; see Käsemann, p. 224)? Paul's verbal inspirations are sometimes too clever by half, but not necessarily difficult. This one means, 'Your body [personal identity] is dead because of Sin's destructive power, but the Spirit brings it to life because of God's saving goodness'.

197. In Rm. 8:13 'put to death the deeds of the body', it is not satisfactory to regard *sōma* ('body') as an equivalent of *sarx* ('flesh'), as does Käsemann, p. 226. Paul has in mind here not the general perversions of fallible humanity (as in Gal 5:19–21) but individual actions arising from the demands of an ego dominated or formerly dominated by Sin (as in Rm 6:12). He now takes it for granted that Christians can kill off such destructive demands.

198. For the translation 'witness to' (against the English versions), see Cranfield, vol. 1 p. 403.

199. Ps 44:26 (MT), *ḥasdeka;* 43:27 (LXX), 'for thy name's sake'.

200. On the meaning of *paraclete* see Grayston, *Johannine Epistles,* pp. 57–58.

201. Räisänen (*Paul and the Law,* chap. 2) argues against Cranfield that for Paul the law was abolished; to the contrary Badenas, *Christ, the End of the Law.*

202. Dunn, *Christology,* pp. 184–87 examines this passage for evidence of the descent of a preexistent Christ and finds none. Even if such ideas were present, they would contribute nothing to Paul's intention.

203. As does Caird, but see Martin's full discussion, which leads to no decision. In studying the epistle I receive the strong impression that Paul is not only bearing up against possibly indefinite imprisonment and the awkwardness of fellow Christians and the thought that he may be put to death but that at last he is willing to withdraw from his apostolic responsibility and hand it over to Timothy. No doubt the *praetorium* and Caesar's household could be related to an Ephesian imprisonment; but mention of them would be very effective in relation to a Roman imprisonment, a brave attempt to make something out of a rather desperate situation, which an Ephesian imprisonment would not be.

204. Recent phases of this discussion are admirably presented and discussed by Martin. Some critics think that Phil 4:10–23, in which Paul thanks his readers for their gift, 'comes unbelievably late in the sequence of the verses and chapters as we have them in our canonical epistle' (Martin, p. 15) and therefore regard this section as a first letter from Paul to Philippi. But that view is probably based on a cultural misunderstanding. In antiquity Jews did not thank donors for their gifts: they blessed God for the gift and called down his blessing on the benefactor. Hebrew has no equivalent for the *eucharisteō* word-group; and *todāh,* in the so-called thanksgiving psalms, is a praise word. Thanks offered to human benefactors occurs first in 2 Mc 12:31 (see TWNT 9 pp. 397–98); and Rm 16:4 is the only passage in NT and early Christian literature that deals with thankfulness to human benefactors (AGB, s.v., 1). Paul begins Philippians by thanking God for the community's remembrance of him (if that is the correct interpretation of *mneia hymōn* in 1:3, cf. Martin, p. 64) and by assuring them of God's continued blessing (1:6) and praying for its extension (1:9–11). When in 4:10–19 he specifically mentions their help and seconding of Epaphroditus to his service, he is almost effusive in his gratitude and characteristically ends with another blessing from God for them. But he has to frame his gratitude so it does not sound like a reproach. He can scarcely say, 'Thank you for your kindness and help— a pity it went wrong'! At the end of his letter, the significant place for making a final impression, he says, in effect, 'You have always helped me, not because I made demands but because you were generous. Even this latest ill-fated attempt gave me enormous pleasure; and as it is I can manage. God is pleased with your sacrifice and will bless you in all your need.'

205. Eschatological language is important in the epistle: 'day of Christ' occurs again in 2:16, destruction and salvation in 1:28, and 'The Lord is at hand' at 4:5. These are standard but unemphatic references. But in 3:1b–4:1 eschatology contributes vigorously to the argument: looking forward to the resurrection from the dead (3:11), the image of the contest (3:14), the destruction of opponents (3:19), the coming of the deliverer and the great transformation (3:20–21), and perhaps even the crown in 4:1. Clearly the eschatological expectation is still vigorous, but Paul seems content for the outcome no longer to depend solely on him.

206. In Phil 2:17 Paul applies sacrificial imagery to his prospective death: 'Even if I am to be poured as a libation upon the *sacrificial offering* of your faith' (*spendomai—* elsewhere in 2 Tm 4:6, which, with *analysis* [cf. Phil 1:23] looks like a reminiscence of

Philippians—and *thysia kai leitourgia*). This can only be described (without denying Paul's serious intent) as a conceit. The faith(fulness) of the community is described as a pious offering and public service for God, and Paul (somewhat affectedly) underplays the significance of his death by comparing it to the accompanying drink offering. No serious theological consequences can be drawn from this language. Equally uninformative is Paul's self-description in Rm 15:16 as 'a *leitourgos* of Christ Jesus to the Gentiles in the priestly service [*hierourgounta*] of the gospel, so that the offering [*prosphora*] of the Gentiles may be acceptable, sanctified by the Holy Spirit.' Paul imagines himself as carrying out a public duty by acting like a hellenistic priest when he presents the devout Gentiles as a sacred offering to God. It is acceptable rhetoric but no more. The same rhetoric, however, is used to save the reputation of Epaphroditus, who can be called 'your apostle and *leitourgos* to my needs' (Phil 2:25; for *leitourgia* as Christian aid, see Rm 15:27; 2 Cor 9:12). He had risked his life trying to complete their *leitourgia* to Paul (Phil 2:30), who gracefully acknowledged having received from Epaphroditus the gifts they sent, a fragrant offering, a sacrifice *(thysia)* acceptable to God (Phil 4:18; for the 'fragrant offering', see on 2 Cor 2:14–15). Thus, Paul uses sacrificial language occasionally as a decorative form of writing when referring to himself and his supporters but very seldom of the death of Christ. Cf. p. 53.

207. For the discussion, see Martin, *Carmen Christi,* supplemented by his commentary; also Dunn, *Christology in the Making,* pp. 114–21.

208. The verb *kenoō* occurs only in Paul: it means to make faith, the cross, and Paul's confidence ineffective or powerless (Rm 4:14; 1 Cor 1:17, 9:15; 2 Cor 9:3). The adjective *kenos* is used when Paul is hoping to make sure that his missionary work, God's grace, and Christian faith have not become ineffective (1 Cor 15:10, 14:58; 2 Cor 6:1; Gal 2:2; Phil 2:16; 1 Thes 2:1, 3:5).

209. See Hengel, *Crucifixion,* pp. 51–63 ('The "Slaves" Punishment').

210. Martin, *Carmen Christi,* p. 227: 'It is not said to whom he was disobedient. Indeed it seems to be unimportant. . . . The fact of his obedience is simply reported'. If this means that instead of telling others what to do, he had to do what he was told (i.e., *obedient* simply means that he was a slave), there is no objection. But if *obedience* is used in its normal sense, it must imply obedience to someone or something. For some interpreters, obedience of any kind is regarded as a good thing. In the first volume of Bultmann's *Theology of the New Testament* obedience appears in the index forty-nine times, love only thirty-eight times. If *obedience* is being used to mean fidelity, it can pass; but strict obedience ought always to be regarded with suspicion, sometimes with dismay. See Grayston, 'Obedience Language in the New Testament'.

211. For attempts at identifying this opposition to the cross, see Martin, ad loc. and pp. 33–34; also Grayston, 'The Opponents in Phil. 3'.

212. There is nothing to suggest that 'their god is the belly' refers to gluttony or food tabus (*koilia* = stomach) or to sexual promiscuity (*koilia* = location of the female—and occasionally of the male—sexual organs; cf. Philo, who can say that the belly signifies *epithumia;* see Behm, TWNT 3). *Koilia* here means 'bowels' (as in Mk 7:19) or even 'excrement' (LSJ, s.v. 1.3), just as Paul has previously referred to *skybala* in Phil 3:8. Out of an apostolic insult have been conjured the Philippian libertines, those comic all-purpose villains who, in modern criticism, rush on to the stage at the slightest provocation and perform shameful debauchery. But when Paul imagines the opponents glorying in their shame he means that they are delighted by what he regards as despicable, namely their promotion of circumcision.

213. Paul Tortelier, *Paul Tortelier—A Self-Portrait in Conversation with David Blum* (London 1984), pp. 149–50. An exceptionally gifted student performed the first move-

ment of the Boccherini Concerto in B-flat major. He played with a lovely tone, pure into-nation, good taste, and respect for the score; but Tortelier and fellow students were unmoved. So Tortelier himself played the piece, and the student himself said, 'It's like night and day'. Tortelier's playing, no better than his student's, had communicated in a way that his student's had not. 'When you play this work', asked Tortelier, 'what does your imagination tell you?' 'I'm busy enough thinking about my cello playing.' 'That's just the trouble. The notes must be taken for granted. Do you know what I did to make it different? I told a story as I played.'

214. *Ei pōs katantēsō,* and cf. *ei kai katalabō,* '[to see] whether I can capture'. In both sentences it is probably *ei* with aorist subjunctive, possible in classical and hellenistic Greek (cf. Lk 9:13), an extension of the deliberative subjunctive in indirect questions.

215. But *salvation (sōtēria)* and the verb *to save (sōzō)* are not uncommon. *Sōtēria* means preservation in a dangerous personal or communal situation (Phil 1:19, 2:12) or in the supremely dangerous situation when destruction is likely (Phil 1:28). When the verb is used, the danger is not always made clear, but the context is usually eschatological: from destruction (1 Cor 1:18, 3:15; 2 Cor 2:15), from death (Rm 10:9, 11:14; 1 Cor 5:5), from disobedience and rebellion (Rm 11:26), from the wrath (Rm 5:9–10), and from bond-age to corruption (Rm 8:24). This last reference is nearest to Phil 3:20.

Chapter 3

1. Authorities are not agreed on the authorship of Colossians. Kümmel, *Introduction to the New Testament,* reviews the discussion: on the basis of language and style he finds no reason to doubt Pauline authorship and attributes theological differences between Colossians and other Pauline writings to the polemical intention of the epistle (pp. 340–46). Houlden (pp. 134–39) cautiously and Caird (pp. 155–57) rashly agree. The argument for authenticity is commonly negative; it might be more successful if it could be put in the form of positive reasons why Colossians *should* be by Paul. Against the Pauline authorship are ranged Marxsen, *Introduction to the New Testament,* pp. 184–85; Vielhauer, *Geschichte der urchristlichen Literatur,* pp. 196–200; Lohse, pp. 177–83; Ernst, pp. 150–52; Schweizer, pp. 15—24. The choices are between attributing Colossians to, say, Timothy (while Paul was still alive but incapacitated) or to someone of Paul's entourage (shortly after his death) or attributing part of it to Paul and part to a later reviser (as in the elab-orate scheme of Schmithals, 'The *Corpus Paulinum* and Gnosis').

2. In other prescripts the recipients are not called *adelphoi* and, except in Ephesians, are not described as *pistoi;* and grace and peace are offered from 'God our father *and the Lord Jesus Christ*' (for the variants, see Metzger, ad loc.).

3. In Col 1:7 it is not easy to decide whether Epaphras—who belonged to the Colossian community (Col 4:12) and at some stage shared Paul's imprisonment (Phlm 23)—was a faithful minister of Christ on *our* or on *your* behalf (cf. Metzger, ad loc.). In either case this letter gives him the approval of the Pauline mission while justifying the visit of Tychi-cus, a member of Paul's entourage (Col 4:7). The whole greeting section, which ends with Paul's own signature (4:7–18) looks like an attempt to make personal links between mem-bers of the Pauline mission and the Colossian community, to bind together the commu-nities at Colossae, Laodicea and Hierapolis; and to limit any damage caused by Paul's condition and imprisonment. After Paul's own dealing with his imprisonment in Philip-pians, this is the next piece of evidence that steps were being taken to reinforce and hold on to the missionary empire.

4. See chap. 2, n. 33.

5. Cf. Col 3:24, where inheritance language is used to persuade slaves to be obedient. For a close parallel, see Paul's speech to Agrippa (Acts 26:18).

6. 'The kingdom of his beloved Son' does not readily convey the correct impression. For one thing, *basileia* means ruling authority, and *beloved* indicates not affection ('dear' NEB) but the delegation of authority. The divine being is *theos* and *kyrios,* creator and sovereign. The Father delegates sovereignty to the Son. Hence, the relevance of verses 15–20.

7. For the meaning of *apolytrōsis,* see the discussion of Rm 3:24 and chap. 2 n. 161. Some late witnesses add *through his blood* from Eph 1:7, but Colossians suggests neither atonement nor ransom.

8. *Aphesis hamartiōn* (forgiveness of sins) is mainly Lukan: (1) in Lk 3:3, Mk 1:4 John Baptist proclaims a baptism of repentance for forgiveness of sins and in Acts 2:38 Peter urges baptism in the name of Jesus Christ for the forgiveness of sins; (2) in Lk 1:77 John Baptist is to give knowledge of salvation to God's people in the forgiveness of their sins and in Lk 24:47, Acts 5:31, 10:43, 13:38, 26:18 apostolic proclamation offers forgiveness of sins by faith in the risen Lord; (3) in Mt 26:28 Jesus' blood of the covenant is poured out for many for the forgiveness of sins; (4) Eph. 1:7 and (5) cf. Heb 9:22, 10:18.

9. The verb *paristēmi* (present) is, of course, a technical term of the sacrificial cultus (as in Rm 12:1), and may be a legal term for presenting someone for judgement (AGB, s.v.). Neither the present passage nor any Pauline use suggests the legal meaning. The cultic meaning may be suggested by *holy,* but scarcely by *blameless* or *irreproachable* (Lohse). At the most this could be only a residual sacrificial metaphor, with little influence on the character of Christ's death. It is better to interpret in agreement with verse 28: 'to present every man mature in Christ', presumably to carry out God's purposes, as in Rm 6:13, 16, 19.

10. Cf. Martin, *Reconciliation,* p. 41.

11. The evidence is most clearly displayed by Lohse, pp. 41–46. Schweizer's discussion (pp. 55–88) is fuller but less easy to grasp. Add Sanders, *The New Testament Christological Hymns,* chap. 4; and Hooker, 'Were There False Teachers at Colossae?' Hooker disbelieves in the 'hymn'.

12. The translation here printed is RSV modified to accord more closely with the Greek, especially in verse 20.

13. So Käsemann, 'A Primitive Christian Baptismal Liturgy'.

14. There are several instances of supplying what is lacking, whether a need or a defect (*hysterēma*). The familiar verb is *anaplēroō* or *prosanaplēroō;* but here *antanaplēroō* presumably suggesting completing something in the place of someone else or for someone else's benefit.

15. Paul never conceals the likelihood of Christian suffering, nor is he reticent in mentioning his own (see 1 Thes 1:6, 2:14–16, 3:3; 1 Cor 4:9–13; 2 Cor 1:8–11, 4:7–8, 6:3–10, 11:23–33, 12:9–10, 13:4; Gal 6:17; Rm 8:35–39; Phil 3:10). But the remarkable statements of Col 1:24 suggest that his bodily sufferings at that time were being taken as damaging evidence against the truth of his gospel.

16. Philo, *L.A.* 2. 55. Colossians uses *apekdusis,* Philo *ekduo.* Ordinances are *dogmata,* cf. Col 2:14. For an actual cult, see Apuleius, *Golden Ass* 11.22–26, quoted in Barrett, *The New Testament,* pp. 127–30.

17. Presumably the conceit of a Christian kind of spiritual circumcision comes from Phil 3:3. That the writer does not take it very seriously is suggested by Col 3:11, though others may have done so. See Grayston, 'The Opponents in Phil 3'.

18. On *syn Christo* see TWNT 7, pp. 780–95. Paul has been 'raised with Christ' and 'with Christ', i.e., covered by his protective power.

19. After 'baptism', *en hō* is naturally taken to mean 'in which', as Schweizer (pp. 145–46 rightly (though not lucidly) argues (so Bruce). But since in this group of verses *en hō* frequently means 'in him' it is possible (so Lohse)—though not elegant—to read it in that sense: 'in whom you have been raised with him'.

20. For *paraptōma* see chap. 2, n. 156.

21. The verb is *charizomai*, which means to behave generously in giving a favour—so to remit a debt that the debtor cannot pay (Lk 7:42–43), to forego the punishment of a repentant offender (2 Cor 2:7, 10—sarcastically in 2 Cor 12:13). In this part of its range, the Greek verb behaves like the Latin *condonare*. In Col 2:13, 3:13 and Eph 4:32 it is used of God and consequently of Christians. It conveys the simplest of all forgiveness ideas: out of goodheartedness, disregarding wrongful activity.

22. Houlden (p. 187–88) thinks there are hints of this deception theory, though he probably finds it distasteful. But embarrassment has been unnecessary since Aulen's *Christus Victor*, pp. 71, 126–7, where he draws attention to the religious conceptions underlying this realistic imagery: 'Good in itself can have the effect of hardening, and can provoke evil, and increase the malice of those who are evil. So the evil assails the good, but it is to its own undoing. Evil over-reaches itself; its power is broken when it seems to have prevailed'.

23. This translation depends on the decisions that (1) God is the subject in verses 13–15, (2) *apekdysamenos* is a middle voice used in the active sense (BDF 316.1), and (3) *en autō* means 'in him' and not 'in the cross' (cf. Bruce, pp. 239–40). Since Lightfoot, some have adopted the interpretation, 'On the cross he discarded the cosmic powers and authorities like a garment' (NEB); but can the angelic powers be both a discarded robe and captive enemies? (Scott). Could Christ really be depicted as celebrating a triumph over a cast off suit of clothes? (Caird). See also Carr, *Angels and Principalities*.

24. Delling, TWNT 7, s.v. *stoicheion*.

25. See also Schweizer, 'Christianity of the Circumcised and Judaism of the Uncircumcised'. For neo-Pythagorean infuence, see Köster, *Introduction to the New Testament*, vol. 1, pp. 374–76; Hengel, *Judaism and Hellenism*, vol. 1, pp. 245–47.

26. Cf. Leaney, *The Rule of Qumran*, pp. 25–30. Cf. 'living according to the truth' in 1 Jn 1:6, discussed in Grayston, *Johannine Epistles*, pp. 49–50.

27. On the virtual absence of *Spirit* from Colossians, see Schweizer, p. 38.

28. The technical discussion is in Percy, *Die Probleme der Kolosser-und Epheserbriefe*; Mitton, *The Epistle to the Ephesians*; Roon, *The Authenticity of Ephesians*. See also Murphy-O'Connor, *Paul and Qumran* for articles by Kuhn and Mussner; and Käsemann, 'Ephesians and Acts'. The following books on particular features of the epistle contain surveys of recent study: Kirby, *Ephesians, Baptism and Pentecost*; Sampley, *'And the Two Shall Become One Flesh'*; Caragounis, *The Ephesian Mysterion*; Rader, *The Church and Racial Hostility*.

29. Doty, *(Letters in Primitive Christianity)* discusses Paul's coworkers: 'A very common feature of Hellenistic letters, mention of the carrier established the carrier's relation to the writer, and guaranteed that what he had to say in interpreting the letter was authorized by the writer. The feature was especially important in Hellenistic letters where the actual information to be conveyed was trusted (only) to the messenger' (p. 30, also pp. 46–47).

30. Could Paul have composed Ephesians? If he could have written to no particular community, having discovered the art of theological reflexion without the stimulus of a theological need; if he had become a late convert to the Qumran type of semi-Gnostic writing, and was content to plaigarise his own Colossians; if he could invent a new view of the Gentiles' relation to the Jewish inheritance and could thrust himself forward as the

special guardian of this mystery, perhaps he could have written it—but it is scarcely likely. It is much more likely that the epistle was composed when it still seemed plausible and effective to recall Paul's authority, in one of the first attempts to rescue his crumbling missionary empire.

31. This rough-and-ready quotation of verses 3–14 (mainly RSV, adapted, shortened and rearranged) is intended to indicate the structure of the passage and to hint at the awkward syntax.

32. *Apolytrōsis* is a neighbour of 'expiation by his blood' in Rm 3:24–25; and Heb 9:15 mentions 'a death for the redemption of transgressions'. *Aphesis* of sins by the shedding of blood appears in Heb 9:22; and Mt 26:28 has the 'blood of the covenant poured out for many for the forgiveness of sins'. More commonly forgiveness comes by repentance and in the name of Christ. Cf. 1 Jn 1:7, 9, where both the blood of Christ and confession of sin cleanse from sins. Cf. Barth, p. 83: 'Freedom is the clear purpose and result of redemption'.

33. Mitton, Bruce, Schnackenburg (ad loc.).

34. Cf. Schnackenburg, ad loc.: 'whereas Paul stresses the ethical requirement (Rm 6:12–14, 15–21) and the eschatological reservation (Rom. 6:8, 22f.; 8:17), the shift of emphasis here is unmistakable' (p. 95). 'Salvation is identified with the already accomplished feat of the resurrection of the saints' (Barth, p. 232, and his excursus 'From Death to Life', pp. 232–36).

35. For recent study of Eph 2:12–19, see Martin, *Reconciliation*, pp. 167–93. And on verses 11–13, see Barth, pp. 298–305.

36. Aboth. 1.1, 'Make a fence to the Torah'. The well-known view that reference is intended to the wall in the Jerusalem Temple prohibiting the passage of Gentiles from the first court into the second would be plausible if readers of the epistle might be supposed familiar with descriptions in Josephus in *War* 5.193–94 and 6.124–25 (published 77–78 C.E.) and Ant. 15.417 (published c.94 C.E.). That and other suggestions seem improbable. Nor is it easy in this context to introduce the barrier dear to hellenistic religion of supernatural powers standing betwen mankind and the supreme deity. They would indeed liven up the passage, but the writer seems to have overlooked the opportunity. See Ernst, ad loc.

37. For further discussion of this view see Rader, chap. 8; also Bruce, pp. 296–97.

38. Cf. Rm 11:14, *mou tēn sarka* means 'my own Jewish race'; and Eph 6:12, 'We are not contending against flesh and blood' because Christ has destroyed the enmity in his flesh. Cf. Schnackenburg, ad loc.

39. But Schnackenburg (p. 117) says that the event of the cross in Ephesians neither achieves creative clarity (as it does in Col 2:14–15) nor gives a shock of offence (as elsewhere in Paul), but is simply the place of reconciliation, a theme of Christian preaching.

40. With Jeremias TWNT, s.v. *akrogōniaios* is the keystone; so Caird, Ernst, Houlden, and Bruce against Mitton, who (in my opinion) overlooks the fabulous nature of the building, though he is supported by Schnackenburg. Cf. also Gärtner, *The Temple and the Community in Qumran and the New Testament,* pp. 60–66, McKelvey, *New Temple,* chap. 8.

41. God's announcement in Is. 57:19 of 'peace to the far and to the near' is sensibly interpreted by rabbinic teachers as referring to those who have made themselves far from God and to those who *were* far from God but have repented and so are near. That view is akin to the estimate of Gentiles in Eph 2:1–10. But some rabbis applied *near* to Jews and *far* to Gentiles, and not always with hostility. A proselyte is 'one who comes near'. In a saying attributed to Rabbi Eliezer (80–120 C.E.): 'When a man comes to you to be made a proselyte, and he comes to you with pure intent, draw him near, and do not keep

him at arm's length. And learn that if you repel with your left hand, you must draw near with your right'. See *Rab Anth*, pp. 323, 325, 331, 568, SB 3 pp. 585–87.

42. The interpretation of 'the breadth and length and height and depth' in Eph 3:18 as a reference to the cosmic cross is as old as Irenaeus, (Danielou, *A History of Early Christian Doctrine*, I pp. 279–92). It is somewhat favoured by Houlden and discussed and rejected by Ernst. It is no more than a device adopted by preachers who know that the cross is theologically important but can only make it rhetorically impressive. The four words belong to the stock language of magical texts (Dibelius, p. 77). Cf. Schnackenburg, p. 154 and Barth, pp. 395–97.

43. Caird (ad loc) may be right in suggesting that the writer (Paul, in his view) is adapting the Shema (Dt 6:4–9). It looks as if an original 'one body, one hope, one Spirit' has been expanded through the writer's conviction that his readers, originally without hope, had been called to a splendid hope (Eph 1:18, 2:12). Cf. Col 3:15, 'Let the peace of Christ rule in your hearts to which indeed you were called in one body'. Then the two triads would be a reasonable match: the one body of the one Lord, the one hope calling for one faith, and the one Spirit received in the one baptism. *One faith* certainly does not mean 'one body of theological propositions' but the set of attitudes and assumptions in a community where God is to be trusted and members can be introduced to one another as beloved and faithful. See Grayston, 'What is Faith?'. On the credal character of the formula, see Bruce, ad loc. and Barth, pp. 462–72.

44. This passage is treated in detail by Sampley, *"And the Two Shall Become One Flesh'*. On verse 23, see Bruce.

45. See Hooke, 'The Myth and Ritual Pattern in Jewish and Christian Apocalyptic'. Cf. Ringgren, *Religions of the Ancient Near East*, pp. 148–54. Barth luxuriantly describes traditions about the Bridegroom and the Bride, pp. 668–99.

46. See Foerster, *Gnosis*, vol. 1, pp. 139 (Iren. *Haer* 1.7.1), 152 (Clem. Alex. *Exc. e Theod.* 63.2), 176 (fr. 38 of Heracleon); 219 (Iren. *Haer* 1.21.3). These are mid-second-century Valentinian sources. For *Acts of Thomas*, 5ff., see Foerster, vol. 1, p. 344 ff. and see Bornkamm in Hennecke, Schneemelcher, and Wilson (London 1965) pp. 432–33 on the wedding hymn in the first Praxis.

47. Rudolph, *Gnosis*, p. 120.

48. For the discussion of this passage and possible parallels, see Dibelius, Ernst, and Bruce (ad loc.) and Kuhn, pp. 125–31.

Chapter 4

1. Thus, conceding something to the plea for a canonical appraisal of NT writings, as suggested by Childs *(The New Testament as Canon)*, chap. 21. In general, the introduction to Houlden's commentary (though in my view placing the epistles too late) is dispassionately effective. To his support for Lukan influence, add Wilson, *Luke and the Pastoral Epistles*.

2. See Dibelius and Conzelmann; and for 'Saviour', see Brox, pp. 232–33.

3. This *pistos logos* is elaborately examined in Knight, *Faithful Sayings in the Pastoral Epistles*, chap 6.

4. For various interpretations in early Christianity, see Spicq, ad loc. Brox (ad loc.) favours a more Gnostic interpretation than I do; cf. Peel, *Epistle to Rheginos*, pp. 140–41. I find support for my own view in Lane, '1 Tim. iv. 1–3: An Early Instance of Over-realized Eschatology?' and Moule, 'St. Paul and Dualism', pp. 111–112.

5. See the excursus on soteriological terminology in Dibelius and Conzelmann, pp. 143–46—hieratic language of the Greeks with a new tone and weight through the emperor cult.

6. 'Redeem us from all iniquity' is *lytrōsetai hēmas apo pasēs anomias.*

7. For its extent, see Knight, *Faithful Sayings,* chap. 5.

Chapter 5

1. Books on the study of Mark: Lightfoot, *History and Interpretation;* idem, *Locality and Doctrine;* idem, *Gospel Message;* Evans, *Beginning of the Gospel;* Marxsen, *Mark the Evangelist;* Martin, *Mark—Evangelist and Theologian;* Donahue, *Are You the Christ?;* Kee, *Community of the New Age;* Kelber, *Kingdom in Mark;* idem, *The Passion in Mark;* idem, *Mark's Story of Jesus;* Hooker, *Message of Mark;* Best, *Mark—The Gospel As Story;* Hengel, *Studies in the Gospel of Mark;* Telford, *Interpretation of Mark.*

2. The *titloi* of the sections *(periochai)* are given by Swete p. liv. See Metzger, *Text of the New Testament,* p. 23.

3. Haenchen, p. 34, n. 9; so Rawlinson, p. xx.

4. E.g., Burkitt, *Gospel History and Its Transmission,* p. 102; Montefiore *Synoptic Gospels,* vol. 1, p. xlii; though Turner had already said in Hastings' *Dictionary of the Bible* i (1898), pp. 406, 410 that Mark groups events by subject matter rather than by time.

5. Taylor, pp. 107–11; similarly Cranfield.

6. Compare the proposals of Pesch and Gnilka.

7. Neirynck, *Duality in Mark.* See Kelber, *Mark's Story,* pp. 43–46.

8. For Kingdom of God the landmarks are Dalman, *Words of Jesus* 1902, TWNT 1 of 1933, and Schürer, Vol. 2, pp. 531–37, though the last is something of an olio. Among recent discussions, reference should perhaps be made to the enthusiastic trilogy by Perrin, *The Kingdom of God in the Teaching of Jesus;* idem, *Rediscovering the Teaching of Jesus;* idem, *Jesus and the Language of the Kingdom;* balanced by Ladd, *The Presence of the Future.* For a recent survey Chilton, *The Kingdom of God in the Teaching of Jesus;* idem, *God in Strength,* pp. 283–88; also Marcus, *The Mystery of the Kingdom of God,* pp. 51–57; Beasley-Murray, *Jesus and the Kingdom of God.* In Mark references to the kingdom of God are much of a kind. Despite Jesus' proclamation in 1:15, everyone in the Gospel is in the position of Joseph of Arimathea, 'looking for the kingdom of God' (15:43), i.e., looking for God's royal power to have proper effect. In 4:11 those around Jesus, including the Twelve, are expected to know the secret of the Kingdom of God. When the parable of the sower is interpreted, the secret turns out to be an excuse for the delayed arrival of the expected great harvest. The parables of the seed growing secretly and the mustard seed have a similar purpose (4:26, 30). In 9:1 some are promised that they will see that the Kingdom of God has come with power, implying that the divine power had not yet been properly displayed. Four sayings indicate qualifications for entering the Kingdom of God (i.e., coming within the scope of its benefits): removing what causes you to sin (9:47), behaving like a child (10:14–15), renouncing wealth (10:23–25), and agreeing with Jesus about the primary commandments (12:34). It could be supposed that such qualifications gained admission to a kingdom already present, but it is more likely that they promised admission to a kingdom yet to come—and that interpretation is confirmed by Jesus' view that he would drink no more the fruit of the vine until the day when he would drink it new in the Kingdom of God (14:25). Mark thus shows a clear perception that the initiative announced in Galilee had not succeeded, but (in still recording qualifications for entry to the kingdom) he has clearly not abandoned that initiative.

9. Cf. Johnson, *Sacral Kingship in Ancient Israel* and Pss 47, 68, 93, 96, 145; Dt 32:39.

10. The formula *ti hēmin kai soi,* (see BDF, sec. 299 [3]), translated 'What have you to do with us', is intended to repel an attack or intrusion. The meaning is, 'What [room is there] for both us and thee [in this situation]?' On the silencing of demons and other magical techniques, see Hull, *Hellenistic Magic and the Synoptic Tradition,* pp. 69–70; Smith, *Jesus the Magician,* pp. 127–28.

11. The silence theme is not confined to pt 1 of the Gospel. In pt. 2 it is attached to the cure of the deaf-mute in 7:31–37, perhaps type 2, though clearly not to the two feedings and the walking on the water nor to the exorcism of the Syro-Phoenician woman's daughter (7:24–30) (type 1). In pt. 3, the blind man of Bethsaida is taken outside the village to have his sight restored, and afterwards he is told not to enter the village, but there is no command of silence (8:22–26); nor could there be for blind Bartimaeus (10:46–52). In 9:14–29 a deaf and dumb spirit is exorcised with much publicity. Instead, the disciples are commanded to be silent about Peter's confession of Jesus as the Christ (8:30) and, until the Son of Man had risen from the dead, about what they had seen on the mount of transfiguration (9:9). Possibly a suggestion of secrecy is intended by words introducing the second prediction of the passion, namely, that Jesus 'would not have anyone know' that he was passing through Galilee because he was teaching his disciples (9:30–31). But when Jesus gives his approval to a rival exorcist by saying, 'No one who does a mighty work in my name will be able soon after to speak evil of me' (9:38–40), secrecy and silence are out of the question. Nor do they appear in pts. 4 and 5 of the gospel. Summarizing, in type 1, where no silence is required, there are five healings, three exorcisms, two feedings and the walking on the water; in type 2, where silence is required but not possible, there are three healings; and in type 3, where silence is commanded there are two exorcisms, with the implication that this was true of very many others, and the two prohibitions laid on the disciples. Clearly, then, the command of silence has a relative importance. It can scarcely be treated as Mark's invention for insisting that only secret knowledge of the great powers of Jesus—and therefore of his status—was available during his lifetime.

Since Wrede's book *Das Messiasgeheimnis in den Evangelien* in 1901 (Kümmel, *The New Testament,* pp. 284–87) attention has constantly been directed to the status of Jesus as Messiah, whether before or after his resurrection (for criticism of such enquiries see Kümmel, *Introduction to the New Testament,* pp. 89–94). One source of confusion has been the indiscriminate use of the adjective *messianic* (despite the rare use of *Christos* in Mark) to imply anything associated with any kind of divinely approved status for Jesus. This will require discussion when we consider the first Passion prediction; but it may here be proposed that Mark in effect says, 'Jesus was indeed Messiah, but that title confines him to the Jews and angers them; therefore it is better to listen to him as Son of Man and confess him as Son of God'. For traditional views, cf. Tuckett, *Messianic Secret.*

12. Satan appears in Mark's formal opening of the Gospel (1:13); elsewhere in words of Jesus: in the parabolic refutation of the scribal accusation that Jesus is possessed by Beelzebul or the prince of the demons (3:23, 26); in a conventional usage, parallel to persecution, hardships and anxieties in the interpretation of the sower (4:15); and in Jesus' rebuke of Peter (8:33) to represent human forces opposed to God. This is in accord with a common trick of language by which violent terms become domesticated and familiar. Unclean spirits and demons occur in 1:23, 26, 27, 32, 34, 39; 3:11–12, 15, 22, 30; 5:2, 8*, 13, 15, 16, 18; 6:7, 13. In pt. 2 only 7:25, 26, 29*, 30 and in pt. 3 9:25*, 38 (* = a saying of Jesus).

1 3. Cf. Schürer vol. 2, sec. 26; Bowker, *Jesus and the Pharisees,* pp. 38–42; Rivkin, *A Hidden Revolution,* chap. 2. For the view that neither Mark nor his sources can be reliably used to confirm that disputations took place between Jesus and the Jewish authorities or as a basis for reconstructing their substance, see Cook, *Mark'sTreatment of the Jewish Leaders.*

1 4. The conflict is kept in mind in subsequent parts of the Gospel. In pt. 2 there is a remarkable passage about purity rules (7:1–23), initiated, according to Mark by questions from the Pharisees and some scribes who had come down from Jerusalem. Jesus' reply is strongly polemical, though he does not explicitly name the rival authorities whom (for the only time in Mark) he calls hypocrites (but see *hypokrisis* in 12:15). In pt. 3 Jesus warns against the leaven of the Pharisees and the leaven of Herod (8:15). *Leaven* means dangerous contamination. The Pharisees (who now appear for the first time in a saying of Jesus) are attacked for their association with the activities of Herod (cf. the two references to the hostility of Pharisees and Herodians in 3:6 and 12:13). In 9:11 Jesus accepts scribal interpretation (some witnesses say Pharisaic and scribal) of the return of Elijah but counters it with a qualifying scriptural reference. In 9:14 scribes argue with disciples about a case for exorcism and in 10:2 Pharisees put a question about divorce. In pt. 4 some Pharisees and Herodians try to trap Jesus by asking him about tribute to Caesar (12:13); a scribe agrees with Jesus on the primary commandments and is himself commended (12:28, 32, 34). Two sayings of Jesus express hostility towards the scribes as regards their messianic teaching and their ritual and monetary excesses (12:35, 38–40). Most of this sounds like disputes about rival customs, irritating but not violent, the result of a sectarian division. But in pt. 3 sayings of Jesus show a new development: he is to be rejected by 'elders, chief priests and scribes', which is Mark's way of indicating the chief influences in the Jerusalem Sanhedrin (mentioned in 14:55 and 15:1); though 'elders', who would mostly have been Pharisees, are sometimes omitted (e.g., 10:33). In pts. 4 and 5 the chief priests and fellow-members of the Sanhedrin are relentless enemies of Jesus until Joseph of Arimathea, a respected member of the council, shows different sympathies. Also in 12:18–27 Jesus adopts the Pharisaic position on resurrection, against the Sadducees (their only explicit mention in Mark).

1 5. Dewey, 'The Literary Structure of the Controversy Stories', pp. 113–15. The subject is given full-length treatment in idem, *Markan Public Debate.*

1 6. In addition to the commentaries, especially Pesch and Gnilka, see also Martin, *Mark,* pp. 184–88. The questions discussed are when the Pharisees came into the saying, whether 19a can be regarded as a genuine saying of Jesus, whether 19b is redactional, whether the bridegroom is metaphorical or allegorical, and whether 20 refers to the death of Christ and is intended to justify the Christian practice of fasting. It is universally agreed that *bridegroom* was not a known description of the Messiah. See Ziesler, 'The Removal of the Bridegroom'.

1 7. On fasting, see TWNT 4 s.v.; Marshall, *Luke ad* Lk 5:33–39; Abrahams, *Studies in Pharisaism and the Gospels,* vol. 1, chap. 91; Pedersen, *Israel,* p. 12. In Mark, fasting occurs only here (‖Matthew, Luke) and in a variant reading in 9:29 (‖Matthew, v.1.): 'This kind (of demon) cannot be driven out by anything but prayer *and fasting'*. The added words may not be original, but they display exactly the purpose of fasting: to induce a perhaps hesitant or reluctant deity to use his power. In Luke and Matthew fasting is an act of piety, intended to persuade God to favour the worshipper. The prophetess Anna constantly frequented the temple and reinforced her prayers with fasting, in the hope of Israel's redemption (Lk 2:37); the self-conscious Pharisee not only prays in the temple but also mentions his fasting twice a week (Lk 18:12). The Sermon on the Mount is addressed to a community that fasts to obtain a reward from God (Mt 6:16–18). The only obligatory

Jewish fast was prescribed for the Day of Atonement, where fasting again is used to rein-force the community's plea for forgiveness. Mt 9:15 interprets Mark's reference to fasting as 'mourning' (probably under the influence of the Q saying in Mt 11:17). But the relation between fasting and dying must be carefully understood. The relatives and friends of a dying person may fast to prevent his death (if God can be persuaded to spare him) and may continue to fast after his death to induce God to remove the resulting uncleanness. (King David had robust ideas on these matters, 2 Sm 12:15–23.) Hence, fasting is an apotropaic feature of mourning but seems to play no part in the NT. Finally, fasting may be used to persuade God to disclose his will (in the temptations of Jesus, Mt 4:2‖Luke; Acts 13:2–3; 14:23 in the appointment of missionaries and elders). This function of fast-ing is piously justified as making the petitioner more receptive to the divine will.

18. The passive *aparthē* may be read as a divine reflexive. Cf. BDF, 313 'Aram. gen-erally uses the pass. for actions of a celestial being'.

19. Cf. abandoning family for the gospels' sake and the recompense in the age to come (Mk 10:29–30). Elsewhere Mark shows Jesus giving strong support to the fifth com-mandment 7:10–13, 10:19; and he does not reproduce the Q logion Mt 10:37‖Lk 14:26.

20. On the problems of feeding, see Fowler, *Loaves and Fishes.*

21. 'The leaven of the pharisees and the leaven of Herod' in 8:15 suggests that (1) Pharisees and Herodians were still troublesome when these words entered the tradition, (2) they were troublesome in different ways (note the repetition: *apo tēs zymēs . . . kai tēs zymēs*), (3) they were joined together because both the divergent attitudes were to be rejected, and (4) what was in mind was their view of Gentiles (cf. Pesch, Gnilka [ad loc.]). In 8:11–13 the Pharisees ask Jesus for a sign from heaven and are denied it. A sign of what? In Mark's view, I suspect, it was a sign that God's benefits were sufficient for Gen-tiles as well as Jews.

22. Fowler, *Loaves and Fishes,* p. 86.

23. Of the fourteen uses of *Kingdom of God (basileia),* seven occur in pt. 3: 9:1, 47; 10:14, 15, 23, 24, 25—distributed among four episodes. See n. 8.

24. The commentaries are unhelpful until Haenchen and Lane; Pesch and Gnilka are especially useful both for bibliography and criticism. Also see Higgins, *Jesus and the Son of Man,* pp. 30–75; Tödt, *The Son of Man in the Synoptic Tradition,* pp. 141–221, 275–77; Hooker, *The Son of Man in Mark,* pp. 103–47; Hahn, *The Titles of Jesus in Chris-tology,* pp. 37–67; Jeremias, *New Testament Theology,* vol. 1, pp. 276–86; Kümmel, *The-ology of the New Testament,* pp. 85–90.

25. TWNT 5, s.v. *paschō.* On p. 914 Michaelis tries hard to derive *polla pathein* from Is 53:3, 4, 11; Pesch (p. 50) disagrees, though he thinks that an influence of the suffering servant on the suffering godly man in Wisdom cannot be ruled out.

26. 2 Mc 6:30, 7:18, 32; 4 Mc 4:25, 9:8, 10:10, 13.17 *(si vera lectio).* Michaelis's argument that these do not prove that *paschō* was a technical term for martyrdom may be right, but it provides the context.

27. TWNT 2, s.v. *dei,* pp. 21–25, though lacking reference to Mk 8:31.

28. Tödt, pp. 188–94; Bennett, 'The Son of Man'.

29. For Isaiah 53, see the discussion of Mk 10:45. Nickelsburg (*Resurrection,* pp. 58–62) regards Wis 2:12–20 as a development of Isaiah 53.

30. The many troubles of the godly man are frequently rehearsed, e.g., Pss 31:9–18, 34:19–22, 37:32–34, 54:3–5, 63:9–11, 69:16–28, 86:14–17, 109:1–29. As for the prophets, they were killed by Jezebel (1 Kgs 18:4, 13); Zachariah was stoned by Joash (2 Chr 24:21); they were killed by disloyal Israelites (Neh 9:26, Jer 2:30), and Jeremiah him-self was constantly in trouble (Jer 11:18–20; 20:2; 26:8–11, 20–23). Cf. Acts 7:52 and 1 Thes 2:15 (and my comments thereon).

31. Bennett, 'Son of Man', p. 129.

32. The order p*resbyteroi*, a*rchiereis*, g*rammateis* only in Mk 8:31. Elsewhere *agp* (2 times), *apg* (2 times), *a* Sanhedrin (1 time), *ag* (4 times), *a* (4 times), and 'the high priest' (3 times). Matthew and Luke have their own variations, but neither puts *presbyteroi* first except when copying Mt. 8:31.

33. Lf. Jeremias (*Jerusalem in the Time of Jesus,* chap. 9) is right; cf. Schürer, vol. 2, pp. 210, 218.

34. Cf. *Treis*, TWNT 8; the threefold blessing (Nm 6:24–26), the threefold curse on Canaan (Gn 9:25–27); Paul's threefold blessing (2 Cor 13:13) and his threefold request for a blessing (2 Cor 12:8). The closest formal parallel to Mark is Jn 21:15–17: the Lord's threefold enquiry and commissioning of Peter.

35. Cf. Marcus, *Mystery*, p. 147: 'God intended the outsiders to be blinded by Jesus' parables and his parabolic actions (4:11–12), so that they oppose him and eventually bring about his death; in his death, however, the new age of revelation will dawn.'

36. The verb *paradidōmi* can express handing over property or authority to another; transmitting memories, records, customs, and teaching to another; and handing over a person into the power of another, whether for security, justice, punishment or safety.

1. John Baptist is *handed over* (Mk 1:14 = Mt 4:12), no doubt to Herod's officers or to death.

2. Those who oppose the followers of Jesus will *hand them over* to Jewish courts for punishment and to civil authorities Mk 13:9–12‖Matthew, Luke.

3. Judas Iscariot is uniquely remembered as the one who betrayed Jesus *(handed him over)* to his enemies (Mk 3:19, 14:44). Mark describes the betrayal in 14:10–11, and reports sayings of Jesus that obliquely refer to Judas: 14:18, 21, 41, 42. There are parallels in Matthew and Luke. In addition Mt 26:23, 25 reinforces references to Judas in the context of Mk 14:19–21, and Mt 27:3–10 inserts the story of Judas's confession that he had betrayed innocent blood and his death. Lk 22:48 adds to the Gethsemane scene 'Judas, would you betray the Son of Man with a kiss?' John has a comprehensive set of references: (6:64, 71; 12:4; 13:2, 11, 21; 18:2, 5; 21:20).

4. The Jewish authorities *hand Jesus over* to the Romans (Mk 10:33b; 15:1, 10 with parallels in Matthew. Luke makes his contribution in 20:20, 24:20; Acts 3:13, 7:52; and John in 18:30, 35 and perhaps 19:11.

5. Pilate *handed him over* to be scourged and crucified (Mk 15:15 and parallels; Jn 19:16).

6. In Gal 2:20 *(paradontos heauton)* and Eph 5:2, 25 *(paredōken heauton)* Jesus *handed himself over* for our benefit.

7. In Mk 9:31 the Son of Man *is handed over* (in Matthew and Luke he 'is to be handed over') by God. So also in Mk 10:33a par. Matthew (Luke has 'everything that is written of the Son of man will be accomplished'); and in Mt 26:2, 'After two days the Passover is coming, and the Son of Man will be delivered up to be crucified'. Cf. Jn 18:36: 'If my kingship were of this world, my servants would fight, that I might not be handed over to the Jews', implying that this situation is due to God; cf. therefore Jn 19:11. In Rm. 4:25 it is said almost explicitly—and in Rm. 8:32 quite explicitly—that God handed over his Son for our benefit.

37. *Eis (tas) cheiras* is not common in the NT. Apart from Mk 9:31 and parallels, there is Mk 14:41 = Mt 26:45, 'The Son of Man is betrayed into the hands of sinners'; Lk 24:7, 'the Son of Man must be delivered into the hands of sinful men', which complements Lk 23:46, 'Father, into thy hands I commit my spirit'. It appears in Acts 21:11, 28:17 of Paul's arrest; in Jn 13:3 of Jesus' knowledge that the Father had given all things

into his hands; and in Heb 10:31 of the conviction that it is a fearful thing to fall in to the hands of the living God. On the general use of *hand* see Grayston, 'The Significance of the Word *Hand* in the New Testament; and TWNT 9.

38. The common translation of *apechei*, namely, 'It is enough', comes from the Latin *sufficit* without proper support in Greek usage. The solitary verb should neither be dismissed as a conventional remark nor as a novel theological statement. The verb occurs only once more in Mark at 7:6 in a quotation from Is 29:13, 'Their heart is far from me' *(porrō apechei)*. Cf. Job's plea in 13:21, 'Withdraw thy hand from me' *(tēn cheira ap emou apechou)*, which is similar to Jesus' entreaty; also Prv 15:29, 'The Lord is far from the wicked' *(makran apechei ho theos apo asebōn)*, for in Gethsemane God is far from the *dikaios*. The threefold sleeping of the disciples is taken as a clear indication that God will do nothing to remove the cup from Jesus. Cf also Is 55:9 where heaven *apechei* from earth and 'my way *apechei* from your ways'.

39. According to Mk 15:43 Joseph of Arimathea, a respected member of the council, a man of substance and influence, was well disposed towards the followers of Jesus. Mt. 27:57 calls him a rich man and a disciple; Lk 23:50–51 describes him as a good and godly member of the council who had withheld his consent to their purpose and deed.

40. The exception is Mk 3:4.

41. 'It is a curse to spit in the face of another (Num 12:14; Dt 25:9), to mock and debase, to make mouths (Isa 57:4), and to point the finger (Isa 58:9), all because it creates evil in the soul, counteracts the blessing and debases the honour' (Pedersen, *Israel*, I–II, p. 441). Cf. also Job 30:10; 1 QS 7.13. Also LSJ, s.v. *ptyō* 4, 'to avert a bad omen, disarm witchcraft, and the like'. For God's faithful people mocked by enemies Pss 44:13, 79:4, 80:6. For the humiliation of the godly man Pss 22:17, 35:16; Jer 20:7.

42. At Mk 9:31 and 10:34 the majority text has 'on the third day', with support from several important witnesses in each place.

43. At Mt 16:21 D(Gk)bo conform to Mark; at Mt 17:23 several witnesses read 'after three days', and several different witnesses have *anastēsetai*; at Mt 20:19 *anastēsetai* appears in the majority text, including important witnesses.

44. But D it have *meth hēmeras treis,* and several important witnesses read *anastēnai*.

45. Cf. McArthur 'On the Third Day'; Schillebeeckx, *Jesus*, pp. 526–32.

46. The *hiphil* of *qwm*.

47. The verb *epitimaō* is used three times in Mark (8:30, 32, 33). In verse 32 Peter rebuked Jesus; and in verse 33 Jesus rebuked Peter, and said 'Get behind me Satan! For you are not on the side of God, but of men.' Yet in verse 30 (as in Mk 3:12) *rebuked* is avoided by English versions and replaced by *charged, gave strict orders,* and the like. Presumably, the translators were influenced by the *hina mē* clause introducing what the censure is to prevent (cf. AGB, s.v.), but they may not have fully understood the significance of a biblical rebuke. It is not simply an expression of disapproval but an exercise of power, often in a conflict of wills. When Jacob rebukes Joseph for reciting his dream, he is exerting his paternal authority (Gn 37:10). When God rebukes he shows strength (Ps 68:30): the sea is rebuked and driven back (Ps 18:15, 104:7, 106:9); God's enemies are destroyed and their name blotted out (Ps 9:5, 76:6, 80:16). In Zec 3:2 the Lord rebukes Satan for accusing Joshua the high priest. Hence, the exchange of rebukes between Jesus and Peter is a trial of strength. Mark makes ample use of that idea. He has Jesus rebuking demon-possessed persons to silence them and expel the demons. In Mk 9:25 the demon is already dumb, so the rebuke expels it (cf. the overpowering of wind and sea in Mk 4:39 and of a fever in Lk 4:39). But in Mk 1:25 and 3:11–12 the possessed persons are claiming to know him as the Holy One of God or the Son of God, and they must be stopped, not because they are giving away a secret but because they hope to use knowledge of his name

to thwart his power. Hence, the misguided rebukes in 10:13, 48 to people who were asking a blessing or an act of power (cf. Lk 9:55, 19:39, 23:40). Hence, Jesus' rebuke in Mk 8:30 is his way of preventing the disciples from naming him as the Christ in order to use that name to their own advantage. Cf. Mk 10:35–45.

48. *Son of God*. First, at the baptism and the transfiguration God identifies Jesus as his son (1:11, 9:7), which ought perhaps to settle the matter. He is acknowledged by Jewish and Gentile demoniacs (3:11, 5:7), which, although less persuasive, implies that the bearer of this title can overcome the demonic forces. And the centurion in charge of the crucifixion unit says, 'Truly this man was [a] son of God' (15:39), whatever the overtones of that remark may have been. Thus, in Mark the title Son of God is satisfactory to God, to Jews and Gentiles, and the Roman army. Second, nowhere does Jesus explicitly call himself Son of God. When the high priest asks 'Are you the Christ, the Son of the Blessed?'—according to Mk 14:61–62—Jesus replies, 'I am [Son of the Blessed], and you will see the Son of Man [without naming the Christ] seated at the right hand of the power,' etc. (taking the view that Mark has artfully constructed a double question for the high priest to ask, which is answered chiastically and ambiguously). In the parable of the wicked husbandmen Mk 12:1–9, the owner's son is surely intended by Mark to represent Jesus and, since a son is the most natural and authoritative agent of a land-owning father, is not implausible on the lips of Jesus. In Mk 13:32 Jesus says that the decisive hour in this generation is known only to God, not to the angels in heaven, nor to the son, i.e. it cannot be argued that Jesus' ignorance of the timing tells against his sonship, because a father keeps some things for his own decision even when his son is acting as his chief agent. Third, In Gethsemane—and significantly there—Jesus addresses God as *Abba, Father* and, as a dutiful son, resigns everything to his Father's will (14:36), perhaps in the expectation that the Son of Man would come in the glory of his Father with the holy angels' (8:38). That is the evidence and it is clear that *Son of God* had not been in the forefront of Jesus' teaching, though Mark thought it should be the matter of Christian confession. He may or may not have announced that conviction in the opening words of his Gospel (see Metzger on 1:1).

49. *Christos* is (1) used editorially at 1:1 and in the rejected variant at 1:34; (2) attributed to Jesus by others—Peter (8:29), high priest (14:61), and chief priests and scribes (15:32); (3) in sayings of Jesus—'Whoever gives you a cup of water to drink because you belong to Christ, amen I say to you he will certainly not lose his reward' (9:41; the parallel in Mt 10:42 has 'because he is a disciple'). The Markan form may be secondary—an attempt to reassure readers that reticence about *Christos* will not diminish the benefits of loyalty to that name. Also 12:35–37 and 13:21–23.

50. In addition to the books mentioned in n.24, there is a confident example of the formerly dominant critical view in C. Colpe, TWNT 8, pp. 403–81, with extensive bibliography. But the situation had already been changed by 'the use of bar nash/bar nasha in Jewish Aramaic' printed as Appendix E in the third edition of Black, *An Aramaic Approach to the Gospels and Acts,* after which scholars have moved rapidly towards ideas held at the beginning of the century. For a selection, see Moule, *The Origins of Christology;* Hooker, 'Is the Son of Man Problem Really Insoluble?'; Casey, *Son of Man;* Lindars, *Jesus and the Son of Man,* Older views on Aramaic scholarship, expressed by Black ('Aramaic *Barnashā* and the "Son of Man"')are ruthlessly rejected by Casey ('Aramaic Idiom and Son of Man Sayings'). The internal debate by Aramaic specialists is not yet sufficiently resolved for confident conclusions to be drawn about interpretation of the Gospels; but interpreters need no longer be overanxious about Christology, they can exorcise the Son of Man expectations from early Judaism and show indifference to the Similitudes of 1 Enoch, and they can stop sorting and classifying Son of man sayings. See also n. 57.

51. In biblical Hebrew *'adam* and *'noš* were almost always collective expressions meaning mankind. Plurality of individual men was expressed by *b'ne 'adam* or *b'ne ha' adam*, lit., 'sons of mankind'. A single man was normally *'iš*. The expression *ben 'adam* was uncommon and normally poetic.

ben 'adam	‖ *'iš* Nm 23:19; Jer 49:18, 33, 50:40, 51:43; Ps 80:17
ben 'adam	‖ *'noš* Pss 8:4 (What is *'noš* that thou art mindful of him, and *ben 'adam* that thou dost care for him?), 144:3 (the same thought, but with *'adam* and *ben 'noš*); Job 25:6; Is 51:12, 56:2
ben 'adam	‖ *geber* = (strong) man Job 16:21

The plural *b'ne 'adam* may be parallel to *'našim* (2 Sm 7:14), *'išim* (Prv 8:4), and *'noš* (Ps 90:3). In LXX the article occurs sometimes with the plural *(hoi huioi tŏn anthrōpōn)*— as at Pss 36(35):7, 49(48):3, 62(61):10—but *never* with the singular.

52. Dn 7:13 belongs to the Aramaic part of Daniel: 'And behold with the clouds of heaven came *k'bar 'naš* ('one like a son of man'). At 8:17 in the Hebrew part of Daniel, God addresses Daniel (in the style of Ezekiel) as 'son of man'.

53. It may be Mark's community feared they would lose Jesus as a source of forgiveness and *halachah* if Mark persuaded them to replace both *prophet* and *Christos* with 'Son of man'.

54. 'Before the angels of God'; *not* Mt 10:33, 'before my Father who is in heaven', which is a strongly Matthean phrase.

55. The rejected reading omitting *logous* (Metzger, *ad loc*) may have been a deliberate attempt to turn a surprising reference to the words of Jesus into an appropriate reference (in view of verses 34–37) to his disciples.

56. Perhaps Mk 14:62 is another saying where Son of man is equivalent to the first personal pronoun, though that could not be argued for Mk 13:26.

57. Vermes, *Jesus the Jew*, p. 168; Casey, 'General, Generic and Indefinite'.

58. Also Gospel of Thomas 55, presumably in a Gnostic sense.

59. Manson, *Sayings of Jesus*, p. 132.

60. Hengel, *Crucifixion*, p. 87; see also remarks on pp. 38, 50, 79, 83.

61. Josephus, *War* 2.75, *Ant.* 17.295; Schürer, vol. 1, p. 332. Other Josephus references in Hengel, Crucifixion, pp. 26, 40.

62. The saying is usually said to be paradoxical. So it is in form, but not in substance. It is no more than a pithy reflexion on reputation, standing or identity. If you have to defend it, you do not confidently possess it and every defensive act weakens it. If you can let it take its chance, you are displaying it confidently.

63. It is often said that *kai elegen autois* is the sign of a Markan addition. Perhaps so, but very often as a climax, (See Mk 2:23–27, 4:3–11, 6:1–4, 6:7–10, 7:1–14, 8:14–21, 11:15–17; cf. Lane, ad loc.)

64. 'They shall see the men that have been taken up, who have not tasted death from their birth' comes in the Second Vision of 2(4) Esdras. See the commentary by Oesterley (ad loc), which identifies the 'immortals' as Enoch, Elijah and others, including Moses according to Josephus, *Ant.* 4.326. It will not do to say that 'to taste death' is a common Jewish expression: the only occurrences are Mk 9:1 and parallels, Heb 2:9, Jn 8:52, and 2(4) Ezr 6:26, all from the same period. The rabbinic occurrences in SB are far later, later even than the probable date of the Gospel of Thomas (*Nag Hammadi Library in English*, p. 117). GT1, extant in Greek in Pap.Oxyr. 654, says, 'Whoever finds the interpretation of these sayings will not taste death' and the sayings immediately following are much to the point. See also GT18 and 19 which repeat the phrase in question.

65. The words 'suffer many things and be treated with contempt' (*exoudenēthē*) obviously echo 8:31, 'suffer many things and be rejected' (*apodokimasthēnai*). The earlier passage recalls Ps 118:22, quoted at Mk 12:10, 1 Pt 2:7, and Acts 4:11 (where the verb *exoudeneō* is used). The godly sufferer in the Psalms is frequently aware of contempt, e.g., Pss 22:6, 89:38–45, 119:22, 123:3. The servant of God in Isaiah is despised and rejected (Is 53:3) and is part of the general background; but the emphasis on Elijah scarcely directs attention to that description of the godly sufferer.

66. For Elijah expectations, see TWNT 2, pp. 930–43—interesting enough but not relevant. The exegesis of this passage has been hampered by failure to discover the question that Mark is trying to answer and by following Matthew's false trail of identifying Elijah redivivus with John Baptist in Mt 17:13 (not Mark) and in 11:14 (not Luke). In the Q logion Mt 11:10‖Lk 7:27 Jesus himself applies Mal 3:1, though not 4:5–6, to the Baptist; and in Mt 10:35‖Lk 12:52–53 by quoting Mi 7:6 he assigns to himself a destabilising, not a restoring, role. But Q logia need not determine the meaning of a Markan debate. If in Mark Jesus, for the sake of argument, applies Elijah's role to himself, there is no need to search anxiously for an OT Elijah prediction to explain Mk 9:13, 'as it is written of him'. What Elijah was expected to achieve has been attempted by the Son of man, but it did not work; so they did to the Son of man what scripture said they would.

67. The marginal reference to *scribes* in 9:14 echoes 9:11; the conjunction of *disciples* and *a great crowd* echoes 8:34; and *faithless generation* echoes *adulterous and sinful generation* in 8:38.

68. In Mark *egeirō* can be (1) intransitive—to get up, arise or come on the scene, rise against or attack; on (2) transitive—to rouse or raise up, particularly from the dead (6:14, 16; 12:26; 14:28; 16:6). *Anistēmi* is intransitive—to get up, rise against, and rise from the dead (as in 8:31; 9:9, 10, 31; 10:34; 12:23, 25). Mark has used verbs that belong naturally both to the story he is telling and to the Christian paradigm.

69. Only these two passages in Mark have *diakonos*.

70. Cf. Mt 13:41, 16:28, 25:31 for the kingdom of Christ.

71. Temple (1 Kgs 7:21, 39, 49; 2 Chr 3:17, 4:6–8); guards (2 Kgs 11:11; 2 Chr 23:10); divine court (1 Kgs 22:19; 2 Chr 18:18); law (Neh 8:4; 1 Esd 9:44); anointed (Zec 4:3, 11); David (2 Sm 16:6).

72. Jn 14:13–14, 16:23; 1 Jn 3:22, 5:14–15. See Grayston, *Johannine Epistles*, pp. 116–17.

73. Did Mt 20:20 give the request to the *mother* of James and John not to save apostolic reputations (if so, why fail to amend verses 22–24) but to recall the female request in Mt 14:7–8?

74. For information, see commentaries and TWNT 6, s.v. *pinō*, pp. 148 ff. and 1, s.v. *baptō, baptizo*.

75. *Doulos* here for the first time in Mark. It reappears in Mk 12:2, 4 of the servants who are ill treated by the tenants of the vineyard and in Mk 13:34 of the servants left in charge by the absent householder.

76. C. S. Lewis, reviewing a book by Austin Farrer, said, 'To talk to us thus Dr. Farrer makes himself almost nothing, almost nobody; to be sure in the event his personality stands out from the pages as clearly as any author—but this is one of heaven's jokes—nothing makes a man so noticeable as vanishing'. Quoted by P. Curtis, *A Hawk among Sparrows: A Biography of Austin Farrer* (London 1985).

77. Cf. also Grayston, 'The Empty Tomb'.

78. 'The Son of man came' is not past narrative about the expected Son of man but very similar to 'I came not' in Mk 2:17, though expressed more circumspectly—a statement of being under orders; cf. Mt 8:8–9. For the date and theological importance of 4

Maccabees, see Anderson in OTP, vol. 2, in pp.533, 542. He says, 'Christianity laid great stress on the saving or redemptive efficacy of the death of Jesus; it was picking up and adapting to its own new faith a doctrine that already enjoyed at least a limited currency in Judaism'. He may be right in citing Isaiah 53, but his other examples from Test Benj. and 1QS are not convincing.

79. Cf. Hill, *Greek Words and Hebrew Meanings*, pp. 77–81; Barrett, 'The Background of Mark 10:45'; and for a different view Moulder, 'The Old Testament Background and Interpretation of Mark X.45:

80. I claim no more for this rendering of the Hebrew than that it may be rather like what readers in early Judaism heard. The targum solves all problems of translation and interpretation by rewriting the text (Stenning, *Targum of Isaiah*). See Westermann, *Isaiah 40–66* and the varying translations of RSV, NEB, and GNB. The LXX translator was often baffled but generally offered a meaningful sense.

81. The word *anti* occurs twice in Is 53:9: 'I will give the wicked *anti* his burial, and the rich *anti* his death', presumably meaning 'in requital for' (so Otley, *Isaiah according to the Septuagint*). It also occurs in the third component of verse 12, '*Anth'ōn* his soul was handed over to death', where *anti* plus the genitive plural *hōn* corresponds to *taḥath ašer* in Hebrew. Both can mean 'instead of', but they can equally mean 'because', as they do here. It would be grotesque to suppose that the servant was allowed to die *in place of* the spoils or the inheritance at the beginning of the verse.

82. There is an administrative compensation in Mt 9:27–31, where there are *two* blind men, partly modelled on blind Bartimaeus.

83. See Marshall, *Gospel of Luke,* ad loc.

84. Howard in MHT, vol. 2, p. 472. Winer adopted the translation much earlier (p. 562), though the translator and editor, W. F. Moulton, disapproved.

85. Cf. Sanders, *Jesus and Judaism;* Rivkin, *What Crucified Jesus?*

86. It is often suggested that the destruction of the fig tree was developed from the parable in Lk 13:6–9 where a fruitless fig tree has a reprieve. It is more likely that Luke, with his sympathy for Jewish piety, moderated Mark's severity and, to give Judaism another chance, made use of an Ancient Near Eastern wisdom story about an unfruitful tree; see Lichtenberger, 'Ahiqar' in OTP, vol. 2, p. 487.

87. It is commonly said that the sayings on faith, prayer, and forgiveness in Mk 11:22–25 were thoughtlessly added when the destruction of the fig tree ceased to be read as an omen for Judaism and became merely an example of miraculous power. Pesch is right to suggest a meaning properly related to the context (not temple worship but prayer in the Christian community); but my own proposal offers an even closer relation.

88. Jeremias (*Parables of Jesus,* pp. 70–76) admits the allegory but argues that originally the parable was quite different. Thereafter, see the bibliography in Pesch, to which add Drury, *Parables in the Gospels*. The discovery—and (it seems) subsequent loss—of the performative function of some parables may be seen in Perrin, *Jesus and the Language of the Kingdom.*

89. For a generous list of parables, see Drury, *Parables in the Gospels,* pp. 171–73. Mk 2:19 has been ruled out above. The only exception is the shepherd allegory in John 10.

90. Weiser, *The Psalms.*

91. It is usually taken for granted that the stone (*lithos,* sometimes *petra*) used as a cornerstone (*akrogōniaios* or *kephalē gōnias*) was a stock symbol, in the primitive church, for Christ's resurrection. Apart from Mk 12:10 par. there are three passages that exploit this imagery: (1) 1 Pt 2:4–8 collects the rejected stone (cornerstone of Ps 118), the stone of offence and rock of stumbling of Is 8:14, and the precious cornerstone of Is 28:16 (both

Isaianic stones appear in Rm 9:32–33) and calls Christ a living stone, which suggests the resurrection by virtue of the word *living,* not by virtue of the stone imagery; (2) Eph 2:20 calls Jesus the cornerstone, which is a structural image—the passage refers not to resurrection but to spirit; (3) Acts 4:11, which plausibly but not necessarily refers to the resurrection. The evidence is meagre, and Barn. 6.4 does not improve it. Cf. Black, 'The Christological Use of the Old Testament in the New Testament; pp. 13–14.

92. In Mk 12:8 the son is first killed, then cast out of the vineyard. In Matthew and Luke he is first cast out, then killed—perhaps to hint that Jesus was first excluded from Israel and then put to death by the Romans. It may be true that he was crucified outside the city (Jn 19:20, Heb 13:12–13), but the evangelists make nothing of it. See Marshall, *Gospel of Luke,* ad loc.

93. Cf. Pesch and Gnilka and Jeremias, *New Testament Theology,* p. 184, n. 3.

94. Cf. Cavallin, *Life after Death,* pt. 1, pp. 23, 70.

95. Cf. Grayston, 'The Study of Mark XIII'.

96. On anointing, see the biblical dictionaries; also SB vol. 1, 427, 986 and M. Taanith 1.4–7.

97. Derrett, *Law in the New Testament,* pp. 266–78 is illuminating. It seems to me unlikely that an anointing of the messianic king was intended (Manson, *Servant-Messiah,* p. 85; Dibelius, *Jesus,* p. 88): both the response of the company and the variety of applications are against it.

98. The literature on this is enormous. No one can have read Jeremias, *Eucharistic Words of Jesus* without profit and some admiration. But the author greatly overplayed his hand in producing some of the historical and theological conclusions—as is plainly evident in the compact, instructive treatment by Haenchen (though in the end I prefer Gnilka). Marshall, *Last Supper and Lord's Supper* is a good account of recent discussion, though the author's historical and theological conclusions are too confident for the uncertain state of our knowledge (which he fully admits). Cf. also Robbins, 'Last Meal: Preparation, Betrayal and Absence'; and now see Lèon-Dufour *Sharing the Eucharistic Bread.*

99. But not by G. D. Kilpatrick, *The Eucharist in Bible and Liturgy.*

100. The alleged grammatical difficulty in Aramaic can apparently be overcome (McNamara, *Targum and Testament,* pp. 127–28); but the awkwardness in Greek remains.

101. There is an apparent exception. The first part of the Hallel psalms (113–14), which were sung before the blessing of the bread, was concluded with a short prayer. For R. Tarfon it should be a sparse statement of redemption, for R. Akiba a fulsome prayer for the enjoyment of future festivals: 'May we eat there of the sacrifices and of the passover-offerings whose blood has reached with acceptance the wall of the altar, and let us praise thee for our redemption and for the ransom of our soul'. Tarfon and Akiba were third-generation Tannaitic teachers (120–140 C.E.), and this addition to the Haggadah comes from that period. Akiba's generalized piety throws no light on Passover a century earlier—against Marshall (*Last Supper,* p. 168, n. 2), whose attribution of 'redemptive and expiatory associations' to Passover (p. 77) is regrettably imaginative.

102. 'It is still rather remarkable that He should have taken for granted that His allusion to that narrative, which seems to have but little occupied the Jewish mind at that period, would be grasped at once by the Disciples' (Dalman, *Jesus-Jeshua,* p. 166). Dalman persuades himself that the Jewish ideas attaching to Zec 9:11, though not referred to by Jesus, are valuable; but he discounts the later Jewish identification of 'the blood of the covenant' with the blood of circumcision. He adds, 'Although the blood of the Paschal lamb . . . was now ordinary offering blood, without any propitiatory effect, yet the dutiful remembrance (expressed in words) of the Egyptian Passover in itself awakened the atten-

tion to the application of blood which once had prevented a threatening destruction, and had been the necessary condition for the redemption from Egypt.' The embarrassment is obvious, but nobody has done better since.

103. *Assigned* (RSV) represents *diatithēmi*, from which *diathēkē* derives.

104. Mt 9:2–6 (Mark); 12:31–32 (Mark); 6:12–15 (Q, Mark); 18:23–35; 18:15–18 (cf. Jn 20:23), 21–22 (Q).

105. In Matthew the verb *save* (*sōzō*) is almost always from Mark. It refers to preserving life, restoring to health, helping in danger of death, especially in eschatological peril. It is never related to *sins*.

106. Luke is more interested than Matthew in forgiveness: Lk 1:77 (the Baptist) and 24:47—when Christ has suffered and risen from the dead, 'repentance and forgiveness of sins should be preached in his name to all nations', as it is in Acts 2:38, 5:31, 10:43, 13:38, 26:18, though associated less with the death and more with the resurrection and exaltation of Christ. In Lk 7:47–49 forgiveness is available in response to love (against Marshall). Eph 1:7 connects redemption, blood and forgiveness; but *blood* is absent from the parallel in Col 1:14; and cf. Heb 9:22, 10:18, 'Without the shedding of blood there is no forgiveness of sins'.

107. I am not persuaded by Neyrey, *The Passion according to Luke.*

108. Luke took trouble to reassure his readers that Christians were unlikely to receive harm from the Roman administration. See Walaskay, *'And So We Came to Rome.'* That theme of Acts is as plausible here as elsewhere. To interpret *hikanon estin* in Lk 22:38 as 'Enough of that talk', a testy rebuke, is incredible. It *was* Jesus who said, 'Go and buy a sword'.

109. I agree (against Schweizer) with Anderson, Lane, Pesch and Gnilka that Mk 16:8 was the intended conclusion of the Gospel. It has to be remembered that Mark's readers already knew that Jesus had died and risen again and did not stand in need of resurrection stories to instil confidence in the resurrection.

110. CD 7 (DSSE, p. 88); Rabin (*Zadokite Documents*, pp. 30–31) suggests, perhaps implausibly, that the shepherd may have been the Teacher of Righteousness.

111. The view that *Abba* suggests baby talk is rightly rejected by Haenchen. As Jeremias says, (*Central Message of the New Testament*, p. 21) 'Grown-up sons and daughters called their fathers *Abba* as well'.

112. See n. 38.

113. See Metzger, *Textual Commentary,* ad loc.

114. Merkel, 'Peter's Curse.' Cf. Gnilka, ad loc.

115. See Vermes, *Jesus the Jew,* pp. 197–99. The discussion has possibly been advanced by Fitzmeyer, 'The Contibution of Qumran Aramaic to the Study of the New Testament', pp. 391–94. Cf. Dunn, *Christology in the Making,* pp. 15–16 (rather uncritical!); Lohse, TWNT, pp. 361–63. See n. 48.

116. Cf. Schürer, vol. 2, sec. 29 ('Messianism'). De Jonge, 'The Earliest Christian Use of Christos' is less eclectic and more tentative.

117. Hart, 'The Crown of Thorns in Jn 19:2–5'; also TWNT 8 pp. 631–32.

118. On the problems of this section, see Schweizer, *The Good News according to Matthew,* pp. 515–517. In general, cf. Senior, *Passion Narrative according to Matthew,* esp. pp. 335–39.

119. Daylight hours run roughly from 6 o'clock A.M. to 6 o'clock P.M.: (1) Mk 15:1, *euthus prōi,* 'as soon as it was morning'—Jesus taken before Pilate (see Sherwin-White, *Roman Society and Roman Law in the New Testament,* p. 45) and led to Golgotha; (2) 15:25, 'the third hour' = 9 o'clock A.M.—the crucifixion and mocking; (3) 15:33, 'the sixth hour' = noon—'darkness over the whole land until the ninth hour' (4) 15:34, 'the

ninth hour' = 3 o'clock P.M.—the dying cries and following events; (5) 15:42, *opsias genomenēs,* 'when evening came' (but just before the end of daylight because it was the day before sabbath)—Joseph of Arimathaea places the body in the tomb.

120. Mt 27:43 quotes a form of Ps 22:8; and in Lk 23:35 *exemyktērizon* and *Sōsatō heauton* reflect Ps 22:7–8.

121. If, as some think, Mark used both Matthew and Luke, he ignored Luke and revised Matthew.

122. The Hebrew/Aramaic words in their Greek transliteration vary in the different witnesses. I have quoted the forms given by Nestle-Aland 26. Matthew has *eli eli lamah ᵉzabtani.* Thus, *eli* is Hebrew, and *elohi* is Aramaic *ᵉlahi, Lama* is Hebrew 'why?' (read by a number of witnesses), and *lema* or *lima* are Aramaic. The verb *ᵉzabtani* is Hebrew *ʾazab,* 'abandon' (which may be present in the D text of Matthew and Mark *zaphthani,* unless it derives from the Hebrew/Aramaic *zaʾap.,* 'to be enraged'); *sabachtani* is Aramaic *sbq,* 'to let alone, leave'. Cf. Pesch, *ad loc.* and Jeremias, *New Testament Theology,* p. 5. LXX has *ho theos ho theos mou, prosches moi. hina ti egkatelipes me?* 'God, my God, give heed to me, why has thou forsaken me?' It looks as if this quotation came from the Aramaic-speaking rather than the Greek-speaking Church. See also the careful study by Léon-Dufour, 'Le dernier cri de Jésus'.

123. Jeremias, *New Testament Theology,* p. 189.

124. Pedersen, *Israel,* vols. 1–2, pp. 437–52.

125. Sifre Deut. (*Rab Anth,* p. 498).

126. Luke abandons Mark's three-hour scheme, retaining only the period between the sixth and ninth hours for the great darkness when 'the sun's light failed, and the curtain of the temple was torn in two'. He omits the wine and myrrh and transfers the mocking from Jews to soldiers when they offer sour wine.

127. See Metzger, *Textual Commentary* and Marshall, ad loc. The passages are Lk 2:43, 49; 9:33, (55); 11:44; 12:39, 46, 48, 56; 18:34; 19:42, 44; 20:7; 24:18. Cf Acts 3:17, 13:27 and the D text of Lk 6:5: 'On that day Jesus saw somebody working on sabbath and said to him: Man, if you know what you are doing, God give you happiness; but if you do not know, you are cursed as a law-breaker'. Cf. also Jn 16:3; 1 Cor 2:8; 1 Tm 1:13.

128. SB, vol. 2, p. 269.

129. Joseph of Arimathaea is called *euschēmōn bouleutēs,* i.e., a well-respected member of the *boulē* (council). Was the Sanhedrin meant or some local council, such as the Arimathea Sanhedrin (if it had one)? This question, conscientiously raised by Nineham and others, was much discussed at the beginning of the century by Bruce, Loisy, Lagrange, and Montefiore. Surely Mark would expect his readers to think, as Luke did, of the high council of the Jews and its hierarchy.

130. Matthew makes Joseph a wealthy disciple (Mt 27:57) so missing Mark's irony. Luke calls him a good and respectable man (*dikaios*—as Jesus was to the centurion) and excuses his share in the Sanhedrin's proceedings by saying that he 'had not consented to their purpose and deed'.

131. See Grayston, 'The Empty Tomb'.

132. For a feature of this episode, see Grayston 'The translation of Matthew 28.17'.

133. On the retention of *ouk estin hōde, alla ēgerthē,* see Metzger, ad loc.

134. On which, see Marshall, ad loc.

135. On the retention of verse 12, see Metzger, ad loc.

136. For the details, see Marshall, ad loc.; also Leaney, pp. 56–57.

137. The material I need to discuss may be displayed according to the convenient

divisions of Marshall's commentary, with my own indication of the audiences to which the speeches were addressed (J = Jews, G = Gentiles, C = Christians).

1.	1:1–2.47	*The Beginning of the Church*	
1.a	1:1–5	Prologue	
		*1:3 Jesus alive after his passion	
1.d	1:15–26	The twelfth apostle	
		*1:16 mention of Judas	C
1.f	2:14–42	Peter preaches the gospel	
		*2:23–36 crucified and risen	J
2.	3:1–5:42	*The Church and the Jewish Authorities*	
2.b	3:11–26	Peter explains the healing	
		*3:13–18,26 killed and raised	J
2.c	4:1–22	The arrest of Peter and John	
		*4:2,10–11 crucified and raised	J
2.d	4:23–31	The disciples pray for further boldness	
		*4:27–28 prayer	
2.h	5:17–42	The second arrest of the apostles	
		*5:28,30–31 killed and raised	J
3.	6:1–9.31	*The Church Begins to Expand*	
3.c	7:1–53	Stephen's speech in court	J
3.d	7:54–8:1a	The death of Stephen	
3.g	8:26–40	The conversion of an Ethiopian	?J ?G
		*8:32–35 explanation of scripture	
3.h	9:1–19a	Conversion and call of Paul	
4.	9:32–12:25	*The Beginning of the Gentile Mission*	
4.b	10:1–11:18	The conversion of Cornelius	G and C
		*10:39–43 life, death and resurrection of Jesus	
5.	13:1–15:35	*The Mission to Asia Minor and its Aftermath*	
5.c	13:13–52	Evangelism in the synagogue at Pisidian Antioch	
		*13:27–41 death and resurrection	J and godfearers
6.	15:36–18:17	*Paul's Missionary Campaign in Macedonia and Achaia*	
6.3	17:1:15	Thessalonica and Beroea	
		*17:3 the necessity of death and resurrection	J and godfearers
6.f	17:16–34	Athens: the Areopagus Address	
		*17:31 raised from the dead	G
7.	18:18–20:38	*Paul's Missionary Campaign in Asia*	
7.h	20:17–38	Paul's farewell address at Miletus	
		*20:20 the blood of his own Son	C
8.	21:1–28:31	*Paul's Arrest and Imprisonment*	

8.c	21:37–	Paul's defence before the crowd	
	22:29	*22:7–10 the risen Lord	J
8.d	22:30–	Paul appears before the Sanhedrin	
	23:10	*23:6 the hope of resurrection	J
8.f	24:1–27	Paul appears before Felix	
		*24:15,21 resurrection	G
8.h	25:13–	Paul appears before Festus and Agrippa	
	26:32	*26:15,22–23 suffer and rise from the dead	G and J
8.j	28:17–31	Paul and the Jews in Rome	
		*28:23,31 the kingdom of God	J

138. Cf. Conzelmann, p. 84 on 13:29.

139. Cf. Haenchen, p. 181.

140. In Acts 4:25 David is God's servant *(pais)*, and in verses 27 and 30, so is Jesus. When Peter says that God has 'glorified his *pais* Jesus' (Acts 3:13), it is difficult to ignore Is 52:13 'Behold my *pais* . . . will be greatly glorified'. But what then? In Acts God glorifies, raises up and anoints his *pais* (even in the presence of enemies) and performs signs and wonders through his name (Acts 3:13, 26; 4:30). If in writing these verses Luke was conscious of Isaiah 52–53, he avoided every suggestion of vicarious suffering. Lk 22:37 quotes Is 53:12, 'he was reckoned with transgressors', a phrase that refers more to the disciples' prospective dangers than to the theological interpretation of Christ's death (see p. 206). The textually uncertain verse Lk 23:34, 'Father forgive them', might correspond to the final phrase of Isaiah 53, where the servant 'made intercession for the transgressors', except that LXX (which Luke uses) says, 'because of their sins he was handed over'. Jeremias, TWNT 5, s.v. *pais theou* overplayed his hand, but he modified his position in Zimmerli and Jeremias, *Servant of God*. The designation of Jesus as *pais* in Acts is no longer derived from Deutero-Isaiah but from the custom of qualifying eminent men as God's servants.

141. See Lambrecht, 'Paul's Farewell-Address at Miletus'. The structure may be set out thus: 18–21 Apologia, 22–25 Paul's Departure and Suffering, 26–27 Apologia, 28–30 Paul's Departure and the Community's Danger, 31 Apolopgia, 32 Commendation of the Community, 33–35 Apologia.

142. Bovon, *Luc le théologian* reviews studies from 1950 to 1975. Conzelmann, *Theology of Saint Luke* briefly discusses the death of Christ, pp. 199–202. There is a desire to save Luke's theological reputation in Kümmel, 'Luc en accusation dans la théologie contemporaine', p. 103; George, *Etudes sur l'Oeuvre de Luc,* which gives the most comprehensive collection of references; and Marshall, *Luke: Historian and Theologian*, pp. 169–75. I find Marshall's argument not persuasive, and even he is modest in his conclusion: that Luke's presentation of the saving work of Jesus is one-sided, that salvation is bestowed by virtue of his position as Lord and Messiah, and that what is lacking is a full understanding of the significance of the cross as the means of salvation (p. 175). See also Fitzmyer, *Gospel according to Luke* and Büchele, *Der Tod Jesu im Lukasevangelium*, pp. 193–96: 'In an original manner Lk represents the death of Christ as a saving event but—in distinction from other New Testament writings—not regarded simply in itself but in the larger context of the complete saving activity of Jesus on his total course, whereby he maintained and carried through the offering of love and forgiveness to its ultimate consequences in suffering and death.' It would be simpler to say that Luke had little idea what Mark and Paul were talking about. C. K. Barrett ('Theologia Crucis—in Acts?') says that Luke's *theologia crucis* is not a doctrine of the atonement but a way of life (p. 75). Indeed,

Luke was not sufficiently interested in theology to be a theologian of any kind; but he has a strictly practical *theologia crucis* of taking up the cross daily (p. 84).

Chapter 6

1. Elliott, *A Home for the Homeless.*

2. Silvanus and Timothy are mentioned together in 2 Cor 1:19 and along with Paul in the addresses to 1 and 2 Thessalonians. In Acts Silas (the Greek form of the latinised Silvanus; see excursus in Windisch, pp. 80–81) begins as a leading member of the community at Antioch and then is exclusively associated with Paul (Acts 15:40; 16:19, 25, 29; 17:4, 10, 14–15 [with Timothy at Thessalonica]; 18:5 [with Timothy]). John Mark is associated with Peter in Acts 12:12 and with Paul and Barnabas in Acts 12:25 and 15:37, 39. He joins with Paul and others in sending greetings to Philemon's house church (Phlm 24). In Col 4:10, described as the cousin of Barnabas, he joins in the greetings to Colossae, which he may possibly visit. Timothy is asked to travel to Rome and take Mark to look after the affairs of the imprisoned Paul (2 Tm 4:11). Thus, in the period after Paul had withdrawn from direct control of his missionary work, Mark was known to some of the Asian communities and may have gone to Rome on Paul's affairs. He turns up in 1 Peter, alongside Silvanus, immediately after a mention of the community in Babylon (a code name for Rome), in a circular letter directed to Christians in areas where Paul has preached the gospel. It is relevant that the pseudonymous second epistle, directed to the same communities as 1 Peter, specifically mentions Paul's writings and difficulties caused by them (2 Pt. 3:15–16).

3. Death and resurrection of Christ: 1:2, 3, 11, 21; 2:21, 24; 3:18; 4:1, 13; 5:1. The verb *paschō*: 2:19–23; 3:14, 17, 18; 4:1, 1, 15, 19; 5:10. The noun *pathēma*: 1:11; 4:13; 5:1, 9.

4. 1 Pt 1:1 calls the readers *parepidēmoi* (RSV 'exiles'); 1 Pt 2:11 calls them *paroikoi* and *parepidēmoi* (RSV 'aliens and exiles'). 1 Pt. 4:12 uses *xenos* ('stranger'). Cf. Eph 2:19, 'You are no longer *xenoi* and *paroikoi* (RSV 'strangers and sojourners'); Heb 11:13, 'They were *xenoi* and *parepidēmoi*' (RSV 'strangers and exiles').

5. Suffering for the name is discussed in the commentaries and needs no attention here; but a suggestion may be offered about 1 Pt. 4:14b. The syntax is notoriously perplexing: *hoti to tēs doxēs kai to tou theou pneuma eph hymas anapauetai.* The words following *kai*, 'The spirit of God rests upon you', come from Isa 11:2; but what could the preceding words mean? It is customary to read them as if *to [pneuma] tes doxēs* were intended ('the spirit of glory and of God' RSV, 'that glorious Spirit which is the Spirit of God' NEB). But that would be surprisingly clumsy syntax. *To tēs doxēs* could sensibly mean 'what is appropriate to or associated with the glory' (cf. Selwyn, p. 222). The clue may perhaps be found in the Passover Haggadah. In answer to the son's question to his father 'Why is this night . . .?' the father instructs him. He begins with the shame and ends with the glory; and he expounds, 'A wandering Aramean was my father . . . until he finishes the whole section (Pes. 10.4). I am not persuaded by Cross, *1 Peter—A Paschal Liturgy* or by Leaney, '1 Peter and the Passover' that the epistle reproduces a paschal liturgy but I think it does contain some fragmentary Passover memories. Perhaps, then, verse 14b implies 'Happy are you for the shame that ends in the glory and the spirit of God rests upon you.'

6. NIV here slightly modernizes and varies RV. Other versions interpret variously. *Prognōsis* becomes *destined* in RSV, *purpose* in NEB, GNB. *Hagiasmos* is *sanctified* in RSV, *hallowed* in NEB, *made a holy people* in GNB. *Hypakoē* is *obedience* in RSV, *obey*

in GNB, and perhaps *to his service* in NEB. Finally *rhantismos* is *sprinkling* in RSV, *puri-fied* in GNB, and *consecrated with the sprinkled blood* in NEB.

7. The formula requires the exegete to be syntactically alert. *Pneumatos* is a subjective genitive after *hagiasmō; hypakoēn* takes *Christou* as an objective genitive, and *haimatos* (itself a descriptive genitive after *rhantismon*) takes *Christou* as a possessive genitive. Those who object to this promiscuity of the genitive try to make *hypakoēn* stand on its own, but that would require another *eis* following *kai*. The formula works quite well when recited. Goppelt is so impressed by the parallel between the formula and 1QS 3.6–8 that he thinks 1 Pt 1:2 derives from a Palestinian baptismal form. But would it not be an eccentric baptism that had the order (1) consecration by the Spirit, (2) promise of obe-dience, (3) sprinkling with blood (or what represents blood)? Standard discussions of blood sprinkling and its meaning in 1 Peter have been collected by Olson *(Atonement in 1 Peter)*, pp. 268–80, though his conclusions are inadequate.

8. *ei deon* sc. *estin* in 1 Pt. 1:6 means 'if such things must happen [as they must]'. Thus, it corresponds nicely to the present popular use of *predictably.* The writer knows that disabilities are often imposed but somewhat discounts them—as he does by the opta-tives in 3:14, 17, which, in effect, say 'By doing good you will not suffer, though if you do. . . .'

9. But AGB, s.v. *eis* 4h says that the preposition can be used for the genitive.

10. "Girding up the loins of your mind' in 1 Pt 1:13 looks like an obvious reminis-cence of the Exodus, though the wording does not exactly reproduce Ex 12:11. But in the epistle it metaphorically implies a mental shift that, already begun, is to be carried through *(teleiōs)*.

11. If Exodus imagery is present, perhaps the paschal lamb is in mind (though in Ex 12:5 it is a *probaton teleion,* not an *amnos amōmos* as in 1 Pt 1:19). The blood of the paschal lamb was apotropaic, not expiatory, as everyone admits (see on 1 Cor 5:7). Noth-ing in the special pleading of Jeremias *(Eucharistic Words* pp. 225–26) was intended to upset that statement (the reference to Billerbeck, Vol. 4 is bluff). There can be no objection to the view that the blood of the original Passover sacrifice came to have saving signifi-cance, which is a sufficiently general statement. If the blood of Christ is a pledge of God's liberation and protection, it bears a certain resemblance to the means of Israel's liberation from Egypt. If, however, it is felt that the verb *elytrōthēte* ought to be rendered 'You were ransomed', which active subject does that passive presume: God (So Goppelt) or Christ? Kelly (p. 84) says the wording 'clearly implies that God is the recipient of the ransom, accepting Christ's obedient surrender of his life as an offering which once for all abolishes the disobedience of sinful mankind.' It is difficult to give the second part of that remark a precise meaning. The first—far from obvious—part may perhaps be a form of the Jewish hope that God will accept the sacrifice of one, instead of demanding the death of many: but such matyrdom theology is not appropriate to this context nor to a conjectured orig-inal setting. Finally, whereas most commentators favour the paschal lamb, some (e.g., Spicq) insist on the lamb in Is 53:7; quoted in the episode of Philip and the Ethiopian eunuch in Acts 8:32: 'As a sheep [*probaton*] led to the slaughter or a lamb [*amnos*] before its shearer is dumb, so he opens not his mouth'. It is indeed true that Isaiah 53 is quoted in 1 Pt. 2:21–25 to identify Christ's suffering and forbearing, for which the lamb in verse 7 would be equally appropriate. But the word *blood* does not appear in Isaiah 53, the lamb is not said to be *amōmos,* and the slaughtered beast is a *probaton.* In other words, the writer has various biblical tags in mind but not a specific passage. He is using the general image of sacrificial slaughter with primary emphasis on the power of blood.

12. RSV still renders *phobos* with *fear* though other translations use *awe* (NEB), *rev-erence* (GNB), *reverent fear* (NIV). In modern speech none of these gives the right impres-

sion, which is something like the attitude of trained workers handling a dangerous but necessary radioactive source. I Peter tries to create that response, rather than 'awed thankfulness' (Kelly) of which there is no hint.

1 3. The evidence is given in Balch, *Let Wives Be Submissive*. The two short quotations are from pp. 63 and 109.

1 4. Selwyn (p. 430) makes a tabular comparison of 'slaves and masters' in I Pt 2:18–25 with corresponding sections in Col 3:22–4:1; Eph 6:5–9; I Tm 6:1–2; Ti 2:9–10 According to Spicq (p. 107) the household servant of I Peter (*oiketēs*, the *doulos* born in the household rather than the purchased slave) includes all servants, male or female, in the personal service of the master of the house, from cooks and porters to stewards, teachers and so on, but not specifically agricultural and industrial labourers.

1 5. Not a hymn, despite Bultmann's proposal: see Goppelt, pp. 204–7. Add Boismard, *Quatre Hymnes Baptismales*, chap. 3.

1 6. The divine calling in I Peter is as important as anywhere in Paul, and rather more explicit. The community has been called by God from its former ignorance to a holy life corresponding to his holiness (1:15); from darkness to his marvellous light (2:9); to a practice of doing what is right (2:21); to possession of unity of spirit, sympathy, love of the brethren, a tender heart and a humble mind, i.e., the divine blessing (3:8–12); to a share in his eternal glory in Christ (5:10).

1 7. The *hypogrammos* was a pattern to be copied in writing or drawing, so, 'an example'. The words, 'follow in his steps [*ichnesin*]' imply a marked path. In modern usage, an example may indicate the kind of action required; 'following an example' may suggest hints for one's own initiative. Here, however, it seems to imply tracing the master copy, treading the same path.

1 8. For *suffered* in I Pt. 2:21 and 3:18 instead of the variant *died* see Metzger, *Textual Commentary*, ad loc. Paul seldom says that Christ suffered; Peter never says that he died (except in the variant text of 3:18).

1 9. But see, now, Mettinger, *A Farewell to the Servant Songs*.

20. It is instructive to compare the LXX text of Is 52:13–53:12 with I Pt 2:21–25 and to notice what is and what is not picked out. The epistle says that Christ suffered for our sins, not his own, and so reflects Is 53:4a–b, 5, 8d, 12e–f. It doubtless implies that this was God's doing, so reflecting Is 53:6c. Under suffering he uttered neither abuse nor threats, perhaps like the voiceless sufferer of Is 53:7. He was free from sin and deceit, as in Is 53:9c,d (the replacement of LXX *anomia* by *hamartia* is not significant, for *hamartia* is used in Is 53:12). The healing effect of Christ's suffering comes from Is 53:5d, and the waywardness of the sheep from Is 53:6a. The attempt to squeeze I Pt 2:23c out of *dikaiōsai* and *paredothē* (Is 53:11–12) is a failure. On the other hand, the epistle does not mention that the sufferer is God's *pais* (Is 52:13a) or that the Gentiles regard him with shocked surprise (Is 52:15)—both of which would have suited the writer's argument. It does not mention his *tapeinōsis* (Is 53:8)—though that would have strengthened one of the writer's themes (see p. 240)—nor his sickness or despicable appearance (Is 52:14, 53:2–4), nor does it mention his restoration by God, the gift of light and understanding and his forthcoming revenge (Is 53:10–12). All in all, it is easier to suppose that the writer of the epistle used apt and moving phrases to give scriptural force to his argument than that he studied the Isaianic poem to discover a key to the atoning power of Christ's death. His key was the social necessities of a slave's life. (Cf. Olson, *Atonement in I Peter* pp. 84–101.)

2 1. Since Christ is not compared to a goat intended for Azazel (a desert demon?) and since the *xylon* is not 'a solitary land' and since the goat's bearing of iniquity is expressed by *lambanō*, not *anapherō*, it is incredible that Gentile readers would detect any reference to the scapegoat (Lv 16:20–22).

22. *Cross* renders *xylon* (which properly means wood, hence 'a gallows' in LXX, once in Paul (Gal 3:13, referring to Dt 21:23), three times in the storytelling tradition of Acts, where it is conventionally translated 'tree'. The reference in 1 Peter has nothing to do with Paul's complex scriptural argument; more probably it comes from popular storytelling.

23. On *Apogenomenoi*, see Kelly, Best.

24. There are obvious coincidences: 2:18 *phobos* (3:14, 16); 2:19 *syneidesis* (3:16); 2:20 *paschō, agathopoieō, hamartano-kakopoieō* (3:14, 17); 2:23 *loidoreō* (3:9); 2:24 *hamartiai, dikaiosynē* (3:18).

25. The optatives *paschoite* and *theloi* are dismissive, i.e., they are not indicating a remote possibility (as against 1 Pt. 4:12) but admitting the likelihood and dismissing its importance.

26. Literature and discussion in Goppelt, pp. 239–42.

27. For the variety of readings (*epathen/ apethanen, peri/ hyper,* hymōn/hēmōn in several combinations), see Metzger, *Textual Commentary* p. 692. The variants show how scribes, when copying formulae, were apt to be influenced by wording familiar to them.

28. Angels appear in a rhetorical conceit, 1 Pt 1:12, the devil as a roaring lion (from Ps 22:13) in a homiletical warning, 1 Pt 5:8. The epistle is more concerned with the malignity of neighbours than of demons.

29. Literature and discussion in Goppelt, pp. 246–60. The problems may be noted with reference to the Greek text:

> 19 *en hō kai tois en phylakē pneumasin poreutheis ekēryxen*
> Problems are the reference of *en hō* ('in which circumstances' or 'in which spirit'?) whether the *phylakē* was a prison beneath the earth or somewhere on the way to heaven? the identity of the *pneumata* (spirits of dead human beings or rebellious angelic beings?) and what Christ proclaimed—his resurrection triumph and hearers' condemnation, the nearness of judgement, or the gospel and a chance of repentance?
>
> 20 *apeithēsasin pote hote apexedecheto hē tou theou makrothumia en hēmerais Nōe kataskeuazomenēs kibōtou eis hēn oligoi, tout estin octō psychai, diesōthēsan di hydatos*
> Problems are the understanding of *eis hēn oligoi* (presumably, 'into which a few [went]') and the meaning of *diesōthēsan di hydatos* ('were brought safely through the water' or 'were rescued by means of water' or even 'were saved [from the divine judgement effected] by water'?).
>
>> 21 *ho kai hymas antitypon nyn sozei baptisma,*
>> *ou sarkos apothesis rhypou*
>> *alla syeidēseōs agathēs eperōtēma eis theon*
>
> Problems are the grammatical difficulty of *ho,* the meaning and references of *antitypon,* the awkward placing of *baptisma* and the meaning of *eperotema* and its relation to the 'good conscience' and to God.

In all this the only fluent, unambiguous phrase comes at the beginning of verse 20: 'When the patience of God was eagerly expectant in the days of Noah, as an ark was being prepared'. This observation suggests that the writer, who normally wrote very competent Greek, at this point had one clear image in his mind but was unhandy in his attempt to link it to an existing formula and to present it as an analogue of baptism. If anything, this is an additional argument against the view that the epistle contains the major parts of a baptismal liturgy or homily.

The translation indicates the decisions I have taken. Verse 19: because of *pneumati/ pneumasin* it is difficult to avoid the reference of *en ho* to 'spirit'; the earnest debate whether Jesus went upwards or downwards is inappropriate and comical because it ignores the advice against visualising Hebrew symbols; the 'spirits' are the dead human beings of Noah's day (as in 1 Pt 4:6), and there is sufficient evidence for this loose usage of *pneumata* for *psychai* (Goppelt; the imprisoned angels of Jude 1:6 and 2 Pt 2:4 are a separate myth, developed from Gn 6:1–4 and passages in 1 Enoch); Christ proclaimed, 'The end of all things is at hand' (as in 1 Pt 4:7). Verse 20: the few who were saved provides the clue to the intention of the paragraph; the flood is an end-time symbol. Verse 21: the baptismal reference can best be disentangled (though the precise significance is not important for my purposes) if my view of 19–20 is accpted.

30. On the addition *for us* or *for you* (*hyper hēmon/ hymōn*) after the first *suffered in the flesh*, see Metzger ad loc, who also deals with the genitive or dative case of *sin* at the end of the verse. The preferred reading is *pepautai hamartias*, with some Greek witnesses writing more explicitly *apo hamartias*; a few witnesses read the dative *hamartiais*. See Metzger, ad loc. The translation given above differs from RSV, NEB, GNB, JB, TNT, all of which treat the second half of the verse as a general statement (which Kelly rightly opposes). My translation is close to NIV and RV and takes up Kelly's observation that *hoti* can be either *for* (thus introducing a reason why they should so arm themselves) or *that* (describing the thought which they are to share).

31. The statement that 'he who suffered in the flesh has ceased from sin' (*ho pathōn sarki pepautai hamartias*) achieves conciseness, ambiguity and implausibility. It sounds something like Rm 6:7, 'He that hath died is justified from sin' (*ho gar apothanon dedikaiotai apo tēs hamartias);* but that comes, as every one agrees, from another world of discourse. It might originate from a proverbial saying, on the lines of Jewish teaching that a repentant death atones for one's own sin and a martyr death atones for the sins of others—with the proviso that Christians attached atonement to the death of Christ. But the writer of the epistle is not now concerned with atonement but with dissuading his readers from sinning. If the words *pepautai hamartias* mean 'has ceased from sinning', the statement becomes implausible in any possible application.

32. Believing as I do that the Petrine writer is using traditional language in 1 Pt 4:5–6, I do not think it necessary to enquire anxiously about the meaning of 'the dead' in verse 6. If it is not a reference back to verse 19, it may perhaps (though not very relevantly) mean all the dead of mankind or Christians who have died after accepting the gospel (but how odd to bring them in as a rhetorical afterthought!).

Chapter 7

1. Aland, *Kurzgefasste Liste,* pp. 31–32.

2. Kümmel, *Introduction to the New Testament,* pp. 394–97.

3. Jones, 'The Epistle to the Hebrews and the Lucan Writings'.

4. Bruce, "'To the Hebrews' or 'To the Esssenes'?", p. 216.

5. Kummel, *Introduction to the New Testament* pp. 399–400; Vielhauer, *Geschichte der urchristlichen Literatur,* p. 247; Hughes, *Hebrews and Hermeneutics,* pp. 2–3; Moffatt. Misreadings: Heb 7:27 thinks that high priests made daily offerings for their own sins; 9:3 places the altar of incense in the Holy of Holies; and 9:22 says that 'without the shedding of blood there is no forgiveness of sins', which is at least misleading if not mistaken.

6. In *Redating the New Testament,* chap. 7, Robinson enthusiastically insisted on a date between the deaths of Peter and Paul and the death of Nero. In *The Birth of the New Testament* Moule urges a date just before 70 C.E. when the political crisis had decisively separated Jews and Christians (pp. 59–60, 68, 97, 174). A later date is demanded by Kümmel, *Introduction to the New Testament* p. 403.

7. In order to preserve in English translation the deliberate order of the Greek phrasing I have taken a small liberty with *palin anakainizein eis metanoian* in 6b, which probably means '[for someone] to give them a renewed opportunity of repentance'. In 6c, although *anastauroō* normally means nothing more than 'crucify', I incline to agree with Michel (against many modern commentators) in following the ancient fathers and the Vulgate, which give 'crucify again' *(rursum crucifigentes).* Otherwise, it is difficult to make sense of the writer's rather savage irony.

8. In both Heb 6:4–6 and 10:26–29 contempt for the Son of God and his death is associated with disregard of the Holy Spirit. But Hebrews does not make the Holy Spirit a leading feature of Christian existence (perhaps implied in the foundation clauses of 6:1, but not mentioned). There are two main uses of *pneuma:* (1) generally, the plural *pneumata,* meaning angelic beings (1:7, 14), the spirits of the godly dead (12:23), own our spirits (12:9 cf. soul and spirit 4:12); (2) the Holy Spirit in which we participate (2:4, 6:4; presumably also the Spirit of grace in 10:29) and the voice that speaks in scripture (3:7, 9:8, 10:15). In addition, *pneuma* may mean the 'eternal spirit' in which Christ offered himself (9:14). It may be that some of the intended readers of Hebrews were tempted to ignore Jesus in favour of the Spirit.

9. Vanhoye *(La structure litteraire)* made a great contribution to our understanding of the plan of Hebrews and of its author's style and method of composition; but his insistence that formal indications must always override subject matter produces some odd results. Vanhoye's rigidity prevents him from recognizing some obvious units (e.g., 2:1–4), commits him to the view that the writer made strict use of formal devices except when he failed to use them at all, and demands considerable artifice in the section 8:1–9:28.

10. Moffatt on 1:3; recently Dunn, *Christology in the Making* pp. 166, 206–9.

11. Dunn, *Christology in the Making* p. 158 says, 'Any attempt to set the exalted Christ merely on the level of the angels was resisted with great vigour, particularly . . . by the writer to the Hebrews'. That, however, implies that the attempt *was* made and suggests hesitations about the immediately preceding sentence: 'There is no clear evidence that any first-century Christian community actually thought of *Christ* as an angel'. Moreover, the midrash in Hebrews 7 suggests that Melchizedek may have been a source of interest in the community. Hence, it is significant that the Melchizedek Document (11QMelch) apparently treats Melchizedek as the head of the 'sons of Heaven', identical with the archangel Michael, who comes to preside over the final judgement, to overcome Satan, and to perform a great act of deliverance on the Day of Atonement. So DSSE, pp. 265–68; Vermes, *The Dead Sea Scrolls,* pp. 82–83 for literature; Dunn, pp. 152–53. For this scroll's hearing on Hebrews 1–2, see de Jonge and van der Woude, "11QMelchizedek and the New Testament', pp. 314–18 (opposed by Carmignac). It is equally significant that Philo can identify Melchizedek as 'A priest, namely *logos,* having as his portion Him that is' (Philo *L.A.* 3.82). See Michel, pp. 131–133; and the explanation by Dey, *Intermediary World* p. 127.

12. The problems occur in verse 9. How is the last clause related to previous clauses? Some make it depend on 'crowned', which immediately precedes it in the Greek; but the resulting sense is odd. Others, therefore, make it depend on 'the suffering of death'. 'For every one' represents *hyper pantos* where the plural *pantōn* would be more familiar. Hence, presumably, 'for every person', not for people as a whole. Finally, 'by the grace of

God' *(chariti theou)* is read by the majority of good Alexandrian and Western witnesses, including papyrus 46 (about 200 C.E.); but two tenth-century mss. and the corrector of an eleventh-century ms. read 'without God' *(chōris theou)*. But the surprising reading was known to Origen (third century), who found it in the majority of mss. known to him; and the reading was much discussed by Eastern and Western fathers, some of whom regarded it as correct. Since it is a very surprising reading and since Hebrews mentions 'grace' on seven more occasions, it is easy to see why the perplexing *chōris* would be changed to the familiar *chariti;* and much less easy to see how the familiar *chariti* could be changed to the surprising *chōris.* Hence, some authorities regard 'apart from God' as the original reading (Montefiore, Michel; 'God leads Christ into suffering [2:10] but increases this suffering by the torment of divine abandonment [Mk 15:34]). But the enigmatic brevity of *chōris theou* is scarcely a satisfactory introduction for such a complex thought; and the author of Hebrews is likely to be closer to the Lukan Passion than the Markan. In any case, the writer is arguing for the crucifixion as an object of belief and veneration. Would he sensibly describe it as apart from God? It is more likely that he simply meant to say, 'by the generosity of God he was allowed to be mankind's representative'? How, then, did the variant arise? Metzger sets out the theory of a pedantic gloss mistakenly included and corrected. It is implausible but the best we can do.

13. This line is an attempt to render the interesting aorist participle in *pollous huious eis doxan agagonta.*

14. 'To equip . . . completely' represents *teleiōsai.* The usual translation, 'to perfect', is avoided because it is notoriously difficult to guide the reader's understanding if the great variety of ideas associated with 'perfection' is introduced. There is no difficulty about the adjective *teleios* and the verb *teleioō:* they mean 'completion' and 'bringing to completion'. The context decides what sort of completion is in mind. In this passage (to my mind) the writer means that God equipped Jesus with everything necessary for his high priestly task. The writer of Hebrews often uses *completion* in relation to the Jewish cult: the priesthood and sacrifices of the Law can bring nothing to completion (7:11, 19; 9:9; 10:1), but Christ as high priest, having passed through the greater and 'more perfect' tent (9:11), by a single offering, has brought to completion those whom he consecrated (10:14). Christians can be described as mature people *(teleioi)* in contrast to babies, and it is their business to press on to completion (5:14–6:1; cf. 11:40, 12:23); and Christ himself, as well as being fully equipped for the work of high priest, must learn obedience through suffering (5:9, 7:28) and so become the pioneer and complete exponent of our faith (12:2). What kind of religious completion the author proposes is discussed in the commentaries and in special studies: recently, Peterson, *Hebrews and Perfection* and more excitingly, Dey, *Intermediary World,* where 'perfection' means approaching, and drawing near to, God (pp. 203–4).

15. Holiness is an important cultic word in Hebrews: (1) *ta hagia* (as in LXX) means the sanctuary (8:2; 9:1(sg.)–3, 12, 24–25; 10:19; 13:11); (2) it means the Holy Spirit (2:4, 3:7, 6:4, 9:8, 10:15); (3) Christians are *hoi hagioi* (3:1, 6:10, 13:24); (4) Christians are consecrated by the shedding of Christ's blood (9:13–14; 10:10, 14, 29; 13:12). Christians are urged to share God's holiness, and without holiness they cannot see the Lord (12:10, 14). Thus, being made holy, being brought to completion (see n. 14) is the condition required for the vision of God. As in Nm 16:5, to consecrate is to bring near to God (Moffatt, p. 32).

16. 'Put to the test' represents *peirastheis,* usually translated too weakly as 'tempted'. In NT, *peirazō* is often an eschatological verb meaning 'to put to the [final] test'. That meaning consorts more readily with the eschatology of Hebrews than the meaning that looks towards moral temptation. The author presumably introduces the verb here because

he intends to launch his next appeal with Psalm 95, in which Israel tested God and put him to the proof (where God had put them to the proof by the prospect of death, Ps. 81:7, Nm 20:4). In 4:15 our great high priest can sympathise with our weaknesses, having been tested *(pepeirasmenon)* in all respects, in similar fashion, without sin. If the statement implies that Jesus, as a particular person, experienced all the temptations of human beings, it is a grotesque exaggeration (e.g., he could not have experienced the temptations of a pregnant mother or a Roman governor). But if it means that like all of us, he was put to the test of death (cf. Abraham, Heb 11:17), it coheres with 2:18 and 5:7: 'Jesus offered up prayers and supplications, with loud cries and tears, to him who was able to save him from death'.

17. See Grayston, 'HILASKESTHAI. 'The essential conception is that of altering that in the character of an object which necessarily excludes the action of the grace of God, so that God cannot (as we speak) look on it with favour. The "propitiation" acts on that which alienates God and not on God whose love is unchanged throughout' (Westcott, ad loc.). 'This removal of sins as an obstacle to fellowship with God comes under the function of *ho hagiazōn*' (Moffatt, ad loc.).

18. Cf. Dey, *Intermediary World*, p. 219; 'The bold and revolutionary thesis of the author of Hebrews . . . is that Jesus has entered and participated in the realm of imperfection (flesh, blood, and temptation) and has accomplished perfection within this realm and thereby has opened the way for others to participate in perfection within this realm of creation and not outside of it'.

19. In disagreement with Peterson, *Hebrews and Perfection*, pp. 75–76 and others. I think it unlikely that the writer was inventing a new *homologia*. He was probably basing his adventurous expositions on an existing formula. For the highpriesthood of Moses in Philo, see Dey, *Intermediary World*, pp. 157–61, with references to Hebrews 3:1–6.

20. This rendering follows the order of Greek phrases. For proposals to identify hymnic or credal formula, of, Michel, p. 223–24 and Brandenburger (in Peterson, p. 235, n. 90), who is followed by Buchanan ad loc. There are many different ways of translating this passage, of relating its clauses, and of understanding the words it contains, especially *eulabeia* in verse 7 ('godly fear' or 'anxiety'?)—all discussed by Peterson, pp. 84–96.

21. For Hebrews 7, see Demarest, *History of Interpretation of Hebrews 7:1–10;* Thompson, *Beginnings of Christian Philosophy—The Epistle to the Hebrews,* chap. 7; Dey, *Intermediary World* chap. 6.

22. The verb *anistēmi* is twice used intransitively of a priest arising (i.e., coming on the scene) (Heb 7:11, 15). The noun *anastasis* appears among the foundational instructions of the community (Heb 6:1–2), which may perhaps be understood as a set of three rituals with corresponding divine counterparts: (1) a repentance ritual with transfer of allegiance to God ('repentance from dead works and faith toward God'); (2) cleansing rituals with imposition of hands that conveys the heavenly gift ('with instruction about ablutions, the laying on of hands'); (3) a new life ritual and anticipated final verdict ('resurrection of the dead and eternal judgement'). The third item looks like standard Jewish eschatology; but the writer of Hebrews probably regarded it metaphorically; for in 11:19 the sparing of Isaac is a parable of resurrection, and in 11:35 the restorations performed by Elijah and Elisha are called resurrections. (For the doxology, see p. 271.) These rituals may have been the substructure, as it were, of *ho tēs archēs tou Christou logos,* 'the account of the beginning of Christ', i.e., the statements about Christ that were taught first.

23. 'Once for all' is *ephapax,* appearing here for the first time and again at 9:12, 10:10. It means the same as *hapax* in 9:26, 28. In 9:7 *hapax* refers to the high priest's entry once a year (on the Day of Atonement) into the inmost sanctuary. In 9:27 it is attached to the truism that we die only once, in 6:4 and 10:2 to our enlightenment and

cleansing, and in 12:26–7 to God's intention of shaking heaven and earth for one last time. The self-offering of Christ is referred to in 9:12, 14, 15, 25, 26, 28; 10:10, 12, 14.

24. Grayston, 'HILASKESTHAI' pp. 649–651.

25. Examination of the *'Loci citati vel allegati'* in Nestle-Aland 26 shows fourteen verses in Hebrews with recognizable references to Isaiah. Only one is a direct quotation (Heb 2:13 quotes Is 8:17, 18; cf. 12:12); the remainder are memorable phrases. Is 53:10 has *ean dōte peri hamartias* ('if you give [an offering] for sin'); but that is echoed neither in Heb 7:27 nor 9:28. The final clause of Is 53:11 is *kai tas hamartias autōn autos anoisei* ('and he indeed shall bear their sins', where *anoisei* is the future of *anapherō*); and the penultimate clause of Is 53:12 is *kai autos hamartias pollōn anēnegken* ('and he indeed bore the sins of many', where *anēnegken* is the aorist of *anapherō*, as in Heb 7:27 and 9:28). But when the object of *anaphero* is *sins,* it cannot mean to offer up a sacrifice; it must mean to bear the consequences of sins, either one's own or someone else's (as in Nm 14:33 LXX).

26. With the high priest's cultic duties in mind, it is improbable that the plural has the extended meaning found elsewhere in the NT; see AGB, s.v., 1b.

27. This is against Peterson, pp. 117, 188–90; see n. 16.

28. Cf. Buchanan, ad loc. The sinlessness of Jesus is well discussed from a modern awareness of sin by Pannenberg, *Jesus—God and Man,* pp. 354–64; and Macquarrie, *Principles of Christian Theology,* pp. 301–3. Both theologians advance a view akin to this interpretation of Heb 7:27.

29. Vanhoye, *La structure litteraire,* pp. 138–72 is instructive but not always persuasive.

30. For the older (1902) traditional description of the Tabernacle, see Kennedy in *Hastings Dictionary of the Bible* 4, pp. 653–68. For a more critical view, Vaux, *Ancient Israel* pp. 294–97.

31. For the magnitude of sacrificial slaughter, see Jeremias, *Jerusalem in the Time of Jesus,* pp. 56–57.

32. 'By this the Holy Spirit indicates . . ., which is a *parabolē* for the present age' (Heb 9:8–9). *Parabolē* means 'an enigmatic indication', called in later writers a *typos.*

33. Cf. Thompson, *Beginnings of Christian Philosophy* on the hellenistic concepts of sacrifice in Hebrews 9 (chap. 6), against the protests of Williamson (*Philo and the Epistle to the Hebrews,* pp. 561–70), who seems too anxious to resist all hellenistic understanding of Hebrews.

34. *Paragenomenos* ('appeared') means 'came on the scene'. In Lk 12:51 the verb is used of Jesus' having come not to bring peace but a sword; in Is 56:1, 62:11, 63:1 of God's arriving as saviour and avenger.

35. RSV translates *genomenōn* 'that have come'; RV translates *mellontōn* 'to come' (see Metzger, ad loc.). In 10:1, which has *mellontōn*, the good things to come were to be made available by the death and ascent of Christ; here, they are available now.

36. I have argued against this insensitive assumption in my article 'Salvation Proclaimed III Hebrews 9:11–14'.

37. Heb 12:9 contrasts fathers of our flesh with the Father of spirits. In 9:10 'regulations of the flesh' are rules for earthly life until the time of the new order. In 5:7 and 10:20 the flesh of Jesus is his temporary earthly existence, as against his permanent spiritual existence. This use of *flesh* comes from the standard Hebraic 'flesh and blood' (2:14).

38. See Michel, p. 311; Peterson, *Hebrews and Perfection,* pp. 140–44.

39. The noun *diathēkē* and the verb *diatithēmi,* though appearing in one or two important passages, are uncommon outside Hebrews, which speaks of the old covenant six times (8:9; 9:4, 15, 20), of the new or better covenant nine times (7:22; 8:6, 8, 10; 9:15; 10:16,

29; 12:24; 13:20), and of covenants in general twice (9:16, 17). Elsewhere, the old covenant or covenants seven times (Lk 1:72; Acts 3:25, 7:8; Rm. 9:4; 2 Cor. 3:14; Gal 3:17; Eph 2:12), a new covenant three times (Lk 22:29; Rm 11:27; 2 Cor 3:6); of two covenants in Gal 3:17; and of covenants in general Gal 3:15. Mk 14:24 (= Mt 26:28) has 'my blood of the covenant', which is close to the language of Heb 10:29, 13:20, and Lk 22:20, 1 Cor 11:25 have 'new covenant in my blood'. It is plain that the NT is not dominated by covenant ideas.

40. Hebrews uses blood of Abel for his death (12:24), and refers to the blood of sacrificial animals (13:11). It uses Passover blood (11:28), blood of the covenant (9:18-21, 10:29, 13:20), blood of the Day of Atonement (9:7, 13, 25; 10:4), and Christ's blood (9:12, 14;10:19; 13:12); and blood as a means of cleansing (9:22).

41. Peterson, *Hebrews and Perfection,* pp. 160-66 mentions the numerous interpretations of Heb 12:22-23, but his discussion of them is rambling.

42. See Moffatt and Bruce for Philo's interpretation of Moses' move outside the camp as transfer from the bodily to the spiritual. Also Thompson *Beginning of Christian Philosophy,* pp. 128-40 ('Outside the Camp'). I do not agree with Thompson's view that Hebrews is not polemical, but he says useful things about this passage.

43. The word is *exousia.* See AGB, s.v. for its range of meaning.

44. The *homologia* references are Heb 3:1, 4:14, 10:23 (for the verb, 11:13 of publicly admitting, 13:15 of liturgical praise). See Michel in TWNT p. 216.

45. The *parrhēsia* references are Heb 3:6; 4:16; 10:19, 35. See Schlier in TWNT pp. 882-83.

46. See Barrett, 'The Eschatology of the Epistle to the Hebrews'. I would not go so far as to say, with Barrett, that in the thought of Hebrews 'the eschatological is the determining element' (p. 366), though I draw closer to him when he says, 'The heavenly tabernacle and its ministrations are from one point of view eternal archetypes, from another, they are eschatological events' (p. 385).

47. He passed through the heavens (Heb 4:14) and was exalted above them (7:26). He went ahead as a forerunner (6:20), and opened up a new and living way (10:20). He entered the holy place in heaven (9:12, 24), was crowned with glory and honour (2:9), and sat down at the right hand of the Majesty on high (1:3, 8:1, 10:12, 12:2).

Chapter 8

1. See Grayston, *Johannine Epistles,* esp. pp. 7-14.

2. See Grayston, *Johannine Epistles,* p. 56.

3. For a technical investigation of the group of words to which *hilasmos* belongs, see Grayston HILASKESTHAI. In LXX *hilasmos* and *exilasmos* seem to be used without distinction: (1) referring to cultic matters (Ex 30:10; Lv 23: 27-28; 25:9; Nm 5:8; 1 Chr 28:11, (20); Ez 43:23; 44:27; 45:19; 2 Mc 3:33; 12:45—all with explicit cultic contexts); and (2) referring to God's forgiveness (Ps 129 [130]:4; Ez 7:25 [LXX]; Dn [Th] 9:9, 'To the Lord our God belong acts of mercy and forgiveness'; Wis 18:21; Sir 5:5; 16:11; 17:29; 18:12, 20; 35:3).

4. See Grayston, 'The Meaning of PARAKLĒTOS', in which I examine the word's nineteen extant occurrences (including ten from Philo). The earliest, from the fourth century B.C.E., one from the first century B.C.E. and possibly another from the second century C.E. occur in legal contexts (though not as technical legal terms). None of the rest do. Franck (*Revelation Taught,* p. 18) gives a qualified welcome to my contentions but says that it should

be embarrassing for me that 'all the earliest instances occur in legal contexts'. The remark is both tendentious and invalid.

5. Cf. Smith, 'Johannine Christianity': 'Despite some residual disagreement on the question of Johannine and Synoptic relationships, few would today defend the view that John's Gospel is based principally upon the Synoptics or that they were his chief written sources' (pp. 226–27 and n. 1). In contrast Barrett (p. 45): 'The fact is that there crops up repeatedly in John evidence that suggests that the evangelist knew a body of traditional material that was either Mark, or was something much like Mark.' Robinson (*The Priority of John*, pp. 10–23) denies John's dependence on the Synoptic Gospels or any other source.

6. These figures are arrived at by counting pages and lines in Nestle-Aland 26.

7. Cf. Harvey, *Jesus on Trial*.

8. Also Barrett, *The Prologue of St. John's Gospel*.

9. But cf. other possible translations of 1:9.

10. See Schweizer, *The Good News according to Mark*, pp. 32–33.

11. Vermes, *Scripture and Tradition in Judaism*. Schoeps, (*Paul*, p. 133) regards Jn 1:29 as an allusion to Passover, not to the Aqedah (pp. 141–49).

12. Forcefully argued by Dodd, *The Interpretation of the Fourth Gospel*, pp. 230–38.

13. Lightfoot, 'Appended Note', pp. 349–56. Cf. also hyssop in Jn 19:29 and Ex 12:22.

14. Other possible identifications, e.g., the lamb of the daily sacrifices, the guilty offering, the scapegoat, are discussed in the commentaries; cf. Morris.

15. Braun, *Jean le Théologien*, vol. 3, p. 164.

16. On the address, see AGB, s.v. *gynē*. The oddity is that anyone should address his *mother* as 'woman'. The English language lacks a word for the formal manner of address (like the French use of *Madame*). Hence 'woman' in English sounds brutally rude when Jesus addresses his mother or the Samaritan woman (Jn 4:21) or Mary Magdalene (Jn 20:13, 15); but it is intended as a civil and formal address. It was doubtless chosen by the Evangelist because in his day the family of Jesus (cf. the rebuke to his brothers in Jn 7:3–9 and the formula 'My time has not yet come') were pressing for recognition that the Johannine community was unwilling to concede.

17. On this phrase, see n. 10. The general meaning is 'What room is there for your concern and mine in this situation?', with the implication that the speaker feels threatened. Whether Jesus regards the threat as improper (e.g., if it diverted him from God's instructions) or proper but premature (e.g., if it made demands on his miraculous power) is a matter of interpretation. Translations vary: 'What have I to do with thee? (RV) is modernised in NEB to 'Your concern . . . is not mine' or, equally dismissive, to 'Why do you involve me?' (NIV) and 'Why turn to me?' (JB). More aggressive are 'What do you want from me?' (NJB), 'Why are you interfering with me?' (TNT), and 'You must not tell me what to do' (GNB). If we leave aside what Jesus himself may have meant (taking the story as a historical account), the Johannine implication is that Jesus indeed performed a miracle but not because his mother hinted that he should.

18. In fact, *hē hōra mou* is unique: elsewhere when Jesus says 'My time has not yet come' (Jn 7:6, 8), the phrase is *ho kairos ho emos* or *ho emos kairos* (the only uses of *kairos* in the Gospel), and in the first place the meaning is 'My time has not come to go to the festival'. The Evangelist uses the idea editorially when he says that Jesus' opponents wanted to arrest him but failed 'because his hour had not yet come' (Jn 7:30, 8:20). The impression is given that God would decide the hour and that Jesus would recognize it when it came and accept it. From chap. 12 onwards Jesus says that the hour of his death and resurrection has come (12:23–25, 27; (editorial 13:1); 17:1). In chap. 16 he says that

an hour is coming that will affect the disciples (16:2, 4, 25, 32). Earlier in the Gospel he says that an hour is coming (and now is) when worship of God will be transformed and when the dead will hear the voice of the Son of God and will live (Jn 4:21, 23; 5:25, 28). Hence *the hour* contains the pivotal events of Jesus' death and resurrection; and *an hour* brings the consequences upon the disciples and mankind.

19. Jn 5:14; 7:14, 28; 8:20, 59; 10:23; 11:56; 18:20.

20. See Abrahams, 'The Cleansing of the Temple' in his *Pharisaism and the Gospels,* vol. 1.

21. The evidence for these statements is best set out systematically.

1. Jesus is asked for a sign in Jerusalem (2:18) and Galilee (6:30).
2. Jesus regrets that they will not believe unless they see signs and wonders (4:48) and distinguishes between receiving benefit from a miraculous supply of bread and perceiving a *sēmeion* (6:26).
3. Unlike John Baptist (10:41), Jesus performed what the Evangelist calls *sēmeia:* (a) water/wine 2:11, (b) recovery of sick child 4:54, (c) bread from heaven 6:14, (d) sight to man born blind 9:16, (e) Lazarus (11:45–46) 12:18. The cripple's cure in 5:2–8 is not called a sign and only obliquely an *ergon* (work) in 5:20, 36 and 7:21. No miracle is directly called an *ergon,* though indirectly sign d is so called in 9:3, and sign e in 5:20–21. Signs a and b are not elaborated; sign c is extended by discussion of bread and scripture; sign d is immersed in debate within the community; and sign e is prefaced by discussion and followed by lengthy exploration of consequences. After the cripple's cure in John 5, the healing is ignored and discussion fixes on Sabbath and the relation of Jesus to God.
4. The Evangelist's intention in recording signs is disclosed in 20:30. The effect of signs in general is mentioned in 2:23, 3:2, 6:2, 7:31, 12:37; and the effect of specific signs in 2:11, 4:53, 6:14, 9:16, 11:45, 47.

22. See Grayston, *Johannine Epistles,* intro.

23. Among modern studies see Sidebottom, *The Christ of the Fourth Gospel* pp. 69–136 Schnackenburg in Excursus 5 of vol. 1 of his commentary; Borsch, *The Son of Man in Myth and History,* pp. 257–313; Smalley, 'Johannine Son of Man', cf. idem, *John—Evangelist and Interpreter,* pp. 212–14; Lindars, 'The Son of Man in Johannine Christology, idem, *Jesus—Son of Man,* pp. 145–147; Moloney, *The Johannine Son of Man;* Coppens, 'Les logia johanniques'; Barrett, *Essays on John,* pp. 37–49.

24. On the surprising form *Son of man* and not the standard form *the Son of the man* (i.e. *huios anthrōpou,* not *ho huios tou anthrōpou*), see Lindars, *Jesus—Son of Man,* pp. 154–55).

25. According to Lindars, Jesus—Son of Man, p. 157, since the verb *hypsoō,* 'to lift up', is not the same as 'to set on a pole' (Nm 21:9) or 'fix to a cross' (*stauroō*), it may possibly derive from either (1) Is 52:13 (LXX), 'My servant . . . shall be exalted and lifted up' (*hypsōthēsetai kai doxasthēsetai*) or (2) the Syriac verb *zeqaph,* which is equivalent to *hypsoō* and can mean both 'to crucify' and 'to exalt'. I doubt that the Evangelist needed such learned stimulus.

26. Cf. Bultmann, in TWNT 2 832–43; Dodd, *Interpretation* 2.

27. Cf. Borgen, *Logos Was the True Light,* pp. 121–32, 'God's Agent on the Fourth Gospel'.

28. Schnackenburg is very good on these verses.

29. Words (12:49, 17:8); deeds (5:36, 17:4); judgement (5:22, 27); all things (3:35, 13:3, 17:2); those given to him (6:37, 39, 65; 10:29; 17:6, 7, 9, 24; 18:9); name (17:11, 12); glory (17:22, 24); cup (18:11).

30. Also Jn 8:5–52; 10:10, 28; 14:19.

31. See Bultmann on 5:26 (ET, p. 260).

32. Putting chapter 6 before chapter 5 is a favourite transposition. Schnackenburg (vol. 2, pp. 5–9) gives the arguments that persuade him to adopt it in his commentary. He tentatively attributes it 'to editors of an organizing but not overly careful turn of mind'. Lindars (pp. 207–209, 234–35) rejects it but accounts for the problems of chap. 6 by arguing that it was inserted in its present position at the second stage of composition. Barrett rejects any displacement, and Haenchen vol. 1, p. 51 says that 'the time of theories of displacement is gone'. For this enquiry they do not greatly matter, and an attempt must be made to explain the text as it stands.

33. Perhaps the prophet king, (as in Meeks, *Prophet-King*), but not the messianic king. If there is historical reminiscence behind this verse, it refers to Galilean separatism; cf. Jn 4:45, where the Galileans welcome Jesus, having seen his attack on the Jerusalem temple.

34. Schnackenburg (vol. 2, pp. 56–59) put the arguments for and against regarding the verses as editorial expansion and concludes, 'The hypothesis of an author other than the Evangelist has arguments in its favour, but since the section is not alien to his thought his authorship can be defended'. In addition to the commentaries, see also Dunn, 'John VI—A Eucharistic Discourse?' and Barrett, *Essays on John.*

35. As Barrett thinks (*Church, Ministry and Sacraments in the New Testament,* p. 74); cf. also idem, *Essays on John,* chap. 2 ('"The Flesh of the Son of Man" John 6:53') and chap. 6 (Sacraments). In passing, I voice a little disquiet about speaking too easily of the eucharist in the Johannine community (knowing full well that *eucharisteō* is used at Jn 6:11, 23 and that *eucharistia* is used in Didache, Ignatius, and Justin): in unguarded moments we may read back later assumptions into John.

36. See BDF, sec. 101 under *esthiein;* TWNT 8, s.v. *trōgō.* For the verb *to eat* Jn uses aorist *phag—,* present *trōgō,* not *esthiō* (cf. Jn 13:18, where a psalm quotation uses *trōgōn* for LXX *esthiōn*).

37. Also Jn 12:35, 13:33, 14:19, 16:16–18.

38. On *egō eimi,* see Barrett, p. 342; Brown, vol. 1, App. #4, Schnackenburg, vol. 2, pp. 79–89. The words are used (1) to identify oneself: 'I am (not) the person concerned', or 'it's me'; (2) to explain what Jesus is and does by a series of images; and (3) to assert the heavenly origin of Jesus, with a possible reference to the divine self-disclosure in, e.g., Is 43:10 (LXX).

39. In verse 7, *egō eimi hē thura tō probatōn* may mean 'I am the door for access to the sheep within the fold' or 'I am the door by which sheep enter the fold'. The second meaning would suit verse 9, 'If anyone enters through me [as the door] he will be saved'. I suspect these two verses of being fragments of the controversy disclosed in 1 John (see p. 277), about whether Jesus was or was not essential for entry to the community and for continuing life within it.

40. It is generally acknowledged that this imagery is closely related to the polemic in Ez 34 and Zec 10:3, 11:4–17.

41. 'Risk his life' translates *tēn psychēn autou tithesin* (Bultmann, pp. 370–71, n. 5). No shepherd in his senses would surrender his life for the sheep (if he did the wolves would indeed grow fat), and any competent shepherd could drive off a wolf (in Palestine they came singly, not in packs). Anyone who uses a *paroimia* can become a victim of his chosen image. If the Evangelist wants to say that Jesus surrended his life (as he certainly does later), he has only said 'risk it' here.

42. Derrett, 'The Good Shepherd' is fanciful (as Barrett says) and incoherent.

43. When Jesus saw them weeping, he *enebrimēsato tō pneumati kai etaraxen heauton* (verse 33), burst into tears (*edakrusen,* verse 35), and again *embrimōmenos en heautō* as

he went to the tomb. The verb *embrimaomai* means 'to snort, to show anger' (AGB). The commentators vary in explaining the anger of Jesus: it was caused by their unbelief or by the presence of Death, or it was necessary for thaumaturgic technique. It is more plausible to say that Jesus was not resigned to the death of his friend but angered by it. Even if the death was required by God, it was upsetting and infuriating.

44. The risk to Jesus is signalled in verse 2, where Mary is identified as the woman who anointed the Lord, an action that belongs to his burial, as Jn 12:7 requires.

45. Cf. Jn 9:3: the blindness from birth occurs 'that the *erga* of God might be made manifest in him'.

46. I have written out the passage in this manner not in order to quarrel with the familiar translation (see especially the note in Sanders and Mastin), but to suggest the economy of the Greek and to indicate the structure.

47. For this interpretation and also for the omission of *and the life* by some witnesses, see Haenchen, ad hoc.

48. Cf. Dodd, 'The Prophecy of Caiaphas'.

49. But see Lindars (p. 420), who thinks that the temple episode of Jn 2:14–22 originally followed the entry into Jerusalem, so that the two events interpret one another. The death and resurrection of Jesus are then the key to understanding both stories. If so, the Evangelist was rather easily prepared to hide the key.

50. The simile of verse 24 requires this kind of application since (1) it is not obviously true of all *individual* believers, and (2) a buried seed that actually dies does not germinate, and that is the end of that. The use of the dying seed in 1 Cor 15:36–37 belongs to another kind of argument, namely, that the bodily form of the fruit is different from the bodily form of the seed that 'died' to produce it.

51. *Diakoneō* only here except Jn 12:2 of Martha's serving; *diakonia* and *diakonos* are absent.

52. The crowd who heard the sound from heaven said that it had thundered ('The adversaries of the Lord shall be broken in pieces; against them will be thunder from heaven', 1 Sm 2:10) or that an angel had spoken to him. So, say the commentators, if the crowd misunderstood, how could Jesus assert that the voice came for their sake, not his? But the commentators assume that the voice came for their *information;* the Evangelist means that it came for their *benefit,* whether they understood or not.

53. A coherent presentation of the view I am opposing is given by Richter, 'Die Fusswaschung'.

54. On the text of this verse, see Metzger, who retains *ei mē tous podas* on sound principles. The majority of exegetes omit these words to support their (to my mind impossible) interpretation of the pericope.

55. Similar statements are in Jn 8:50, 54; 12:28; 14:13—with a climax in Jn 17:1, 4–5, 10, 22, 24. In Jn 7:39 and 12:16, 23 it is taken for granted that the death of Jesus is implied.

56. The commentaries consider the several ways (indicative or imperative) of translating *pisteuete.* If Jn 13:36–14:31 explains the consequences of 13:33–35, it is proper to regard 14:1–3 as appearing in response to 13:38. 'Believe in God' is not to be taken in the weak modern sense (that theistic propositions are generally acceptable) but in the older sense of 'Trust in God' (and therefore in me, as his Agent).

57. The *pollai monai* of Jn 14:2 are now often translated 'rooms'; but that sounds like bed and breakfast accommodation with exactly the wrong impression of impermanence. Schnackenburg, vol. 3, p. 60 compares heavenly dwellings in Jewish and non-Jewish (particularly Gnostic texts): it is only necessary to read the passages quoted to see how little they have to do with the thought of the Gospel.

58. See App. 5 in Brown, vol. 2, pp. 1135–44; Excursus 16 in Schnackenburg, vol. 3, pp. 138–54; and the bibliography in Haenchen, vol. 2, pp. 120–22—to which add Grayston, 'The meaning of PARAKLETOS and Burge, *The Anointed Community*.

59. The *hithpael* (a primarily reflexive verb form) of *qdš* is often translated 'sanctify yourselves', though LXX uses the passive. It means 'put yourself in the proper condition for performing sacred duties and for encountering God'.

60. Such illumination would forewarn their intended victims. Perhaps they were drunk! The commentaries discuss the facts and improbabilities.

61. This (and much else in the trial narrative) can only be a guess. It does not imply that I am reading the narrative as if it were a good historical report (see the commentaries for the numerous problems); but presumably the story must have been intended to be dramatically convincing. Fortunately, I am concerned less with judicial reporting and more with theological perusasion.

62. Many commentaries are entranced by the seamless tunic and refer it to the high priest's vestment (Ex 28:32), comparing the description in Josephus and the compulsive allegorising of Philo. So Jesus is allusively represented as a high priest, or the tunic represents the unity of the Church. If that is so, why is the symbol of Christ's priesthood or Christian unity in the hands of Roman soldiers who will sell it to advantage in the market? (Schnackenburg, vol. 3, p. 224 is sound on this). The other four instances of not unfavourable Roman actions are in verses 29, 33, 34, 38.

63. Interpreters are divided about the Evangelist's intention: Is he referring to the godly sufferer of Psalms, or the Passover lamb of Exodus and Numbers, or does he combine them both? The interpretation of the Lamb of God (see p. 287), the sequence of Passover references, and the timing of the death of Jesus push the reference of 19:36 in the direction of Passover—as Jews no doubt would recognize. But would it seem religiously right to Jews that the Passover offering of a lamb without spot or blemish should provide the symbolic justification for the death of a man who had been scourged and derided and who still bore the marks of crucifixion when next seen (Jn 20:20)?

64. The purport of the question is much discussed. *Who* looks on him? The obvious answer is, Those who pierced him, i.e., the Romans. What is the meaning of *opsontai eis hon* (an unusual but not unknown construction; see LSJ and MM)? Probably it is 'They shall look to him for salvation' (not in repentance as in Zechariah) in the sense that they put him to death but will have to come back to him for succour.

65. In this respect it is a matter of indifference whether medical persons think they can or cannot account for the phenomenon.

66. This of course implies that in Jn 20:19–29 the disciples experienced *appearances* of Jesus, i.e., visual, auditory, and possibly tactile impressions that made possible genuine communication with the ascended Lord. In principle, an appearance that stimulates the sense of touch is no more suspect than one that stimulates vision and hearing.

Chapter 9

1. Cf. Lambrecht, *L'apocalypse johannique et l'Apocalyptique*, in which Vanni provides a survey of recent scholarship (pp. 21–46), and Lambrecht himself is illuminating on the structure of the Apocalypse (pp. 77–104); also Court, *Myth and History in the Book of Revelation*.

2. *Machinery* in the sense of a 'group of contrivances, esp. supernatural persons and incidents, used in literary work' (COD)—as in e.g., Alexander Pope's *Rape of the Lock*.

3. For the prophet's concern with accusation and indictment, see McKane, 'Prophecy and Prophetic Literature'. The prophetic qualities of Revelation are examined in Hill, *New Testament Prophecy,* pp. 70:93.

4. Ford, ad loc.

5. Of Antipas, the faithful witness (Rv 2:13); of Christ (3:14); of the two witnesses who are killed (11:3, 7–8); and of the blood of the martyrs (17:6); see discussion in Court, pp. 88–90.

6. Ps 89:38 (LXX) reads, 'The witness in heaven is faithful'; Ps 89:28 reads, 'I will make him the first-born, the highest of the kings of the earth'. Zec 4:2 with its seven lamps and 4:10 with its seven eyes probably suggested the seven spirits of Rv 1:4; and Zec 12:10, 12–13 certainly provided most of the composite quotation in Rv 1:7.

7. The aorist *agapēsanti* is read by 2053, 2062 and the A majority group in Nestle-Aland 26.

8. See Metzger, ad loc. *Lysanti* probably has the better ms. support, though P and 2053 which normally go with AC here read *lousanti*. For *lyein ek,* see Rv 20:7; for *lyein apo* Lk 13:16; 1 Cor 7:27. For *lyein hamartian* Jb 42:9 (LXX), Is 40:2 (LXX). Acts 16:33, 'washed from [the effect of] the blows', gives a sort of parallel to *louein apo;* and the verb *katharizō* is used of cleansing from sin, e.g., Ps 50:4; Heb 1:3, 9:14, 22–23; 10:2; 1 Jn 1:7. The washing is sometimes explained as a failure to understand '*in* his blood' (as a Hebrew idiom meaning 'at the cost of his blood' and sometimes defended by reference to Zec 13:1, the fountain opened to cleanse from sin and uncleanness (Cowper, 'There Is a Fountain Filled with Blood')—but not so in LXX.

9. Since commentators discern new exodus imagery here, they are tempted to explain Rv 1:5d—6 by reference to the Passover sacrifice. So Sweet, ad loc.: 'In Jewish thought a martyr death could win atonement for the sins of the people, but the primary references is to the Passover Lamb', presumably because Jesus appears as the Lamb in Revelation 5. So on p. 61 Sweet says, 'As God rescued Israel from Egyptian bondage by the blood of the Passover lamb—which would be a good comment if the analogy were completed—so God rescued his people from Roman oppression by the blood of his Son'. But what has that to do with sins? Osten-Sacken argues for a baptismal context ("Christologie Taufe, Homologie").

10. See above, chap. 2, p. 39 and n. 77). Cf. Rordorf, *Sunday,* pp. 207ff.

11. Cf. Rordorf, Sunday, pp. 90–95.

12. So that the phrase 'in the midst of the throne' means 'within the ambience of Majesty'.

13. This observation is well known or ought to be. C. E. Raven used to lecture on it in the late 1930s. Cf., now, Sweet, p. 14 and (a book much criticised but sound in this respect) Boman, *Hebrew Thought Compared with Greek,* chap. 2.

14. This is one of the four uses of *arnion* in LXX. Also Jer 11:19 'I was like a guileless lamb led to the slaughter'. Ps 113:4, 6 are irrelevant as is Is 40:11 (Aquila).

15. So the daily burnt offering, morning and evening, in Ex 29:38–41; Nm 28:3–8; Ez 46:13–15 (morning only), as well as numerous other occasions (Nm 28–29; Lv 1:10; 3:6; 9:3; 12:6, 8; 14:10–25; 23:18–20.

16. Lohmeyer drew attention to 1 En 89–90, Test. Jos. 19 (now conveniently available in OTP, pp. 67–70, 324), and that origin is favoured by Beasley-Murray and Ford. The apocalyptic ram became well-known when C. H. Dodd used it to explain the Lamb of God in Jn 1:29, 36: (*Interpretation of the Fourth Gospel,* pp. 230–38).

17. Ex 12:3–5 and Noth, *Exodus,* p. 91.

18. van Unnik, ('Worthy Is the Lamb') argued for the meaning 'worthy to receive

secret information'. But the whole aim of Revelation is to disclose concealed information; see esp. 22:10.

19. Cf. Hanson, *Wrath of the Lamb*, chap. 7. See Rm, n. 5.

20. Apart from Revelation, *agorazō* occurs in 1 Cor 6:20; 7:23; 2 Pt 2:1, 'the master who bought them'—a stock reference in a document that otherwise lacks all reference to the death and resurrection of Christ. *Exagorazō* in Gal 3:13, 4:5. See chap. 2, p. 32, and p. 80 and n. 142.

21. This occurrence corresponds well to the vivid description by Pliny the Younger of the eruption of Vesuvius in 79 C.E. (Ep. 6.16.20).

22. Cf. Court, chapt. 4.

23. The familiar perplexity about Rev 19:12c can be dispelled if *onoma* is given the meaning 'authority, office'. For evidence. MM; AGB, s.v. *onoma* 5; cf. Phil 2:9, Heb 1:4.

24. See Metzger, ad loc.

25. With Sweet, against Beasley-Murray.

26. The *go' el haddam* (avenger of blood) appears in Nm 35:19–27; Dt 19:6, 12; Jos 20:3, 5; 2 Sm 14:11. Cf. Vaux, *Ancient Israel*, pp. 10–12; Pedersen, *Israel*, vol. 1–2, pp. 378–92. According to Hertzberg, *I and II Samuel*, p. 331, the practice of blood vengeance prevailed among Bedouin until modern times.

27. The parody of the marriage supper of the Lamb is taken from Ez 39:17–20, but John suppresses the drinking of blood.

Bibliography

STANDARD SOURCES

The English translation normally used is the Revised Standard Version, though sometimes I give my own rendering. Other translations are sometimes quoted.

AV	Authorized (King James) Version
GNB	Good News Bible
NEB	New English Bible
NIV	New International Version
NJB	New Jerusalem Bible
RV	Revised Version
TNT	Translator's New Testament (British and Foreign Bible Society, London 1973)

New Testament

The Greek text used is the 26th edition of Nestle-Aland *Novum Testamentum Graece* (Stuttgart 1979).

Occasional use is made of Η ΚΑΙΝΗ ΔΙΑΘΗΚΗ (Ancient text with modern Greek translation) published by ΒΙΒΛΙΚΗ ΕΤΑΙΡΕΙΑ, Athens 1967.

Textual information is taken from K. Aland, *Kurzgefasste Liste der Griechischen Handschriften des Neuen Testaments,* vol. 1 (Berlin 1963).

Other standard works are:

AGB	Arndt, Gingrich, and Bauer, *A Greek-English Lexicon of the New Testament and Other Early Christian Literature* (Chicago and Cambridge, rev. 1979 by Gingrich and Danker)
BDF	Blass, Debrunner, and Funk, *A Greek Grammar of the New Testament* (Cambridge and Chicago 1961)
Field	F. Field, *Notes on the Translation of the New Testament* (Cambridge 1899)
LSJ	Liddell, Scott, and Jones, *Greek-English Lexicon* (Oxford 1968).
Metzger	B. M. Metzger, *A Textual Commentary on the Greek New Testament* (London and New York 1971)

MHT   Moulton, Howard, and Turner, *A Grammar of New Testament Greek*, 4 vols. (Edinburgh 1900–1976)

MM   Moulton and Milligan, *Vocabulary of the Greek Testament* (London 1930)

Moule   C.F.D. Moule, *An Idiom Book of New Testament Greek* (Cambridge 1953)

TWNT   Kittel and Friedrich, eds. *Theologisches Wörterbuch zum Neuen Testament*, vols. 1–10 (Stuttgart 1932–1979)

Winer   Winer and Moulton, *A Treatise on the Grammar of New Testament Greek* (Edinburgh 1882)

Old Testament

BDB   Brown, Driver, and Briggs, *A Hebrew and English Lexicon of the Old Testament* (Oxford 1907)

LXX   A. Rahlfs, ed., *Septuaginta*, (Stuttgart 1950) H. B. Swete, ed., *The Old Testament in Greek*, 3 vols. (Cambridge 1887–1894)

MT   Masoretic Text (the standard Hebrew text published by the British and Foreign Bible Society, London 1950)

Otley   R. R. Otley, *Isaiah according to the Septuagint* (Cambridge 1904–1906)

Other Jewish Writings

APOT   R. H. Charles, *Apocrypha and Pseudepigrapha of the Old Testament*, 2 vols. (Oxford 1913)

AOT   H. F. D. Sparks, ed., *The Apocryphal Old Testament* (Oxford 1984)

Ber Rab   Bereshith Rabba, midrash on Genesis (see H. L. Strack, *Introduction to the Talmud and Midrash*, New York and Philadelphia 1959)

DSSE   G. Vermes, *The Dead Sea Scrolls in English* (Harmondsworth 1987) with references to IQS, CD, IQSa, IQM, IQpHab, and IIQMelch

Josephus   *Against Apio, The Jewish War, The Jewish Antiquities* by Flavius Josephus, Loeb Classical Library (London and Cambridge, Mass. 1926)

M   *The Mishnah*, trans. H. Danby (Oxford 1933), with reference to the following Tractates: Aboth, Baba Metzia, Berakoth, Kerithoth, Nedarim, Pesachim, Sanhedrin, Yoma, and Zebahim

Midr Ps   Midrash Tehillim on the Psalms (see Bereshith Rabba above)

Oesterley   *The Sayings of the Jewish Fathers*, trans. W. O. E. Oesterley (London 1919)

OTP   *Old Testament Pseudepigrapha*, 2 vols. J. H. Charlesworth, ed., (London 1983, 1985) with reference to: Ascension of Isaiah, 2 Baruch, 4 Baruch, 1 Enoch, Joseph and Asenath, Jubilees, 3 Maccabees, 4 Maccabees, Psalms of Solomon, and Testaments of the Twelve Patriarchs (specially Benjamin, Gad, Joseph, Levi, Reuben)

Pesiqta   Midrashim on pentateuchal and prophetic lessons (see Bereshith Rabba above)

Philo   Philo Judaeus, Works in 10 vols., Loeb Classical Library (London and Cambridge, Mass. 1929–1962), with reference to Decal., de Fug., Her., L.A., Mos., Quod omn. prob. liber sit, Virt.

Rab Anth   *A Rabbinic Anthology* by Montefiore and Loewe (London 1938)

Rabin   C. Rabin, ed. and trans., *The Zadokite Documents* 2nd ed. (Oxford 1958)

Roth   C. Roth, ed. and trans., *The Haggadah* (London 1934)

Sanh.	Tractate Sanhedrin in the Babylonian Talmud
SB	Strack and Billerbeck, *Kommentar zum Neuen Testament aus Talmud und Midrash,* 6 vols. (Munich 1922–1961)
Sifre Deut	Midrash on Deuteronomy (see Bereshith Rabba above)
Stenning	J. F. Stenning, ed., *The Targum of Isaiah* (Oxford 1949)
Targ.PsJon	Targum Pesudo Jonathan, an Aramaic paraphrase of the Pentateuch (see M. McNamara, *Targum and Testament,* Shannon, Ireland 1972)
Tosephta	Tosephta Sanhedrin, in *Tractate Sanhedrin,* trans. H. Danby (London 1919)

Greek Writers

Aesch.	Aeschylus, *Persae* (5th century B.C.E.)
Dem.	Demosthenes, *Orations* (4th century B.C.E.)
Dion. Hal.	Dionysius of Halicarnassus, *On the character of Thucydides* (1st century B.C.E.)
Pliny	Pliny the Younger, *Epistles* (1st century C.E.)

Early Christian Writings

Barnabas	*Epistle of Barnabas* (between 70 and 100)
1 Clement	*First Epistle of Clement* (c.96)
Clem. Alex.	Clement of Alexandria (c.150–c.215)
Didache	The so-called 'Teaching of the Twelve Apostles,' an early Church Order
Eusebius HE	Eusebius of Caesarea, *Ecclesiastical History* (c.260–c.340)
Hermas	*The Shepherd of Hermas,* a writing from the second century containing Visions, Mandates, and Similitudes
Ignatius	Letters of Ignatius of Antioch (c.35–c.107), with references to Ephesians, Magnesians, Philadelphians, Polycarp, Romans, and Trallians
Irenaeus	Irenaeus of Lyons, *Against Heresies* (c.130–c.200)
Justin	Justin Martyr, *Dialogue with Trypho* (c.100–c.165)
Papias	Papias of Hierapolis (c.60–130)
Polycarp	Polycarp of Smyrna (c.69–155)
Tertullian	Tertullian, *Against All the Heresies* and *On Chastity* (c.160–c.225)

Goodspeed	*Index apologeticus* (Leipzig 1912)
	Index patristicus (Leipzig 1907)
Hennecke, Schneemelcher, and Wilson	*New Testament Apocrypha,* 2 vols. (London 1963, 1965)

Gnostic Writings

Förster, W	*Gnosis,* 2 vols. (Oxford 1972, 1974)
GT	Gospel of Thomas
Nag Hammadi	J. M. Robinson, ed., *The Nag Hammadi Library in English* (Leiden 1977)
Rheginos	M. L. Peel, ed., *The Epistle to Rheginos* (London 1969)

Journals

ExT	*Expository Times*
HTR	*Harvard Theological Review*
JQR	*Jewish Quarterly Review*
NT	*Novum Testamentum*
NTS	*New Testament Studies*
RQ	*Revue de Qumran*
ZNTW	*Zeitschrift für die neutestamentliche Wissenschaft*

Dictionaries

COD	*Concise Oxford Dictionary*
SOED	*Shorter Oxford English Dictionary*

COMMENTARIES

Matthew

In addition to the earlier commentaries of W. C. Allen (1907), A. Plummer (1915), A. H. McNeile (1915, repr. 1928), and M.-J. Lagrange (1927), see the following:

Beare, F. W. *The Gospel according to Matthew* (Oxford 1981).
Bonnard, P. *L'Evangile selon Matthieu* (Neuchatel 1963).
Filson, F. V. *The Gospel according to St. Matthew* (London 1960).
Gundry, R. H. *Matthew: A Commentary on His Literary and Theological Art* (Grand Rapids 1982).
Green, H. B. *The Gospel according to Matthew* (Oxford 1975).
Hill, D. *The Gospel of Matthew* (London 1972).
Lohmeyer, E., and W. Schmauch. *Das Evangelium des Matthäus* (Göttingen 1958).

Mark

In addition to the earlier commentaries of E. P. Gould (1896), H. B. Swete (1908), C. G. Montefiore (1909), M.-J. Lagrange (1911), A. B. Bruce (1917), A. E. J. Rawlinson (1925), and B. H. Branscombe (1937), see the following:

Anderson, H. *The Gospel of Mark* (London 1967).
Cranfield, C. E. B. *The Gospel according to Saint Mark* (Cambridge 1959).
Gnilka, J. *Das Evangelium nach Markus* (Zurich-Neukirchen-Vluyn 1978–79).
Haenchen, E. *Der Weg Jesu* (Berlin 1968). With synoptic parallels.
Johnson, S. E. *The Gospel according to St. Mark* (London 1960).
Klostermann, E. *Das Markusevangelium,* 4th ed. (Tübingen 1950).
Lane, W. L. *The Gospel of Mark* (London 1974).
Montefiore, C. G. *The Synoptic Gospels* vol. 1 (London 1909).
Moule, C. F. D. *The Gospel according to Mark* (Cambridge 1965).
Nineham, D. E. *Saint Mark* (Harmondsworth 1963).
Pesch, R. *Das Markusevangelium,* vols. 1–2 (Freiburg 1977).

Schweizer, E. *The Good News according to Mark* (London 1971).
Taylor, V. *The Gospel according to St. Mark* (London 1952).

Luke

In addition to the earlier commentaries of A. Plummer (1896), B. S. Easton (1926), M.-J. Lagrange (1927), and J. M. Creed (1930), see the following:

Ellis, E. E. *The Gospel of Luke* (London 1966)
Fitzmyer, J. A. *The Gospel according to Luke,* vols. 1–2 (New York 1981, 1985).
Leaney, A. R. C. *A Commentary on the Gospel according to St. Luke* (London 1958).
Marshall, I. H. *The Gospel of Luke* (Exeter 1978).
Schweizer, E. *The Good News according to Luke* (London 1984).

John

In addition to the earlier commentaries of B. F. Westcott (1882), W. Bauer (1925), M.-J. Lagrange (1927), G. H. C. Macgregor (1928), and J. H. Bernard (1928), see the following:

Barret, C. K. *The Gospel according to St. John* 2d ed. (London 1978).
Brown, R. E. *The Gospel according to John,* vols. 1–2 (New York 1966, London 1971).
Bultmann, R. *The Gospel of John* (Oxford 1971).
Fenton, J. C. *The Gospel according to John* (Oxford 1970).
Haenchen, E. *A Commentary on the Gospel of John,* vols. 1–2 (Philadelphia 1984).
Hoskyns, E. C. *The Fourth Gospel,* ed. F. N. Davey (London 1940).
Lightfoot, R. H. *St. John's Gospel,* ed. C. F. Evans (Oxford 1956).
Lindars, B. *The Gospel of John* (London 1972).
Morris, L. *The Gospel according to John* (London 1971).
Sanders, J. N. *A Commentary on the Gospel according to St. John,* ed. B. A. Mastin (London 1968).
Schnackenburg, R. *The Gospel according to St. John,* vols. 1–3 (London 1968, 1980, 1982).

Acts

Bruce, F. F. *The Acts of the Apostles,* 2d ed. (London 1952). The Greek text with introduction and commentary.
———. *The Book of the Acts,* (Grand Rapids 1988). Previously published as *Commentaryon the Book of the Acts.*
Conzelmann, H. *Die Apostelgeschichte, 2d ed.* (Tübingen 1972).
Foakes Jackson, F. J., and Kirsopp Lake eds. *The Beginnings of Christianity* (London 1920–33). (Vols. 4 and 5 ed. Kirsopp Lake and H. J. Cadbury.)
Haenchen, E. *The Acts of the Apostles* (Oxford 1971).
Hanson, R. P. C. *The Acts* (Oxford 1967).
Marshall, I. H. *Acts* (Leicester 1980).

Romans

Barrett, C. K. *A Commentary on the Epistle to the Romans* (London 1957).
Best, E. *The Letter of Paul to the Romans* (Cambridge 1967).

Black, M. *Romans* (London 1973).
Cranfield, C. E. B. *A Critical and Exegetical Commentary on the Epistle to the Romans,*
    vols. 1–2 (Edinburgh 1975, 1979).
Dodd, C. H. *The Epistle of Paul to the Romans* (London 1932).
Gifford, E. H. *The Epistle of St. Paul to the Romans* (London 1886).
Käsemann, E. *Commentary on Romans* (London 1980).
Leenhardt, F.-J. *L'Epitre de Saint Paul aux Romains* (London 1967).
Lietzmann, E. *Römerbrief,* 4th ed. (Tübingen 1933).
Lightfoot, J. B. 'The Epistle to the Romans: Analysis and Commentary,' in his *Notes on*
    the Epistles of St. Paul (London 1895).
Nygren, A. *Commentary on Romans* (London 1952).
Sanday, W., and Headlam, A. C. *A Critical and Exegetical Commentary on the Epistle to*
    the Romans (Edinburgh 1895).
Wilckens, U. *Der Brief an die Römer,* 3 vols. (Zürich, Köln, Neukirchen-Vluyn 1978,
    1980, 1982).

1 Corinthians

Allo, E. B. *Première Epitre aux Corinthiens* (Paris 1956).
Barrett, C. K. *The First Epistle to the Corinthians* (London 1968).
Bruce, F. F. *1 and 2 Corinthians* (London 1971).
Conzelmann, H. *A Commentary on the First Epistle to the Corinthians* (Philadelphia
    1975).
Lietzmann, H., and W. G. Kümmell. *An die Korinther,* I–II (Tübingen 1949).
Senft, C. *La Première Epitre de Saint Paul aux Corinthiens* (Neuchatel 1979).

2 Corinthians

Allo, E. B. *Second Epitre aux Corinthiens* (Paris 1956).
Barrett, C. K. *The Second Epistle to the Corinthians* (London 1973).
Bruce, F. F. *1 and 2 Corinthians* (London 1971).
Bultmann, R., *Der zweite Brief an die Korinther,* ed. E. Dinkler (Göttingen 1976).
Collange, J.-F. *Enigmes de la Deuxième Epitre de Paul aux Corinthiens* (Cambridge 1972).
Hèring, *La Seconde Epitre de Saint Paul aux Corinthiens* (Neuchatel and Paris 1958).
Lietzmann, H., and W. G. Kümmel. *An die Korinther,* I–II, (Tübingen 1949).
Plummer, A. *Second Epistle of St Paul to the Corinthians* (Edinburgh 1915).

Galatians

Betz, H. D. *Galatians* (Philadelphia 1979).
Bonnard, P. *L'Epitre de Saint Paul aux Galates* (Neuchatel 1972).
Bruce, F. F. *The Epistle to the Galatians* (Exeter 1982).
Burton, E. de W. *A Critical and Exegetical Commentary on the Epistle to the Galatians*
    (Edinburgh 1921).
Guthrie, D. *Galatians* (London 1969).
Lietzmann, H. *An die Galater,* 4th ed. (Tübingen 1971).
Lightfoot, J. B. *Saint Paul's Epistle to the Galatians* (London 1896).
O'Neill, J. *The Recovery of Paul's Letter to the Galatians* (London 1972).

Ephesians

Abbott, T. K. *A Critical and Exegetical Commentary on the Epistles to the Ephesians and to the Colossians* (Edinburgh 1987).

Barth, M. *Ephesians,* vols. 1–2 (New York 1974).

Bruce, F. F. *The Epistles to the Colossians, to Philemon and to the Ephesians* (Grand Rapids 1984).

Caird, G. B. *Paul's Letters from Prison* (Oxford 1976).

Dibelius, M. and H. Greeven. *An die Kolosser Epheser an Philemon* (Tübingen 1953).

Ernst, J. *Die Briefe an die Philipper, an Philemon, an die Kolosser, an die Epheser* (Regensburg 1974).

Houlden, J. L. *Paul's Letters from Prison* (Harmondsworth 1970).

Masson, C. *L'Epitre de Saint Paul aux Ephésiens* (Neuchatel and Paris 1953).

Mitton, C. L. *Ephesians* (London 1974).

Rendtorf, H. *Die Kleineren Briefe des Apostels Paulus* (Göttingen 1949).

Robinson, J. Armitage. *St. Paul's Epistle to the Ephesians* (London 1903).

Schnackenburg, R. *Der Brief an die Epheser* (Zürich 1982).

Scott, E. F. *The Epistles of Paul to the Colossians, to Philemon and to the Ephesians* (London 1930).

Philippians

Beare, F. W. *A Commentary on the Epistle to the Philippians,* 3d ed. (London 1973, repr. 1988 with rev. bibl.).

Caird, G. B. *Paul's Letters from Prison* (Oxford 1976).

Collange, J.-F. *L'Epitre de Saint Paul aux Philippiens* (Neuchatel 1973).

Dibelius, M. *An die Thessalonicher I, II; An die Philipper* (Tübingen 1937).

Ernst, J. *Die Briefe an die Philipper, an Philemon, an die Kolosser, an die Epheser* (Regensburg 1974).

Grayston, K. *The Letters of Paul to the Philippians and to the Thessalonians* (Cambridge 1967).

Heinzelmann, G. *Die Kleineren Briefe des Apostels Paulus* (Göttingen 1949).

Houlden, J. L. *Paul's Letters from Prison* (Harmondsworth 1970).

Lightfoot, J. B. *St. Paul's Epistle to the Philippians* (London 1896).

Martin, R. P. *Philippians* (London 1976).

Scott, E. F. 'Philippians,' in *The Interpreter's Bible,* ed. G. A Buttrick, vol. II (New York 1955).

Colossians

Abbott, T. K. *A Critical and Exegetical Commentary on the Epistles to the Ephesians and to the Colossians* (Edinburgh 1987).

Bruce, F. F. *The Epistles to the Colossians, to Philemon and to the Ephesians* (Grand Rapids 1984).

Caird, G. B. *Paul's Letters from Prison* (Oxford 1976).

Dibelius, M. and H. Greeven. *An die Kolosser Epheser an Philemon* (Tübingen 1953).

Ernst, J. *Die Briefe an die Philipper, an Philemon, an die Kolosser, an die Epheser* (Regensburg 1974).

Houlden, J. L. *Paul's Letters from Prison* (Harmondsworth 1970).
Lightfoot, J. B. *Saint Paul's Epistles to the Colossians and Philemon* (London 1892).
Lohse, E. *Colossians and Philemon* (Philadelphia 1971).
Martin, R. P. *Colossians: The Church's Lord and the Christian's Liberty* (Exeter 1972).
Masson, C. *L'Epitre de Saint Paul aux Colossiens* (Neuchatel and Paris 1950).
Moule, C. F. D. *The Epistles of Paul the Apostle to the Colossians and to Philemon* (Cambridge 1957).
O'Brien, P. T. *Colossians, Philemon* (Waco 1982, Milton Keynes 1987).
Schweizer, E. *The Letter to the Colossians* (London 1982).

Thessalonians

Best, E. *A Commentary on the First and Second Epistles to the Thessalonians* (London 1972).
Bruce, F. F. *1 and 2 Thessalonians* (Waco 1982, Milton Keynes 1986).
Dibelius, M. *An die Thessalonicher I, II; An die Philipper* (Tübingen 1937).
Grayston, K. *The Letters of Paul to the Philippians and to the Thessalonians* (Cambridge 1967).
Marshall, I. H. *1 and 2 Thessalonians* (Grand Rapids and London 1983).
Rigaux, B. *Les Epitres aux Thessaloniciens* (Paris and Gembloux 1956).

The Pastoral Epistles

Barrett, C. K. *The Pastoral Epistles* (Oxford 1963).
Brox, N. *Die Pastoralbriefe* (Regensburg 1969).
Dibelius, M., and H. Conzelmann. *The Pastoral Epistles* (Philadelphia 1972).
Easton, B. S. *The Pastoral Epistles* (London 1948).
Hanson, A. T. *The Pastoral Letters* (Cambridge 1966).
Houlden, J. L. *The Pastoral Epistles* (Harmondsworth 1976).
Kelly, J. N. D. *The Pastoral Epistles* (London 1963).
Lock, W. *A Critical and Exegetical Commentary on the Pastoral Epistles* (Edinburgh 1924).
Scott, E. F. *The Pastoral Epistles* (London 1936).
Simpson, E. K. *The Pastoral Epistles* (London 1954).
Spicq, C. *Les Epitres Pastorales*, vols. 1–2 (Paris 1969).

Hebrews

Bruce, F. F. *Commentary on the Epistle to the Hebrews* (Edinburgh 1964).
Buchanan, G. W. *To the Hebrews* (New York 1972).
Héring, J. *L'Epitre aux Hebreux* (Neuchatel 1954).
Michel, O. *Der Brief an die Hebraer* (Göttingen 1975).
Moffatt, J. *A Critical and Exegetical Commentary on the Epistle to the Hebrews* (Edinburgh 1924).
Montefiore, H. W. *The Epistle to the Hebrews* (London 1964).
Spicq, C. *L'Epitre aux Hebreux*, vols. 1–2 (Paris 1952–1953).
Westcott, B. F. *The Epistle to the Hebrews* (London 1906).

1 Peter

Beare, F. W. *The First Epistle of Peter,* 2d ed. (Oxford 1958).
Best, E. *1 Peter* (London 1971).
Bigg, C. *A Critical and Exegetical Commentary on the Epistles of St. Peter and St. Jude,* 2d ed. (Edinburgh 1902).
Goppelt, L. *Der Erste Petrusbrief* (Göttingen 1978).
Hort, F. J. A. *The First Epistle of St. Peter I. 1–II. 17* (London 1898).
Kelly, J. N. D. *A Commentary on the Epistles of Peter and Jude* (London 1969).
Schrage, W. 'Der erste Petrusbrief,' in *Die "Katholischen" Briefe,* ed. H. Balz and W. Schrage (Göttingen 1973).
Selwyn, E. G. *The First Epistle of St. Peter* (London 1946).
Spicq, C. *Les Epitres de Saint Pierre* (Paris 1966).
Windisch, H., and H. Preisker. *Die Katholischen Briefe,* 3d ed. (Tübingen 1951).

Johannine Epistles

Brown, R. E. *The Epistles of John* (London 1983).
Bruce, F. F. *The Epistles of John* (London 1970).
Bultmann, R. *The Johannine Epistles* (Philadelphia 1973).
Dodd, C. H. *The Johannine Epistles* (London 1946).
Grayston, K. *The Johannine Epistles* (Grand Rapids and London 1984).
Houlden, J. L. *A Commentary on the Johannine Epistles* (London and New York 1973).
Lieu, J. *The Second and Third Epistles of John* (Edinburgh 1986).
Marshall, I. H. *The Epistles of John* (Grand Rapids and London 1978).
Schnackenburg, R. *Die Johannesbriefe,* 6th ed. (Freiburg 1979).
Smalley, S. S. *1, 2, 3 John* (Waco 1984).
Westcott, B. F. *The Epistles of St. John* (London 1905).

Revelation

Allo, E.-B. *Saint Jean–L'Apocalypse* (Paris 1921).
Beasley-Murray, G. R. *The Book of Revelation* (London 1974).
Bousset, W. *Die Offenbarung Johannis* (Göttingen 1906).
Caird, G. B. *The Revelation of St. John the Divine* (London 1966).
Charles, R. H. *The Revelation of St. John,* vols. 1–2 (Edinburgh 1920).
Farrer, A. *The Revelation of St. John the Divine* (Oxford 1964).
Ford, J. M. *Revelation* (New York 19;75).
Lohmeyer, E. *Die Offenbarung des Johannes* (Tübingen 1953).
Sweet, J. *Revelation* (London 1979).
Swete, H. B. *The Apocalypse of St. John* (London 1909).

GENERAL STUDIES

Abrahams, I. *Studies in Pharisaism and the Gospels,* vol. 1 (Cambridge 1917).
Allison, D. C. *The End of the Ages Has Come* (Edinburgh 1987).
Angus, S. *The Mystery Religions and Christianity* (London 1925).
Ashton, J., ed. *The Interpretation of John* (Philadelphia and London 1980).

Atkinson, J. 'Blood of Christ,' in *A New Dictionary of Christian Theology*, ed. A. Rich-
    ardson and J. Bowden (London 1983).
Aulen, G. *Christus Victor* (London 1931).
Badenas, R. *Christ, the End of the Law* (Sheffield 1985).
Balch, D. L. *Let Wives Be Submissive: The Domestic Code in 1 Peter* (Chico, Texas 1981).
Barbour, R. S. 'Wisdom and the Cross in 1 Corinthians 1 and 2,' in *Theologia crucis—
    signum crucis*, ed. C. Andresen and G. Klein (Tübingen 1979).
Barr, J. *Escaping from Fundamentalism* (London 1984).
Barrett, C. K. 'The Eschatology of the Epistle to the Hebrews,' in *The Background of the
    New Testament and Its Eschatology*, ed. W. D. Davies and D. Daube. (Cambridge
    1956).
————. 'The Background of Mark 10:45,' in *New Testament Essays*, ed. A. J. B. Higgins
    (Manchester 1939).
————. *The Prologue of St. John's Gospel* (London 1971).
————. 'Theologia Crucis—in Acts?' in *Theologia Crucis–Signum Crucis*, ed. C. Andre-
    sen and G. Klein (Tübingen 1979).
————. *Essays on John* (London 1982).
————. *Essays on Paul* (London 1982).
————. *Church, Ministry and Sacraments in the New Testament* (Exeter 1985).
————. *The New Testament Background: Selected Documents*, rev. ed. (London 1987).
Beasley-Murray, G. R. *Jesus and the Kingdom of God* (Grand Rapids and Exeter 1986).
Beker, J. C. *Paul the Apostle* (Edinburgh 1980).
Bennett, W. J. 'The Son of Man must . . .' *Novum Testamentum 17* (1975): 113–129.
Best, E. *One Body in Christ* (London 1955).
————. *Mark—The Gospel As Story* (Edinburgh 1983).
Betz, H. D. *Der Apostel Paulus und die Sokratische Tradition: Eine exegetische Untersu-
    chung zu eine 'Apologie' 2 Korinther 10–13* (Tübingen 1972).
Betz, O. 'Rechtfertigung in Qumran,' in *Rechtfertigung*, ed. J. Friedrich, W. Pöhlmann
    and P. Stuhlmacher (Tübingen 1976).
Black, M. *An Aramaic Approach to the Gospels and Acts* (Oxford 1967).
————. Aramaic *Barnashā* and the "Son of Man," *ExT*95(1984): 200–206.
Boers, H. 'The Form-critical Study of Paul's Letters—1 Thessalonians As a Case Study'
    NTS 22(1976): 140–58.
Boismard, M.-E. *Quatre hymnes baptismales dans La Première Épitre de Pierre* (Paris
    1961).
Boman, T. *Hebrew Thought Compared with Greek* (London 1960).
Bömer, F. *Untersuchungen über die Religion der Sklaven in Griechenland und Rom*, pt. 2,
    'Die Sogenannte sacrale Freilassung in Griechenland und die *(douloi) hieroi'*
    (Mainz 1960).
Borgen, P. *Logos Was the True Light* (Trondheim 1983).
Bornkamm, G. 'Lord's Supper and Church in Paul,' in *Early Christian Experience* (London
    1969).
————. 'Sin, Law and Death—an Exegetical Study of Romans 7,' in *Early Christian
    Experience* (London 1969).
Borsch, F. H. *The Son of Man in Myth and History* (London 1967).
Bouttier, M. *En Christ* (Paris 1962).
Bovon, F. *Luke, le théologien* (Neuchatel 1978).
Bowker, J. *Jesus and the Pharisees* (Cambridge 1973).
Braun, F. M. *Jean le théologien*, vol. 3 (Paris 1966).
Bruce, F. F. "'To the Hebrews' or 'To the Essenes'?" *NTS* 9(1962–1963): 217–32.

Büchele, A. *Der Tod Jesu im Lukasavengelium* (Frankfort 1978).

Bultmann, R. *Theology of the New Testament* (London 1952, 1955).

———. 'Das Verhältnis des urchristlichen Botschaft zum historischen Jesus,' Abhandlungen der Heidelberger Akademie der Wissenschaften (Phil.-hist. Klasse, H 3) (1960): 111.

———. 'Romans 7 and the Anthropology of Paul,' trans. and ed. S. Ogden, in *Existence and Faith* (London 1961).

Burge, G. M. *The Anointed Community: The Holy Spirit in the Johannine Literaure* (Grand Rapids and Exeter 1987).

Burkitt, F. C. *The Gospel History and Its Transmission* 2d ed. (Edinburgh 1907).

Caird, G. B. *Principalities and Powers* (Oxford 1956).

Caragounis, C. C. *The Ephesian Mysterion* (Lund 1977).

Carmignac, J. *RQ* 7(1970): 343–78.

Carr, W. 'The Rulers of This Age' *NTS* 23(1976–1977): 20–35.

———. *Angels and Principalities* (Cambridge 1981).

Casey, M. *Son of Man* (London 1979).

———. 'Aramaic Idiom and Son of Man Sayings' *ExT* 96(1985): 233–36.

———. 'General Generic and Indefinite: The Use of the Term "Son of Man" in Aramaic Sources and in the Teaching of Jesus' *JSNT* 29(1987): 21–56.

Cavallin, H. C. C. *Life after Death*, pt. 1 (Lund 1974).

Charlesworth, J. H., ed. *The Old Testament Pseudepigrapha*, 2 vols. (London 1983, 1985).

Childs, B. S. *Exodus* (London 1974).

———. *The New Testament As Canon–An Introduction* (London 1984).

Chilton, B., ed. *God in Strength: Jesus' Announcement of the Kingdom* (Freistadt 1979).

———. *A Galilean Rabbi and His Bible* (London 1984).

———. *The Kingdom of God in the Teaching of Jesus* (London 1984).

Conzelmann, H. *An Outline of the Theology of the New Testament* (London 1969.

———. *The Theology of Saint Luke* (London 1970).

Cook, M. J. *Mark's Treatment of the Jewish Leaders* (Leiden 1978).

Coppens, J. 'Les logia johanniques du fils de l'homme,' in *L'Evangile de Jean,* ed. M. de Jonge (Gembloux and Leuven 1977).

Court, J. M. *Myth and History in the Book of Revelation* (London 1979).

Cross, F. L. *1 Peter–A Paschal Liturgy* (London 1957).

Dalman, G. *The Words of Jesus* (Edinburgh 1902).

———. *Jesus–Jeshua* (London 1929).

Danielou, J. *A History of Early Christian Doctrine*, vol. 1, *The Theology of Jewish Christianity* (London 1964).

David, P. *The Epistle of James* (Exeter 1982).

Davies, W. D. *Paul and Rabbinic Judaism* (London 1948).

De Jonge, M. 'The Earliest Christian Use of *Christos*' *NTS* 32(1968), 321–43.

De Jonge, M, and A. S. Van der Woude. 'IIQ Melchizedek and the New Testament' *NTS* 12(1956–66): 301–26.

Demarest, B. *A History of Interpretation of Hebrews 7:1–10 from the Reformation to the Present* (Tübingen 1976).

Derrett, J. D. M. *Law in the New Testament* (London 1970).

———. 'The Good Shepherd: St. John's Use of Jewish Halakah and Haggadah,' in *Studies in the New Testament,* vol. 2 (Leiden 1978).

Dewey, J. 'The Literary Structure of the Controversy Stories in Mark 2:1–3:6' *JBL* 92(1973): 394–401. Also in *The Interpretation of Mark,* ed. W. Telford (Philadelphia and London 1985).

————. *Markan Public Debate–Literary Technique, Concentric Structure, and Theology in Mark 2:1–3:6* (Chico, Texas 1980).

Dey, L. K. K. *The Intermediary World and Patterns of Perfection in Philo and Hebrews* (Missoula, Montana 1975).

Dibelius, M. *Die Geisterwelt im Glauben des Paulus* (Göttingen 1909).

————. *Jesus* (London 1963).

Dodd, C. H. *The Interpretation of the Fourth Gospel* (Cambridge 1953).

————. 'The Prophecy of Caiaphas (Jn. xi 47–53),' in *Neo Testamentica et Patristica*, (Leiden 1962).

Donahue, J. R. *Are you the Christ? The Trial Narrative in the Gospel of Mark* (Missoula, Montana 1973).

Doty, W. G. *Letters in Primitive Christianity* (Philadelphia 1973).

Drain, J. W. *Paul—Libertine or Legalist?* (London 1975).

Driver, S. R. *Notes on the Hebrew Text . . . of the Books of Samuel* (Oxford 1913).

Drury, J. *The Parables in the Gospels* (London 1985).

Dunn, J. D. G. 'John VI—A Eucharistic Discourse?' *NTS* 17 (1970–1971): 328–38.

————. *Christology in the Making: An Inquiry into the Origins of the Doctrine of the Incarnation* (London 1980).

Dupont, J. SYN CHRISTO: *L'union avec le Christ suivant saint Paul*, pt. 1, '"Avec le Christ" dans la vie future' (Bruges 1952).

Eichrodt, W. *Theology of the Old Testament* (London 1967).

Elliott, J. H. *A Home for the Homeless: A Sociological Exegesis of I Peter—Its Situation and Strategy* (London 1982).

Ellis, E. E. *Prophecy and Hermeneutic in Early Christianity* (Tübingen 1978). 'Wisdom' and 'Knowledge' in 1 Corinthians (1975).

Enslin, M. S. 'How the Story Grew: Judas in Fact and Fiction,' *Festchrift to Honour F. Wilbur Gingrich*, ed. E. H. Barth and H. E. Cocroft (Leiden 1972).

Evans, C. F. *The Beginning of the Gospel . . .* (London 1968).

————. *Resurrection and the New Testament* (London 1970).

Feldman, A. *The Parables and Similes of the Rabbis* (Cambridge 1927).

Fischel, H. A. 'Martyr and Prophet,' *JQR* 37(1946): 265–80, 363–86.

Foerster, W. *Palestinian Judaism in New Testament Times* (Edinburgh and London 1964).

Fowler, R. M. *Loaves and Fishes* (Chico, Texas 1981).

Franck, E. *Revelation Taught: The Paraclete in the Gospel of John* (Uppsala, Sweden 1985).

Gärtner, B. *The Temple and the Community in Qumran and the New Testament* (Cambridge 1965).

Gaster, T. H. *Passover: Its History and Traditions* (London and New York 1958).

George, A. *Etudes sur l'oeuvre de Luc* (Paris 1978).

George, A. R. *Communion with God in the New Testament* (London 1953).

Goppelt, L. *Theology of the New Testament*, vol. 2 (Grand Rapids 1982).

Grant, R. M. *A Historical Introduction to the New Testament* (London 1963).

Gray, J. *I and II Kings* (London 1964).

Grayston, K. 'The Significance of the Word *Hand* in the New Testament,' in *Mélanges Bibliques* Rigaux FS (Gembloux 1970).

————. 'The Study of Mark XIII,' *Bulletin of the John Rylands University Library of Manchester* 56(1974): 371–87.

————. 'Obedience Language in the New Testament,' *Epworth Review* 2:(1975): 72–80.

————. 'Not with a Rod,' *ExT* 88(1976): 13–16.

————. 'A Problem of Translation,' *Scripture Bulletin* 11(1980): 27–31.

―――. 'The Empty Tomb,' *ExT* 92(1981): 263–67.

―――. 'HILASKESTHAI and Related Words in LXX,' *NTS* 27(1981): 640–56.

―――. 'The Meaning of PARAKLETOS,' *JSNT* 13(1981): 67–82).

―――. 'Salvation Proclaimed III Hebrews 9:11–14,' *ExT* 93(1982): 164–68.

―――. 'What Is Faith? (1) Biblical Faith,' *Epworth Review* 10(1983): 67–72.

―――. *The Johannine Epistles* (London 1984).

―――. 'The Translation of Matthew 28.17,' *JSNT* 21(1984): 105–9.

―――. 'The Opponents in Philippians 3,' *ExT* 97(1986): 170–72.

Gubler, M.-L. *Die Frühesten Deutungen des Todes Jesu* (Göttingen 1977).

Hahn, F. *The Titles of Jesus in Christology* (London 1969).

Hanson, A. T. *Grace and Truth: Study in the Doctrine of the Incarnation* (London 1975).

―――. *The Wrath of the Lamb* (London 1957).

Hart, H. S.-J. 'The Crown of Thorns in John 19:2–5' *JTS*, n.s., 3(1952): 66–75.

Harvey, A. E. *Jesus on Trial: A Study in the Fourth Gospel* (London 1976).

―――. *Jesus and the Constraints of History* (London 1982).

Hengel, M. *Judaism and Hellenism* (London 1974).

―――. *Crucifixion* (London 1977).

―――. *The Atonement* (London 1981).

―――. *Studies in the Gospel of Mark* (London 1985).

Hertzberg, H. W. *I and II Samuel* (London 1964).

Higgins, A. J. B. *Jesus and the Son of Man* (London 1964).

Hill, D. *Greek Words and Hebrew Meanings* (Cambridge 1967).

―――. *New Testament Prophecy* (London 1979).

Hooke, S. H. 'The Myth and Ritual Pattern in Jewish and Christian Apocalyptic,' in *The Labyrinth*, ed. S. H. Hooke (London 1935): 213–33.

Hooker, M. D. *Jesus and the Servant* (London 1959).

―――. *The Son of Man in Mark* (London 1967).

―――. 'Interchange in Christ,' *JTS* 22(1971): 349–61.

―――. 'Were There False Teachers at Colossae?,' in *Christ and Spirit in the New Testament*, ed. B. Lindars, and S. S. Smalley (Cambridge 1973).

―――. 'Interchange and Atonement,' *Bulletin of the John Rylands University Library of Manchester* 60(1978): 462–81.

―――. 'Is the Son of Man Problem Really Insoluble?,' in *Text and Interpretation*, ed. E. Best and R. Wilson (Cambridge 1979).

―――. 'Paul and Covenantal Nomism,' in *Paul and Paulinism*, ed. M. D. Hooker and S. G. Wilson (London 1982).

―――. *The Message of Mark* (London 1983).

Hughes, G. *Hebrews and Hermeneutics* (Cambridge 1979).

Hull, J. M. *Hellenistic Magic and the Synoptic Tradition* (London 1974).

Hunter, A. M. *Paul and His Prececessors*, 2d ed. (London 1961).

Hurd, J. C. *The Origin of 1 Corinthians* (London 1965).

Jeremias, J. *The Central Message of the New Testament* (London 1965).

―――. *The Eucharistic Words of Jesus* (London 1966).

―――. *Jerusalem in the Time of Jesus* (London 1969).

―――. *The Parables of Jesus*, rev. ed. (London 1970).

―――. *New Testament Theology vol. 1, The Proclamation of Jesus* (London 1971).

Jewett, R. *Paul's Anthropological Terms* (Leiden 1971).

Johnson, A. R. *The Vitality of the Individual in the Thought of Ancient Israel* (Cardiff 1949).

————. *Sacral Kingship in Ancient Israel* (Cardiff 1955).

Jones, C. P. M. 'The Epistle to the Hebrews and the Lucan writings,' in *Studies in the Gospels,* ed. D. E. Nineham (Oxford 1955).

Käsemann, E. 'The Pauline Doctrine of the Lord's Supper,' in *Essays on New Testament Themes* (London 1964).

————. 'A Primitive Christian Baptismal Liturgy' in *Essays on New Testament Themes* (London 1964).

————. 'Ephesians and Acts,' in *Studiesin Luke-Acts,* ed. L. E. Keck and J. L. Martyn (London 1968).

————. '"The Righteousness of God" in Paul,' in *New Testament Questions of Today* (London 1969).

————. *Perspectives on Paul* (London 1971).

Kee, H. C. *Community of the New Age* (London 1976).

Kelber, W. H. *The Kingdom in Mark—A New Place and a New Time* (Philadelphia 1974).

————. *Mark's Story of Jesus* (Philadelphia 1979).

————. ed. *The Passion in Mark* (Philadelphia 1976).

Kennedy, A. R. S. 'Tabernacle,' in *Hastings Dictionary of the Bible* (Edinburgh 1902) 4, pp. 653–68.

Kilpatrick, G. D. *The Eucharist in Bible and Liturgy* (Cambridge 1983).

Kirby, J. C. *Ephesians, Baptism and Pentecost* (London 1968).

Klijn, A. F. J. '1 Thessalonians 4:13–18 and Its Background in Apocalyptic Literature,; in *Paul and Paulinism,* ed. M. D. Hooker and G. Wilson (London 1982).

Knight, G. W. *The Faithful Sayings in the Pastoral Epistles* (Kampen 1968).

Knox, R. A. *On Englishing the Bible* (London 1949).

Koch, K. *The Prophets* (London 1982, 1983).

Koester, H. '1 Thessalonians—Experiment in Christian Writing,' in *Continuity and Discontinuity in Church History,* ed. F. F. Church and T. George (Leiden 1979).

————. *Introduction to the New Testament* (New York 1982).

Köhler, L. *Old Testament Theology* (London 1957).

Kramer, W. *Christ, Lord, Son of God* (London 1966).

Kuhn, K. G. 'New Light on Temptation, Sin and Flesh in the New Testament,' in *The Scrolls and the New Testament,* ed. K. Stendahl (London 1958).

————. 'The Epistle to the Ephesians in the Light of the Qumran Texts,' in *Paul and Qumran,* ed. J. Murphy-O'Connor (London 1968).

Kümmel, W. G. *Römer 7 und die Bekehrung des Paulus* (Leipzig 1929).

————. *Man in the New Testament* (London 1963).

————. 'Luc en accusation dans la théologie contemporaine,' in *L'Évangile de Luc,* ed. F. Neirynck (Gembloux 1973).

————. *The New Testament–The History of the Investigation of Its Problems* (London 1973).

————. *The Theology of the New Testament according to Its Major Witnesses, Jesus–Paul–John* (London 1974).

————. *Introduction to the New Testament,* rev. ed. (London 1975).

Ladd, G. E. *A Theology of the New Testament* (Guildford and London 1975).

————. *The Presence of the Future* (Grand Rapids 1974, London 1980).

Lambrecht, J. 'Paul's Farewell–Address at Miletus (Acts 20:17–38)' in *Les Actes des Apôtres,* ed. J. Kremer (Gembloux 1979).

————. ed. *L'apocalypse johannique et l'apocalyptique dans le Nouveau Testament* (Gembloux 1980).

Lampe, G. W. H. *The Seal of the Spirit* (London 1951).

————. *God as Spirit* (Oxford 1977).

Lane, W. L. '1 Tim, iv.1–3: An Early Instance of Over-realized Eschatology?' *NTS* 11(1964–1965): 164–67.

Laws, S. *A Commentary on the Epistle of James* (London 1980).

Leaney, A. R. C. '1 Peter and the Passover: An Interpretations; *NTS* 10(1963–1964):244–48.

————. *The Rule of Qumran and Its Meaning* (London 1966).

Léon-Dufour, X. 'Le dernier cri de Jésus,' *Études* 348/5 (1978): 666–82.

————. 'Jésus face à la mort menaçante,' *Nouvelle Revue Théologique* 100(1978): 802–21.

————. *Sharing the Eucharistic Bread: The Witness of the New Testament* (New York 1987).

Lightfoot, R. H. *History and Interpretation in the Gospels* (London 1935).

————. *Locality and Doctrine in the Gospels* (London 1936).

————. *The Gospel Message of St. Mark* (Oxford 1950).

Lindars, B. 'The Son of Man in the Johannine Christology,' in *Christ and Spirit in the New Testament,* ed. B. Lindars and S. S. Smalley (Cambridge 1973).

————. *Jesus Son of Man* (London 1983).

Lohse, E. *Grundriss der neutestamentlichen Theologie* (Stuttgart 1974).

Louth, A. 'Mysticism,' in *A Dictionary of Christian Spirituality,* ed. G. S. Wakefield (London 1983).

Lüdemann, G. *Paul, Apostle to the Gentiles* (London 1984).

Lyonnet, S. 'L' emploi paulinien de *exagorazein* au sens de 'redimere' est-il attesté dans la litérature grecque?' *Biblica* 42(1961): 85–89.

Macquarrie, J. *Principles of Christian Theology,* rev. ed. (London 1977).

Manson, T. W. *The Sayings of Jesus* (London 1949).

————. *The Servant-Messiah* (Cambridge 1953).

Marcus, J. *The Mystery of the Kingdom of God* (Atlanta 1976).

Marshall, I. H. *Luke: Historian and Theologian* (Exeter 1970).

————. *Last Supper and Lord's Supper* (Exeter 1980).

Martin, R. P. *Carmen Christi: Philippians ii 5–11 in Recent Interpretation and in the Setting of Early Christian Worship* (Cambridge 1967).

————. *Mark–Evangelist and Theologian* (Exeter 1972).

————. *Reconciliation: A Study of Paul's Theology* (London 1981).

Marxsen, W. 'Erwägungen zum Problem des verkündigten Kreuzes,' in *Der Exeget als Theologe* (Gütersloh 1968).

————. *Introduction to the New Testament* (Oxford 1968).

————. *Mark the Evangelist* (New York 1969).

Meeks, W. A. *The Prophet-King* (Leiden 1967).

Merkel, H. 'Peter's Curse,' in *The Trial of Jesus,* ed. E. Bammel (London 1970).

Mettinger, T. N. D. *A Farewell to the Servant Songs* (Malmö 1983).

Metzger, B. M. *The Text of the New Testament* (Oxford 1964).

Mitton, C. L., *The Epistle to the Ephesians: Its Authorship, Origin and Purpose* (Oxford 1951).

McArthur, H. K. 'On the Third Day,' *NTS* 18(1971–1972): 81–86.

McKane, W. 'Prophecy and Prophetic Literature,' in *Tradition and Interpretation,* ed. G. W. Anderson (Oxford 1979).

McKelvey, R. J. *The New Temple* (Oxford 1969).

McNamara, M. *Targum and Testament* (Shannon 1972).

Moloney, F. J. *The Johannine Son of Man* (Rome 1976).

Morris, L. *The Apostolic Preaching of the Cross* (London 1955).
———. *The Cross in the New Testament* (Exeter 1965).
———. *The Atonement—Its Meaning and Significance* (Leicester 1983).
Moulder, W. J. 'The Old Testament Background and Interpretation of Mark X.45' *NTS* 24(1977): 120–27.
Moule, C. F. D. 'St. Paul and Dualism: The Pauline Conception of the Resurrection,' *NTS* 12(1965–1966): 107–23.
———. *The Origins of Christology* (Cambridge 1977).
———. *The Birth of the New Testament*, 3d ed. (London 1981).
———. 'Death "to Sin," "to Law," and "to the World": A Note on Certain Datives,' in *Essays in New Testament Interpretation* (Cambridge 1982).
Munck, J. *Paul and the Salvation of Mankind* (London 1959).
———. 'I Thess. 1.9–10 and the Missionary Preaching of Paul' *NTS* 9(1962–1963): 95–110.
Murphy-O'Connor, J. ed. *Paul and Qumran* (London 1968).
Mussner, F. 'Contributions Made by Qumran to the Understanding of the Epistle to the Ephesians,' in *Paul and Qumran*, ed. J. Murphy-O'Connor (London 1968).
Neirynck, F. *Duality in Mark* (Leuven 1972).
Neugebauer, F. Das paulinische "in Christo" *NTS* 4(1957–1958): 124–38.
———. *In Christus* (Göttingen 1961).
Neyrey, J. *The Passion according to Luke* (New York 1985).
Nicholson, E. W. *Preaching to the Exiles* (Oxford 1970).
Nickelsburg, G. W. E. *Resurrection, Immortality, and Eternal Life in Intertestamental Judaism* (Cambridge, MA 1972).
———. *Jewish Literature between the Bible and the Mishnah* (London 1981).
Noth, M. *Exodus* (London 1962).
Nygren, A. *Agape and Eros*, pt. 1 (London 1932).
Oesterley, W. O. E. *II Esdras* (The Esra Apocalypse) (London 1933).
Olson, U. S. 'The Atonement in 1 Peter,' Ph.D. diss., Union Theological Seminary (Virginia), 1979.
Osten-Sacken, P. von der. "'Christologie, Taufe Homologie"—ein Beitrag zu Apc Joh 1:5f.', *ZNTW* 58(1967): 255–66.
Pannenberg, W. *Jesus—God and Man* (London 1968).
Pearson, O. A. '1 Thess. 2:13–16: A Deutero-Pauline Interpolation?' *HTR* 64(1971): 79–94.
Pedersen, J. *Israel*, Vols. I–II and III–IV (London and Copenhagen 1926, 1940).
Percy, E. *Die Probleme der Kolosser- und Epheserbriefe* (Lund 1946).
Perrin, N. *The Kingdom of God in the Teaching of Jesus* (London 1963).
———. *Rediscovering the Teaching of Jesus* (London 1967).
———. *Jesus and the Language of the Kingdom* (London 1976).
Peterson, D. *Hebrews and Perfection* (Cambridge 1982).
Pierce, C. A. *Conscience in the New Testament* (London 1955).
Price, J. L. 'Light from Qumran upon Some Aspects of Johannine Theology,' in *John and Qumran*, ed. J. H. Charlesworth (London 1972).
Rad, G. von *Old Testament Theology*, vol. 2 (Edinburgh 1965).
———. *Deuteronomy* (London 1966).
Rader, W. *The Church and Racial Hostility* (Tübingen 1978).
Räisänen, H. *Paul and the Law* (Tübingen 1983).
Ramsey, A. M. *The Glory of God and the Transfiguration of Christ* (London 1949).

Reitzenstein, R. *Hellenistic Mystery-Religions,* trans. J. E. Steely (Pittsburgh 1978).

Reumann, J, *Righteousness in the New Testament* (Philadelphia 1982).

Richter, G. 'Die Fusswaschung Joh 13.1–20,' with responses by J. A Fitzmyer and J. D. Quinn, in *Studien zum Johannesevangelium* (Regensburg 1977).

Ridderbos, H. *Paul* (London 1977).

Ringgren, H. *Religions of the Ancient Near East* (London 1973).

Rivkin, E. *A Hidden Revolution* (Nashville 1978).

———. *What Crucified Jesus?* (London 1986).

Robbins, V. K. 'Last Meal: Preparation, Betrayal and Absence,' in *The Passion in Mark* ed. W. H. Kelber (Philadelphia 1976).

Robinson, J. A. T. *The Human Face of God* (London 1973).

———. *Redating the New Testament* (London 1976).

———. *The Priority of John* (London 1985).

Roon, A. van. *The Authenticity of Ephesians* (Leiden 1974).

Rordorf, W. *Sunday* (London 1968).

———. *Sabbat et dimanche dans l'Eglise ancienne* (Neuchâtel 1972).

Rudolph, K. *Gnosis* (Edinburgh 1983).

Russell, D. S. *The Method and Message of Jewish Apocalyptic* (London 1964).

Sampley, J. P. *'And the Two Shall Become One Flesh'* (Cambridge 1971).

Sanders, E. P. *Paul and Palestinian Judaism* (London 1977).

———. *Paul, the Law and the Jewish People* (Philadelphia 1983).

———. *Jesus and Judaism* (London 1985).

Sanders, J. T. *The New Testament Christological Hymns* (Cambridge 1971).

Schillebeeckx, E. *Jesus* (London 1979).

Schippers, R. 'The Pre-Synoptic Tradition in 1 Thessalonians II 13–16' NT 8(1966): 222–34.

Schmithals, W. *'The Corpus Paulinum* and Gnosis,' in *The New Testament and Gnosis,* ed. A. H. B. Logan and A. J. M. Wedderburn (Edinburgh 1983).

Schnackenburg, R. 'Der Menschensohn im Johannesevangelium,' NTS 11(1964): 123–37.

Schoeps, H. J. *Aus frühchristlicher Zeit* (Tübingen 1950). 'Die jüdischen Prophetenmorde'

———. *Paul* (London 1961).

Schürer, E. *The History of the Jewish People in the Age of Jesus Christ* Geza Vermes et al., vols. 1–2 and vol. 3, pts. 1–2 (Edinburgh 1973, 1979, 1986, 1987).

Schürmann, H. *Jesu ureigener Tod: Exegetische Besinnungen und Ausblick* (Freiburg 1975).

Schweizer, E. 'Dying and Rising with Christ,' NTS 14(1967–1968): 1–14.

———. 'Christianity of the Circumcised and Judaism of the Uncircumcised,' in *Jews, Greeks and Christians,* ed. R. Hamerton-Kelly and R. Scroggs (Leiden 1976).

Schweitzer, A. *The Mysticism of Paul the Apostle* (London 1931).

Scroggs, R. 'Paul: *Sophos* and *Pneumatikos,'* NTS 14(1967–1968): 33–55.

Senior, D. P. *The Passion Narrative according to Matthew: A Redactional Study* (Louvain 1975).

Sherwin-White, A. N. *Roman Society and Roman Law in the New Testament* (Oxford 1963).

Sidebottom, E. M. *The Christ of the Fourth Gospel* (London 1961).

Smalley, S. S. 'The Johannine Son of Man Sayings,' NTS 15(1968): 278–301.

———. *John—Evangelist and Interpreter* (Exeter 1978).

Smith, D. M. 'Johannine Christianity: Some Reflections on Its Character and Delineation' NTS 21(1975): 222–48.

Smith, M. *Jesus the Magician* (London 1978).

Spicq, C. *Notes de lexicographie Néo-Testamentaire*, vol. 1 (Fribourg and Göttingen 1978).

Steck, O. H. *Israel und das gewaltsame Geschick der Propheten* (Neukirchen-Vluhn 1976).

Stendahl, K. *Paul among Jews and Gentiles* (London 1977).

Tannehill, R. C. *Dying and Rising with Christ* (Berlin 1967).

Taylor, V. *The Virgin Birth* (London 1920).

––––––. *Jesus and His Sacrifice* (London 1937).

––––––. *The Atonement in New Testament Teaching* (London 1940).

––––––. *Forgiveness and Reconciliation* (London 1941).

––––––. *The Names of Jesus* (London 1953).

––––––. *The Life and Ministry of Jesus* (London 1954).

––––––. *The Cross of Christ* (London 1956).

––––––. *The Person of Christ in New Testament Theology* (London 1958).

Telford, W. ed. *The Interpretation of Mark* (Philadelphia and London 1985).

Theissen, G. *The Social Setting of Pauline Christianity* (Edinburgh 1982).

Thompson, J. W. *The Beginnings of Christian Philosophy—The Epistle to the Hebrews* (Washington 1982).

Tinsley, E. J. 'Mysticism, in *A New Dictionary of Christian Theology*, ed. A. Richardson, and S.Bowden (London 1983).

Tödt, H. E. *The Son of Man in the Synoptic Tradition* (London 1965).

Tuckett, C. ed. *The Messianic Secret* (Philadelphia and London 1983).

Turner, C. H. 'Chronology of New Testament' in Hastings' *Dictionary of the Bible* (Edinburgh 1898;).

Unnik, W. C. van '"Worthy Is The Lamb": The Background of Apoc 5,' in *Mélanges Bibliques* Rigaux FS (Gembloux 1970).

Vanhoye, A. *La Structure Litteraire de l'Epitre aux Hebreux* (Paris and Bruges 1963).

Vaux, R. de *Ancient Israel* (London 1961).

Vermes, G. *Jesus the Jew* (London 1973).

––––––. *The Dead Sea Scrolls: Qumran in Perspective*, 2d ed. (London 1982).

Vielhauer, P. *Geschichte der Urchristlichen Literatur* (Berlin 1975).

Wainwright, G. *Eucharist and Eschatology* (London 1971).

Walaskay, P. W. *'And So we Came to Rome'* (Cambridge 1983).

Weber, H.-R. *The Cross: Tradition and Interpretation* (London 1979).

Weiser, A. *The Psalms* (London 1959).

Westermann, C. *Isaiah 40–66* (London 1969).

White, J. L. *The Body of the Greek Letter* (Missoula, Montana 1972).

Wilckens, U. *Weisheit und Torheit* (Tübingen 1959).

Williamson, R. *Philo and the Epistle to the Hebrews* (Leiden 1970).

Wilson, R. M. *Gnosis and the New Testament* (Oxford 1968).

––––––. 'How Gnostic Were the Corinthians?' *NTS* 19(1972–1973) 64–74.

Wilson, S. G. *Luke and the Pastoral Epistles* (London 1979).

Ziesler, J. A. *The Jesus Question* (Guildford and London 1950).

––––––. *The Meaning of Righteousness in Paul* (Cambridge 1972).

––––––. 'The Removal of the Bridegroom–A note on Mark II 18–22 and Parallels,' *NTS* 19(1972–1973): 190–94.

––––––. 'SOMA in the Septuagint,' *NT* 25(1983): 133–45.

Zimmerli, W., and J. Jeremias. *The Servant of God*, rev. ed. (London 1964).

Index of Citations

Note: This index provides citations for biblical and other ancient writings. The references are complete for all but the New Testament writings, for which only the main citations are given.

OLD TESTAMENT

Genesis

1:3	53
1:4–25	105
1:26	53
1:31	105
2:7	75
2:8	225
2:17	106
2:24	31, 152, 406n.33
3:4–5	106, 125
6:1–4	447n.29
9:4	408n.71
9:25–27	432n.34
15:6	77, 89, 91, 377, 380
22:2	263
22:8,13	286
22:16	113
28:12	292, 319
37:10	433n.47
37:27	407n.49

Exodus

7:17–21	333
12:3–5	458n.17
12:5	287, 444n.11
12:11	444n.10
12:15	29
12:22	453n.13
12:46	246, 287
19:5	160
19:10	329
19:11,16	410n.85
21:32	396
24:1–2,3– 8,9–11	208, 241, 268
25:17	94
26:33	221
28:32	457n.62
29:38–41	458n.15
30:6	265
30:10	452n.3
32:11–14	193
34:20	32
34:27–35	53

Leviticus

1:10; 3:6; 9:3	458n.15
11:24– 25,28,32,40	329
11:44	315
12:6,8; 14:10–25	458n.15
15	329

16:20–22	445n.21
16:34	47
17:11,14	408n.71
18:5	78, 115
18:6	407n.49
18:8,29	29
20:7–8	315
23:18–20	458n.15
23:27–28; 25:9	452n.3
27	32

Numbers

3:12	192
5:8	452n.9
6:24–26	432n.34
9:12	246, 287
12:14	433n.41
16:5	449n.15
20:4	450n.16
21:9	454n.25
21:6–9	293
23:19	435n.51
24:3	200
24:8	208
28:3–8,29	458n.15
35:19–27	459n.26
35:33	37

OTHER JEWISH WRITINGS

NEW TESTAMENT

EARLY CHRISTIAN WRITINGS

Subject Index